PRINCIPLES OF CHEMICAL REACTOR ANALYSIS AND DESIGN

Uzi Mann

Department of Chemical Engineering
Texas Tech University

Plains Publishing Company
Lubbock, Texas

ISBN 0-9673761-0-6

Library of Congress Cataloging-in-Publication Data:

Mann, Uzi
Principles of Chemical Reactor Analysis and Design / Uzi Mann

Includes Index.
ISBN Number 0-9673761-0-6
1. Chemical reactors. 2. Chemical engineering. 3. Process Design
I. Mann, Uzi.   II. Title.

Plains Publishing Company
P. O. Box 6311
Lubbock, TX 79493-6311

In memory of my sister, Meira Lavie

To Helen, and to David, Amy, and Joel

"Discovery consists of looking at the same thing as everyone else and thinking something different."

Albert Svent-Gyorgyi
Nobel Laureate, 1937

PREFACE

The objective of this book is to present a new **fundamental** approach and methodology to analyze and design chemical reactors. The new approach overcomes the deficiencies of the current design methodology (dimensional, case-specific formulation of operations with **single** reactions), and provides a general unified methodology to design chemical reactors with **multiple** reactions in terms of **dimensionless** equations. The ultimate goal is to provide engineers with the tools needed to handle contemporary, **industrial** reactor operations, and, at the same time, simplify their work.

The main advantages of the new approach (reaction-based design formulation, instead of the conventional species-based formulation) are:

- A **unified**, general methodology to design chemical reactors with **multiple** reactions (reactors with single reactions are merely simple special cases).
- Reactor operations are described in terms of generic **dimensionless** design equations that generate dimensionless **operating curves** (a new concept in chemical reaction engineering). These curves provide complete information on the progress of the chemical reactions, the formation (or depletion) of all chemical species, and temperature variation. Reactor sizing is represented by a "point" on these curves (similar to the way a point on friction factor versus Reynolds number curves represents a specific operation in fluid flow).
- Reactor operations are described in terms of the **most robust** set of design equations (**smallest** number of design equations and the **least** number of terms in each).
- New design **capabilities** for reactor configurations whose design equations are not currently available (filling reactor, plug-flow reactor with distributed feed, etc.).
- A framework for developing **economic-based** optimization tools as well as for system-based analysis of integrated chemical processes.

In order to provide a clear and thorough presentation of the new reactor design methodology and its applications, the text is divided into two parts comprising of nine chapters. It begins with the Introduction Chapter that provides an overview of the new approach and the tenets on which it is based. Part One (Chapters 1 through 4) covers the four fundamental concepts used in the analysis of reactor operations (i.e., stoichiometry, chemical kinetics, species balances, and energy balance) and the manner in which they are used in the new methodology. A fifth fundamental concept, momentum balance, which is applicable only in tubular and packed-bed reactors, is covered in Chapter 6. In an effort to present a structured methodology and to introduce each concept and definition in a convenient way for later reference and review, the contents of each chapter are organized in sections that focus on specific topics. At the end of each chapter, a Summary section provides a list of the key concepts and topics covered in the chapter. Each chapter contains numerous examples that elucidate theoretical concepts and illustrate their applications.

Part Two (Chapters 5 through 9) covers the applications of the new approach to different reactor configurations and to various operating modes. Chapter 5 deals with ideal batch reactor operations, Chapter 6 with plug-flow reactors, and Chapter 7 with continuous stirred-tank reactors. Each of these chapters begins with an overview of the key assumptions and a brief discussion when these assumptions are applicable. The first section summarizes the design and energy balance equations as well as the auxiliary relations to express species concentrations. These equations are then applied, in increased degree of complexity, in the following sections. The next section covers isothermal operations with single reactions, followed by a section on isothermal operations with multiple reactions, and finally by a section on non-isothermal operations with multiple reactions. Chapter 8 deals with other reactor configurations such as semi-batch reactors, distillation reactors, etc. In this chapter, general non-isothermal operations with multiple reactions are covered. Chapter 9 provides a brief discussion of economic-based optimization of reactor operations.

The book is written for two groups of readers: (i) chemical engineering students who are exposed to analysis and design of chemical reactors for the first time, and (ii) practicing engineers who have learned the subject using the conventional (species-based) approach. An attempt has been made to present the material in a way that is appealing to both groups. However, each may prefer to learn the material in a different way. Instructors who use the text in an undergraduate course may prefer to cover the material by the "horizontal approach" as described by the arrows in the diagram on the facing page. This way, the application of the methodology is learned slowly with increasing degree of difficulty. On the other hand, practicing engineers may cover the material by chapters, according to their interest and need. Both approaches require a thorough understanding of the fundamental concepts of the new approach, covered in the first four chapters.

A knowledge of formulating and solving (numerically) simultaneous first-order differential equations (initial value problems) and multiple nonlinear algebraic equations is required. The use of a mathematical software that provides numerical solutions to these type of equations (HiQ, Matlab, Mathematica, Maple, Mathcad, Polymath, etc.) is strongly recommended.

As the title indicates, this book deals with the **principles** of analyzing and designing chemical reactors. In order to concentrate on the main objective of the book (introducing the new approach and methodology), some topics that are commonly covered in chemical reaction engineering texts (chemical kinetics, catalysis, effect of diffusion and mass transfer, etc.) are not covered here. In many instances, determining the effects of transport limitations on the reaction rates is the most challenging task in the design process. These effects and the methods to describe them mathematically are discussed in detail in many texts and articles but are beyond the scope of this text. The reader is expected to have the knowledge of how to obtain the **global** rate expressions of the chemical reactions. Also, advanced topics related to special reactor types (fluidized-bed, trickle-bed, etc.) are not covered in the text.

The problems at the end of each chapter are categorized by their level of difficulty, indicated by a subscript next to the problem number. Subscript 1 indicates problems whose solutions require substitution of numerical values into equations provided in the text. Subscript 2 indicates problems whose solutions require some more in-depth analysis and modifications of given equations. Subscript 3 indicates problems whose solutions require more complex analysis and involve application of several equations. Subscript 4 indicates problems that require the use of a mathematical software or writing of a computer code to obtain numerical solutions.

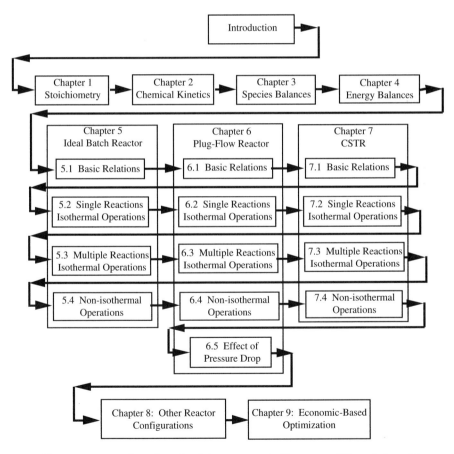

The new approach and methodology presented here are still evolving. New and effective methods for estimating kinetic parameters of multiple simultaneous chemical reactions from reactor operating data are being developed, and additional economic-based optimization procedures are being prepared. Although further refinements may be necessary, the development of the new approach has reached the stage that a unified methodology can be presented. As more teachers and practicing engineers become familiar with it and recognize its advantages and power, new applications will be developed. To inform readers on new developments and to provide new examples and case studies, a web site is being prepared (www.che.ttu.edu/faculty/mann/reactor.htm). The site will also provide additional resource material for teaching and learning the new approach.

The lack of a unified methodology to analyze and design chemical reactors with multiple reactions and of dimensionless design formulation have hindered the work of chemical engineers and limited the application of chemical reaction engineering concepts. Hopefully, this text will influence the way chemical reaction engineering is taught and practiced. Readers are invited to communicate to me their experiences in applying the new methodology and any suggestion on how to improve its presentation. Please contact me by e-mail at: umann@ttu.edu.

I am indebted to many individuals for their encouragement and help during the development of this text. Michael O'Leary and Steve Wolfe materially supported this project and suggested the title of the text. M. D. Morris encouraged me to develop this book. Sandeep Lal provided constructive criticism. Amy and Helen Mann edited the text, in which they put not only their skills, but their hearts. I also express my thanks to Oscar Marrero, Joel Mann, and Mahesh Iyer for proof reading the final version.

Lubbock, Texas   Uzi Mann
June, 1999

CONTENTS

INTRODUCTION

THE NEW APPROACH

This chapter provides an overview of the basic concepts that form the foundation of the new approach to analyze and design chemical reactors presented in this book. The chapter introduces the framework of what follows in the rest of the text. We first review the major deficiencies of the current reactor design methodology and list the advantages and new capabilities provided by the new approach. We then discuss certain fundamental characteristics of reacting systems that have not been recognized and exploited in the conventional reactor design methodology. We also compare the structure of reactor design formulation to the designs of other operations in the process industries and show how the new approach fits the common process design framework. Upon completing this short chapter, the reader should be convinced of the validity and importance of the new approach and recognize the obstacles that have inhibited its development.

Despite the key role chemical reactors play in many industrial processes, the current design formulation of chemical reactors is limited in scope, cumbersome and inefficient. Most procedures employed to design processes and operations in the chemical industry usually involve two stages: (i) generic dimensionless description of the operation and its features, and (ii) sizing of the equipment to meet a specified production rate. (For example, in binary distillation, the McCabe-Thiele diagram is used first to describe the level of separation that can be achieved with a given number of trays, but the sizing of the column diameter is determined on the basis of specified capacity and flooding consideration. In fluid flow, the operation is described by a dimensionless plot of friction factor versus Reynolds number, but the sizing of the tube diameter is carried out by selecting a point on the curve and using the specified flow rate.) However, the current methodology to design chemical reactors skips the first step and is concerned solely with the reactor sizing needed to meet specified production rates. Consequently, some characteristic features of reactor operations are overlooked in the design. Further, useful tools that are commonly derived from the generic dimensionless description of the operation (e.g., economic-based optimization) have not been developed for chemical reactors.

Specifically, the current reactor design methodology has three major deficiencies:

- No systematic methodology is available to design chemical reactors with **multiple** reactions. By and large, the current methodology has been developed for reactors with **single** chemical reactions, and the formulation is usually in terms of the **conversion** of a reactant. The design of reactors with **multiple** chemical reactions is formulated by writing species balance equations and expressing them in terms of either the species concentration or their molar flow rates. This "brute-force" approach results, in many cases, in design formulations that consist of a number of equations larger than the minimum necessary to describe the operation, often leading to unstable or inaccurate numerical solutions.

- Chemical reactor operations are described in terms of **system-specific, extensive** quantities (reactor volume, species flow rate, operating time, etc.) and **dimensional** design equations, instead of **intensive** quantities and **dimensionless** design equations (as done in essentially all other operations in the process industries). As a result, little, if any, attention is given to the underlying behavior of the reactor, and little insight into its operation is gained.

- The design formulation is structured to provide information for **specific cases** — required reactor volume to obtain a specified conversion for a given flow rate, etc. Consequently, the design formulations should be "tailored" to each specific case. (It is analogous to conducting separate sets of pressure-drop calculations instead of using dimensionless friction factor and the Reynolds number).

These deficiencies have implications beyond the design and operation of chemical reactors. The genesis of all system-based modeling and optimization procedures is the generic, **dimensionless** design formulations of the operation without regard to equipment sizing. (For example, the optimization of flow systems is based on the dimensionless friction factor versus Reynolds number description of flow in tubes.) Such generic description is not currently available for reactor operations, and this has prevented the development of system-based design methodologies for integrated chemical processes.

This book describes a new fundamental approach to analyze and design chemical reactors that overcomes all the deficiencies described above. The new approach — reaction-based design formulation instead of the conventional species-based formulation — enables us to describe the generic features of reactor operations by dimensionless equations and consider the required size separately. Specifically, the new reactor design methodology provides the following features:

- A **unified**, general methodology to analyze and design chemical reactors with **multiple** reactions (reactors with single reactions are merely simple, special cases). It is applicable to **all** reactor configurations, **any** complex stoichiometry, and **all** forms of reaction rate expressions.

- Reactor operations are described in terms of generic **dimensionless** design equations that generate dimensionless **operating curves** (a new concept in chemical reaction engineering). These curves provide complete information on the progress of the chemical reactions, the formation (or depletion) of the species, and changes in temperature. Specific cases (designs) are "points" on these curves (similar to the way a "point" on dimensionless friction factor versus Reynolds number charts is used).

- Reactor operations are described in terms of the **most robust** set of design equations (**smallest** number of design equations and the **least** number of terms in each). Since, in many instances, the design equations are stiff differential equations, the new formulation is not merely a matter of mathematical elegance, but it enables us to obtain accurate numerical solutions.
- **New** design **capabilities** for reactor configurations whose design equations are not currently available in the literature (filling reactor, plug-flow reactor with distributed feed, etc.).
- A framework for developing new **economic-based** optimization tools for chemical reactor operations as well as for integrated chemical processes.

The new approach to analyze and design chemical reactors is based on basic physical principles and known relations, but it applies them in a **different way**. The new approach (reaction-based design formulation instead of the conventional species-based formulation) is founded on the following tenets:

- The core of reactor operations is the **chemical reactions** that are taking place; the species balance equations serve merely as the means to describe reactor operations. Therefore, the design equations should be formulated in terms of quantities related to chemical reactions (e.g., extent of reaction) rather than on quantities related to species (conversion, concentration, etc.). An examination of the structure of the design equations of rate processes (see Table 1) indicates that the final design equations are expressed in terms of quantities that affect the operation, not the basic quantity used in formulating them. For example, the genesis of the design equation for fluid flow operations is the momentum balance equation, but the final design equation is expressed in terms of flow-related quantities (velocity, pressure, etc.), not momentum. The energy balance equation is used to describe heat-transfer operations, but the final design equation is expressed in terms of temperature and other quantities related to heat-transfer, not energy. By using reaction-based quantities, we describe reactor operations more accurately and obtain more insight into the process.
- The **extent of reaction** is the most suitable reaction-related quantity to describe reactor operations and should be used in the design formulation. (The extent is a measure of "units" of a chemical reaction that have been transformed.) The reluctance to use the extent might be because it is a **calculated** (not-measurable) quantity. At first, it seems counterproductive to introduce and use calculated quantities. However, this is routinely done in thermodynamics, where calculated quantities (enthalpy, free-energy, etc.) are used to simplify the formulations. Here, too, by expressing the design equations in terms of extents of reactions rather than measurable quantities (concentrations, species flow rates, etc.), we simplify the design formulation – we formulate the design with the smallest number of design equations and obtain an insight into the operation.
- There are two "types" of chemical reactions — **independent** reactions and **dependent** reactions. All **state** variables of the reactor (composition, temperature, enthalpy, etc.) are determined by a set of **independent** chemical reactions. On the other

hand, the species formation rates depend on **all** the chemical reactions that **actually** take place in the reactor, including the dependent reactions. The two reaction "types" are analogous to the two types of thermodynamic quantities — those related only to

Table 1: Structure of Design Equations of Rate Processes

Process/ Operation	Fundamental Principle	Variables in Final Design Equation	Dimensionless Quantities
Fluid Flow (Laminar, Turbulent)	Momentum Balance	Viscosity Density Velocity Pressure Elevation	Reynolds Number Friction Factor
Heat-Transfer (conduction, Convection)	Energy Balance	Thermal Conductivity Heat Capacity Heat-Transfer Coefficient Temperature Density Velocity Viscosity	Nusselt Number Reynolds Number Prandl Number Biot Number
Mass-Transfer (Diffusion, Convection)	Species Balance	Diffusion Coefficient Mass-Transfer Coefficient Species Concentrations Density Velocity Viscosity	Sherwood Number Reynolds Number Schmidt Number Lewis Number Biot Number
Chemical Reactors	Species Balances, Stoichiometry, Chemical Kinetics	Reaction Rate Constants Species Concentrations Temperature Reactor Volume Feed Flow Rate Feed Composition Reactant Conversion	

the "state" of the system (temperature, enthalpy, etc.) and those related to the process (work, heat, etc.). Each species balance equation (the genesis of the reactor design equation) consists of two distinct constituent parts. One is related to the amount of the species in the reactor (or, for flow reactors, species flow rate), thus depending only on the independent reactions. The other is related to the species formation rates, thus depending on **all** the chemical reactions that **actually** take place. Since all state variables are determined by the extents of the independent reactions, we have to derive a relationship that enables us to express the rates of the **dependent** reactions in terms of the **independent** reactions. This known, but rarely used, stoichiometric relationship is the "Rosetta Stone" that enables us to formulate reaction-based design equations.

- The operation of a chemical reactor is, basically, a rate process and, therefore, has a characteristic time constant. We select a **reaction-based** time constant to formulate the design equation in terms of dimensionless time. By doing so, we express the reactor operation on the time scale of the chemical reactions.

- Careful selection of a reference state (or reference stream) serves as a basis for a scaling factor for **all** the design calculations. Beside selecting a reference flow rate, we define a reference reaction rate, reference heat capacity, etc., that enable us to formulate the design in terms of **dimensionless** quantities and describe the reactor operations in terms of dimensionless operating curves. Dimensionless design equations not only provide insight into the behavior of the reactor but also simplify the calculations (no need to select appropriate time increments for numerical solution).

Based on these concepts, we develop here a methodology that leads to reaction-based, dimensionless design formulations of chemical reactors. The incorporation of the different elements that constitute the design equation is shown schematically in Figure 1. The new approach applies certain known stoichiometric relations that have not been used in the context of reactor design formulation. The new methodology sorts

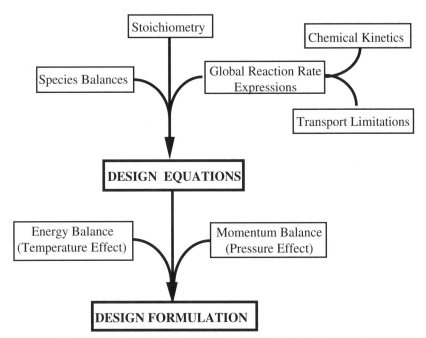

Figure 1: Schematic Diagram of Reactor Design Formulation

all the stoichiometric relations between species and the chemical reactions prior to combining them with kinetics and material balance in the framework of the design equation. As a result, we are able to discern the various factors affecting the reactor operation and the interrelationships between them.

The derivation of the design equation is only the first step in the design formulation. To design chemical reactors, we have to consider the effects of temperature and pressure variations on the operations (see Figure 1). We present here a procedure to incorporate the energy balance equation and the momentum balance equation into the design formulation. By sorting all the stoichiometric relations first, we are able to tie the chemical reactions directly to the energy balance and the momentum balance equations and combine them in a unified, dimensionless design formulation. The end result is a generic, complete detailed description of the reactor operation, new design capabilities, a foundation for economic-based optimization, and the framework for ssyten-based analysis of integrated chemical processes.

PART ONE

THE FUNDAMENTALS

As a discipline, chemical reaction engineering (CRE) is built on four fundamental concepts:
* stoichiometry,
* chemical kinetics providing the rate expression,
* species mass balance leading to the design equation, and
* thermodynamics.

Part One covers these concepts in some detail and develops the tools needed to analyze and design chemical reactors.

Stoichiometry is used in essentially all phases of reactor design and analysis. Primarily, it is an accounting system that enables us to determine what species are being consumed, what species are being formed and in what amounts. Stoichiometry plays a similar role in reaction engineering as financial accounting does in business. Furthermore, stoichiometry also provides important relationships between the various chemical reactions. It enables us to select a set of independent reactions, and to express the dependent reactions in terms of the independent reactions. Stoichiometry also provides tools to identify independent species compositions and to define other quantities such as yield and selectivity to characterize reactor operations.

Chemical kinetics is the branch of chemistry that deals with reaction mechanisms, the rates of chemical reactions and their dependence on the operating conditions (temperature, concentrations, etc.). In reaction engineering, we consider the rate expressions (sometimes called "rate law"), their various forms, and how to obtain their parameters from experimental data.

Material balance is the physical principle used to derive the design equations of chemical reactors. Since reaction engineering concerns with chemical transformations, for convenience we write the material balance for individual species and express the them in terms of moles. Species balances over the reactors or portions of them form the species-based design equations. By applying stoichiometric relations, we convert the species-based design equations to reaction-based design equations that completely describe the reactor operations.

Thermodynamics provides the reaction engineer with information on the limits of the chemical transformations at given operating conditions (equilibrium). The first law of thermodynamics — the conservation of energy — serves as the physical principle that enables us to determine temperature variations during the reactor operation.

Those four fundamental concepts are the foundation of chemical reaction engineering and appear in the formulation of all reactor operations. It is therefore essential to develop a good understanding of these concepts and to be familiar with their key definitions. The remainder of the text deals with the applications of those concepts to different reactor operations. We apply one additional fundamental concept, the conservation of momentum, in Chapter 6 to describe the effect of pressure drop on gaseous plug-flow reactor operations.

CHAPTER ONE

STOICHIOMETRY

In this chapter, we discuss a tool used to analyze processes involving chemical transformations — stoichiometry. Literally, stoichiometry means measurement of elements. In practice, stoichiometry is an accounting system that keeps track of what species are being formed, what species are being consumed, and in what amounts. It enables us to calculate the composition of chemical reactors. Furthermore, when dealing with chemical reactors with multiple reactions, stoichiometry also indicates how many independent chemical reactions are required to determine all the state quantities and the number of design equations necessary to describe the reactor operation. Stoichiometry provides a method to select sets of independent chemical reactions and a procedure to relate dependent and independent reactions. It also indicates what set of independent reactions is most suitable for the design formulation of chemical reactors. We develop in this chapter a simple stoichiometric methodology that enables us to handle any conceivable complex reacting system. Stoichiometry is being applied in all phases of reactor analysis and design; therefore, a good knowledge of its key definitions and structure is essential.

We start the chapter with a brief discussion of selecting a "basis" for analyzing reacting systems and define the stoichiometric coefficients of individual species in a chemical reaction. We then define the extent of a chemical reaction and discuss its relation to the formation (or depletion) of individual species. Next, we relate the extent to changes in the composition of batch and steady flow reactors with single chemical reactions and describe a procedure to identify the limiting reactant. We then expand the discussion to reactors with multiple chemical reactions and define dimensionless extent and the conversion of a reactant. Next, we discuss the role of independent and dependent reactions and describe a procedure to determine the number of independent reactions. We also describe how to select a set of independent reactions and determine the relationships between the dependent reactions and the independent reactions. Next, we discuss how to determine what set of species compositions can be specified to determine the state variables of the reactor. We close the chapter by defining other stoichiometric quantities commonly used in reactor analysis such as product yield and selectivity.

1.1 STOICHIOMETRIC COEFFICIENTS

The first step in analyzing any chemical engineering problem is to define a system and select the framework for solving the problem. This step is commonly referred to as selecting a "basis" for the calculation. A similar step is required when dealing with processes involving chemical reactions. We can express chemical transformations in many forms and have to select one of them as the "basis" for the analysis. To illustrate this point, consider, for example, the reaction between oxygen and carbon monoxide to form carbon dioxide. We can describe the reaction by one of many chemical equations, such as

$$CO + \frac{1}{2}O_2 \rightarrow CO_2, \tag{1.1-1}$$

$$2\,CO + O_2 \rightarrow 2\,CO_2, \tag{1.1-2}$$

or, if you wish to do so,

$$20\,CO + 10\,O_2 \rightarrow 20\,CO_2. \tag{1.1-3}$$

We select one of these chemical reactions as the "system" and relate all relevant quantities (heat of reaction, rate of reaction, etc.) to that chemical equation. Note that the numerical values of the species' coefficients of the reaction depend on the specific chemical equation selected, but the ratio of any two coefficients is constant.

We adopt common conventions concerning chemical equations. Each chemical equation has an arrow indicating the direction of the chemical transformation and clearly defines the reactants and products of the reaction. Reversible reactions are treated as two distinct reactions, one forward and one backward. The arrow also serves as an equality sign for the total mass represented by the chemical equation (44 mass units in reaction (1.1-1), 88 mass units in reaction (1.1-2), etc.). Thus, each chemical equation should be balanced: an unbalanced chemical equation violates the conservation of mass principle and does not make physical sense. Usually, for convenience, we write chemical equations such that either the largest species' coefficient in the equation is one, as in reaction (1.1-1), or the smallest species' coefficient in the equation is the smallest integer, as in reaction (1.1-2).

Once we select a specific chemical equation, the stoichiometric coefficients of the individual species can be defined. Consider the general chemical reaction

$$aA + bB \rightarrow cC + dD. \tag{1.1-4}$$

We define the stoichiometric coefficients of the different chemical species in the reaction as follows: for each product species, the stoichiometric coefficient is identical to the coefficient of that species in the chemical equation. For each reactant, the stoichiometric coefficient is the negative value of the coefficient of that species in the chemical equation. If a species does not participate in the reaction, by definition its stoichiometric coefficient is zero. Thus, for chemical reaction (1.1-4), the stoichiometric coefficients are: $s_A = -a$, $s_B = -b$, $s_C = c$, $s_D = d$, and, for any inert species, say I, $s_I = 0$. By defining the stoichiometric coefficients in this manner, we express the chemical reactions as homogeneous algebraic equations. For example, we write chemical reaction (1.1-4) as

$$-aA - bB + cC + dD = 0.$$

The advantage of doing so will become clear later when we consider multiple simultaneous chemical reactions. Using the definition of the stoichiometric coefficients, the mathematical condition for a balanced chemical reaction is

$$\sum_{j}^{all} s_j \cdot MW_j = 0, \qquad (1.1-5)$$

where s_j and MW_j are, respectively, the stoichiometric coefficient and the molecular mass of species j, and the summation is over all chemical species participating in the reaction.

Each chemical reaction is characterized by the sum of its stoichiometric coefficients,

$$\Delta = \sum_{j}^{all} s_j = s_A + s_B + ... \qquad (1.1-6)$$

The parameter Δ indicates the change in the number of moles per "unit" of the chemical reaction selected. For example, for chemical reaction (1.1-4), $\Delta = (-a) + (-b) + c + d$. This parameter is especially useful in determining changes in either the total mole of a batch reactor or the total molar flow rate of a flow reactor.

Example 1-1 Consider the reaction between nitrogen and hydrogen to produce ammonia. The chemical reaction can be written as

$$N_2 + 3 H_2 \rightarrow 2 NH_3, \qquad (a)$$

or

$$\frac{1}{2} N_2 + \frac{3}{2} H_2 \rightarrow NH_3. \qquad (b)$$

For chemical reactions (a) and (b) above, define the stoichiometric coefficients, the value of D, and in each case calculate the ratio of the stoichiometric coefficients of ammonia and hydrogen.

Solution For chemical reaction (a), the stoichiometric coefficients are:

$$s_{N2} = -1; \quad s_{H2} = -3; \quad s_{NH3} = 2; \quad \text{and} \quad \Delta = -2.$$

The ratio of the stoichiometric coefficients of NH_3 and H_2 is $s_{NH3}/s_{H2} = 2/(-3) = -0.667$.

For chemical reaction (b), the stoichiometric coefficients are:

$$s_{N2} = -\frac{1}{2}; \quad s_{H2} = -\frac{3}{2}; \quad s_{NH3} = 1; \quad \text{and} \quad \Delta = -1.$$

The ratio of the stoichiometric coefficients of NH_3 and H_2 is $s_{NH3}/s_{H2} = (1)/(-1.5) = -0.667$. Note that the ratio of any two stoichiometric coefficients is independent of the specific chemical equation (or "system") selected.

1.2 EXTENT OF A CHEMICAL REACTION

Once we select a specific form to represent the chemical transformation, we consider next how to express the progress of the reaction. For the chemical reaction between oxygen and carbon monoxide, we can do it in one of three ways: by expressing the consumption of oxygen, the consumption of carbon monoxide, or the formation of carbon dioxide. To avoid confusion, we develop a methodology that relates the amount of individual species formed or depleted to the chemical reaction selected as the "basis." We define the extent of a chemical reaction as one "unit" of the chemical equation selected. For example, for chemical reaction (1.1-1), one extent is the reaction of one mole of CO with half a mole of O_2 to form one mole of CO_2. For chemical reaction (1.1-2), one extent is the formation of two moles of CO_2 by the reaction of one mole of O_2 with two moles of CO.

The extent of a chemical reaction relates to the change in the number of moles of any species, say species j, taking part in the reaction by

$$X \equiv \frac{\text{moles of species j formed by the reaction}}{\text{stoichiometric coefficient of species j}} = \frac{(n_j - n_{j0})}{s_j}. \qquad (1.2\text{-}1)$$

The units of extent are moles of extent. (Note that the heat of reaction of a chemical reaction is expressed in terms of energy per mole extent.) For chemical reaction (1.1-4), the extent relates to the number of moles of each species formed (or consumed) by the reaction by

$$X \equiv \frac{n_A - n_{A0}}{-a} = \frac{n_B - n_{B0}}{-b} = \frac{n_C - n_{C0}}{c} = \frac{n_D - n_{D0}}{d}.$$

Note that the stoichiometric coefficient of species j, s_j, is a dimensionless parameter indicating the moles of species j formed per mole extent.

Next we want to relate the composition of chemical reactors to the reactions taking place in them. To do so, we derive a relationship between the mass basis selected for calculating the amount of a species in the reactor and the chemical reactions selected to represent the chemical transformations. The extents of the chemical reactions are the tying elements between these bases. For convenience, we distinguish between two modes of reactor operations: batch operation (closed reactors) and steady continuous operation (flow reactors), shown schematically in Figure 1-1. In batch reactors, reactants are charged into the reactor and, after a certain time, the products are discharged from the reactor; hence, the chemical transformations take place over time. In steady flow reactors, reactants are continuously fed into the reactor, and products are continuously withdrawn from the reactor; hence, the chemical transformations take place over space.

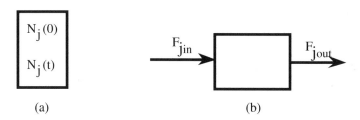

(a)   (b)

Figure 1-1: Modes of Reactor Operations: (a) Closed Reactor; (b) Flow Reactor

We first consider batch (closed) reactors. Let $N_j(t)$ denote the total number of moles of species j in the reactor at time t. When a **single** chemical reaction takes place, the change in the number of moles of species j in the reactor at time t is equal to the number of moles of species j generated by the reaction,

$$N_j(t) - N_j(0) = (n_j - n_{j0}).$$

Using (1.2-1), we relate $N_j(t)$ to the reaction extent at time t, $X(t)$, by

$$N_j(t) = N_j(0) + s_j \cdot X(t). \tag{1.2-2}$$

By writing (1.2-2) for any two species, say A and j, we obtain

$$N_j(t) = N_j(0) + \frac{s_j}{s_A}\left[N_A(t) - N_A(0)\right]. \tag{1.2-3}$$

Eq. (1.2-3) provides a relation between the number of moles of any two species in the reactor at time t without calculating the extent itself. It is an algebraic expression of elementary stoichiometric calculations we commonly do in our heads. To determine the total number of moles in a batch reactor at time t, $N_{tot}(t)$, we sum the terms in (1.2-2) over all the species in the reactor and obtain

$$N_{tot}(t) = N_{tot}(0) + \Delta \cdot X(t), \tag{1.2-4}$$

where parameter Δ, indicating the change in the number of moles per unit extent of the reaction, is defined by (1.1-6).

To express the total mass of species j in the reactor at time t, $W_j(t)$, we multiply (1.2-2) by the molecular mass of j, MW_j,

$$W_j(t) = W_j(0) + MW_j \cdot s_j \cdot X(t). \tag{1.2-5}$$

Note that by summing (1.2-5) over all species and using (1.1-6), we obtain, as expected,

$$W_{tot}(t) = W_{tot}(0).$$

Next, we consider a steady flow reactor with a **single** chemical reaction and conduct a species balance over the reactor. At steady state, the molar flow rate of species j at the reactor outlet is equal to the molar flow rate of species j at the reactor inlet plus the rate species j is being generated inside the reactor by the reaction, G_j,

$$F_{jout} = F_{jin} + G_j. \tag{1.2-6}$$

We discuss the generation term, G_j, in detail in Chapter 3; here, we concentrate on its relation to the extent of the reaction. Using (1.2-1), the moles of species j formed by the reaction per unit time is

$$G_j = \frac{d(n_j - n_{j0})}{dt} = s_j \cdot \dot{X}, \tag{1.2-7}$$

where $\dot{X}$ is the units of reaction extent converted per unit time in the reactor. Hence, we can write (1.2-6) as

$$F_{jout} = F_{jin} + s_j \cdot \dot{X}. \tag{1.2-8}$$

Eq. (1.2-8) relates the molar flow rate of any species at the reactor outlet and inlet to the rate of the chemical transformation in the reactor. We can write (1.2-8) for any two species, say A and j, and obtain,

$$F_{jout} = F_{jin} + \frac{s_j}{s_A}(F_{Aout} - F_{Ain}). \tag{1.2-9}$$

Eq. (1.2-9) provides, for flow reactors with single reactions, the molar flow rate of any species at the reactor outlet in terms of the molar flow rate of any other species without calculating the extent itself.

To obtain the total molar flow rate at the reactor outlet, $(F_{tot})_{out}$, we sum (1.2-8) over all the species in the reactor and obtain,

$$(F_{tot})_{out} = (F_{tot})_{in} + \Delta \cdot \dot{X}, \tag{1.2-10}$$

where Δ, defined by (1.1-6), denotes the change in the number of moles per unit extent.

To express the mass flow rate of species j, we multiply (1.2-8) by the molecular mass of species j, MW_j,

$$\dot{W}_{jout} = \dot{W}_{jin} + MW_j \cdot s_j \cdot \dot{X}. \tag{1.2-11}$$

Note that by summing (1.2-11) over all the species in the reactor and using (1.1-5), we obtain

$$(\dot{W}_{tot})_{out} = (\dot{W}_{tot})_{in}.$$

Example 1-2 A batch reactor contains 4 moles of CO and 1 mole of O_2. Calculate the extent of the reaction and the composition of the reactor when the reaction goes to completion.
a. Use the chemical equation (1.1-1) and employ (1.2-2).
b. Use the chemical equation (1.1-2) and employ (1.2-3).
Solution We select chemical reaction (1.1-1) as the chemical equation,

$$CO + \frac{1}{2} O_2 \rightarrow CO_2,$$

and its stoichiometric coefficients are

$$s_{CO} = -1; \quad s_{O2} = -\frac{1}{2} ; \quad s_{CO2} = 1; \quad \text{and} \quad \Delta = -\frac{1}{2} .$$

a. The reaction reaches completion when one of the reactants is depleted. Assuming that CO is depleted, $N_{CO}(t) = 0$, and we write (1.2-2) to determine the extent of the reaction,

$$X(t) = \frac{N_{CO}(t) - N_{CO}(0)}{s_{CO}} = \frac{0 - 4.0}{(-1)} = 4 \text{ mole extent}.$$

Now that we know the extent of the chemical reaction, we can calculate the moles of O_2 and CO_2 in the reactor at the end of operation by (1.2-2),

$$N_{O_2}(t) = N_{O_2}(0) + s_{O_2} \cdot X(t) = 1 + (-0.5) \cdot (4) = -1 \text{ mole}$$
$$N_{CO_2}(t) = N_{CO_2}(0) + s_{CO_2} \cdot X(t) = 0 + (1) \cdot (4) = 4 \text{ mole}.$$

Since the number of moles of any species cannot be negative, obtaining such values is an indication that there is an error in the calculation, or that it is based on incorrect information. The error here is, of course, the assumption that the CO is depleted before the oxygen, while actually the oxygen is the limiting reactant. Hence, to determine the extent at completion, we write (1.2-2) for oxygen with $N_{O2}(t) = 0$,

$$X(t) = \frac{N_{O_2}(t) - N_{O_2}(0)}{s_{O_2}} = \frac{0 - 1.0}{(-0.5)} = 2 \text{ mole extent}.$$

We calculate the moles of CO and CO_2 in the reactor at the end of operation by (1.2-2),

$$N_{CO}(t) = N_{CO}(0) + s_{CO} \cdot X(t) = 4 + (-1) \cdot (2) = 2 \text{ mole}$$
$$N_{CO_2}(t) = N_{CO_2}(0) + s_{CO_2} \cdot X(t) = 0 + (1) \cdot (2) = 2 \text{ mole}.$$

The total number of moles in the reactor at time t is

$$N_{tot}(t) = N_{O_2}(t) + N_{CO}(t) + N_{CO_2}(t) = 0 + 2 + 2 = 4 \text{ mole}.$$

We can also calculate $N_{tot}(t)$ by (1.2-4)

$$N_{tot}(t) = N_{tot}(0) + \Delta \cdot X(t) = 5 + \left(-\frac{1}{2}\right) \cdot (2) = 4 \text{ mole}.$$

b. In this case, we select chemical equation (1.2-2) as the chemical equation

$$2\,CO + O_2 \rightarrow 2\,CO_2,$$

and its stoichiometric coefficients are

$$s_{CO} = -2; \quad s_{O2} = -1; \quad s_{CO2} = 2; \quad \text{and} \quad \Delta = -1.$$

Using the given initial composition, $N_{O2}(0) = 1$ mole, and that at completion $N_{O2}(t) = 0$, we write (1.2-2) for oxygen with $N_{O2}(t) = 0$,

$$X(t) = \frac{N_{O_2}(t) - N_{O_2}(0)}{s_{O2}} = \frac{0 - 1.0}{(-1)} = 1 \text{ mole extent}.$$

We calculate the moles of CO and CO_2 in the reactor at the end of operation by (1.2-3),

$$N_{CO}(t) = N_{CO}(0) + \frac{s_{CO}}{s_{O2}}\left[N_{O_2}(t) - N_{O_2}(0)\right] = 4 + \frac{(-2)}{(-1)}(0-1) = 2 \text{ mole}$$

$$N_{CO_2}(t) = N_{CO_2}(0) + \frac{s_{CO2}}{s_{O2}}\left[N_{O_2}(t) - N_{O_2}(0)\right] = 0 + \frac{(2)}{(-1)}(0-1) = 2 \text{ mole}.$$

Note that, as expected, the final reactor composition does not depend on the specific chemical equation selected. Also, note that the extent in part (b) is one half of the extent in part (a).

Example 1-3 Ethylene oxide is produced by catalytic oxidation on a silver catalyst according to the chemical reaction:

$$C_2H_4 + \frac{1}{2}O_2 \rightarrow C_2H_4O.$$

A gaseous stream consisting of 60% C_2H_4, 30% O_2, and 10% N_2 (by mole) is fed at a rate of 40 mole/min into a flow reactor operating at steady state. If the mole fraction of oxygen in the reactor effluent stream is 0.08, calculate the production rate of ethylene oxide.

Solution We select the written chemical reaction as the chemical equation; hence, the stoichiometric coefficients are:

$$s_{C2H2} = -1; \quad s_{O2} = -\frac{1}{2}; \quad s_{C2H2O} = 1; \quad s_{N2} = 0; \quad \text{and} \quad \Delta = -\frac{1}{2}.$$

We use (1.2-8) to express the oxygen molar flow rate at the reactor exit,

$$(F_{O_2})_{out} = (F_{O_2})_{in} + s_{O_2} \cdot \dot{X} = (0.3) \cdot (40) + \left(-\frac{1}{2}\right) \dot{X}. \tag{a}$$

We use (1.2-10) to express the total molar flow rate of the effluent stream,

$$(F_{tot})_{out} = (F_{tot})_{in} + \Delta \cdot \dot{X} = 40 + \left(-\frac{1}{2}\right) \dot{X}. \tag{b}$$

Thus, the oxygen mole fraction in the effluent stream is

$$(y_{O_2})_{out} = \frac{(F_{O_2})_{out}}{(F_{tot})_{out}} = \frac{(0.3) \cdot (40) + \left(-\frac{1}{2}\right) \dot{X}}{40 + \left(-\frac{1}{2}\right) \dot{X}} = 0.08. \tag{c}$$

Solving (c) for $\dot{X}$, we obtain $\dot{X} = 19.13$ mole/min. Now that we know the value of $\dot{X}$, we use (1.2-10) to calculate the production rate of ethylene oxide,

$$(F_{C_2H_4O})_{out} = (F_{C_2H_4O})_{in} + s_{C_2H_4O} \cdot \dot{X} = 0 + (1) \cdot (19.13) = 19.13 \text{ mole/min}.$$

1.3 LIMITING REACTANT

Example 1-2 illustrated the advantage of identifying the limiting reactant of a chemical reaction and describing the reaction progress in terms of its depletion. Here, we describe how to identify the limiting reactant. Let A and B be two reactants of a chemical reaction; using (1.2-2) or (1.2-8), we relate the number of moles and the molar flow rate between them, respectively,

$$\frac{N_A(t) - N_A(0)}{s_A} = \frac{N_B(t) - N_B(0)}{s_B} \quad \text{or} \quad \frac{F_{A_{out}} - F_{A_{in}}}{s_A} = \frac{F_{B_{out}} - F_{B_{in}}}{s_B}.$$

If species A is indeed the limiting reactant, $N_A(t)$ vanishes before $N_B(t)$ or $F_{A_{out}}$ vanishes before $F_{B_{out}}$. Hence, one of the following respective relations should be satisfied,

$$\left|\frac{N_A(0)}{s_A}\right| \leq \left|\frac{N_B(0)}{s_B}\right| \quad \text{or} \quad \left|\frac{F_{A_{in}}}{s_A}\right| \leq \left|\frac{F_{B_{in}}}{s_B}\right|. \tag{1.3-1}$$

Note that because the stoichiometric coefficients of both reactants are negative, we should use the absolute value of each term. When A and B are in stoichiometric proportion, (1.3-1) reduces to

$$\frac{N_A(0)}{s_A} = \frac{N_B(0)}{s_B} \quad \text{or} \quad \frac{F_{A_{in}}}{s_A} = \frac{F_{B_{in}}}{s_B}. \quad (1.3\text{-}2)$$

This is the mathematical condition for stoichiometric proportion of the reactants.

The procedure for identifying the limiting reactant of a chemical reaction is simple and goes as follows: for each reactant calculate

$$\left|\frac{N_j(0)}{s_j}\right| \quad \text{or} \quad \left|\frac{F_{j_{in}}}{s_j}\right|.$$

The reactant with the smallest value is the limiting reactant. Since very rarely a chemical reaction has more than three reactants, the procedure is rather short. When the reactants are in stoichiometric proportion, any reactant can be used as the limiting reactant.

In the remainder of the monograph, the subscript "A" denotes the limiting reactant; other species, either reactants or products, are labeled by other letters. In the two examples below, we identify the limiting reactant and show how to relate the extent to different measurable quantities.

Example 1-4 The gas-phase reaction

$$2\,CO + O_2 \rightarrow 2\,CO_2,$$

takes place in a closed, constant-volume batch reactor at isothermal conditions. Initially, the reactor contains 1.5 kmole of CO, 1 kmole of O_2, 1 kmole of CO_2, and 0.5 kmole of N_2, and its pressure is 5 atm. At time t, the reactor pressure is $P(t) = 4.5$ atm. Assuming ideal gas behavior, find:
a. the reactor pressure when the reaction goes to completion,
b. the extent of reaction at time t, and
c. the reactor composition at time t.

Solution We select the given chemical reaction as the chemical equation and the stoichiometric coefficients are:

$$s_{CO} = -2;\ s_{O2} = -1;\ s_{CO2} = 2;\ s_{N2} = 0;\ \text{and}\ \Delta = -1.$$

First, we identify the limiting reactant using (1.3-1) for each reactant,

$$\left|\frac{N_{CO}(0)}{s_{CO}}\right| = \left|\frac{1.5}{-2}\right| = 0.75 \quad \text{and} \quad \left|\frac{N_{O2}(0)}{s_{O2}}\right| = \left|\frac{1.0}{-1}\right| = 1.0. \quad (a)$$

Hence, CO is the limiting reactant.
a. At completion, $N_{CO}(t) = 0$, and, using (1.2-2), the extent of the reaction is

$$X(t) = \frac{N_{CO}(t) - N_{CO}(0)}{s_{CO}} = \frac{0 - 1.5}{-2} = 0.75 \text{ kmole}. \quad (b)$$

To determine the pressure, we derive a relationship between P(t) and X(t). For ideal gas,

$$P(t) = \frac{R \cdot T}{V} N_{tot}(t),$$
(c)

where the total number of moles in the reactor at any time t, $N_{tot}(t)$, is given by (1.2-4). Hence, for isothermal operation,

$$\frac{P(t)}{P(0)} = \frac{N_{tot}(t)}{N_{tot}(0)} = \frac{N_{tot}(0) + \Delta \cdot X(t)}{N_{tot}(0)},$$
(d)

and, in this case,

$$N_{tot}(t) = N_{tot}(0) + \Delta \cdot X(t) = N_{tot}(0) + (-1) \cdot X(t),$$

and, at t = 0, $N_{tot}(0) = 4$ kmoles, so

$$\frac{P(t)}{P(0)} = \frac{4 - X(t)}{4}.$$
(e)

The reactor pressure at completion is

$$P = (5 \text{ atm}) \frac{4 - 0.75}{4} = 4.06 \text{ atm}.$$

b. Rearranging (e), the reaction extent at time t is

$$X(t) = \left(1 - \frac{P(t)}{P(0)}\right) \cdot N_{tot}(0) = \left(1 - \frac{4.5}{5}\right) \cdot (4) = 0.4 \text{ kmole}.$$
(f)

c. Now that the extent at time t is known, we determine the amount of each species using (1.2-2),

$$N_{CO}(t) = N_{CO}(0) + s_{CO} \cdot X(t) = 1.5 + (-2) \cdot (0.4) = 0.7 \text{ kmole},$$
$$N_{O_2}(t) = N_{O_2}(0) + s_{O_2} \cdot X(t) = 1.0 + (-1) \cdot (0.4) = 0.6 \text{ kmole},$$
$$N_{CO_2}(t) = N_{CO_2}(0) + s_{CO_2} \cdot X(t) = 1.0 + (2) \cdot (0.4) = 1.8 \text{ kmole},$$
$$N_{N_2}(t) = N_{tot}(0) + \Delta \cdot X(t) = N_{tot}(0) + (0) \cdot (0.4) = 0.5 \text{ kmole},$$

and, using (1.2-4),

$$N_{tot}(t) = N_{tot}(0) + \Delta \cdot X(t) = 4.0 + (-1) \cdot (0.4) = 3.6 \text{ kmole}.$$

The species mole fractions at time t are: $y_{CO}(t) = 0.194$, $y_{O2}(t) = 0.167$, $y_{CO2}(t) = 0.5$, and $y_{N2}(t) = 0.139$.

Example 1-5 An aqueous solution with a concentration of 0.8 mole/liter of reactant A is fed into a flow reactor at a rate of 150 liter/min. The chemical reaction $2A \rightarrow B$ takes place in the reactor. Conductivity cells determine the compositions of the inlet and outlet streams. The conductivity reading at the inlet is 140 units and at the outlet is 90 units. If the conductivity is proportional to the sum of the concentrations of A and B, calculate

a. the extent of the reaction and
b. the production rate of B.

Solution The chemical reaction is

$$2\,A \rightarrow B,$$

and its stoichiometric coefficients are $s_A = -2$, $s_B = 1$, and $\Delta = -1$.

a. To determine $\dot{X}$, we have to derive a relationship between $\dot{X}$ and the conductivity. Since the conductivity is proportional to $(C_A + C_B)$, we can write

$$\frac{\lambda_{out}}{\lambda_{in}} = \frac{C_{A_{out}} + C_{B_{out}}}{C_{A_{in}} + C_{B_{in}}} = \frac{90}{140}. \tag{a}$$

The concentrations are related to the species molar flow rates by

$$C_{A_{out}} = \frac{F_{A_{out}}}{v_0} \qquad \text{and} \qquad C_{B_{out}} = \frac{F_{B_{out}}}{v_0}, \tag{b}$$

where v_0 is the volumetric flow rate. Using (1.2-8), the molar flow rates of A and B at the reactor outlets are

$$F_{A_{out}} = F_{A_{in}} + s_A \cdot \dot{X} = v_0 \cdot C_{A_{in}} + s_A \cdot \dot{X} = 120 - 2 \cdot \dot{X} \tag{c}$$

$$F_{B_{out}} = F_{B_{in}} + s_B \cdot \dot{X} = v_0 \cdot C_{B_{in}} + s_B \cdot \dot{X} = \dot{X}. \tag{d}$$

Noting that $C_{B_{in}} = 0$, we substitute (c) and (d) into (b) and the latter into (a) to obtain

$$\frac{\lambda_{out}}{\lambda_{in}} = \frac{F_{A_{in}} - \dot{X}}{F_{A_{in}}} = \frac{120 - \dot{X}}{120} = \frac{90}{140}, \tag{e}$$

or

$$\dot{X} = 120 \cdot \left(1 - \frac{90}{140}\right) = 42.86 \text{ mole/min}. \tag{f}$$

b. Using (d), the production rate of B is 42.86 mole/min.

1.4 MULTIPLE CHEMICAL REACTIONS

So far we have considered only reactors with single chemical reactions. However, in most applications, several simultaneous reactions take place. Below, we derive stoichiometric relations for reactors with multiple chemical reactions.

First, we consider a closed (batch) reactor where n simultaneous chemical reactions take place and focus our attention on one species, say species j. The total number of moles of species j in the reactor, $N_j(t)$, is related to the number of moles of species j formed by each of the individual chemical reactions

$$N_j(t) - N_j(0) = (n_j - n_{j0})_1 + \ldots + (n_j - n_{j0})_i + \ldots + (n_j - n_{j0})_n, \quad (1.4\text{-}1)$$

where $(n_j - n_{jo})_i$ is the number of moles of species j formed by the i-th chemical reaction at time t. Note that, in general, species j may be formed in some reactions and consumed in others. As will be discussed below, to determine the species composition (and, in general, all other state quantities), we have to consider only a set of **independent** reactions and not all the chemical reactions that take place. Hence, substituting (1.2-2), (1.4-1) becomes

$$N_j(t) = N_j(0) + \sum_{m}^{n_{ind}} (s_j)_m \cdot X_m(t), \quad (1.4\text{-}2)$$

where $(s_j)_m$ is the stoichiometric coefficient of species j in the m-th reaction, and $X_m(t)$ is the extent of the m-th independent reaction at time t. Eq. (1.4-2) relates the composition in a batch reactor to the extents of the independent chemical reactions. To express the mass of species j in the reactor at time t, $W_j(t)$, we multiply (1.4-2) by the molecular mass of species j, MW_j,

$$W_j(t) = W_j(0) + MW_j \sum_{m}^{n_{ind}} (s_j)_m \cdot X_m(t). \quad (1.4\text{-}3)$$

To obtain the total number of moles in the reactor at time t, $N_{tot}(t)$, we sum the terms of (1.4-2) for all species in the reactor and collect them by the individual extents,

$$N_{tot}(t) = N_{tot}(0) + \sum_{m}^{n_{ind}} \Delta_m \cdot X_m(t), \quad (1.4\text{-}4)$$

where Δ_m is the change in the number of moles per unit extent of the m-th independent reaction, defined by (1.1-6), and $N_{tot}(0)$ is the total number of moles initially in the reactor. Note that the relations for batch reactors with single reactions, (1.2-2) and (1.2-4), are special cases of (1.4-2) and (1.4-4), respectively. Also note that by summing (1.4-3) over all species and using (1.1-5), we obtain, as expected,

$$W_{tot}(t) = W_{tot}(0).$$

Similarly, for flow reactors operating at steady state, we express the generation of species j by n simultaneous reactions,

$$F_{j_{out}} = F_{j_{in}} + \sum_{m}^{n_{ind}} G_{jm}, \quad (1.4\text{-}5)$$

where G_{jm} indicates the moles of species j formed per unit time by the m-th independent chemical reaction in the reactor. Using (1.2-8), (1.4-5) becomes

$$F_{j_{out}} = F_{j_{in}} + \sum_m^{n_{ind}} (s_j)_m \cdot \dot{X}_m ,$$

(1.4-6)

where $\dot{X}_m$ is the units of extent of the m-th independent reaction converted per unit time in the reactor. Eq. (1.4-6) relates the molar flow rate of any species at the outlet of a steady flow reactor to the extents of the individual chemical reactions taking place in the reactor. To express the mass flow rate of species j at the reactor exit, we multiply (1.4-6) by the molecular mass of species j, MW_j, and obtain

$$\dot{W}_{j_{out}} = \dot{W}_{j_{in}} + MW_j \sum_m^{n_{ind}} (s_j)_m \cdot \dot{X}_m .$$

(1.4-7)

To obtain the total molar flow rate at the reactor outlet, $(F_{tot})_{out}$, we sum (1.4-6) for all species in the reactor and collect terms by the individual extents,

$$(F_{tot})_{out} = (F_{tot})_{in} + \sum_m^{n_{ind}} \Delta_m \cdot \dot{X}_m ,$$

(1.4-8)

where Δ_m is the change in the number of moles per unit extent of the m-th independent reaction, defined by (1.1-6), and $(F_{tot})_{in}$ is the total molar flow rate at the reactor inlet. Note that for steady flow reactors with single reactions, (1.2-8) and (1.2-10), are special cases of (1.4-6) and (1.4-8), respectively. Also note that by using (1.1-5) and summing (1.4-7) over all species, we obtain, as expected,

$$\dot{W}_{out} = \dot{W}_{in} .$$

Example 1-6   Consider a batch reactor where the two simultaneous chemical reactions

$$CO + \frac{1}{2} O_2 \rightarrow CO_2$$

(a)

$$C + \frac{1}{2} O_2 \rightarrow CO,$$

(b)

take place. Initially, the reactor contains 4 mole of CO, 4 mole of O_2 and 2 mole of C. At the end of the operation, the reactor contains 2 mole of CO, and 2 mole of O_2. Determine:

a. the composition of the reactor at the end of the operation and
b. the portion of O_2 that reacts in each reaction.

Solution   We select chemical equations (a) and (b) as the "system," and their stoichiometric coefficients are, respectively:

$$(s_{CO})_1 = -1; \quad (s_{O_2})_1 = -\frac{1}{2}; \quad (s_{CO_2})_1 = -1; \quad (s_C)_1 = 0; \quad \Delta_1 = -\frac{1}{2};$$

$$(s_{CO})_2 = 1; \quad (s_{O_2})_2 = -\frac{1}{2}; \quad (s_{CO_2})_2 = 0; \quad (s_C)_2 = -1; \quad \Delta_2 = -\frac{1}{2}.$$

Since each reaction has a species that does not participate in the other, the two reactions are independent.

a. We write (1.4-2) for CO,

$$N_{CO}(t) = N_{CO}(0) + (s_{CO})_1 \cdot X_1(t) + (s_{CO})_2 \cdot X_2(t) = 2 \text{ mole}.$$

Substituting the numerical values for the stoichiometric coefficients and the initial composition, we obtain

$$-X_1(t) + X_2(t) = -2. \tag{c}$$

We write (1.4-2) for O_2,

$$N_{O_2}(t) = N_{O_2}(0) + (s_{O_2})_1 \cdot X_1(t) + (s_{O_2})_2 \cdot X_2(t) = 2 \text{ mole},$$

and obtain

$$X_1(t) + X_2(t) = 4. \tag{d}$$

We solve (c) and (d) and obtain $X_1(t) = 3$ mole and $X_2(t) = 1$ mole. Now that the extents of the two independent chemical reactions are known, we can calculate the amount of C and CO in the reactor. Using (1.4-2),

$$N_C(t) = N_C(0) + (s_C)_1 \cdot X_1(t) + (s_C)_2 \cdot X_2(t) = 1.0 \text{ mole},$$

and

$$N_{CO_2}(t) = N_{CO_2}(0) + (s_{CO_2})_1 \cdot X_1(t) + (s_{CO_2})_2 \cdot X_2(t) = 3.0 \text{ mole}.$$

With the given values $N_{CO}(t) = 2$ mole, $N_{O_2}(t) = 2$ mole, and the calculated values, $N_{tot}(t) = N_{CO}(t) + N_{O_2}(t) + N_C(t) + N_{CO_2}(t) = 8$ mole. We can determine the total number of moles in the reactor at time t by using (1.4-4),

$$N_{tot}(t) = N_{tot}(0) + \Delta_1 \cdot X_1(t) + \Delta_2 \cdot X_2(t) =$$

$$= 10.0 + \left(-\frac{1}{2}\right) \cdot (3.0) + \left(-\frac{1}{2}\right) \cdot (1.0) = 8.0 \text{ mole}.$$

b. We use (1.2-1) to determine the number of moles of oxygen reacted by each chemical reaction,

$$(n_{O_2} - n_{O_20})_1 = (s_{O_2})_1 \cdot X_1(t) = \left(-\frac{1}{2}\right) \cdot (3.0) = -1.5 \text{ mole}$$

$$(n_{O_2} - n_{O_20})_2 = (s_{O_2})_2 \cdot X_2(t) = \left(-\frac{1}{2}\right) \cdot (1.0) = -0.5 \text{ mole}.$$

The minus sign indicates that oxygen is being consumed in the chemical reactions. Thus, 75% of the oxygen reacts by reaction (a) and 25% by reaction (b).

Example 1-7 Wafers for integrated circuits are made of pure silicon that is produced by reacting raw silicon with HCl to form silicon trichloride, $SiHCl_3$. (The silicon trichloride is later reduced with hydrogen to provide pure silicon.) At the reactor operating conditions, silicon tetrachloride, $SiCl_4$, is also formed. Molten raw silicon is fed into a flow reactor operated at 1250°C at a rate of 80 lbmole/hr, and gaseous HCl is fed in proportion of 4 mole HCl per mole silicon. The following reactions take place in the reactor

$$Si + 3\ HCl \rightarrow SiHCl_3 + H_2 \tag{a}$$

$$Si + 4\ HCl \rightarrow SiCl_4 + 2\ H_2. \tag{b}$$

All the silicon fed into the reactor is reacted, and the effluent stream contains 40% H_2 (by mole). Calculate:
a. the production rate of $SiHCl_3$ and
b. the amount of HCl reacted.

Solution The stoichiometric coefficients of chemical reactions (a) and (b) are:

$$(s_{Si})_1 = -1; \ (s_{HCl})_1 = -3; \ (s_{SiHCl3})_1 = 1; \ (s_{SiCl4})_1 = 0; \ (s_{H2})_1 = 1; \ \Delta_1 = -2;$$

$$(s_{Si})_2 = -1; \ (s_{HCl})_2 = -4; \ (s_{SiHCl3})_2 = 0; \ (s_{SiCl4})_2 = 1; \ (s_{H2})_2 = 2; \ \Delta_2 = -2.$$

Since each reaction has a species that does not participate in the other, the two chemical reactions are independent.

a. We select the inlet stream as a basis for the calculation; thus, $(F_{Si})_{in} = 80$ lbmole/hr and $(F_{HCl})_{in} = 320$ lbmole/hr. Using (1.4-6), the production rate of $SiHCl_3$ is

$$(F_{SiHCl_3})_{out} = 0 + (s_{SiHCl_3})_1 \cdot \dot{X}_1 + (s_{SiHCl_3})_2 \cdot \dot{X}_2 = \dot{X}_1. \tag{c}$$

To obtain $\dot{X}_1$, we first use the fact that all the silicon is consumed in the reactor; hence $(F_{Si})_{out} = 0$. Using (1.4-6),

$$(F_{Si})_{out} = 80 + (-1) \cdot \dot{X}_1 + (1) \cdot \dot{X}_2,$$

or

$$\dot{X}_1 + \dot{X}_2 = 80 \text{ lbmole/hr}. \tag{d}$$

To obtain $\dot{X}_2$, we use the given composition of H_2 in the product stream,

$$(y_{H_2})_{out} = \frac{(F_{H_2})_{out}}{(F_{tot})_{out}} = 0.4. \tag{e}$$

Using (1.4-6),

$$(F_{H_2})_{out} = 0 + \dot{X}_1 + 2 \cdot \dot{X}_2, \tag{f}$$

and using (1.4-8),

$$(F_{tot})_{out} = (400) - 2 \cdot \dot{X}_1 - 2 \cdot \dot{X}_2. \tag{g}$$

Substituting (f) and (g) into (e), we obtain

$$(1.8) \cdot \dot{X}_1 + (2.8) \cdot \dot{X}_2 = 160 \text{ lbmole/min} . \tag{h}$$

Solving (h) and (d), we obtain $\dot{X}_1 = 71.11$ lbmole/hr and $\dot{X}_2 = 8.89$ lbmole/hr, and from (c), the production rate of $SiHCl_3$ is $(F_{SiHCl3})_{out} = 71.11$ lbmole/hr.
b. Using (1.4-6), the flow rate of HCl at the reactor exit is

$$(F_{HCl})_{out} = 320 + (-3) \cdot (71.11) + (-4) \cdot (8.89) = 71.11 \text{ lbmole/hr} . \tag{i}$$

The amount of HCl reacted in the reactor is

$$(F_{HCl})_{in} - (F_{HCl})_{out} = 320 - 71.11 = 248.89 \text{ lbmole/hr} . \tag{e}$$

1.5 DIMENSIONLESS EXTENT AND REACTANT CONVERSION

The stoichiometric relations derived above provide a glimpse at the key role the extent plays in the analysis of chemical reactors; whenever the extents of the independent reactions are known, we can readily determine the reactor composition and all other state variables (temperature, enthalpy, etc.). Unfortunately, the extent has two deficiencies:
* it is not a measurable quantity, and consequently must be related to other measurable quantities (concentrations, pressure, etc.), and
* it is an extensive quantity depending on the amount of reactants initially in the reactor or on the inlet flow rate into the reactor.

While the use of calculated quantities rather than measurable quantities seems, at first, cumbersome and even counter productive, it actually simplifies the analysis of chemical reactors with multiple reactions. This approach is similar to the use of calculated quantities such as enthalpy and free energy in thermodynamics that results in simplified expressions. Here too, by using the extents of independent reactions, we formulate the design of chemical reactors by the smallest number of design equations. Examples 1-4 and 1-5 illustrated how to determine the reaction extent from measured quantities.

To characterize the generic behavior of chemical reactors, we would like to describe their operations in terms of intensive quantities. To convert the extents to extensive quantities, we define dimensionless extents. For chemical reactors with single reactions, it is common to define and use the conversion of the limiting reactant. We define the dimensionless extent, Z_m, of the m-th independent reaction by selecting a convenient reference state or a reference stream. For batch reactors, it is defined by

$$Z_m = \frac{X_m}{(N_{tot})_0} , \qquad \left[\frac{mole}{mole} \right] \tag{1.5-1}$$

where $(N_{tot})_0$ is the total number of moles initially in the reactor. For flow reactors, Z_m is defined by

$$Z_m = \frac{\dot{X}_m}{(F_{tot})_0} \qquad \left[\frac{mole/time}{mole/time}\right] \qquad (1.5\text{-}2)$$

where $(F_{tot})_0$ is the total molar flow rate of a reference stream, which is, in many instances, the inlet stream to the reactor. The numerical values of dimensionless extents depend on the reference state or reference stream selected as well as on the stoichiometry of the chemical reactions. In most cases, they vary between -1 and 1, and very rarely, their absolute value is greater than 3.

For batch reactors, using (1.5-1), stoichiometric relation (1.4-2) reduces to

$$N_j(t) = (N_{tot})_0 \left(y_j(0) + \sum_{m}^{n_{ind}} (s_j)_m \cdot Z_m(t) \right), \qquad (1.5\text{-}3)$$

where $y_j(0) = N_j(0)/(N_{tot})_0$ is the molar fraction of species j in the reference state. Similarly, stoichiometric relation (1.4-4) reduces to

$$N_{tot}(t) = (N_{tot})_0 \left(1 + \sum_{m}^{n_{ind}} \Delta_m \cdot Z_m(t) \right). \qquad (1.5\text{-}4)$$

For flow reactors, using (1.5-2), stoichiometric relation (1.4-6) reduces to

$$F_j = (F_{tot})_0 \left(y_{j0} + \sum_{m}^{n_{ind}} (s_j)_m \cdot Z_m \right), \qquad (1.5\text{-}5)$$

where $y_{j0} = F_{j0}/(F_{tot})_0$ is the molar fraction of species j in the reference stream. Similarly, (1.4-8) reduces to

$$F_{tot} = (F_{tot})_0 \left(1 + \sum_{m}^{n_{ind}} \Delta_m \cdot Z_m \right). \qquad (1.5\text{-}6)$$

Another dimensionless, intensive quantity commonly used is the conversion of a reactant. It is defined as the fraction of the reactant that has been consumed in the reactor. For batch reactors, we define the conversion of reactant A at time t, $f_A(t)$, by

$$f_A(t) \equiv \frac{\text{moles of reactant A consumed in time t}}{\text{moles of reactant A initially in the reactor}} = \frac{N_A(0) - N_A(t)}{N_A(0)}. \qquad (1.5\text{-}7a)$$

For flow reactors operating at steady state, we define the conversion of reactant A in the reactor by

$$f_A \equiv \frac{\text{rate reactant A consumed inside the reactor}}{\text{rate reactant A fed into the reactor}} = \frac{F_{A_{in}} - F_{A_{out}}}{F_{A_{in}}}. \qquad (1.5\text{-}7b)$$

Note that the conversion is defined only for reactants, and, by definition, it may vary between 0 and 1. Also note that the conversion is related to the composition or flow rate in the reactor, and it is not defined on the basis of a chemical reaction. In general, species A may be a reactant in several chemical reactions. However, if species A is a product in any independent chemical reaction, the conversion of A is not defined.

When a **single** chemical reaction takes place in a reactor, the conversion of a reactant relates to the extent of the reaction. For batch reactors with a single reaction, from (1.2-2) and (1.5-1a),

$$N_A(0) - N_A(t) = -s_A \cdot X_1(t) = -s_A \cdot (N_{tot})_0 \cdot Z(t),$$

and, substituting in (1.5-7), the relationship between the conversion of reactant A and the extent is

$$f_A(t) = -\frac{s_A}{N_A(0)} X(t) = -\frac{s_A}{y_A(0)} Z(t). \qquad (1.5\text{-}8)$$

To express the number of moles of any species in terms of the conversion of reactant A, we substitute $X(t)$ from (1.5-8) into (1.2-3) and obtain

$$N_j(t) = N_j(0) - \frac{s_j}{s_A} N_A(0) \cdot f_A(t). \qquad (1.5\text{-}9)$$

To express the total number of moles in the reactor in terms of the conversion of reactant A, we substitute $X(t)$ from (1.5-8) into (1.4-4) and obtain

$$N_{tot}(t) = N_{tot}(0) - \frac{\Delta}{s_A} N_A(0) \cdot f_A(t), \qquad (1.5\text{-}10)$$

where Δ, defined by (1.2-3), denotes the change in the number of moles per unit extent.

To obtain a relationship between the conversion and the reaction extent in steady flow reactors with single chemical reactions, we write (1.2-8) for reactant A and substitute it in (1.5-8),

$$f_A = -\frac{s_A}{F_{A_{in}}} \dot{X} = -\frac{s_A}{y_{A0}} \dot{Z}. \qquad (1.5\text{-}11)$$

To express the molar flow rate of any species at the reactor outlet in terms of the conversion of reactant A, we substitute (1.5-11) into (1.2-9) and obtain

$$F_{j_{out}} = F_{j0} - \frac{s_j}{s_A} F_{A0} \cdot f_A. \qquad (1.5\text{-}12)$$

To relate the total molar flow rate at the reactor exit to the conversion, we substitute (1.5-11) into (1.2-10) and obtain

$$(F_{tot})_{out} = (F_{tot})_0 - \frac{\Delta}{s_A} F_{A0} \cdot f_A. \qquad (1.5\text{-}13)$$

When species A is a reactant in several chemical reactions, the term $N_A(0) - N_A(t)$ in (1.5-7) should account for the consumption of reactant A by all the independent reactions. Hence,

$$f_A(t) \equiv \frac{N_A(0) - N_A(t)}{N_A(0)} = -\frac{1}{N_A(0)} \sum_m^{n_{ind}} (n_A - n_{A0})_m,$$

or

$$f_A(t) = \sum_m^{n_{ind}} f_{Am}(t), \qquad (1.5\text{-}14)$$

where f_{Am} is the conversion of reactant A by the m-th independent reaction

$$f_{Am}(t) = -\frac{s_{Am}}{N_A(0)} X_m(t) = -\frac{s_{Am}}{y_A(0)} Z_m(t). \qquad (1.5\text{-}15)$$

Similarly, for steady flow reactors

$$f_A = \sum_m^{n_{ind}} f_{Am}, \qquad (1.5\text{-}16)$$

where f_{Am} is the conversion of A by the m-th independent reaction defined by

$$f_{Am} = -\frac{s_{Am}}{F_{A0}} \dot{X}_m = -\frac{s_{Am}}{y_{A0}} Z_m, \qquad (1.5\text{-}17)$$

and Z_m is defined by (1.5-2).

Example 1-8 The selective oxidation of ethylene to produce ethylene oxide is studied in a batch reactor. The following chemical reactions take place in the reactor:

Reaction 1 $2\,C_2H_4 + O_2 \rightarrow 2\,C_2H_4O$

Reaction 2 $C_2H_4 + 3\,O_2 \rightarrow 2\,CO_2 + 2\,H_2O.$

Initially, the reactor contains 4 moles of C_2H_4 and 2 moles of O_2. At time t, the molar fraction of ethylene is 0.434, and the molar fraction of oxygen is 0.0474. Calculate:
a. the conversions of ethylene and oxygen, and
b. the conversion of each reactant by each reaction.
Solution We select the two given chemical equations as the "system," and their stoichiometric coefficients are, respectively:

$(s_{C2H4})_1 = -2;$ $(s_{O2})_1 = -1;$ $(s_{C2H4O})_1 = 2;$ $(s_{CO2})_1 = 0;$ $(s_{H2O})_1 = 0;$ $\Delta_1 = -1;$

$(s_{C2H4})_2 = -1;$ $(s_{O2})_2 = -3;$ $(s_{C2H4O})_2 = 0;$ $(s_{CO2})_2 = 2;$ $(s_{H2O})_1 = 2;$ $\Delta_2 = 0.$

Since each reaction has a species that does not participate in the other, the two reactions are independent. We select the initial conditions as the reference state; hence $N_{tot}(0) = 6$ mole, and $y_{C2H4}(0) = 0.667$, $y_{O2}(0) = 0.333$, $y_{C2H4O}(0) = y_{CO2}(0) = y_{H2O}(0) = 0$.

To determine the conversions of the two reactants, we first determine the dimensionless extent of each chemical reaction. Using (1.5-3) and (1.5-4), the molar fraction of ethylene and oxygen at time t are:

$$y_{C_2H_4}(t) = \frac{N_{C_2H_4}(t)}{N_{tot}(t)} = \frac{0.667 - Z_1(t) - Z_2(t)}{1 - (0.5) \cdot Z_1(t)} = 0.434 \tag{a}$$

$$y_{O_2}(t) = \frac{N_{O_2}(t)}{N_{tot}(t)} = \frac{0.333 - (0.5) \cdot Z_1(t) - (3) \cdot Z_2(t)}{1 - (0.5) \cdot Z_1(t)} = 0.0474. \tag{b}$$

Solving (a) and (b), we obtain $Z_1 = 0.113$ and $Z_2 = 0.060$. Using (1.5-15), the conversion of ethylene by each of the chemical reactions is

$$(f_{C_2H_4})_1 = -\frac{(-1)}{0.667}(0.226) = 0.339 \quad \text{and} \quad (f_{C_2H_4})_2 = -\frac{(-1)}{0.667}(0.060) = 0.090.$$

The conversion of oxygen by each of the chemical reaction is

$$(f_{O_2})_1 = -\frac{(-0.5)}{0.333}(0.226) = 0.339 \quad \text{and} \quad (f_{O_2})_2 = -\frac{(-3)}{0.333}(0.060) = 0.541.$$

Using (1.5-13), the total conversion of ethylene is

$$(f_{C_2H_4})_1 + (f_{C_2H_4})_2 = 0.429,$$

and the total conversion of oxygen is

$$(f_{O_2})_1 + (f_{O_2})_2 = 0.880.$$

Example 1-9 Ammonia is produced in a continuous catalytic reactor according to the reaction

$$N_2 + 3\,H_2 \rightarrow 2\,NH_3.$$

A synthesis gas stream consisting of 24.5% N_2, 73.5% H_2, and 2% argon is available at a rate of 60 mole/min. At the operating conditions, the conversion per pass in the reactor is 12%. To enhance the operation, a portion of the reactor effluent stream is recycled and combined with the fresh synthesis gas as illustrated in the diagram below. If the mole fraction of the argon in the product stream is 3%, determine:
a. the ammonia production rate,
b. the overall nitrogen conversion in the process, and
c. the recycle ratio (F_5/F_4).

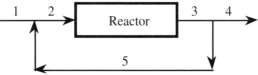

Solution The stoichiometric coefficients of the chemical reaction are:

$$s_{N2} = -1; \quad s_{H2} = -3; \quad s_{NH3} = 2; \quad s_{Ar} = 0; \quad \text{and} \quad \Delta = -2.$$

Note that the argon is an inert species, and its stoichiometric coefficient is zero.

a. We select the entire process as the "system" (see figure) and choose the inlet stream as a basis. Hence, $F_1 = (F_{tot})_0 = 60$ mole/min and $(F_{tot})_{out} = F_4$. We write an argon balance using (1.2-9),

$$(F_{Ar})_4 = (F_{Ar})_1 + s_{Ar} \cdot \dot{X} = (F_{Ar})_1$$

$$(0.03) \cdot (F_{tot})_4 = (0.02) \cdot (F_{tot})_1 = (0.02) \cdot (60 \text{ mole/min}). \tag{a}$$

Solving (a), we obtain $(F_{tot})_4 = 40$ mole/min. Using (1.2-10) to relate the total molar flow rate of the outlet stream to the extent,

$$(F_{tot})_4 = (F_{tot})_1 + \Delta \cdot \dot{X},$$

$$40 = 60 + (-2) \cdot \dot{X}, \tag{b}$$

and $\dot{X} = 10$ mole/min. Now that we know the extent of the reaction, we can calculate the production rate of ammonia by (1.2-8),

$$(F_{NH3})_4 = (F_{NH3})_1 + s_{NH3} \cdot \dot{X} = 0 + (2) \cdot (10) = 20 \text{ mole/min}. \tag{c}$$

b. We calculate the nitrogen molar flow rate at the process product stream by (1.2-8),

$$(F_{N2})_4 = (F_{N2})_1 + s_{N2} \cdot \dot{X} = (0.245) \cdot (60) + (-1) \cdot (10) = 4.70 \text{ mole/min}. \tag{d}$$

The overall nitrogen conversion in the process, calculated by (1.5-7), is

$$f_{N2} = \frac{(F_{N2})_1 - (F_{N2})_4}{(F_{N2})_1} = \frac{14.7 - 4.7}{14.7} = 0.680. \tag{e}$$

Note that we can determine the nitrogen conversion by using (1.5-11),

$$f_{N2} = -\frac{s_{N2}}{(F_{N2})_1} \dot{X} = -\frac{(-1)}{14.7}(10) = 0.680.$$

c. To calculate the recycle ratio, we define the reactor itself as the "system." In this case, $(F_{N2})_{in} = (F_{N2})_2$ and $(F_{N2})_{out} = (F_{N2})_3$. Using the given conversion per pass,

$$f_{N2} = \frac{(F_{N2})_2 - (F_{N2})_3}{(F_{N2})_2} = 0.120,$$

or

$$(F_{N2})_3 = 0.880 \cdot (F_{N2})_2. \tag{f}$$

We use (1.2-8) to express $(F_{N_2})_3$ and $(F_{N_2})_2$ in terms of the extent,

$$(F_{N_2})_3 = (F_{N_2})_2 + s_{N_2} \cdot \dot{X} = (F_{N_2})_2 + (-1) \cdot (10). \tag{g}$$

Substituting (f),

$$(0.880) \cdot (F_{N_2})_2 = (F_{N_2})_2 + (-1) \cdot (10), \tag{h}$$

we obtain $(F_{N_2})_2 = 83.33$ mole/min, and, from (f), $(F_{N_2})_3 = 73.33$ mole/min. Now, conducting a nitrogen balance over the mixer, we obtain

$$(F_{N_2})_2 = (F_{N_2})_1 + (F_{N_2})_5$$

$$83.33 = (0.245) \cdot (60) + (F_{N_2})_5. \tag{i}$$

Solving (i), we obtain $(F_{N_2})_5 = 68.63$ mole/min. Thus, the recycle ratio is

$$R = \frac{(F_{N_2})_5}{(F_{N_2})_4} = \frac{68.63}{4.70} = 14.6.$$

1.6 INDEPENDENT AND DEPENDENT CHEMICAL REACTIONS

In the preceding section, we considered chemical reactors with multiple reactions and derived mathematical expressions relating the reactor composition to the extents of the independent chemical reactions. These relationships deserve a closer examination. As indicated, the summations in (1.5-3) through (1.5-6) should be taken over a set of **independent** chemical reactions and not over **all** the reactions that take place. The concept of independent reactions, or, more accurately, independent stoichiometric relations, is an important concept in stoichiometry and reactor analysis. The number of independent reactions indicates the smallest number of stoichiometric relations needed to describe the chemical transformations that actually take place and to determine the state quantities of a chemical reactor (species composition, temperature, enthalpy, etc.). The number of independent reactions also indicates the smallest number of design equations needed to describe reactors with multiple reactions. Since state quantities are independent of the path, we can select different sets of independent reactions to describe the reactor operation. Below, we discuss the roles of dependent and independent reactions in reactor calculations. We also describe a procedure to determine the number of independent reactions and how to identify a set of independent reactions.

To understand the role of independent reactions, we first develop some insight into the concept. Consider, for example, the reversible chemical reaction

$$CO + H_2O \leftrightarrow CO_2 + H_2. \tag{1.6-1}$$

According to the stoichiometric methodology adopted here, we describe this reversible chemical reaction as two separate reactions: a forward reaction,

$$CO + H_2O \rightarrow CO_2 + H_2, \tag{1.6-1a}$$

whose stoichiometric coefficients are

$$(s_{CO})_1 = -1; \quad (s_{H2O})_1 = -1; \quad (s_{CO2})_1 = 1; \quad (s_{H2})_1 = 1,$$

and a reverse reaction,

$$CO_2 + H_2 \rightarrow CO + H_2O, \tag{1.6-1b}$$

whose stoichiometric coefficients are

$$(s_{CO})_2 = 1; \quad (s_{H2O})_2 = 1; \quad (s_{CO2})_2 = -1; \quad (s_{H2})_2 = -1.$$

Although we have here two chemical reactions, there is only one relationship for the proportions between the individual species participating in these reactions. Hence, only one chemical equation is needed to calculate the reactor composition and we can calculate the composition of the reactor whenever we know how many moles of any single species were formed or consumed. This is because the reverse reaction does not provide any new information on the amounts of individual species that are formed and consumed. Since we write chemical reactions as homogeneous algebraic equations, the second equation is merely the first equation multiplied by -1. In mathematical terms, we say that the two reactions are linearly dependent. For more complex reaction systems, the dependency between the chemical reactions may be due to a linear combination of two (or more) independent reactions.

Consider for example the following simultaneous chemical reactions:

$$C + \frac{1}{2}O_2 \rightarrow CO \tag{1.6-2a}$$

$$CO + \frac{1}{2}O_2 \rightarrow CO_2 \tag{1.6-2b}$$

$$C + O_2 \rightarrow CO_2. \tag{1.6-2c}$$

A close examination of these reactions reveals that Reaction (1.6-2c) is the sum of Reaction (1.6-2a) and Reaction (1.6-2b). Hence, we have here only two independent chemical reactions, and, to determine any state quantity, we have to consider only a set of two independent reactions. In this case, any two reactions among the three form a set of independent reactions.

To determine the number of independent reactions in reactors with multiple chemical reactions, we construct the matrix of the stoichiometric coefficients for the chemical reactions. We designate a row for each chemical reaction and a column for each chemical species and write the stoichiometric coefficient of each species in the respective reaction in the corresponding matrix element. The order that the reactions or the species are assigned in the matrix is not important; however, to avoid forming ill-behaved matrices, it is prudent to consider the important reactions first and write the species in the order they appear in the reactions. Once a column is assigned to a specific species, it should not be changed. For example, for the reversible reaction (1.6-1), we construct a stoichiometric matrix by listing the two individual reactions as they are given and the species in the order they appear in the respective reactions,

$$
\begin{array}{cccc}
\text{CO} & H_2O & CO_2 & H_2 \\
\end{array}
$$
$$
\begin{bmatrix}
-1 & -1 & 1 & 1 \\
1 & 1 & -1 & -1
\end{bmatrix}. \tag{1.6-3}
$$

For chemical reactions (1.6-1a), (1.6-1b), and (1.6-1c), the stoichiometric matrix is

$$
\begin{array}{cccc}
\text{C} & O_2 & \text{CO} & CO_2 \\
\end{array}
$$
$$
\begin{bmatrix}
-1 & -0.5 & 1 & 0 \\
0 & -0.5 & -1 & 1 \\
-1 & -1 & 0 & 1
\end{bmatrix}. \tag{1.6-4}
$$

We exploit the fact that chemical reactions are expressed as homogeneous linear equations and use elementary row operations to reduce the stoichiometric matrix to a diagonal form, using Gaussian elimination procedure. In the reduced matrix, all the elements below the diagonal (elements 1,1; 2,2; 3,3; etc.) should be zero. The number of nonzero rows in the reduced matrix indicates the number of independent chemical reactions, and the nonzero rows in the reduced matrix represent a set of independent chemical reactions for the system.

Elementary row operations are mathematical operations that we can perform on individual equations in a system of linear equations without changing the solution of the system. There are three elementary row operations: (i) interchanging any two rows, (ii) multiplying a row by a nonzero constant, and (iii) adding a scalar-multiplied row to another row. To reduce a matrix to a diagonal form, we use the three elementary row operations to modify the matrix such that all the entries below the diagonal are zeros. First, we check if the first diagonal element (the element in the first row) is nonzero. If it is nonzero, we use the three elementary row operations to convert all the elements in the first column below it to zero. For example, in stoichiometric matrix (1.6-3), the diagonal element in the first row is nonzero, and, to eliminate the entry below it, we add the first row to the second row and obtain

$$
\begin{bmatrix}
-1 & -1 & 1 & 1 \\
0 & 0 & 0 & 0
\end{bmatrix}. \tag{1.6-5}
$$

In general, when the diagonal element is zero, we replace it (if possible) with a nonzero element by interchanging the row with a lower row. We repeat the procedure for the diagonal element in the second column, then the third column, etc. until we obtain a reduced matrix. In the matrix above, all the elements below the diagonal are zero, it is a reduced matrix, and, since it has only one nonzero row, the system has one independent chemical reaction. Note that unlike a conventional Gaussian elimination procedure, we do not convert the elements on the diagonal to 1, since by multiplying a row by a negative constant we change the corresponding chemical reaction.

To determine the number of independent reactions among chemical reactions (1.6-2a, 1.6-2b) and (1.6-2c), we reduce the matrix of the stoichiometric coefficients (1.6-4). Since the diagonal element in the first row is nonzero, we leave the first row unchanged and eliminate the nonzero elements in the first column below it. Similarly, since the first element in the second row is zero, we leave the second row unchanged. To eliminate the nonzero element in the first column of the third row, we subtract the first row from the third row and obtain the following matrix:

$$\begin{bmatrix} -1 & -0.5 & 1 & 0 \\ 0 & -0.5 & -1 & 1 \\ 0 & -0.5 & -1 & 0 \end{bmatrix}. \tag{1.6-6}$$

Now we check the diagonal element of the second column. Since it is nonzero, we leave the second row unchanged and eliminate the nonzero elements in the second column below it. To eliminate the nonzero element in the second column of the third row, we subtract the second row from the third row and obtain the following matrix:

$$\begin{bmatrix} -1 & -0.5 & 1 & 0 \\ 0 & -0.5 & -1 & 1 \\ 0 & 0 & 0 & 0 \end{bmatrix}. \tag{1.6-7}$$

All the elements below the diagonal in this matrix are zero; hence, it is a reduced matrix, and, since it has two nonzero rows, there are two independent reactions.

Once we obtain a reduced stoichiometric matrix, we can identify a set of independent chemical reactions from the nonzero rows of the reduced matrix. In the case of matrix (1.6-5), the nonzero row represents chemical reaction (1.6-1a). In the case of matrix (1.6-6), the two nonzero rows represent chemical reactions (1.6-2a) and (1.6-2b). Note that the set of independent reactions is not unique, and we can generate other sets by replacing one or more reactions in the original set by a linear combination of some or all reactions in the original set. For example, chemical reaction (1.6-1b), the reverse of reaction (1.6-1a), can serve as the independent reaction of reactions (1.6-1a) and (1.6-1b). In matrix (1.6-7), we can replace the second row by the sum of the first and second row and obtain reactions (1.6-2a) and (1.6-2c) as a set of independent reactions. In fact, for this case, any two reactions of the original set form a set of independent reactions for chemical reactions (1.6-2a), (1.6-2b), and (1.6-2c). In principle, the set of independent reactions may include a reaction that does not actually take place in the reactor, yet we can use the set to calculate the reactor composition and other state variables. In practice, to simplify the calculations, we select a convenient set of independent reactions. Below, we discuss how to identify such a set.

As will be discussed later, the rates chemical species are being formed (or depleted) depend on **all** the chemical reactions that actually take place in the reactor. Hence, to design chemical reactors with multiple reactions, we have to consider all the chemical reactions that are taking place, including the **dependent** reactions. Therefore, we have to express the dependent reactions in terms of the independent reactions. Next, we describe how to do so.

Once we select a set of independent reactions, we write each of the dependent reactions as a linear combination of the independent reactions. The total number of reactions is the sum of the dependent and independent reactions

$$n_{reactions} = n_{ind} + n_{dep}. \tag{1.6-8}$$

Let index m denote the m-th independent reaction and index k the k-th dependent reaction. For example, for chemical reactions (1.6-1a) and (1.6-1b), we have two reactions but one independent reaction. We select reaction (1.6-1a) as the independent reaction; thus, m = 1, and the index of the dependent reaction is k = 2. For chemical reactions (1.6-2a), (1.6-2b), and (1.6-2c), we have three reactions and two independent reactions. If we select reactions (1.6-2a) and (1.6-2b) as the set of independent reactions, m = 1, 2, and k = 3. Let α_{km} denote the scalar factor relating the k-th dependent reaction to the m-th independent reaction. Thus, α_{km} is the multiplier of the m-th independent reaction to obtain the k-th independent reaction. To determine the numerical values of the α_{km} factors, we conduct species balances for each dependent reaction. For the k-th dependent reaction, the following set of linear equations should be satisfied for each species

$$\sum_{m}^{n_{ind}} \alpha_{km} \cdot (s_j)_m = (s_j)_k. \tag{1.6-9}$$

Hence, (1.6-9) provides a set of linear equations whose unknowns are α_{km}'s. As will be seen later, it plays an important role in formulating the design equations for chemical reactors with multiple reactions.

To illustrate how the α_{km} factors are determined, consider for example chemical reactions (1.6-1a) and (1.6-1b). We select (1.6-1a) as the independent reaction (m = 1) and reaction (1.6-1b) as the dependent reaction (k = 2). We write (1.6-9) for dependent reaction (1.6-1b) and take j = CO; thus,

$$\alpha_{21} \cdot (-1) = 1,$$

and α_{21} = -1. Indeed, reaction (1.6-1b) is the inverse of reaction (1.6-1a) and is obtained by multiplying the latter by -1. To determine the relationships between dependent and independent reactions among chemical reactions (1.6-2a), (1.6-2b) and (1.6-2c), we select (1.6-2a) and (1.6-2b) as the independent reactions and (1.6-2c) as the dependent reaction. Hence, m = 1, 2 and k = 3. We write (1.6-9) for dependent reaction (1.6-2c) for two species. For j = C,

$$\alpha_{31} \cdot (-1) + \alpha_{32} \cdot (0) = -1,$$

and we obtain α_{31} = 1. For j = CO,

$$\alpha_{31} \cdot (1) + \alpha_{32} \cdot (-1) = 0,$$

and we obtain α_{32} = 1. Hence, dependent reaction (1.6-2c) is the sum of reactions (1.6-2a) and (1.6-2b),

$$\text{Reaction 3} = \text{Reaction 1} + \text{Reaction 2}.$$

Example 1-10 Carbon disulfide is a strong solvent produced by reacting sulfur vapor with methane. The following reactions are believed to take place in the reactor:

$$\text{Reaction 1} \quad CH_4 + 2\,S \rightarrow CS_2 + 2\,H_2$$

$$\text{Reaction 2} \quad CH_4 + 4\,S \rightarrow CS_2 + 2\,H_2S$$

$$\text{Reaction 3} \quad CH_4 + 2\,H_2S \rightarrow CS_2 + 4\,H_2.$$

Methane is fed into a flow reactor at a rate of 80 kg-mole/min and vapor sulfur at a rate of 400 kg-mole/min. The methane conversion is 80%, and the hydrogen mole fraction in the product stream is 10.5%.

a. Identify a set of independent reactions.
b. Determine the production rate of carbon disulfide.
c. Determine the composition of the product stream.
d. Calculate the sulfur conversion.
e. Express each of the dependent reactions in terms of the independent reactions.

Solution   a. To determine the number of independent reactions, we first construct a matrix of stoichiometric coefficients for the given reactions (we list the species in the order they appear in the reactions)

$$\begin{array}{ccccc} CH_4 & S & CS_2 & H_2 & H_2S \end{array}$$

$$\begin{bmatrix} -1 & -2 & 1 & 2 & 0 \\ -1 & -4 & 1 & 0 & 2 \\ -1 & 0 & 1 & 4 & -2 \end{bmatrix}. \qquad (a)$$

The diagonal element in the first column is nonzero, so we do not change the first row. To eliminate the nonzero elements below the diagonal in the first column, we subtract the first row from the second row and the first row from the third row. Matrix (a) reduces to

$$\begin{bmatrix} -1 & -2 & 1 & 2 & 0 \\ 0 & -2 & 0 & -2 & 2 \\ 0 & 2 & 0 & 2 & -2 \end{bmatrix}. \qquad (b)$$

Once all the elements in the first column below the diagonal element are zero, we proceed to the second column. The diagonal element in the second column (in the second row) is nonzero, so we do not change the second row. To eliminate the nonzero elements below the diagonal in the second column in matrix (b), we add the second row from the third row. To simplify the reduced matrix, we also divide the second row by 2, and the matrix reduces to

$$\begin{bmatrix} -1 & -2 & 1 & 2 & 0 \\ 0 & -1 & 0 & -1 & 1 \\ 0 & 0 & 0 & 0 & 0 \end{bmatrix}. \qquad (c)$$

Since all the elements under the diagonal in matrix (c) are zero, this matrix is a reduced matrix. Since matrix (c) has two nonzero rows, the system has two independent chemical reactions. In this case, we can select any two reactions of the original system as a set of independent reactions. The nonzero rows in matrix (c) represent another set of independent reactions,

$$\text{Reaction 1} \quad CH_4 + 2\,S \rightarrow CS_2 + 2\,H_2$$

$$\text{Reaction 4} \quad H_2 + S \rightarrow H_2S.$$

Note that the second row in matrix (c) represents a reaction that is not among the original chemical reactions, and we denote it as Reaction 4.

We select these two reactions as the set of independent reactions; thus, the indices of the independent reactions are $m = 1, 4$, and the indices of the dependent reactions are $k = 2, 3$. The stoichiometric coefficients of the two independent reactions are:

$$(s_{CH4})_1 = -1; \quad (s_S)_1 = -2; \quad (s_{CS2})_1 = 1; \quad (s_{H2})_1 = 2; \quad (s_{H2S})_1 = 0; \quad \Delta_1 = 0;$$

$$(s_{CH4})_4 = 0; \quad (s_S)_4 = -1; \quad (s_{CS2})_4 = 0; \quad (s_{H2})_4 = -1; \quad (s_{H2S})_4 = 1; \quad \Delta_4 = -1.$$

To determine the required quantities, we express the molar flow rates of the individual species in terms of the extents of Reactions 1 and 4.

b. We select the given inlet flow rates as a basis for our calculation. Hence, $(F_{CH4})_{in} = 80$ kmole/min, $(F_S)_{in} = 400$ kmole/min, and $(F_{tot})_{in} = 480$ kmole/min. To express the molar flow rates of the individual species in the product stream, we use (1.4-6) for the individual species:

$$(F_{CH_4})_{out} = 80 + (-1) \cdot \dot{X}_1 + (0) \cdot \dot{X}_4 = 80 - \dot{X}_1, \tag{d}$$

$$(F_S)_{out} = 400 + (-2) \cdot \dot{X}_1 + (-1) \cdot \dot{X}_4 = 400 - (2) \cdot \dot{X}_1 - \dot{X}_4, \tag{e}$$

$$(F_{CS_2})_{out} = 0 + (1) \cdot \dot{X}_1 + (0) \cdot \dot{X}_4 = \dot{X}_1, \tag{f}$$

$$(F_{H_2})_{out} = 0 + (2) \cdot \dot{X}_1 + (-1) \cdot \dot{X}_4 = (2) \cdot \dot{X}_1 - \dot{X}_4, \tag{g}$$

$$(F_{H_2S})_{out} = 0 + (0) \cdot \dot{X}_1 + (1) \cdot \dot{X}_4 = \dot{X}_4, \tag{h}$$

and, using (1.4-8),

$$(F_{tot})_{out} = 480 + (0) \cdot \dot{X}_1 + (-1) \cdot \dot{X}_4 = 480 - \dot{X}_4. \tag{i}$$

Using (d) and (1.5-8), for the given CH_4 conversion,

$$f_{CH_4} = \frac{(F_{CH_4})_{in} - (F_{CH_4})_{out}}{(F_{CH_4})_{in}} = \frac{80 - (80 - \dot{X}_1)}{80} = 0.8,$$

and we obtain $\dot{X}_1 = 64$ kmole/min. Using (g) and (i), the fraction of H_2 in the product stream is

$$(y_{H_2})_{out} = \frac{(F_{H_2})_{out}}{(F_{tot})_{out}} = \frac{2 \cdot \dot{X}_1 - \dot{X}_4}{480 - \dot{X}_4} = 0.105. \tag{j}$$

Substituting $\dot{X}_1 = 64$ kmole/min, we obtain $\dot{X}_4 = 86.7$ kmole/min. Now that we know the extents of the two independent reactions, we can calculate the species molar flow rates of the effluent stream using (d) through (h): $(F_{CH4})_{out} = 16$; $(F_S)_{out} = 185.3$; $(F_{CS2})_{out} = 64$; $(F_{H2})_{out} = 41.3$; $(F_{H2S})_{out} = 86.7$; and $(F_{tot})_{out} = 393.3$ kmole/min. The production rate of CS_2 is 64 kmole/min.

c. Now that we know the molar flow rates of the individual species in the outlet stream, we can calculate the stream's composition using (j). The respective mole fractions are: $(y_{CH4})_{out} = 0.041$, $(y_S)_{out} = 0.471$, $(y_{CS2})_{out} = 0.163$, $(y_{H2})_{out} = 0.105$, and $(y_{H2S})_{out} = 0.220$.

d. Using (1.5-8), the sulfur conversion is

$$f_S = \frac{(F_S)_{in} - (F_S)_{out}}{(F_S)_{in}} = \frac{400 - 185.3}{400} = 0.537. \tag{k}$$

e. To determine the relations between the two dependent reactions (Reactions 2 and 3) and the two independent reactions (Reactions 1 and 4), we write (1.6-9) for each dependent reaction. We start with the first dependent reaction, $(k = 2)$,

$$\sum_m^{n_{ind}} \alpha_{2m} \cdot (s_j)_m = \alpha_{21} \cdot (s_j)_1 + \alpha_{24} \cdot (s_j)_4 = (s_j)_2. \tag{l}$$

We write (l) for $j = CH_4$

$$\alpha_{21} \cdot (-1) + \alpha_{24} \cdot (0) = -1, \tag{m}$$

and obtain $\alpha_{21} = 1$. Next, we write (l) for $j = H_2S$

$$\alpha_{21} \cdot (0) + \alpha_{24} \cdot (1) = 2, \tag{n}$$

and obtain $\alpha_{24} = 2$. Thus, Reaction 2 relates to the two independent reactions by

$$\text{Reaction 2} = \text{Reaction 1} + 2 \text{ Reaction 4}.$$

Next, we consider the second dependent reaction and write (1.6-9) for $k = 3$,

$$\sum_m^{n_{ind}} \alpha_{3m} \cdot (s_j)_m = \alpha_{31} \cdot (s_j)_1 + \alpha_{34} \cdot (s_j)_4 = (s_j)_3. \tag{o}$$

We write (o) for $j = CH_4$

$$\alpha_{31} \cdot (-1) + \alpha_{34} \cdot (0) = -1, \tag{p}$$

and obtain $\alpha_{31} = 1$. Next, we write (o) for $j = H_2S$,

$$\alpha_{31} \cdot (0) + \alpha_{34} \cdot (1) = -2 , \qquad (q)$$

and obtain $\alpha_{34} = -2$. Thus, Reaction 3 relates to the two independent reactions by

Reaction 3 = Reaction 1 - 2 Reaction 4.

The fact that we can determine the composition of a reactor by using a set of independent chemical reactions that includes one or more reactions that do not actually take place in the reactor raises some basic questions regarding stoichiometry. Are the actual chemical reactions taking place in the reactor irrelevant? Can we always substitute some reactions by other reactions? When should we consider the actual reactions that are taking place in the reactor? Also, since we can select different sets of independent chemical reactions, are some sets preferable to other? And, if yes, what is the optimal set? To answer these questions, we have to address the more fundamental issue of what is a chemical reaction, or, more accurately, what a written chemical reaction represents.

When we write a chemical reaction such as $A + B \rightarrow C$, we describe one of two situations. The written chemical reaction may represent a one-step reaction between an atom of reactant A and an atom of reactant B to form an atom of product C without the formation of any intermediate species. Such chemical reactions are called elementary reactions, and, in this case, the written reaction represents the actual interaction between the molecules — the actual chemical transformation. Alternatively, the written chemical reaction may represent a complex series of elementary reactions whose net result is the formation of one atom of species C and the depletion of one atom of A and one atom of species B. In this case, the written reaction is a stoichiometric presentation of the chemical transformations taking place, some of which may even be unknown. The written reaction is merely a description of changes in the species compositions taking place. For example, the reaction between hydrogen and bromine to form hydrogen bromide, $H_2 + Br_2 \rightarrow 2\ HBr$, is a stoichiometric presentation of a complex series of elementary reactions. Hence, with the exception of elementary reactions, what we commonly consider "chemical reactions" are not real chemical transformations taking place in the reactor but rather stoichiometric presentations selected to describe changes in the reactor composition.

Since species compositions are state quantities, when multiple reactions are used to describe the chemical transformations, we can select different sets of independent reactions to calculate them. Thus, to determine any state quantity of chemical reactors (composition, temperature, pressure, enthalpy, entropy, etc.), we can consider any set of independent reactions. It is analogous to thermodynamic calculations, where we can select any convenient path (process) to determine changes in the state of a system. This point was illustrated in Example 1-10 above, where the set of independent reactions selected for calculating the composition of the product stream consisted of a chemical reaction (Reaction 4) that was not among the original reactions describing the chemical transformations. On the other hand, to determine non-state quantities of chemical reactors, we have to consider all the chemical reactions used to describe the actual chemical transformations. For example, to calculate the needed reactor operating time or the needed reactor size, we have to consider the rates of all chemical reactions, including dependent reactions, actually taking place in the reactor. It is similar to calculating thermodynamic quantities that depend on the path, such as work.

What is the optimal set of independent reactions we should use in our calculations? When we have to determine only state quantities of chemical reactors, the optimal set of independent reactions is a set that includes the smallest number of species. This is the set that consists of the reactions corresponding to the nonzero rows of the reduced matrix of stoichiometric coefficients (obtained in the elimination procedure to determine the number of independent reactions). By using this set, we reduce the number of terms in the equations and minimize the number of calculations needed to determine the required quantities. When we have to determine non-state quantities of chemical reactors, the optimal set of independent reactions is a set consisting of reactions whose rates are known. Among those reactions, we select a set that includes the smallest number of species, consequently reducing the number of terms in the equations. This issue is discussed in depth in Chapter 3.

1.7 INDEPENDENT COMPOSITION SPECIFICATIONS

In the preceding section, we discussed how to determine the number of independent chemical reactions and how to select a set of independent reactions. The number of independent reactions indicates the number of equations that should be solved to determine the composition of the reactor and all other state quantities. To solve these equations, a number of conditions, equal to the number of equations, must be specified. However, the specified conditions are not arbitrary; they should provide two types of independent information:
- on all chemical transformations (independent reactions) and
- on the proportions among all individual species.

Below, we describe a method to determine, a priori, what sets of species compositions can be specified and what sets cannot.

Let's examine more closely the nature of the problem. Consider first a chemical reactor with the following two reactions:

$$N_2 + 3 H_2 \rightarrow 2 NH_3$$

$$2 NO + O_2 \rightarrow 2 NO_2.$$

Since each chemical reaction has at least one species that does not participate in the other, both reactions are independent. Hence, we need to specify two conditions to determine the extents of the two reactions. At first glance, it seems that we can specify the composition of any two species. However, a closer examination reveals that these are background reactions (no species participates in both), and, by specifying the amount of two species from the same chemical reaction, we cannot determine the extents of the second reaction and the reactor composition. Rather, we have to specify two independent compositions such that each relates to a distinct independent chemical reaction. Similar situations may arise when we deal with more complex sets of chemical reactions, where the identification of independent compositions is not so obvious. Consider next the hydrogenation of toluene where diphenyl is formed in an undesirable reaction. The following two reactions take place:

$$C_6H_5CH_3 + H_2 \rightarrow C_6H_6 + CH_4, \tag{1.7-1}$$

and

$$2\, C_6H_5CH_3 + H_2 \rightarrow (C_6H_5)_2 + 2\, CH_4. \tag{1.7-2}$$

These two chemical reactions are independent, yet if we specify the amounts of toluene and methane, we cannot determine the composition of other species. The reason is that, while the two reactions are independent, we do not provide independent information on all the species. In this case, compositions of the toluene (T) and methane (M) do not provide us with a relationship on the amount of diphenyl (D) and hydrogen (H). This becomes evident if we multiply reaction (1.7-1) by two and subtract reaction (1.7-2) to obtain

$$(C_6H_5)_2 + H_2 \rightarrow 2\, C_6H_6. \tag{1.7-3}$$

It is clear that specifications on the amounts of toluene and methane do not provide information on the amounts of diphenyl and hydrogen.

The independent composition specifications depend on the relationships among individual species and the chemical reactions taking place in the reactor, but they are invariant of the specific set of independent reactions selected (they apply to any sets of independent reactions selected). Since the set of independent reactions generated by Gaussian elimination consists of chemical reactions with the least number of species, we use it to identify restrictions on composition specifications. To determine the extents of the independent chemical reactions, we usually specify either the amount of individual species (moles of molar flow rates) or a quantity related to all the independent reactions (total number of moles, pressure of the system, etc.) together with compositions of certain species. Note that specifications of species mole fractions contain information on the total number of moles or, for flow reactors, on the molar flow rate.

To identify independent composition specifications, we construct a matrix of independent specifications. It is obtained by taking the reduced matrix (of the Gaussian elimination) and adding to it a column representing information of the total reactor content. Since only independent reactions whose net change in the number of moles (factor Δ defined in (1.1-6)) are not zero affect the total number of moles, the entries in this column are the corresponding Δ factors of each reaction. A set of independent composition specifications is one where each row in this matrix is specified by a non-zero element from a distinct column. When species amounts are specified, we should use information related to distinct species, each participating in a different independent reaction. When information related to the total number of moles is specified, one non-zero element from the last column (of Δ factors) is selected, and we indicate its row. Then, for each of the other rows, we can specify a species corresponding to a nonzero element from distinct columns. For example, for chemical reactions (1.7-1) and (1.7-2), the reduced matrix of stoichiometric coefficients is

$$
\begin{array}{ccccc}
\text{T} & \text{H} & \text{B} & \text{M} & \text{D} \\
\end{array}
$$
$$
\begin{bmatrix}
-1 & -1 & 1 & 1 & 0 \\
0 & 1 & -2 & 0 & 1
\end{bmatrix}. \tag{1.7-4}
$$

The two rows in Matrix (1.7-4) correspond to reactions (1.7-1) and (1.7-3), respectively. The Δ factor of each of them is zero; hence, the matrix of independent specifications is

$$
\begin{array}{cccccc}
\text{T} & \text{H} & \text{B} & \text{M} & \text{D} & \Delta \\
\end{array}
$$
$$
\begin{bmatrix}
-1 & -1 & 1 & 1 & 0 & 0 \\
0 & 1 & -2 & 0 & 1 & 0
\end{bmatrix}. \tag{1.7-5}
$$

It is clear from matrix (1.7-5) that to obtain a nonzero element from each row, we can specify the composition of any two species except methane and toluene.

The conditions of independent composition specifications are demonstrated in the example below. We show that dependent and independent specifications apply regardless of what set of independent reactions is selected.

Example 1-11 The catalytic gas-phase oxidation of ammonia is investigated in an isothermal batch reactor. The following reactions take place in the reactor:

$$\text{Reaction 1} \quad 4\,NH_3 + 5\,O_2 \rightarrow 4\,NO + 6\,H_2O$$

$$\text{Reaction 2} \quad 4\,NH_3 + 3\,O_2 \rightarrow 2\,N_2 + 6\,H_2O$$

$$\text{Reaction 3} \quad 2\,NO + O_2 \rightarrow 2\,NO_2$$

$$\text{Reaction 4} \quad 4\,NH_3 + 6\,NO \rightarrow 5\,N_2 + 6\,H_2O.$$

Initially, the reactor contains 4 mole of NH_3 and 6 mole of O_2. At time t, the reactor contains 0.8 mole of NH_3, 3 mole of O_2, 0.4 mole of NO, 4.8 mole of H_2O, 0.4 mole of NO_2, and 1.2 mole of N_2. Use only three items of the given reactor data at time t in (d) and (e) below.

a. Determine the number of independent chemical reactions.
b. Identify a set of independent chemical reactions.
c. Identify a set of independent reactions among the given reactions.
d. Identify a set of species compositions that can be specified to determine the extents and sets that cannot be specified. Show for one set of each that this is the case using a set of independent reactions from (b) and (c).
e. If the final pressure of the reactor is specified, identify a set of two species compositions that can be specified to determine the extents and sets that cannot be specified. Show for one set of each that this is the case using a set of independent reactions from (b) and (c).

Solution   a. To determine the number of independent reactions, we first construct a matrix of stoichiometric coefficients for the given reactions (we list the species in the order they appear in the reactions)

$$\begin{array}{cccccc} NH_3 & O_2 & NO & H_2O & N_2 & NO_2 \end{array}$$

$$\begin{bmatrix} -4 & -5 & 4 & 6 & 0 & 0 \\ -4 & -3 & 0 & 6 & 2 & 0 \\ 0 & -1 & -2 & 0 & 0 & 2 \\ -4 & 0 & -6 & 6 & 5 & 0 \end{bmatrix}. \tag{a}$$

We conduct a Gaussian elimination procedure and reduce matrix (a) to

$$\begin{array}{cccccc} NH_3 & O_2 & NO & H_2O & N_2 & NO_2 \end{array}$$

$$\begin{bmatrix} -4 & -5 & 4 & 6 & 0 & 0 \\ 0 & 1 & -2 & 0 & 1 & 0 \\ 0 & 0 & -4 & 0 & 1 & 2 \\ 0 & 0 & 0 & 0 & 0 & 0 \end{bmatrix}. \tag{b}$$

Matrix (b) is a reduced matrix, and, since it has three nonzero rows, there are three independent chemical reactions.

b. The nonzero rows in matrix (b) represent the following set of independent reactions,

$$\text{Reaction 1}\quad 4\,NH_3 + 5\,O_2 \rightarrow 4\,NO + 6\,H_2O$$

$$\text{Reaction 5}\qquad 2\,NO \rightarrow O_2 + N_2$$

$$\text{Reaction 6}\qquad 4\,NO \rightarrow N_2 + 2\,NO_2.$$

Note that the second and third row in the matrix represent two chemical reactions that are not in the original set of reactions; we denote them as Reaction 5 and Reaction 6. The stoichiometric coefficients of these three independent chemical reactions are:

$(s_{NH3})_1 = -4;\ (s_{O2})_1 = -5;\ (s_{NO})_1 = 4;\ (s_{H2O})_1 = 6;\ (s_{N2})_1 = 0;\ (s_{NO2})_1 = 0;\ \Delta_1 = 1;$

$(s_{NH3})_5 = 0;\ (s_{O2})_5 = 1;\ (s_{NO})_5 = -2;\ (s_{H2O})_5 = 0;\ (s_{N2})_5 = 1;\ (s_{NO2})_5 = 0;\ \Delta_5 = 0;$

$(s_{NH3})_6 = 0;\ (s_{O2})_6 = 0;\ (s_{NO})_6 = -4;\ (s_{H2O})_6 = 0;\ (s_{N2})_6 = 1;\ (s_{NO2})_6 = 2;\ \Delta_6 = -1.$

To determine all the state quantities of the reactors, we express them in terms of the extents of these three independent reactions, Z_1, Z_5, and Z_6.

c. To select a set of independent reactions among the original reactions, we use the matrix of stoichiometric coefficients (a). Noting that NO_2 participates only in Reaction 3, this chemical reaction should be included in any independent set among the original reactions. In this case, we can construct a set of independent reactions by selecting Reaction 3 and any two of the remaining three reactions. We select Reactions

1, 2, and 3, and the stoichiometric coefficients of these three independent chemical reactions are:

$$(s_{NH3})_1 = -4; \ (s_{O2})_1 = -5; \ (s_{NO})_1 = 4; \ (s_{H2O})_1 = 6; \ (s_{N2})_1 = 0; \ (s_{NO2})_1 = 0; \ \Delta_1 = 1;$$

$$(s_{NH3})_2 = -4; \ (s_{O2})_2 = -3; \ (s_{NO})_2 = 0; \ (s_{H2O})_2 = 6; \ (s_{N2})_2 = 2; \ (s_{NO2})_2 = 0; \ \Delta_2 = 1;$$

$$(s_{NH3})_3 = 0; \ (s_{O2})_3 = -1; \ (s_{NO})_3 = -2; \ (s_{H2O})_3 = 0; \ (s_{N2})_3 = 0; \ (s_{NO2})_3 = 2; \ \Delta_3 = -1.$$

To determine all the state quantities of the reactors, we express them in terms of the dimensionless extents of these three independent reactions, Z_1, Z_2, and Z_3.

d. To identify sets of independent specifications, we construct a matrix of independent specifications by taking the nonzero rows of the reduced matrix (b) and adding to it a column of the corresponding Δ factors,

$$\begin{array}{ccccccc} NH_3 & O_2 & NO & H_2O & N_2 & NO_2 & \Delta \end{array}$$
$$\begin{bmatrix} -4 & -5 & 4 & 6 & 0 & 0 & 1 \\ 0 & 1 & -2 & 0 & 1 & 0 & 0 \\ 0 & 0 & -4 & 0 & 1 & 2 & -1 \end{bmatrix}. \tag{c}$$

To identify sets of species amounts that can be specified to determine the composition of the reactor, for each row in matrix (c) we select a nonzero element from a distinct column among the first six columns. An examination of matrix (c) reveals that since NH_3 and H_2O participate only in Reaction 1, their amount cannot be specified together. Hence, we can specify the amount of any three species as long as the set does not include both NH_3 and H_2O. Let's examine what happens when we specify the amount of NH_3, H_2O, and NO, and use Reactions 1, 5, and 6 as a set of independent reactions. We select the initial state as the reference state; hence, $(N_{tot})_0 = N_{tot}(0) = 10$ moles, $y_{NH3}(0) = 0.4$, $y_{O2}(0) = 0.6$, and $y_{NH3}(0) = y_{H2O}(0) = y_{NO}(0) = y_{NO2}(0) = 0$. Applying (1.5-3) for each of the three species, we obtain

$$N_{NH_3}(t) = N_{tot}(0) \left[y_{NH_3}(0) + (s_{NH_3})_1 \cdot Z_1 + (s_{NH_3})_5 \cdot Z_5 + (s_{NH_3})_6 \cdot Z_6 \right]$$
$$10 \cdot (0.4 - 4 \cdot Z_1) = 0.800 \tag{d}$$

$$N_{H_2O}(t) = N_{tot}(0) \left[y_{H_2O}(0) + (s_{H_2O})_1 \cdot Z_1 + (s_{H_2O})_5 \cdot Z_5 + (s_{H_2O})_6 \cdot Z_6 \right]$$
$$10 \cdot (6 \cdot Z_1) = 4.800 \tag{e}$$

$$N_{NO}(t) = N_{tot}(0) \left[y_{NO}(0) + (s_{NO})_1 \cdot Z_1 + (s_{NO})_5 \cdot Z_5 + (s_{NO})_6 \cdot Z_6 \right]$$
$$10 \cdot (4 \cdot Z_1 - 2 \cdot Z_5 - 4 \cdot Z_6) = 0.400. \tag{f}$$

Rearranging (d) and (e), we discover that these two equations are identical; hence, we cannot solve (d), (e), and (f). Similarly, we use Reactions 1, 2, and 3 as a set of independent reactions and obtain:

$$N_{NH_3}(t) = N_{tot}(0)\left[y_{NH_3}(0) + (s_{NH_3})_1 \cdot Z_1 + (s_{NH_3})_2 \cdot Z_2 + (s_{NH_3})_3 \cdot Z_3\right]$$
$$10 \cdot (0.4 - 4 \cdot Z_1 - 4 \cdot Z_2) = 0.800 \tag{g}$$

$$N_{H_2O}(t) = N_{tot}(0)\left[y_{H_2O}(0) + (s_{H_2O})_1 \cdot Z_1 + (s_{H_2O})_2 \cdot Z_2 + (s_{H_2O})_3 \cdot Z_3\right]$$
$$10 \cdot (6 \cdot Z_1 + 6 \cdot Z_2) = 4.800 \tag{h}$$

$$N_{NO}(t) = N_{tot}(0)\left[y_{NO}(0) + (s_{NO})_1 \cdot Z_1 + (s_{NO})_2 \cdot Z_2 + (s_{NO})_3 \cdot Z_3\right]$$
$$10 \cdot (4 \cdot Z_1 - 2 \cdot Z_3) = 0.400. \tag{i}$$

Rearranging (g) and (h), we discover that these two equations are also identical. It is easy to verify that if we specify the amount of any three species not including both NH_3 and H_2O, we can solve these equations and determine the reactor composition.

e. To identify sets of independent specifications that include information on the total number of moles, we select a row with a nonzero element in the last column of matrix (c), and then, for each of the remaining two rows, we select a nonzero element from a distinct column among the first six columns. An examination of matrix (c) indicates that the second row does not have a nonzero element in the last column. Hence, any specified information on the total number of moles does not relate to Reaction 5. Consequently, the species specifications should include one on a species that participates in Reaction 5 (O_2, N_2 or NO). Let's examine what does happen when we specify the reactor pressure at time t (3.12 atm) and the amounts of NH_3 and NO_2. First, we use Reactions 1, 5, and 6 as a set of independent reactions. Applying (1.5-4), the pressure relates to extents of the independent reactions by

$$\frac{P(t)}{P(0)} = \frac{N_{tot}(t)}{N_{tot}(0)} = 1 + \Delta_1 \cdot Z_1 + \Delta_5 \cdot Z_5 + \Delta_6 \cdot Z_6$$
$$1 + Z_1 - Z_6 = 1.06. \tag{j}$$

Using (1.5-3) for NH_3 and NO_2, we obtain

$$N_{NH_3}(t) = N_{tot}(0)\left[y_{NH_3}(0) + (s_{NH_3})_1 \cdot Z_1 + (s_{NH_3})_5 \cdot Z_5 + (s_{NH_3})_6 \cdot Z_6\right]$$
$$10 \cdot (0.4 - 4 \cdot Z_1) = 0.800 \tag{k}$$

$$N_{NO_2}(t) = N_{tot}(0)\left[y_{NO_2}(0) + (s_{NO_2})_1 \cdot Z_1 + (s_{NO_2})_5 \cdot Z_5 + (s_{NO_2})_6 \cdot Z_6\right]$$
$$10 \cdot (2 \cdot Z_6) = 0.400. \tag{l}$$

Rearranging (j), (k), and (l), we discover that these equations do not include Z_5; therefore, we cannot determine the reactor composition. When we use Reactions 1, 2, and 3 as a set of independent reactions, we obtain:

$$\frac{P(t)}{P(0)} = \frac{N_{tot}(t)}{N_{tot}(0)} = 1 + \Delta_1 \cdot Z_1 + \Delta_2 \cdot Z_2 + \Delta_3 \cdot Z_3$$

$$1 + Z_1 + Z_2 - Z_3 = 1.06. \tag{m}$$

Using (1.5-3) for NH_3 and NO_2, we obtain

$$N_{NH_3}(t) = N_{tot}(0)\left[y_{NH_3}(0) + (s_{NH_3})_1 \cdot Z_1 + (s_{NH_3})_2 \cdot Z_2 + (s_{NH_3})_3 \cdot Z_3\right]$$

$$10 \cdot (0.4_1 - 4 \cdot Z_1 - 4 \cdot Z_2) = 0.800 \tag{n}$$

$$N_{NO_2}(t) = N_{tot}(0)\left[y_{NO_2}(0) + (s_{NO_2})_1 \cdot Z_1 + (s_{NO_2})_5 \cdot Z_5 + (s_{NO_2})_6 \cdot Z_6\right]$$

$$10 \cdot (4 \cdot Z_3) = 0.400. \tag{o}$$

Rearranging (m), (n), and (o), we discover that (m) and (n) reduce to an identical equation; hence we cannot determine the reactor composition. It is easy to verify that if we specify the amount of either NO, N_2, or O_2 instead of NO_2, we can determine the reactor composition.

1.8 OTHER STOICHIOMETRIC QUANTITIES

In the preceding sections, we identified and discussed all the stoichiometric quantities and relationships needed to analyze and design chemical reactors. However, in many applications, it is convenient to define special quantities that characterize the operation or certain conditions. They provide us with a quick measure of important operating or performance parameters. In this section, we define and discuss some stoichiometric quantities that are commonly used in reactor analysis.

1.8.1 Excess Reactant and Excess Air

Excess reactant is a quantity related to the proportion of the reactants. It indicates the surplus amount of a reactant provided over the stoichiometric amount. It is usually used in combustion reactions to indicate the surplus amount of oxygen and to characterize the combustion conditions. The excess amount of reactant B is defined by

$$\left\{\begin{matrix} excess \\ B \end{matrix}\right\} \equiv \frac{(\text{amount of B supplied}) - (\text{stoichiometric amount of B required})}{(\text{stoichiometric amount of B required})}. \tag{1.8-1}$$

Note that, although the excess amount of a reactant applies to the proportion of the reactants, it is defined with respect to a selected set of independent chemical reactions that determines the stoichiometric amount needed. For combustion reactions, the convention is to select chemical reactions that represent complete oxidation of all the fuel components to their highest oxidation level (all carbon atoms to CO_2, all sulfur atoms to SO_2, etc.). Hence, although other chemical reactions may take place during the operation, generating CO and other products, the excess oxygen is defined and calculated on the basis of complete oxidation reactions.

Essentially all combustion operations involve the introduction of an external source of oxygen, usually in the form of air. To characterize the combustion conditions, a quantity related to the **external** oxygen, called Excess Air, is defined by

$$\left\{\begin{array}{c} \text{excess} \\ \text{air} \end{array}\right\} \equiv \frac{(\text{external air supplied}) - (\text{stoichiometric external air required})}{(\text{stoichiometric external air required})} . \quad (1.8\text{-}2)$$

Note that Excess Air is defined on the basis of stoichiometric amount of **external** oxygen at the exclusion of the oxygen fed with the fuel stream.

Example 1-12 A gaseous fuel consisting of 72% CH_4, 24% C_2H_6, 3% N_2, and 1% O_2 (mole %) is fed into a combustion chamber at a rate of 10 g-mole/min. A steam of external air is mixed with the fuel, and the following chemical reactions are believed to take place in the combustion chamber:

$$\text{Reaction 1} \quad CH_4 + 2\,O_2 \rightarrow CO_2 + 2\,H_2O$$

$$\text{Reaction 2} \quad 2\,CH_4 + 3\,O_2 \rightarrow 2\,CO + 4\,H_2O$$

$$\text{Reaction 3} \quad 2\,C_2H_6 + 7\,O_2 \rightarrow 4\,CO_2 + 6\,H_2O$$

$$\text{Reaction 4} \quad 2\,C_2H_6 + 5\,O_2 \rightarrow 4\,CO + 6\,H_2O$$

$$\text{Reaction 5} \quad 2\,CO + O_2 \rightarrow 2\,CO_2.$$

An analysis of the flue gas indicates that all the ethane has been converted and that, on a dry basis, its composition is 83.96% N_2, 7.05% CO_2, 0.18% CO, and 8.69% O_2 (mole percent). Determine

a. the flow rate of the air fed,
b. the conversions of methane, ethane, and oxygen,
c. the excess amount of oxygen fed, and
d. the excess air fed.

Solution First, we determine the number of independent reactions. We construct a matrix of stoichiometric coefficients for the given reactions,

$$\begin{array}{cccccc} CH_4 & O_2 & CO_2 & H_2O & CO & C_2H_6 \end{array}$$
$$\begin{bmatrix} -1 & -2 & 1 & 2 & 0 & 0 \\ -2 & -3 & 0 & 4 & 2 & 0 \\ 0 & -7 & 4 & 6 & 0 & -2 \\ 0 & -5 & 0 & 6 & 4 & -2 \\ 0 & -1 & 2 & 0 & -2 & 0 \end{bmatrix} . \quad (a)$$

Applying Gaussian elimination, matrix (a) reduces to

$$
\begin{bmatrix}
-1 & -2 & 1 & 2 & 0 & 0 \\
0 & 1 & -2 & 0 & 2 & 0 \\
0 & 0 & -5 & 3 & 7 & -1 \\
0 & 0 & 0 & 0 & 0 & 0 \\
0 & 0 & 0 & 0 & 0 & 0
\end{bmatrix}. \qquad \text{(b)}
$$

Since matrix (b) has three nonzero rows, there are three independent chemical reactions. We select Reactions 1, 3, and 5 as a set of independent reactions, and their stoichiometric coefficients are:

$(s_{CH4})_1 = -1$; $(s_{O2})_1 = -2$; $(s_{CO2})_1 = 1$; $(s_{H2O})_1 = 2$; $(s_{CO})_1 = 0$; $(s_{C2H6})_1 = 0$; $\Delta_1 = 0$;

$(s_{CH4})_3 = 0$; $(s_{O2})_3 = -7$; $(s_{CO2})_3 = 4$; $(s_{H2O})_3 = 6$; $(s_{CO})_3 = 0$; $(s_{C2H6})_3 = 0$; $\Delta_3 = 1$;

$(s_{CH4})_5 = 0$; $(s_{O2})_5 = -1$; $(s_{CO2})_5 = 2$; $(s_{H2O})_5 = 0$; $(s_{CO})_5 = -2$; $(s_{C2H6})_5 = 2$; $\Delta_5 = -1$.

Since the flow rate of the air is not given, it is more convenient to use the extents rather than the dimensionless extents of the independent reactions. Hence, we have to determine the extents of Reactions 1, 3, and 5..

a. We select the fuel feed as a basis for the calculation ($F_1 = 10$ mole/min) and denote the air fed by F_2. To use the given dry-basis flue gas compositions, we have to express them in terms of the extents. Using (1.4-6), the total molar flow rate of the flue gas is

$$
(F_{tot})_{out} = (10 + F_2) + \dot{X}_3 - \dot{X}_5. \qquad \text{(c)}
$$

Using (1.4-5), the molar flow rate of H_2O in the flue gas is

$$
(F_{H2O})_{out} = (0) + 2 \cdot \dot{X}_1 + 6 \cdot \dot{X}_3. \qquad \text{(d)}
$$

Combining (c) and (d), the total molar flow rate of the dry flue gas is

$$
(F_{tot})_{dry} = (F_{tot})_{out} - (F_{H2O})_{out} = 10 + F_2 - 2 \cdot \dot{X}_1 - 5 \cdot \dot{X}_3 - \dot{X}_5. \qquad \text{(e)}
$$

Using (1.4-5) and (e) for each of the given dry compositions, we obtain:

$$
y_{N_2} = \frac{0.3 + 0.79 \cdot F_2}{10 + F_2 - 2 \cdot \dot{X}_1 - 5 \cdot \dot{X}_3 - \dot{X}_5} = 0.8396 \qquad \text{(f)}
$$

$$
y_{O_2} = \frac{0.1 + 0.21 \cdot F_2 - 2 \cdot \dot{X}_1 - 7 \cdot \dot{X}_3 - \dot{X}_5}{10 + F_2 - 2 \cdot \dot{X}_1 - 5 \cdot \dot{X}_3 - \dot{X}_5} = 0.0869 \qquad \text{(g)}
$$

$$
y_{CO_2} = \frac{\dot{X}_1 + 4 \cdot \dot{X}_3 + 2 \cdot \dot{X}_5}{10 + F_2 - 2 \cdot \dot{X}_1 - 5 \cdot \dot{X}_3 - \dot{X}_5} = 0.0705 \qquad \text{(h)}
$$

$$
y_{CO} = \frac{-2 \cdot \dot{X}_5}{10 + F_2 - 2 \cdot \dot{X}_1 - 5 \cdot \dot{X}_3 - \dot{X}_5} = 0.0018. \qquad \text{(i)}
$$

Solving (f), (g), (h), and (i), we obtain: $F_2 = 173.5$, $\dot{X}_1 = 7.00$, $\dot{X}_3 = 1.20$, and $\dot{X}_5 = -0.1472$. The molar feed rate of the air stream is 173.5 mole/min. The amount of external oxygen fed is $0.21 \cdot F_2 = 36.44$ mole/min. The total amount of oxygen fed to the combustor is $0.001 \cdot (10) + 36.44 = 36.54$ mole/min.

b. Using the conversion definition (1.5-8) and stoichiometric relation (1.4-6), for a species j that is a reactant in the independent reaction, the conversion is

$$f_j = \frac{\sum\limits_{m}^{n_{ind}} (s_j)_m \cdot \dot{X}_m}{F_{j_{in}}}. \tag{j}$$

Hence, the conversion of methane, ethane, and oxygen are, respectively,

$$f_{CH_4} = -\frac{(-1) \cdot \dot{X}_1}{7.20} = 0.972$$

$$f_{C_2H_6} = -\frac{(-2) \cdot \dot{X}_3}{2.40} = 1.00$$

$$f_{O_2} = -\frac{(-2) \cdot \dot{X}_1 + (-7) \cdot \dot{X}_3 + (-1) \cdot \dot{X}_5}{0.1 + 36.44} = 0.609.$$

c. To determine the excess amount of oxygen, we first have to determine the stoichiometric amount needed. For methane and ethane, the complete combustion reactions are, respectively

$$\text{Reaction 1} \quad CH_4 + 2\,O_2 \rightarrow CO_2 + 2\,H_2O \tag{k}$$

$$\text{Reaction 3} \quad 2\,C_2H_6 + 7\,O_2 \rightarrow 4\,CO_2 + 6\,H_2O. \tag{l}$$

Note that we do not consider Reaction 5 because CO is not a species in the fuel. Using (1.3-2), the stoichiometric amount of oxygen needed for complete combustion is

$$(F_{O_2})_{stoich} = \frac{(s_{O_2})_1}{(s_{CH_4})_1} \cdot (F_{CH_4})_{in} + \frac{(s_{O_2})_3}{(s_{C_2H_6})_3} \cdot (F_{C_2H_6})_{in} = 22.8 \text{ mole/min.}$$

Substituting into (1.8-1), the excess amount of oxygen is

$$\left\{ \begin{array}{c} \text{excess} \\ O_2 \end{array} \right\} = \frac{36.54 - 22.8}{22.8} = 0.602. \tag{m}$$

d. To determine the excess air, we have to determine first the stoichiometric amount of external air needed. The total amount of oxygen needed is 22.8 mole/min, but 0.1 mole/min is supplied by the fuel stream itself. Hence, the external oxygen needed is 22.7 mole/min. Using (1.8-2), the excess air is

$$\left\{\begin{array}{c} \text{excess} \\ \text{air} \end{array}\right\} = \frac{36.54 - 22.7}{22.7} = 0.610. \tag{n}$$

1.8.2 Yield and Selectivity

When several chemical reactions take place simultaneously, producing both de-sired and undesired products, it is convenient to define parameters that indicate what portion of the reactant was converted to valuable products and to useless by-products. Below, we define and discuss two quantities that are commonly used: yield and selec-tivity.

Yield is a measure of the portion of a reactant converted to the desired product by the desirable chemical reaction. It indicates the amount of product V produced relative to the amount of V that could have been produced if only the desirable reaction took place. The yield is defined in two ways; each applies a different basis. In both cases, we define the yield such that it varies between zero and one. The first definition of the yield is based of the amount of reactant A fed to the reactor. It indicates the amount of valuable product, say V, produced relative to the theoretical amount of V that could be produced if all the reactant fed were consumed only by the desirable reaction. For batch reactors, the yield of product V at time t is

$$\eta_V(t) = -\left(\frac{s_A}{s_V}\right)\frac{N_V(t) - N_V(0)}{N_A(0)}, \tag{1.8-3}$$

where s_A and s_V are, respectively, the stoichiometric coefficients of A and V in the desirable chemical reaction. For flow reactors, the yield is

$$\eta_V = -\left(\frac{s_A}{s_V}\right)\frac{F_V - F_{V_0}}{F_{A_0}}. \tag{1.8-4}$$

Using stoichiometric relations (1.5-3) or (1.5-5), the yield relates to the dimensionless extents of the independent chemical reactions by

$$\eta_V = -\left(\frac{s_A}{s_V}\right)\frac{1}{y_A(0)}\left(y_V(0) + \sum_{m}^{n_{ind}}(s_V)_m \cdot Z_m\right). \tag{1.8-5}$$

The second yield definition is based on the amount of reactant A that has been reacted. It indicates the amount of V produced relative to the theoretical amount of V that could be produced if all the consumed A were reacted by the desirable chemical reaction. For batch reactors, the yield of product V at time t is

$$\eta_V^*(t) = -\left(\frac{s_A}{s_V}\right)\frac{N_V(t) - N_V(0)}{N_A(0) - N_A(t)}, \tag{1.8-6}$$

and, for flow reactors, it is

$$\eta_V^* = -\left(\frac{s_A}{s_V}\right)\frac{F_V - F_{V_0}}{F_{A_0} - F_A}. \tag{1.8-7}$$

Using the conversion definition, (1.5-7) or (1.5-8), the relationship between the two definitions of the yield is

$$\eta_V = \eta_V^* \cdot f_A. \tag{1.8-8}$$

Another quantity commonly used to describe reactors with multiple reactions is the selectivity. It is a measure of the amount of desirable product produced relative to the amount of undesired product. Selectivity is usually used to characterize a catalyst or the reactor operating conditions. The selectivity, s, is defined by

$$\sigma = \frac{\text{amount of desired product V formed}}{\text{amount of undesired product W formed}}, \tag{1.8-9}$$

and its numerical value varies between zero and infinity. In some cases, the instantaneous selectivity is calculated. It is defined by

$$\sigma = \frac{\text{formation rate of desired product V}}{\text{formation rate of undesired product W}}. \tag{1.8-10}$$

Note that the above definitions of yield and selectivity are not universal. Some researchers and authors define the term "yield" and "selectivity" differently.

Example 1-13 Ethylene oxide is produced in a catalytic flow reactor. A feed consisting of 70% C_2H_4 and 30% O_2 (mole percent) is fed into the reactor at a rate of 100 mole/sec. The following chemical reactions take place in the reactor:

Reaction 1 $2\,C_2H_4 + O_2 \rightarrow 2\,C_2H_4O$

Reaction 2 $C_2H_4 + 3\,O_2 \rightarrow 2\,CO_2 + 2\,H_2O$

Reaction 3 $C_2H_4 + 2\,O_2 \rightarrow 2\,CO + 2\,H_2O.$

An analysis of the exit stream indicates that its composition is 41.17% C_2H_4, 37.65% C_2H_4O, and 7.06% O_2 (mole percent). Determine
a. the conversions of ethylene and oxygen,
b. the yield of ethylene oxide (by both definitions), and
c. the selectivity of ethylene oxide relative to CO_2.
Solution   Since each chemical reaction has a species that does not participate in the other two, the three reactions are independent, and their stoichiometric coefficients are:

$(s_{C2H4})_1 = -2;\ (s_{O2})_1 = -1;\ (s_{C2H4O})_1 = 2;\ (s_{H2O})_1 = 0;\ (s_{CO2})_1 = 0;\ (s_{CO})_1 = 0;\ \Delta_1 = -1;$

$(s_{C2H4})_2 = -1;\ (s_{O2})_2 = -3;\ (s_{C2H4O})_2 = 0;\ (s_{H2O})_2 = 2;\ (s_{CO2})_2 = 2;\ (s_{CO})_2 = 0;\ \Delta_2 = 0;$

$(s_{C2H4})_3 = -1$; $(s_{O2})_3 = -2$; $(s_{C2H4O})_3 = 0$; $(s_{H2O})_3 = 2$; $(s_{CO2})_3 = 0$; $(s_{CO})_3 = 2$; $\Delta_3 = 1$.

We select the feed stream as the reference stream; hence $(F_{tot})_{in} = 100$ mole/sec and $y_{C2H40} = 0.70$; $y_{O20} = 0.30$; $y_{C2H4O0} = 0$; $y_{CO20} = 0$; $y_{H2O0} = 0$; and $y_{CO0} = 0$. To calculate the required quantities, we first have to determine the dimensionless extents of the reactions, Z_1, Z_2, and Z_3. Using stoichiometric relations (1.5-5) and (1.5-6), the molar fraction of species j in the product stream is

$$y_j = \frac{y_{j0} + \sum\limits_m^{n_{ind}} (s_j)_m \cdot Z_m}{1 + \sum\limits_m^{n_{ind}} \Delta_m \cdot Z_m}.$$ (a)

Writing (a) for C_2H_4, C_2H_4O, and O_2, we obtain

$$y_{C_2H_4} = \frac{0.7 - 2 \cdot Z_1 - Z_2 - Z_3}{1 - Z_1 + Z_3} = 0.4118$$ (b)

$$y_{C_2H_4O} = \frac{2 \cdot Z_1}{1 - Z_1 + Z_3} = 0.3765$$ (c)

$$y_{O_2} = \frac{0.3 - Z_1 - 3 \cdot Z_2 - 2 \cdot Z_3}{1 - Z_1 + Z_3} = 0.0706.$$ (d)

Solving (b), (c), and (d), we obtain $Z_1 = 0.16$, $Z_2 = 0.02$, and $Z_3 = 0.01$. Using (1.5-5), the molar flow rates of the respective species are

$$F_{C_2H_4} = 100 \cdot (0.7 - 2 \cdot 0.16 - 0.02 - 0.01) = 35 \text{ mole/sec}$$ (e)

$$F_{C_2H_4O} = 100 \cdot 2 \cdot 0.16 = 32 \text{ mole/sec}$$ (f)

$$F_{O_2} = 100 \cdot (0.3 - 0.16 - 3 \cdot 0.02 - 2 \cdot 0.01) = 6 \text{ mole/sec}$$ (g)

$$F_{CO_2} = 100 \cdot 2 \cdot 0.02 = 4 \text{ mole/sec}.$$ (h)

a. Using the conversion definition (1.5-8) together with (e) and (g), the conversions of C_2H_4 and O_2 are, respectively,

$$f_{C_2H_4} = \frac{70 - 35}{70} = 0.50$$ (i)

$$f_{O_2} = \frac{30 - 6}{30} = 0.80.$$ (j)

b. The desirable reaction is Reaction 1. Hence, using (1.8-4), the yield of ethylene oxide defined on the basis of the ethylene fed is

$$\eta_{C_2H_4O} = -\left(\frac{-2}{2}\right)\frac{32}{70} = 0.457.\tag{k}$$

Using (1.8-7), the yield of ethylene oxide defined on the basis of the ethylene converted is

$$\overset{*}{\eta}_{C_2H_4O} = -\left(\frac{-2}{2}\right)\frac{32}{70-35} = 0.914.\tag{l}$$

c. Using (1.8-9), the overall selectivity of ethylene oxide with respect to CO_2 is

$$\sigma = \frac{32}{4} = 8.\tag{m}$$

1.9 SUMMARY

In this chapter, we described a methodology to express the reactor composition in terms of the chemical reactions taking place. We covered the following topics:
a. Selection of a chemical reaction as a basis for the calculation and defining its stoichiometric coefficients.
b. Extent of a chemical reaction and how to calculate it.
c. Definition of conversion in batch and flow reactors and its relation to the extent of a chemical reaction.
d. How to identify the limiting reactant of a chemical reaction.
e. Relation between the composition of a reactor and the extents of individual reactions.
f. A method to determine the number of independent chemical reactions and how to select a set of independent reactions.
g. A method to express dependent chemical reactions in terms of the independent reactions.
h. The role of independent and dependent reactions in reactor calculations.
Table 1-1 provides the definitions of the key stoichiometric parameters. Tables 1-2 and 1-3 summarize the stoichiometric relations for batch reactors and for steady flow reactors, respectively.

To simplify the analysis and design of chemical reactors with multiple reactions, we adopt the following procedure:
Step 1: Determine the number of independent reactions, and identify a convenient set of independent reactions.
Step 2: Define the stoichiometric coefficients of the independent reactions. Express the dependent reactions in terms of the independent reactions.
Step 3: Select a reference stream or state, and define the dimensionless extents of the independent reactions.

Step 4: Express all state quantities in terms of the dimensionless extents of the independent reactions. Express the dependent reactions in terms of the independent reactions.

BIBLIOGRAPHY

Other treatments of stoichiometric coefficients, extent of chemical reactions, and independent reactions can be found in:

Aris, R., *Introduction to the Analysis of Chemical Reactors.* Englewood Cliffs, NJ; Prentice-Hall, 1965.

Sandler, S. I., *Chemical and Engineering Thermodynamics*, 2nd Ed. New York: John Wiley & Sons, 1989.

Reklaitis, G. V., *Introduction to Material and Energy Balances.* New York: John Wiley & Sons, 1983.

Table 1-1: Definitions of Stoichiometric Parameters

Definitions of Stoichiometric Coefficients

$$aA + bB \rightarrow cC + dD$$

$$s_A = -a : \quad s_B = -b : \quad s_C = c : \quad s_D = d \tag{A}$$

$$\Delta \equiv \sum_j^{all} s_j = c + d - a - b \tag{B}$$

Mathematical Condition for a Balanced Chemical Reaction

$$\sum_j^{all} s_j \cdot MW_j = 0 \tag{C}$$

Relationship Between the k-th Dependent Reaction and the m-th Independent Reaction

$$\sum_m^{n_{ind}} \alpha_{km} \cdot (s_j)_m = (s_j)_k \tag{E}$$

Table 1-2: Summary of Stoichiometric Relations for Batch Reactors

Definitions	Extent of the i-th Chemical Reaction $$X_i(t) = \frac{(n_j(t) - n_{j0})_i}{(s_j)_i} \quad \text{(A)}$$ Dimensionless Extent of the i-th Chemical Reaction $$Z_i(t) = \frac{X_i(t)}{(N_{tot})_0} \quad \text{(B)}$$ Conversion of Reactant A $$f_A(t) \equiv \frac{N_A(0) - N_A(t)}{N_A(0)} \quad \text{(C)}$$
Stoichiometric Relations for Multiple Reactions	$$N_j(t) = N_j(0) + \sum_m^{n_{ind}} (s_j)_m \cdot X_m(t) \quad \text{(D)}$$ $$N_{tot}(t) = (N_{tot})_0 + \sum_m^{n_{ind}} \Delta_m \cdot X_m(t) \quad \text{(E)}$$ $$N_j(t) = (N_{tot})_0 \left(y_j(0) + \sum_m^{n_{ind}} (s_j)_m \cdot Z_m(t) \right) \quad \text{(F)}$$ $$N_{tot}(t) = (N_{tot})_0 \left(1 + \sum_m^{n_{ind}} \Delta_m \cdot Z_m(t) \right) \quad \text{(G)}$$
Stoichiometric Relations for Single Reactions	$$N_j(t) = N_j(0) + \frac{s_j}{s_A}\left[N_A(t) - N_A(0)\right] \quad \text{(H)}$$ $$f_A(t) = -\frac{s_A}{N_A(0)} X(t) \quad \text{(I)}$$ $$N_j(t) = N_j(0) - \frac{s_j}{s_A} N_A(0) \cdot f_A(t) \quad \text{(J)}$$ $$N_{tot}(t) = N_{tot}(0) - \frac{\Delta}{s_A} N_A(0) \cdot f_A(t) \quad \text{(K)}$$

Table 1-3: Summary of Stoichiometric Relations for Steady Flow Reactors

Definitions	Extent per Time of the i-th Chemical Reaction
	$$\dot{X}_i = \frac{(G_j)_i}{(s_j)_i} = \frac{1}{(s_j)_i} \frac{d(n_j)_i}{dt} \qquad (A)$$
	Dimensionless Extent of the i-th Chemical Reaction
	$$Z_i = \frac{\dot{X}_i}{(F_{tot})_0} \qquad (B)$$
	Conversion of Reactant A
	$$f_A \equiv \frac{F_{A0} - F_A}{F_{A0}} \qquad (C)$$

Stoichiometric Relations for Multiple Reactions	
	$$F_j = F_{j0} + \sum_{m}^{n_{ind}} (s_j)_m \cdot \dot{X}_m \qquad (D)$$
	$$F_{tot} = (F_{tot})_0 + \sum_{m}^{n_{ind}} \Delta_m \cdot \dot{X}_m \qquad (E)$$
	$$F_j = (F_{tot})_0 \left(y_{j0} + \sum_{m}^{n_{ind}} (s_j)_m \cdot Z_m \right) \qquad (F)$$
	$$F_{tot} = (F_{tot})_0 \left(1 + \sum_{m}^{n_{ind}} \Delta_m \cdot Z_m \right) \qquad (G)$$

Stoichiometric Relations for Single Reactions	
	$$F_j = F_{j0} + \frac{s_j}{s_A} (F_A - F_{A0}) \qquad (H)$$
	$$f_A = -\frac{s_A}{F_{A0}} \dot{X} \qquad (I)$$
	$$F_j = F_{j0} - \frac{s_j}{s_A} F_{A0} \cdot f_A \qquad (J)$$
	$$F_{tot} = (F_{tot})_0 - \frac{\Delta}{s_A} F_{A0} \cdot f_A \qquad (K)$$

PROBLEMS

1-1$_2$ The gas-phase decomposition reaction, $C_2H_6 \rightarrow C_2H_4 + 2\ H_2$, is being investigated in a batch, constant-volume isothermal reactor. The reactor is charged with 20 lbmoles of (pure) ethane, and the initial pressure is 2 atm. When the reaction is stopped, the pressure of the reactor is 5 atm. Assuming ideal gas behavior, calculate:
a. the extent of the reaction,
b. the conversion of ethane, and
c. the partial pressure of H_2 and C_2H_2 at the end of the reaction.

1-2$_2$ Repeat Problem 1-1, but, instead of pure ethane, a mixture of 50% ethane and 50% hydrogen is charged into the reactor. The initial pressure is 2 atm, and the final pressure is 3 atm.

1-3$_2$ The gas-phase decomposition reaction, $C_2H_6 \rightarrow C_2H_4 + H_2$, is being investigated in an isobaric, batch reactor operated isothermally. Initially, 10 lbmoles of ethane (pure) are charged into the reactor. If the final volume of the reactor is 80% larger than the initial volume, calculate:
a. the conversion,
b. the extent, and
c. the mole fraction of H_2 at the end of the reaction.
Assume ideal gas behavior.

1-4$_2$ The gas-phase reaction $A \rightarrow 2\ R + P$ takes place in a constant-volume batch reactor. A thermal conductivity detector is used to determine the progress of the reaction. The conductivity reading is proportional to the sum of the concentrations of A and R. At the beginning of the operation, 2 kmoles of A and 1 kmoles of P are charged into the reactor and the conductivity reading is 120 (arbitrary units.) At time t, the conductivity reading is 180. Calculate:
a. the conversion of A at time t and
b. the composition of the reactor at time t.

1-5$_2$ In many organic substitution reactions, the product generated by the reaction is prone to additional substitution. A semi-batch reactor was used to produce monochlorobenzene by reacting benzene with chlorine. The reactor was charged with 20 lbmoles of liquid benzene. A stream of gaseous chlorine bubbled through the liquid, and the unreacted chlorine was recycled. During the operation, monochlorobenzene reacted with the chlorine to produce dichlorobenzene, and the dichlorobenzene reacted with the chlorine to produce trichlorobenzene,

$$C_6H_6\ (l) + Cl_2\ (g) \rightarrow\ C_6H_5Cl\ (l)\ +\ HCl\ (g)$$

$$C_6H_5Cl\ (l) + Cl_2(g) \rightarrow\ C_6H_4Cl_2\ (l)\ +\ HCl\ (g)$$

$$C_6H_4Cl_2\ (l) + Cl_2\ (g) \rightarrow\ C_6H_3Cl_3\ (l)\ +\ HCl\ (g).$$

At the end of the operation, the reactor contained 11 lbmole of benzene and a product mixture in which the amount of monochlorobenzene was three times the amount of dichlorobenzene, and the amount of the latter was twice the amount of trichlorobenzene. The total amount of chlorine fed during the operation was 40 lbmoles. Find:

a. the conversion of benzene,
b. the composition of the reactor (liquid contents),
c. the amount of HCl produced, and
d. the conversion of chlorine.

1-6₂ Dimerization reaction $2 A \rightarrow R$ is taking place in a liquid solution. The progress of the reaction is monitored by an I-R analyzer whose signal is adjusted such that the percent of the I-R absorbed by the solution is proportional to the sum of the concentrations of A and R. Initially, the solution contains 5 kmoles of A and no R, and the reading of the analyzer is 85%. At time t, the analyzer reading is 60%. Calculate:

a. the extent of reaction at time t,
b. the conversion at time t, and
c. the moles of R in the solution at time t.

1-7₂ A gaseous fuel consisting of a mixture of methane (CH_4) and ethane (C_2H_6) is fed into a burner in a proportion of 1 mole of fuel per 20 moles of air. The following reactions are believed to take place in the reactor:

$$CH_4 + 2 O_2 \rightarrow CO_2 + 2 H_2O$$

$$2 C_2H_6 + 7 O_2 \rightarrow 4 CO_2 + 6 H_2O.$$

An analysis of the flue gas (dry basis) indicates that it consists of 83.7% N_2, 7.01% CO_2, 9.15% O_2, and 0.14% methane (% mole). Calculate:

a. the composition of the fuel,
b. the conversion of oxygen,
c. the conversion of methane,
d. the conversion of ethane, and
e. the excess amount of air used.

1-8₁ Ethylene oxide is produced by a catalytic reaction of ethylene and oxygen at controlled conditions. However, side reactions cannot be completely prevented, and it is believed that the following reactions take place:

$$2 C_2H_4 + O_2 \rightarrow 2 C_2H_4O$$

$$C_2H_4 + 3 O_2 \rightarrow 2 CO_2 + 2 H_2O$$

$$C_2H_4 + 2 O_2 \rightarrow 2 CO + 2 H_2O$$

$$2 CO + O_2 \rightarrow 2 CO_2$$

$$CO + H_2O \rightarrow CO_2 + H_2.$$

a. Determine the number of independent reactions.
b. Choose a set of independent reactions among the reactions above.
c. Express the dependent reactions in terms of the independent reactions.

1-9₂ Ammonium nitrate is a raw material used in the manufacture of agricultural chemicals and explosives. The following reactions are believed to take place in the production of ammonium nitrate:

$$4 \, NH_3 + 5O_2 \rightarrow 4 \, NO + 6 \, H_2O$$

$$2 \, NO + O_2 \rightarrow 2 \, NO_2$$

$$4 \, NH_3 + 7 \, O_2 \rightarrow 4 \, NO_2 + 6 \, H_2O$$

$$3 \, NO_2 + H_2O \rightarrow 2 \, HNO_3 + NO$$

$$NH_3 + HNO_3 \rightarrow NH_4NO_3.$$

a. Determine the number of independent reactions.
b. Choose a set of independent reactions among the reactions above.
c. Express the dependent reactions in terms of the independent reactions.

1-10₁ In the reforming of methane, the following chemical reactions may occur:

$$CH_4 + 2 \, H_2O \rightarrow CO_2 + 4 \, H_2$$

$$CH_4 + 2 \, O_2 \rightarrow CO_2 + 2 \, H_2O$$

$$2 \, CH_4 + O_2 \rightarrow 2 \, CH_3OH$$

$$2 \, CH_4 + 3 \, O_2 \rightarrow 2 \, CO + 4 \, H_2O$$

$$CO_2 + 3 \, H_2 \rightarrow CH_3OH + H_2O$$

$$2 \, CH_3OH + 3 \, O_2 \rightarrow 2 \, CO_2 + 4 \, H_2O$$

$$2 \, CO + O_2 \rightarrow 2 \, CO_2$$

$$CO + H_2O \rightarrow CO_2 + H_2.$$

a. Determine the number of independent reactions.
b. Choose a set of independent reactions among the reactions above.
c. Express the dependent reactions in terms of the independent reactions.

1-11₁ Consider the classic mechanism of the reaction between hydrogen and bromine to form hydrogen bromide (the asterisks indicate free radicals):

$$Br_2 \rightarrow 2 \, Br*$$

$$Br* + H_2 \rightarrow HBr + H*$$

$$H^* + Br_2 \rightarrow HBr + Br^*$$

$$H^* + HBr \rightarrow H_2 + Br^*$$

$$2\ Br^* \rightarrow Br_2.$$

a. Determine the number of independent reactions.
b. Identify a set of independent reactions from the reactions above.
c. Express the dependent reactions in terms of the independent reactions.

1-12$_3$ Methanol is being produced according to the reaction

$$CO + 2\ H_2 \rightarrow CH_3OH.$$

A synthesis gas stream consisting of 67.1% H_2, 32.5% CO, and 0.4% CH_4 (by mole) is fed into the process described below at a rate of 100 lbmole/hr. The effluent stream from the reactor is fed into a separator where the methanol is completely removed, and the unconverted reactants are recycled to the reactor. To avoid the buildup of methane, a portion of the recycled stream is purged. At present operating conditions, the CO conversion over the entire process is 90%, and the methane mole fraction at the reactor inlet (Stream 2) is 2.87%. Calculate:
a. The production rate of methanol.
b. The flow rate of the purge stream (Stream 6).
c. The composition of the purge stream.
d. The portion of the recycle stream (Stream 5) that is purged.
e. The CO conversion per pass in the reactor.

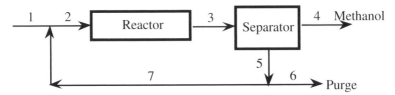

1-13$_3$ Solve Problem 1-12 for the following information. The CO conversion per pass is 15%, and the methane mole fraction at the reactor inlet (Stream 2) is 2.5%. Calculate:
a. The production rate of methanol.
b. The flow rate of the purge stream (Stream 6).
c. The composition of the purge stream.
d. The portion of the recycle stream (Stream 5) that is purged.
e. The CO conversion in the process.

1-14$_2$ The following reactions are believed to take place during the direct oxidation of methane:

$$2\ CH_4 + 3\ O_2 \rightarrow 2\ CO + 4\ H_2O$$

$$CH_4 + 2 O_2 \rightarrow CO_2 + 2 H_2O$$

$$2 CH_4 + O_2 \rightarrow 2 CH_3OH$$

$$CH_4 + O_2 \rightarrow HCHO + H_2O$$

$$CO + H_2O \rightarrow CO_2 + H_2$$

$$CO + 2 H_2 \rightarrow CH_3OH$$

$$CO + H_2 \rightarrow HCHO$$

$$2 CH_4 + O_2 \rightarrow 2 CO + 4 H_2$$

$$2 CO + O_2 \rightarrow 2 CO_2.$$

a. Determine the number of independent reactions.
b. Identify a set of independent reactions from the reactions above.
c. Express the dependent reactions in terms of the independent reactions.

1-15$_2$ Selective oxidation of hydrocarbons is a known method to produce alcohols. However, the alcohols react with the oxygen to produce aldehydes, and the latter react with oxygen to produce organic acids. A 50 g-mole/sec stream consisting of 90% ethane and 10% nitrogen is mixed with a 40 g-mole/sec air stream and fed into a catalytic reactor. The following reactions take place in the reactor:

$$2 C_2H_6 + O_2 \rightarrow 2 C_2H_5OH$$

$$2 C_2H_5OH + O_2 \rightarrow 2 CH_3CHO + 2 H_2O$$

$$2 CH_3CHO + O_2 \rightarrow 2 CH_3COOH.$$

The oxygen conversion is 80%, and the concentration of the ethanol in the product stream is three times that of the aldehyde and four times that of the acetic acid. Calculate:
a. the ethane conversion and
b. the production rate of the ethanol.

1-16$_3$ Ammonia is being used by reacting nitrogen and hydrogen in the system described below. The reaction is $N_2 + 3 H_2 \rightarrow 2 NH_3$, and all the ammonia is removed in the separator. A feed stream consisting of 2% argon (by mole) and stoichiometric proportion of the reactants is fed into the system at the rate of 100 mole/min. The mole fraction of argon in the purge stream (Stream 6) is 5%, and the conversion per pass in the reactor is 10%.
Determine:
a. the production rate of ammonia,
b. the overall (process) conversions of nitrogen and hydrogen, and
c. the flow rate of the recycle stream (Stream 7).

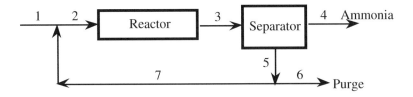

1-17₃ A gaseous fuel consisting of a mixture of methane (CH_4) and ethane (C_2H_6) is fed into a burner in a proportion of 1 mole of fuel per 18 moles of air. The following reactions are believed to take place in the reactor:

$$CH_4 + 2\,O_2 \rightarrow CO_2 + 2\,H_2O$$

$$2\,CH_4 + 3\,O_2 \rightarrow 2\,CO + 4\,H_2O$$

$$2\,C_2H_6 + 5\,O_2 \rightarrow 4\,CO + 6\,H_2O$$

$$2\,C_2H_6 + 7\,O_2 \rightarrow 4\,CO_2 + 6\,H_2O$$

$$2\,CO + O_2 \rightarrow 2\,CO_2.$$

An analysis of the flue gas (dry basis) indicates that it consists of 7.71% CO_2, 0.51% CO, 7.41% O_2, 84.32% of N_2, and a trace amount of methane and ethane (% mole). Calculate:

a. the composition of the fuel,
b. the conversion of oxygen,
c. the conversion of methane,
d. the conversion of ethane, and
e. the excess amount of air used.

1-18₃ Toluene, $C_6H_5CH_3$, is hydrogenated according to the two simultaneous chemical reactions

$$C_6H_5CH_3 + H_2 \rightarrow C_6H_6 + CH_4 \tag{a}$$

$$2\,C_6H_5CH_3 + H_2 \rightarrow (C_6H_5)_2 + 2\,CH_4. \tag{b}$$

Initially, the reactor contains 40% toluene and 60% H_2 (% mole). At the end of the operation, the reactor contains 10% toluene and 30% CH_4 (% mole). Find:

a. the % mole of the diphenyl, $(C_6H_5)_2$, at the end of the operation and
b. the conversion of H_2.
c. Solve (a) and (b) when the following final composition is specified: the reactor contains 10% toluene and 40% H_2 (% mole).

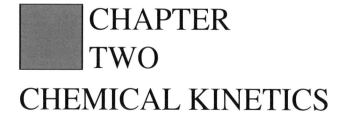

CHAPTER TWO
CHEMICAL KINETICS

In this chapter, we cover the second fundamental concept used in chemical reaction engineering — chemical kinetics. We discuss the kinetic relationships needed to analyze and design chemical reactors. In the first section, we discuss the various definitions of the species formation rates and how they are used. In the second section, we define the rates of chemical reactions and discuss how they relate to the formation rates of individual species. In the third section, we discuss the rate expression indicating the dependency of the reaction rate on the temperature and species concentrations. Without going into the theory of chemical kinetics, we discuss the common forms of the rate expression for homogeneous and heterogeneous reactions. In the last section, we introduce a new measure of the reaction rate — the characteristic reaction time.

2.1 SPECIES FORMATION RATES

Consider a closed (batch) reactor with volume V where species j is formed (or consumed) by chemical reactions. We denote the total moles of species j in the reactor by N_j; hence, the rate species j is being formed in the reactor is dN_j/dt. The magnitude of dN_j/dt depends, of course, on the size of the reactor, and to define an intensive quantity for the formation rate of species j, we consider the nature of the chemical reactions that are taking place. For homogeneous chemical reactions (reactions that take place throughout a given phase), we define a volume-based species formation rate, (r_j), by

$$(r_j) = (r_j)_V = \frac{1}{V} \frac{dN_j}{dt}. \qquad (2.1\text{-}1a)$$

Thus, (r_j) is the formation rate of species j expressed in terms of moles of species j formed per unit time per unit volume. For heterogeneous chemical reactions (reactions that take place on the surface of a solid catalyst or at the interface of two phases, we define a surface-based formation rate of species j, $(r_j)_S$, by

$$(r_j)_S = \frac{1}{S} \frac{dN_j}{dt}, \tag{2.1-1b}$$

where $(r_j)_S$ is expressed in terms of moles of species j formed per unit time per unit surface area. In some cases, it is convenient to define the formation rate of species j on the basis of the mass of solid catalyst:

$$(r_j)_W = \frac{1}{W} \frac{dN_j}{dt}. \tag{2.1-1c}$$

Here, $(r_j)_W$ is the mass-based formation rate of species j expressed in moles of species j formed per unit time per unit mass of catalyst. Note that all these definitions do not relate to any specific chemical reaction but rather to the net formation of species j, which may be formed simultaneously by some reactions and consumed by others. When species j is consumed, (r_j) is negative.

These three definitions of species formation rates relate to each other by

$$\frac{dN_j}{dt} = (r_j)_V \cdot V = (r_j)_S \cdot S = (r_j)_W \cdot W. \tag{2.1-2}$$

Therefore,

$$(r_j) = \left(\frac{S}{V}\right) \cdot (r_j)_S \tag{2.1-3}$$

$$(r_j) = \left(\frac{W}{V}\right) \cdot (r_j)_W. \tag{2.1-4}$$

Hence, when any one of these rates is known, we can determine the other two if the properties of the reactor (mass of solid catalyst per unit volume or catalyst surface per unit volume) are provided.

2.2 RATES OF CHEMICAL REACTIONS

Next, we focus our attention on a specific chemical reaction and consider the rate that it progresses. For a homogeneous chemical reaction, we define a volume-based rate of a reaction by

$$r = \frac{1}{V} \frac{dX}{dt}, \tag{2.2-1a}$$

where r denotes the reaction rate expressed in terms of moles extent per unit time per unit volume. For heterogeneous reactions, we define a surface-based rate of a reaction by

$$r_S = \frac{1}{S}\frac{dX}{dt}, \tag{2.2-1b}$$

where r_s is expressed in terms of moles extent per unit time per unit surface area. Similarly, the mass-based rate of a chemical reaction is defined by

$$r_W = \frac{1}{W}\frac{dX}{dt}, \tag{2.2-1c}$$

where r_w is expressed in terms of moles extent per unit time per unit mass. The relationships between these three rate definitions are

$$\frac{dX}{dt} = r \cdot V = r_S \cdot S = r_W \cdot W. \tag{2.2-2}$$

Therefore,

$$r = \left(\frac{S}{V}\right) \cdot r_S \tag{2.2-3}$$

$$r = \left(\frac{W}{V}\right) \cdot r_W. \tag{2.2-4}$$

Here, too, when any one of the three rates is known, we can determine the other two if the properties of the reactor are provided.

Next, we consider the relationship between the formation rates of individual species and the rates of the chemical reactions. When a **single** chemical reaction takes place, we can easily relate the rate of the chemical reaction, r, to the formation rate of a specific species, say species j, participating in the reaction, (r_j). Differentiating the definition of the extent, (1.2-1),

$$\frac{dX}{dt} = \frac{1}{s_j}\frac{dn_j}{dt},$$

and substituting this into (2.2-1a), we obtain

$$(r_j) = s_j \cdot r. \tag{2.2-5}$$

Thus, for single reactions, when the rate of the chemical reaction is known, we can calculate the rate of formation (or depletion) of any species that participates in the reaction. When **multiple** chemical reactions take place simultaneously, species j may participate in several reactions, formed by some and consumed by others. Differentiating (1.4-1) and substituting into (2.1-1a), the formation rate of species j is

$$(r_j) = \sum_{i=1}^{n_{all}} (s_j)_i \cdot r_i, \tag{2.2-6}$$

where $(s_j)_i$ is the stoichiometric coefficient of species j in the i-th chemical reaction, and r_i is the rate of the i-th reaction. Eq. (2.2-6) relates the formation rate of any species to the rates of the chemical reactions. It is important to note that the summation in (2.2-6) is over **all** chemical reactions that **actually** take place in the reactor (both dependent and independent). Also, note that when a single chemical reaction takes place, (2.2-6) reduces to (2.2-5).

Example 2-1 The heterogeneous catalytic chemical reaction $2\,C_2H_4 + O_2 \rightarrow 2\,C_2H_4O$ is investigated in a packed-bed reactor. The reactor is filled with catalytic pellets whose surface area is 7 m² per gram, and the density of the bed is 1.4 kg/liter. The measured consumption rate of ethylene is 0.35 mole/hr g-catalyst. Determine:
a. the volume-based and surface-based formation rates of ethylene,
b. the volume-based rate of the chemical reaction,
c. the volume-based formation rate of oxygen, and
d. the volume-based formation rate of ethylene oxide.
Solution   The stoichiometric coefficients of the chemical reaction are: $s_{C2H4} = -2$, $s_{O2} = -1$, and $s_{C2H4O} = 2$. The given mass-based formation rate of ethylene is -0.35 mole/hr g (the minus sign indicates that ethylene is consumed).
a. Using (2.1-4), the volume-based formation rate of ethylene is

$$(r_{C_2H_4}) = \left(\frac{W}{V}\right)\cdot (r_{C_2H_4})_W = \rho_{bed}\cdot (r_{C_2H_4})_W =$$

$$= \left(1.4\,\frac{kg}{liter}\right)\cdot\left(-0.35\,\frac{mole\ Et.}{hr\cdot g}\right)\left(\frac{1000\ g}{kg}\right) = -490\,\frac{mole\ Et.}{hr\cdot liter}. \tag{a}$$

To determine the surface-based formation rate of ethylene, we first calculate the characteristic surface of the reactor:

$$\left(\frac{S}{V}\right) = \left(\frac{W}{V}\right)\cdot\left(\frac{S}{W}\right) = \rho_{bed}\cdot\left(\frac{S}{W}\right)$$

$$\left(\frac{S}{V}\right) = (1.4\ kg/lit)\cdot(7.0\ m^2/kg)\cdot(1,000\ g/kg) = 9,800\ m^2/lit. \tag{b}$$

Using (2.1-3), (a), and (b), the surface-based formation rate of ethylene is

$$(r_{C_2H_4})_S = \frac{(-490\ mole\ Et./hr\cdot lit)}{(9.8310^3\,m^2/lit)} = -5.00\cdot10^{-2}\,mole\ Et./hr\cdot m^2. \tag{c}$$

b. Using (2.2-5) and (a), the volume-based rate of the chemical reaction is

$$r = \frac{(r_{C_2H_4})}{s_{C_2H_4}} = \frac{(-490\ mole\ Et./hr\ lit)}{(-2\ mole\ Et./mole\ extent)} = 245\ mole\ extent/(hr\ lit). \tag{d}$$

c. Using (2.2-5) and (d), the volume-based formation rate of oxygen is

$$r_{O_2} = \left(-1 \frac{\text{mole oxygen}}{\text{mole extent}}\right) \cdot \left(245 \frac{\text{mole extent}}{\text{hr} \cdot \text{lit}}\right) = \overline{245} \text{ mole oxygen/(hr} \cdot \text{lit)}. \quad \text{(e)}$$

The negative sign indicates that oxygen is consumed.

d. Using (2.2-5) and (d), the volume-based formation rate of ethylene oxide is

$$r_{C_2H_4O} = \left(1 \frac{\text{mole oxide}}{\text{mole extent}}\right) \cdot \left(245 \frac{\text{mole extent}}{\text{hr} \cdot \text{lit}}\right) = 245 \text{ mole oxide/(hr} \cdot \text{lit)}. \quad \text{(f)}$$

$$490$$

2.3 RATE EXPRESSIONS OF CHEMICAL REACTIONS

Next, we direct our attention to the parameters that affect the rates of chemical reactions. The rate of a chemical reaction is a function of the temperature, the composition of the reacting mixture, and the presence of a catalyst. The relationship between the reaction rate and these parameters is commonly called the "rate expression" or, sometimes, the "rate law." Chemical kinetics is the branch of chemistry that deals with reaction mechanisms and provides a theoretical basis for the rate expression. When such information is available, we use it to obtain the rate expression. However, from an engineering perspective, we consider the rate expression as a mathematical expression (or a mathematical model) that describes the rate of the chemical reaction. In many instances, the rate expression is not available, and we have to obtain it from experimental data obtained on the reactor.

For most chemical reactions, the rate expression is a product of two functions, one of temperature only, $k(T)$, and the second of species concentrations only, $h(C_j\text{'s})$:

$$r = k(T) \cdot h(C_j \text{' s}). \quad (2.3\text{-}1)$$

The function $k(T)$ is commonly called the "reaction rate constant." But, it is not a constant: it depends on the temperature. The term "rate constant" comes about because $k(T)$ is independent of the **composition** and has a constant value at isothermal operations.

Practically for all chemical reactions, $k(T)$ relates to the temperature by the Arrhenius equation,

$$k(T) = k_0 \cdot e^{-E_a/RT}, \quad (2.3\text{-}2)$$

where E_a is a parameter called the activation energy, and k_0 is a parameter called the frequency factor or the pre-exponential coefficient. Both parameters are characteristic of the chemical reaction and the presence (or absence) of a catalyst.. The value of E_a indicates the sensitivity of the chemical reaction to changes in temperature. When the activation energy is large, say 50 to 100 kcal/mole (200 to 400 kJ/mole), the reaction rate is sensitive to temperature. Such values are typical of combustion and gasification reactions that take place at high temperatures and are very slow at room temperature. On the other hand, when the value of E_a is low, say 5 to 10 kcal/mole (or 20 to 40 kJ/mole), the reaction rate is not sensitive to temperature. These values are typical of biological and enzymatic reactions that take place at room temperature. Figure 2-1 shows, schematically, the relation between k and T for large and small activation energies.

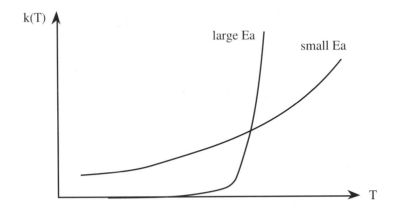

Figure 2-1: Reaction Rate Constant as a Function of Temperature

The value of the rate constant at a given temperature is readily calculated when both parameters in the Arrhenius equation, k_0 and E_a, are known. However, it is convenient to calculate the value of the rate constant at one temperature on the basis of its value at a different temperature using only the activation energy. To obtain a relationship between the values of the reaction rate constant at two temperatures, say T and T_0, we take the log of (2.3-2) for each and combine the two equations to obtain

$$k(T) = k(T_0) \cdot \exp\left[-\frac{E_a}{R}\left(\frac{1}{T} - \frac{1}{T_0} \right) \right]. \qquad (2.3-3)$$

We determine the values of the parameters in the Arrhenius equation, k_0 and E_a, from experimental measurements of the reaction rate constant, k(T), at different temperatures. Using (2.3-3), we plot ln k(T) versus 1/T to obtain a straight line with a slope of $-E_a/R$, as shown schematically in Figure 2-2. In principle, we can determine value of k_0 by extrapolating the line to 1/T → 0 (T → ∞) and reading the value of the intercept. However, this procedure is not accurate because it involves an extrapolation far beyond the range of the experimental data, and a small error in the slope will result in a large error in k_0. Instead, we first determine the slope and then use Figure 2-2 to read the value of the rate constant at the average temperature, T_m. We then calculate k_0

from (2.3-2) using the value of E_a obtained from the slope (see Figure 2-2).

Physically, the activation energy represents an energy barrier that should be overcome as the reaction proceeds. This barrier and its relationship to the heat of reaction is shown schematically in Figure 2-3 for exothermic and endothermic reactions. Note that for reversible chemical reactions, the heat of reaction is the difference between the activation energy of the forward reaction and the activation energy of the backward reaction, $\Delta H_R = (E_a)_{forr} - (E_a)_{back}$.

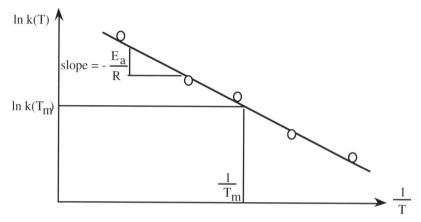

Figure 2-2: Determination of Activation Energy

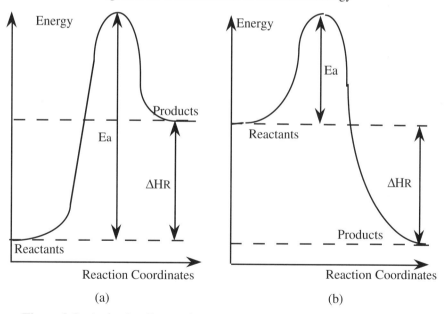

Figure 2-3: Activation Energy for (a) Exothermic Reaction, (b) Endothermic Reaction

As will be discussed later, we formulate the design equations of chemical reactors in terms of dimensionless quantities and would like to express the reaction rate constants in terms of them. We define dimensionless temperature

$$\theta \equiv \frac{T}{T_0}, \tag{2.3-4}$$

where T_0 is a conveniently-selected reference temperature. For most practical situations, θ varies between 0.5 and 2.0. We also define a dimensionless activation energy, γ, by

$$\gamma \equiv \frac{E_a}{R \cdot T_0}, \tag{2.3-5}$$

which is a characteristic of the chemical reaction and the reference temperature. Using (2.3-4) and (2.3-5), the Arrhenius Equation reduces to

$$k(\theta) = k(T_0) \cdot e^{\gamma \frac{\theta-1}{\theta}}, \tag{2.3-6}$$

where $k(T_0)$ is the reaction rate constant at reference temperature T_0.

Example 2-2 The rate constant of a chemical reaction is determined experimentally at two temperatures. Based on the data below, determine
a. the activation energy,
b. the dimensionless activation energy if the reference temperature is 30°C, and
c. the pre-exponential coefficient.

Data:	T (°C)	30	50
	k (min.$^{-1}$)	0.25	1.4

Solution a. To determine the activation energy, we write (2.3-3) for the two given temperatures, $T_1 = 30 + 273.15 = 303.15°K$ and $T_2 = 50 + 273.15 = 323.15°K$,

$$\ln\left(\frac{1.4}{0.25}\right) = -\frac{E_a}{R}\left(\frac{1}{323.15} - \frac{1}{303.15}\right). \tag{a}$$

We obtain $E_a/R = 8,430°K$, and

$$E_a = (8,430°K)(1.987 \text{ cal/mole}°K) = 16.75 \text{ kcal/mole}. \tag{b}$$

b. The reference temperature is $T_0 = 303.15°K$, and, using (2.3-5), the dimensionless activation energy is

$$\gamma = \frac{8,430}{1.987 \cdot 298.15} = 27.81.$$

c. To determine the pre-exponential coefficient k_0, we write (2-12) for one of the given temperatures. For $T_1 = 303.15°K$,

$$(0.25 \text{ min}^{-1}) = k_0 \cdot e^{\frac{-8,430}{1.987 \cdot 303.15}},$$

and we obtain $k_0 = 3.03 \ 10^{11} \text{ min}^{-1}$.

Next, we consider the dependence of the rate expression on the species composition — function $h(C_j\text{'s})$ in (2.3-2). For many homogeneous chemical reactions, $h(C)$ is expressed as a power relation of the species concentrations,

$$h(C_j\text{' s}) = C_A^{\alpha} \cdot C_B^{\beta}, \qquad (2.3\text{-}7)$$

where C_A, C_B, ... are the concentrations of the different species. The powers in (2.3-7) are called the orders of the reaction: α is the order of the reaction with respect to species A, β is the order of the reaction with respect to species B, etc., and $\alpha + \beta + \ldots$ is the overall order of the reaction. The orders can be either integers or fractions and should be determined experimentally. Chemical kinetics provides a theoretical basis for the values of the orders for some reactions. For elementary reactions (reactions that takes place in a single step without formation of intermediates), the orders of the species are identical to the absolute value of their respective stoichiometric coefficients when the reaction is written in its simplest form. However, in general, we consider (2.3-7) to be an empirical expression, and the orders are parameters that should be determined from experimental data. Combining (2.3-1) and (2.3-7), the rate expression for many homogeneous reactions is

$$r = k(T) \cdot C_A^{\alpha} \cdot C_B^{\beta} = k(T_0) \cdot e^{\gamma \frac{\theta-1}{\theta}} \cdot C_A^{\alpha} \cdot C_B^{\beta}. \qquad (2.3\text{-}8)$$

Other forms of composition function $h(C_j\text{'s})$ are used for different types of homogenous and heterogeneous reactions. Recognizing that a written chemical reaction is merely a stoichiometric presentation of many elementary reaction steps, the global rate of a chemical reaction depends on the rates of the individual steps and results in different forms. For example, the rate of many biological and enzymatic reactions is described by an expression of the form

$$r = k(T) \frac{C_A}{K_m + C_A}, \qquad (2.3\text{-}9)$$

where $k(T)$ is the reaction rate constant, and K_m is a parameter to be determined experimentally. The rate expression of the reaction between hydrogen and bromine to form hydrogen bromide is

$$r = k(T) \cdot \frac{C_{H_2} \cdot C_{Br_2}^{0.5}}{1 + K \cdot \dfrac{C_{HBr}}{C_{Br_2}}}. \qquad (2.3\text{-}10)$$

For most heterogeneous catalytic reactions, rate expressions of the following general forms are usually used:

$$r = k(T) \cdot \frac{C_A}{1 + K_A \cdot C_A}, \tag{2.3-11a}$$

$$r = k(T) \cdot \frac{C_A \cdot C_B}{1 + K_A \cdot C_A + K_B \cdot C_B}, \tag{2.3-11b}$$

$$r = k(T) \cdot \frac{C_A \cdot C_B}{(1 + K_A \cdot C_A + K_B \cdot C_B)^2}, \tag{2.3-11c}$$

where K_A, K_B, and K_C, are parameters that should be determined experimentally.

The dimensions and units of the reaction rate constant, $k(T)$, depend on the form of function $h(C_j\text{'s})$. For chemical reactions whose rate expressions are power functions of the species concentrations, we can determine the units of the rate constant. From (2.3-8), we can write the rate as

$$r = k(T) \cdot C^n,$$

where $n = \alpha + \beta + \ldots$ In terms of units,

$$\left(\frac{\text{mole}}{\text{volume} \cdot \text{time}}\right) [=] k \cdot \left(\frac{\text{mole}}{\text{volume}}\right)^n.$$

Thus, the dimensions of the rate constant are

$$k [=] \left(\frac{\text{mole}}{\text{volume}}\right)^{1-n} \left(\frac{1}{\text{time}}\right). \tag{2.3-12}$$

Example 2-3 Determine the overall order of the reaction in Example 2-2 from the units of $k(T)$.
Solution Using (2.3-12) for the given units of k,

$$\left(\frac{1}{\text{min}}\right) [=] \left(\frac{\text{mole}}{\text{volume}}\right)^{1-n} \left(\frac{1}{\text{time}}\right),$$

to obtain consistent units, $1 - n = 0$. Thus, the reaction is first-order.

2.4 EFFECT OF TRANSPORT LIMITATIONS

The rate expressions derived above describe the dependence of the reaction rate expressions on kinetic parameters related to the chemical reactions. These rate expressions are commonly called the "intrinsic" rate expressions of the chemical reactions.

However, in many instances, the local species concentrations depend also on the rate species are transported in the reaction medium. Hence, the actual reaction rates are affected by the transport rates of reactants and products. This is manifested in two general cases: (i) gas-solid heterogeneous reactions, where species diffusion through the pore plays an important role, and (ii) gas-liquid reactions, where interfacial species mass-transfer rate as well as solubility and diffusion play an important role. Considering the effect of transport limitations on the global rates of the chemical reactions represents a very difficult task in the design of many chemical reactors. These topics are beyond the scope of this text, but the reader should remember to take them into consideration.

2.5 CHARACTERISTIC REACTION TIME

In the preceding sections, we characterized the rates of chemical reactions in terms of their sensitivity to changes in temperature and their dependency on species compositions. The progress of chemical reactions is a rate process that can be characterized by a time constant. Reactions with the same rate expression (first-order, second-order, etc.) exhibit a similar behavior with respect to concentration variations — the difference between fast and slow reactions is merely the time scale over which the reaction takes place. To describe the generic behavior of chemical reactions and to derive dimensionless design equations for chemical reactors, we define the characteristic reaction time. The characteristic reaction time is a measure of the time scale over which the reaction takes place (second, minute, etc.).

For volume-based rate expressions, we define the characteristic reaction time, t_{cr}, by

$$t_{cr} = \frac{\text{characteristic concentration}}{\text{characteristic reaction rate}} = \frac{C_0}{r_0}, \qquad (2.5\text{-}1)$$

where both the characteristic concentration and the characteristic reaction rate are conveniently selected. Usually, these two are selected for a reference state or a reference stream. The characteristic concentration is elected as the total concentration of the reference stream (or state). For batch reactors, it is defined by

$$C_0 \equiv \frac{(N_{tot})_0}{V_R(0)}, \qquad (2.5\text{-}2a)$$

where $(N_{tot})_0$ is the total number of moles initially in the reactor, and $V_R(0)$ is the initial volume of the reactor. For flow reactors, it is defined by

$$C_0 \equiv \frac{(F_{tot})_0}{v_0}, \qquad (2.5\text{-}2b)$$

where $(F_{tot})_0$ is the total molar flow rate of the reference stream, and v_0 is the stream's volumetric flow rate. As for the characteristic reaction rate, since, in many instances, the initial (or feed) composition is a parameter of the operation, we would like to select a characteristic rate that is independent of the specific composition of the reference state. Hence, we select a conveniently-defined rate, not necessarily the actual reaction rate. For chemical reactions whose rate expressions are power functions of the concentrations, it is convenient to define a characteristic reaction rate as the rate, at the reference temperature, if the concentration of each species is C_0. The advantage of doing so will be clear when we formulate the dimensionless design equations of chemical reactors. Hence, if the overall order of the reaction is n ($n = \alpha + \beta + ...$), the characteristic reaction time is

$$t_{cr} = \frac{C_0}{k(T_0) \cdot C_0^n} = \frac{1}{k(T_0) \cdot C_0^{n-1}}, \qquad (2.5\text{-}3)$$

where $k(T_0)$ is the reaction rate constant at the reference temperature, T_0. For chemical reactions whose rate expressions have other forms, we define the characteristic reaction rate as follows: we express the species concentrations in terms of the dimensionless extents of the independent reactions and rearrange the rate expression as a product of a dimensional term and a dimensionless term. Then, we select the characteristic reaction rate by equating the dimensionless term to one. This is illustrated in Examples 2-5 and 2-6 below.

Example 2-4 The homogeneous chemical reaction $A + B \rightarrow C$ is carried out in a batch reactor. The reaction is first-order with respect to reactant A, and the order of reactant B is 1.5. Initially, the reactor contains 2 mole of A, 3 mole of B and 0.5 mole of C, and its initial volume is 2 liter. The reaction rate constant at the initial temperature is 0.1 (lit/mole)$^{1.5}$ min^{-1}. Determine the characteristic reaction time.

Solution For chemical reactions whose rate expressions are power functions of the concentrations, we define the characteristic reaction time by (2.5-3). First, we calculate the total concentration. In this case,

$$N_{tot}(0) = N_A(0) + N_B(0) + N_C(0) = 5.50 \text{ mole},$$

and, using (2.5-2),

$$C_0 = \frac{(N_{tot})_0}{V_R(0)} = \frac{5.5}{2.0} = 2.75 \text{ mole/liter}. \qquad (a)$$

Substituting (a) into (2.5-3), the characteristic reaction time is

$$t_{cr} = \frac{1}{(0.1) \cdot (2.75)^{1.5}} = 2.19 \text{ min}. \qquad (b)$$

Example 2-5 A biological waste, A, is decomposed by an enzymatic reaction $A \rightarrow B + C$ in aqueous solution. The rate expression of the reaction is

$$r = \frac{k \cdot C_A}{K_m + C_A}.$$

An aqueous solution with a concentration of 2 mole A/liter is charged into a batch reactor. For the enzyme type and concentration used, $k = 0.1$ mole/lit min and $K_m = 4$ mole/lit. Determine the characteristic reaction time.

Solution Since A is the only species charged into the reactor, $C_0 = C_A(0)$ and using stoichiometric relation (1.5-3),

$$C_A(t) = C_0[1 - Z(t)]. \tag{a}$$

Substituting (a) in the rate expression and rearranging, we obtain

$$r = k \cdot \frac{1 - Z}{\dfrac{K_m}{C_0} + 1 - Z}. \tag{b}$$

The quotient term is dimensionless; hence, the characteristic rate is

$$r_0 = k. \tag{c}$$

Using (2.5-1), the characteristic reaction time is

$$t_{cr} = \frac{C_0}{k} = \frac{(2 \text{ mole/lit})}{(0.1 \text{ mole/lit min})} = 20 \text{ min.} \tag{d}$$

Example 2-6 The heterogeneous, gas-phase catalytic chemical reaction $A + B \rightarrow C$ is carried out in an isothermal isobaric flow reactor. The volume-based rate expression of the reaction is

$$r = k(T) \cdot \frac{C_A \cdot C_B}{1 + K_A \cdot C_A + K_B \cdot C_B},$$

and a gas stream at 2 atmosphere and 230°C is fed into a reactor. At this temperature, $k = 80$ lit/mole min. Determine the characteristic reaction time.

Solution We select the inlet stream as the reference stream. For gas-phase reactions,

$$C_0 = \frac{P_0}{R \cdot T_0} = \frac{(2 \text{ atm})}{(0.08206 \text{ lit} \cdot \text{atm/mole} \cdot °K) \cdot (503°K)} = 4.85 \cdot 10^{-2} \text{ mole/lit}, \tag{a}$$

and the species concentrations are (see Section 6.1)

$$C_A = \frac{F_A}{v} = C_0 \left(\frac{y_{A0} + s_A \cdot Z}{1 + \Delta \cdot Z} \right) \tag{b}$$

$$C_B = \frac{F_B}{v} = C_0 \left(\frac{y_{B0} + s_B \cdot Z}{1 + \Delta \cdot Z} \right). \tag{c}$$

Substituting (b) and (c) into the rate expression, we obtain

$$r = k(T_0) \cdot C_0^2 \cdot \frac{\left(\dfrac{y_{A_0} + s_A \cdot Z}{1 + \Delta \cdot Z}\right)\left(\dfrac{y_{B_0} + s_B \cdot Z}{1 + \Delta \cdot Z}\right)}{(1 + K_A \cdot C_0 + K_B \cdot C_0)^2}. \tag{d}$$

The numerator and denominator of the quotient are dimensionless; hence, the characteristic reaction rate is

$$r_0 = k(T_0) \cdot C_0^2.$$

Using (2.5-1), the characteristic reaction time is

$$t_{cr} = \frac{C_0}{k(T_0) \cdot C_0^2} = \frac{1}{(80 \text{ lit/mole} \cdot \text{min}) \cdot (4.85 \cdot 10^{-2} \text{ mole/lit})} = 0.258 \text{ min}. \tag{e}$$

Example 2-7 Allyl chloride is produced at 400°F by a homogeneous gas-phase reaction between chlorine and propylene,

$$Cl_2 + C_3H_6 \rightarrow C_3H_5Cl + HCl.$$

Based on experimental runs, the chlorine depletion rate is

$$(-r_{Cl_2}) = k(T) \cdot P_{Cl_2} \cdot P_{C_3H_6},$$

where $(-r_{Cl_2})$ is in lb-mole/hr ft³, and P is in psia. A gas stream at 30 psia and 400°F, consisting of 50% propylene and 50% chlorine, is fed into the reactor. The reaction rate constant at 400°F is 0.02 lb-mole/(hr ft³ psia²). Determine the characteristic reaction time.

Solution Since the chlorine depletion rate is provided (rather than the reaction rate), we have to convert it to a volume-based, extension-based reaction rate. Using (2.2-5), the reaction rate is

$$r = -\frac{(-r_{Cl_2})}{s_{Cl_2}} = (-r_{Cl_2}). \tag{a}$$

To calculate the characteristic reaction time, we first have to determine the characteristic reaction rate, r_0. To do so, we express the partial pressures of the reactants in terms of the dimensionless extent of the chemical reaction. Using stoichiometric relations (1.5-5) and (1.5-6), for this case,

$$P_{Cl_2} = P_{tot} \cdot \left(\frac{N_{Cl_2}}{N_{tot}}\right) = P_{tot} \cdot (y_{Cl_2 0} - Z) \tag{b}$$

$$P_{C_3H_6} = P_{tot} \cdot \left(\frac{N_{C_3H_6}}{N_{tot}}\right) = P_{tot} \cdot (y_{C_3H_6 0} - Z). \tag{c}$$

Substituting (b) and (c) into the rate expression,

$$r = k(T) \cdot P_{tot}^2 \cdot \left[(y_{Cl_{2\,0}} - Z) \cdot (y_{C_3H_{6\,0}} - Z) \right].$$ (d)

We select the inlet stream as the reference stream, and, using the procedure described above, since the term inside the bracket is dimensionless, the characteristic reaction rate is

$$r_0 = k(T_0) \cdot P_0^2 = (0.02 \text{ lb-mole hr}^{-1} \text{ ft}^{-3} \text{ psia}^{-2}) \cdot (30 \text{ psia})^2 = 18.0 \text{ lbmole/hr ft}^3. \quad \text{(e)}$$

Assuming ideal-gas behavior, the reference concentration is

$$C_0 = \frac{P_0}{R \cdot T_0} = \frac{(30 \text{ psia})}{(10.73 \text{ psia ft}^3/\text{lbmole}^\circ R) \cdot (860^\circ R)} = 3.25 \cdot 10^{-3} \text{ lbmole/ft}^3. \quad \text{(f)}$$

Substituting (e) and (f) into (2.5-1), the characteristic reaction time is

$$t_{cr} = \frac{C_0}{r_0} = \frac{1}{k(T_0) \cdot P_0 \cdot R \cdot T_0} = 1.81 \cdot 10^{-4} \text{ hr} = 0.650 \text{ sec}. \quad \text{(g)}$$

2.6 SUMMARY

In this chapter, we discussed the main concepts of chemical kinetics related to reactor design. We covered the following topics:

a. Definition of species formation rates on the basis of volume, mass, and surface and the relations among them.

b. Definition of reaction rates and their relation to the species formation rates.

c. General form of the rate expression for many homogeneous reactions (power expression).

d. Activation energy and how to determine it.

e Orders of the reaction.

f. Different forms of the rate expressions.

g. Definition of the characteristic reaction time as a measure of the reaction time scale.

Table 2-1 summarizes the main kinetic definitions and relations.

BIBLIOGRAPHY

More detailed treatments of chemical kinetics, reaction mechanisms, and the theoretical basis for the rate expression can be found in:

K. J. Laidler, *Chemical Kinetics*, Third Ed., Harper & Row, New York, 1987.

S. W. Benson, *The Foundation of Chemical Kinetics*, McGraw-Hill, New York, 1960.

Table 2-1: Summary of Kinetic Relations

Definitions of Species Formation Rates

$$(r_j) \equiv \frac{1}{V}\frac{dN_j}{dt}; \quad (r_j)_S \equiv \frac{1}{S}\frac{dN_j}{dt}; \quad (r_j)_W \equiv \frac{1}{W}\frac{dN_j}{dt} \tag{A}$$

Relations Between the Different Species Formation Rates

$$(r_j) \equiv \left(\frac{S}{V}\right)\cdot(r_j)_S; \quad (r_j) \equiv \left(\frac{W}{V}\right)\cdot(r_j)_W \tag{B}$$

Definitions of Rates of Chemical Reactions

$$r \equiv \frac{1}{V}\frac{dX}{dt}; \quad r_S \equiv \frac{1}{S}\frac{dX}{dt}; \quad r_W \equiv \frac{1}{W}\frac{dX}{dt} \tag{C}$$

Relations Between Species Formation Rates and Rates of Reactions

$$(r_j) = \sum_i^{n_{all}} (s_j)_i \cdot r_i \tag{D}$$

Power Function Rate Expression

$$r = k(T)\cdot C_A{}^{\alpha}\cdot C_B{}^{\beta} \tag{E}$$

where $k(T)$, the reaction rate constant, is expressed by

$$k(T) = k_0 \cdot \exp\left(-\frac{E_a}{R\cdot T}\right) \qquad k(\theta) = k(T_0)\cdot e^{\gamma\frac{\theta-1}{\theta}} \tag{F}$$

α, β order of species A, B, etc.
E_a activation energy
k_0 pre-exponential factor
θ dimensionless temperature, T/T_0
γ dimensionless activation energy, E_a/RT_0

Characteristic Reaction Time

$$t_{cr} \equiv \frac{\text{Characteristic concentration}}{\text{Characteristic reaction rate}} = \frac{C_0}{r_0} \tag{G}$$

For Power Function Rate Expressions

$$t_{cr} = \frac{1}{k(T_0)\cdot C_0{}^{n-1}} \tag{H}$$

J. H. Espenson, *Chemical Kinetics and Reaction Mechanisms*, McGraw-Hill, New York, 1981.

W. C. Gardiner, *Rates and Mechanisms of Chemical Reactions*, Benjamin, New York, 1969.

M. Boudart, *Kinetics of Chemical Processes*, Prentice-Hall, 1968.

O. A. Hougen and K. M. Watson, *Chemical Process Principles III; Kinetics and Catalysis,* Wiley, New York, 1947.

Compilations of experimental and theoretical reaction rate data can be found in:

Bamford, C. H. and C. F. H. Tipper, Eds., "Comprehensive Chemical Kinetics," Elsevier, Amsterdam, 1989

Kerr, J. A. and M. J. Parsonage, "Evaluated Kinetic Data of Gas-Phase Addition Reactions," Butterworth, London, 1972

Kondratiev, V. N. " Rate Constants of Gas-Phase Reactions — Reference Book," Translated by L. J. Holtschlag, Edited by R. M. Fristrom, National Bureau of Standards, NITI Service, Springfield, VA, 1972.

NIST, "NIST Chemical Kinetic Database on Diskette," NIST, Gaithersburg, MD, 1993.

PROBLEMS

2-1$_1$ The rate of a heterogeneous reaction was measured in a rotating basket reactor. The volume of the basket was 100cc and it was filled with 240 g of catalytic particles whose specific surface area was 9.5 m^2 per gram of catalyst. At 100°C, the measured volume-based reaction rate constant was 2.5 sec^{-1}. Determine:

a. the order of the reaction, based on the units of the reaction rate constant,

b. the mass-based rate constant, and

c. the surface-based rate constant.

2-2$_1$ For a first-order homogeneous reaction, the following experimental data were recorded:

T (°C)	0	20	40	60	80
k 10^3 (min^{-1})	0.3	6.18	86.4	879	6900

Determine:

a. the activation energy,

b. the rate constant at 25°C,

c. the frequency factor,

d. the dimensionless activation energy if the reference temperature is 25°C, and

e. the characteristic reaction time at 40°C.

2-3$_1$ A common rule-of-thumb for many chemical reactions states that, at normal conditions, the reaction rate doubles for each increase of 10°C. Assuming a "normal" temperature of 20°C, estimate the activation energy of a "typical" chemical reaction. If we select 20°C as the reference temperature, what is a "typical" dimensionless activation energy?

2-4 Fine and Beall (Chemistry for Engineers and Scientists, p. 854, 1990) provide the kinetic data for a particular proton transfer reaction:

T (°K)	273	283	293	303
k (lit. mole^{-1} s^{-1})	35	150	500	1800

Determine:
a. The overall order of the reaction from the units of k, and
b. the activation energy.
c. If we select 300°K as the reference temperature, what is the dimensionless activation energy?
d. If $C_0 = 2$ mole/lit, what is the characteristic reaction time at 273°K?

2-5 The use of a pressure cooker is a common method to increase the reaction rate and to shorten the cooking time. A typical pressure cooker operates at 15 psig, thus raising the cooking temperature from 212 to 249°F. Neglecting the cooking that occurs during the heating-up period and assuming the rate constant is inversely proportional to the cooking time,
a. estimate the cooking activation energy of the foods listed below.
b. If we select 212°F as the reference temperature, what is the dimensionless activation energy of each food?
c. Assuming first-order reaction, what are the characteristic reaction times at 212°F?

	Normal cooking time	Cooking time in a Pressure Cooker
Green beans	12 min.	3 min
Corn beef	4 hr	60 min
Chicken stew	2.5 hr	20 min
(Data taken from a cook book.)		

CHAPTER THREE

SPECIES BALANCES AND DESIGN EQUATIONS

This chapter covers the fundamental concept that leads to the formulation of the design equations for chemical reactors — conservation of mass. Since we deal here with operations involving changes in composition, we carry out mass balances for individual species. Also, since the operations involve chemical transformations, it is convenient to express the amount of a species in terms of moles rather than mass. In general, we carry out the species balances in one of two ways — as microscopic balances or macroscopic balances. Microscopic species balances, often referred to as the "species continuity equations," are carried out over a differential element and describe what takes place at a given "point" in a reactor. When integrated over the volume of the reactor, they provide the species-based design equations which describe the reactor operation. Macroscopic species balances are carried out over the entire reactor and provide the design equations.

The general conservation statement for species j over a stationary system is

$$
\begin{Bmatrix} \text{molar flow} \\ \text{rate of} \\ \text{species j} \\ \text{into system} \end{Bmatrix} + \begin{Bmatrix} \text{rate moles} \\ \text{of species j} \\ \text{are formed} \\ \text{inside system} \end{Bmatrix} = \begin{Bmatrix} \text{molar flow} \\ \text{rate of} \\ \text{species j out} \\ \text{of system} \end{Bmatrix} + \begin{Bmatrix} \text{rate moles} \\ \text{of species j} \\ \text{accumulate} \\ \text{in system} \end{Bmatrix}. \quad (3.0\text{-}1)
$$

In Section 3.1, we consider a microscopic system and derive the species continuity equation. We integrate the continuity equation over the volume of the reactor to derive the general, species-based design equations. In Section 3.2, we carry out macroscopic species balances to derive the species-based design equation of any chemical reactor. In Section 3.3, we apply the general species-based design equations to reactor configurations commonly used in reactor analysis — ideal batch reactor, continuous stirred tank reactor (CSTR), and plug flow reactor. In Section 3.4, we derive the reaction-based design equations for the three ideal reactor configurations. In Section 3.5, we reduce the reaction-based design equations to dimensionless forms and discuss their solutions — the dimensionless operating curves of chemical reactors.

3.1 MICROSCOPIC SPECIES BALANCES — SPECIES CONTINUITY EQUATIONS

(This section can be skipped on first reading; it is provided here for a complete presentation of the basic principles.)

Consider a stationary volume element, $\Delta x\,\Delta y\,\Delta z$ shown in Figure 3-1, through which species j flows and in which chemical reactions take place. Let J_{jx}, J_{jy}, and J_{jz} be

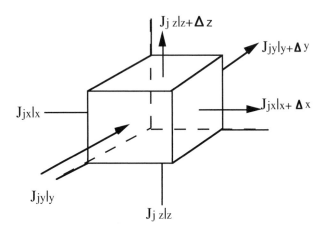

Figure 3-1: Diagram of the Molar Flux Components in a Cartesian Element

the components of the local molar flux of species j, C_j the local molar concentration of species j, and (r_j) the volume-based formation rate of species j defined by (2.1-1a). We write a species balance over a stationary cubic element (see Figure 3-1) in terms of the molar flux of species j through the six surfaces of the element; each bracket corresponds to a term in (3.0-1),

$$\left[(J_{jx})_x \cdot \Delta y \cdot \Delta z + (J_{jy})_y \cdot \Delta x \cdot \Delta z + (J_{jz})_z \cdot \Delta x \cdot \Delta y\right] + \left[(r_j)\,\Delta x \cdot \Delta y \cdot \Delta z\right] =$$

$$\left[(J_{jx})_{x+\Delta x} \cdot \Delta y \cdot \Delta z + (J_{jy})_{y+\Delta y} \cdot \Delta x \cdot \Delta z + (J_{jz})_{z+\Delta z} \cdot \Delta x \cdot \Delta y\right] + \left[\frac{d}{dt}C_j \cdot \Delta x \cdot \Delta y \cdot \Delta z\right].$$

$$(3.1\text{-}1)$$

Dividing both sides by $\Delta x \cdot \Delta y \cdot \Delta z$ and taking the limit, $\Delta x \to 0$, $\Delta y \to 0$, $\Delta z \to 0$, we obtain

$$\frac{\partial C_j}{\partial t} + \frac{\partial J_{jx}}{\partial x} + \frac{\partial J_{jy}}{\partial y} + \frac{\partial J_{jz}}{\partial z} = (r_j). \qquad (3.1\text{-}2)$$

In general, we can write (3.1-2) in vector notation

$$\frac{\partial C_j}{\partial t} + \nabla \cdot \mathbf{J}_j = (r_j),$$

(3.1-3)

where ∇ is the divergence operator of the molar flux of species j. Eq. (3.1-3) is commonly called the species continuity equation. It provides a relation between the time variations in the species concentration at a fixed point, the local motion of the species, and the rate the species is formed by chemical reactions. The species continuity equations for cylindrical and spherical coordinates are given in Table 3-1.

Table 3-1: Species Continuity Equations

In general vector notation

$$\frac{\partial C_j}{\partial t} + \nabla \cdot \mathbf{J}_j = (r_j)$$

(A)

For rectangular coordinates,

$$\frac{\partial C_j}{\partial t} + \frac{\partial J_{jx}}{\partial x} + \frac{\partial J_{jy}}{\partial y} + \frac{\partial J_{jz}}{\partial z} = (r_j)$$

(B)

For cylindrical coordinates,

$$\frac{\partial C_j}{\partial t} + \frac{1}{r}\frac{\partial}{\partial r}(r \cdot J_{jr}) + \frac{1}{r}\frac{\partial J_{j\theta}}{\partial \theta} + \frac{\partial J_{jz}}{\partial z} = (r_j)$$

(C)

For spherical coordinates,

$$\frac{\partial C_j}{\partial t} + \frac{1}{r^2}\frac{\partial}{\partial r}(r^2 \cdot J_{jr}) + \frac{1}{r \cdot \sin\theta}\frac{\partial}{\partial \theta}(J_{j\theta} \cdot \sin\theta) + \frac{1}{r \cdot \sin\theta}\frac{\partial J_{j\phi}}{\partial \phi} = (r_j)$$

(D)

To describe the operation of chemical reactors, we integrate the species continuity equation over the reactor volume. Multiplying each term in (3.1-3) by dV and integrating, we obtain

$$\int_{V_R} \frac{\partial C_j}{\partial t} \cdot dV + \int_{V_R} (\nabla \cdot \mathbf{J}_j) \cdot dV = \int_{V_R} (r_j) \cdot dV.$$

(3.1-4)

The first term reduces to

$$\int_{V_R} \frac{\partial C_j}{\partial t} dV = \frac{dN_j}{dt},$$

where N_j is the total number of moles of species j in the reactor. Applying Gauss' divergence theorem, the second term reduces to

$$\int_{S_R} (\nabla \cdot \mathbf{J}_j) \cdot dV = \int_{S_R} (\mathbf{J}_j \cdot \mathbf{n}) \cdot dS = F_{j_{out}} - F_{j_{in}},$$

where $\mathbf{n}$ is the outward unit vector on the boundaries of the reactor, and $(F_{jout} - F_{jin})$ is the net molar flow rate of species j through the boundaries of the reactor. Thus, (3.1-4) reduces to

$$\frac{dN_j}{dt} + F_{j_{out}} - F_{j_{in}} = \int_{V_R} (r_j) \cdot dV, \qquad (3.1\text{-}5)$$

which is the general, integral, species-based design equation of chemical reactors, written for species j. The application of this equation is discussed in Section 3.3 after we derive it from a macroscopic species balance.

3.2 MACROSCOPIC SPECIES BALANCES — GENERAL SPECIES-BASED DESIGN EQUATIONS

To perform a macroscopic species balance, we conduct a species balance over an entire reactor or a well-defined system. Consider an arbitrary system with one inlet and one outlet and volume V_R, shown schematically in Figure 3-2. Fluid flows through the system, and chemical reactions take place inside the system. We impose no restrictions on the system except the assumption that the chemical reactions are homogeneous

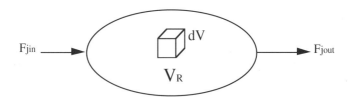

Figure 3-2: An Arbitrary Flow System

(i.e., they take place throughout the system).

$$F_{j_{in}} + G_j = F_{j_{out}} + \frac{dN_j}{dt}, \qquad (3.2\text{-}1)$$

where G_j is the rate species j is formed inside the system by chemical reactions. To obtain useful relations, we should express the generation term, G_j. When the same conditions (temperature and concentrations) exist throughout the system (due to good mixing), the volume-based formation rate of species j, (r_j), is the same everywhere in the system and

$$G_j = (r_j) \cdot V_R,$$

where (r_j) is defined by (2.1-1a). When different conditions exist at different points in the system, (r_j) varies from point to point in the system. In this case, we consider a differential volume element, dV, and express the rate species j is generated in the element by

$$dG_j = (r_j) \cdot dV.$$

Therefore, the rate species j is being generated in the entire system is

$$G_j = \int_{V_R} dG_j = \int_{V_R} (r_j) \cdot dV, \qquad (3.2\text{-}2)$$

where V_R is the volume of the system. Substituting (3.2-2), (3.2-1) reduces to

$$F_{jin} - F_{jout} + \int_{V_R} (r_j) \cdot dV = \frac{dN_j}{dt}. \qquad (3.2\text{-}3)$$

Eq. (3.2-3) is identical to (3.1-5) and is the general, integral, species-based design equation of chemical reactors, written for species j. We derive specific reactor design equations by imposing certain conditions on the flow conditions inside the reactor.

The differential form of the general, species-based design equation can be derived either by a formal differentiation of (3.2-3) or by conducting a balance for species j over a differential reactor. For a differential macroscopic reactor, the terms in (3.2-3) are: $V_R = dV_R$, $F_{jin} = F_j$, $F_{jout} = F_j + dF_j$, $G_j = (r_j) dV_R$, and the design equation reduces to

$$(r_j) \cdot dV_R = dF_j + \frac{dN_j}{dt}. \qquad (3.2\text{-}4)$$

Eq. (3.2-4) is the differential form of the general, species-based design equation for all chemical reactors.

For reactors with **single** chemical reactions, it has been customary to write the species-based design equation for the limiting reactant A. Hence, for limiting reactant A, (3.2-4) becomes

$$F_{A_{in}} = F_{A_{out}} + \int_{V_R} (-r_A) \cdot dV + \frac{dN_A}{dt}, \qquad (3.2\text{-}5)$$

where $(-r_A)$ is the depletion rate of reactant A, defined by (2.1-1a). Note that $(-r_A)$ is a positive quantity. Eq. (3.2-5) is the integral form of the general, species-based design equation, written for the limiting reactant. Similarly, from (3.2-4), we can readily obtain the general differential, species-based design equation, written for limiting reactant A,

$$-dF_A = (-r_A) \cdot dV + \frac{dN_A}{dt}. \tag{3.2-6}$$

To obtain useful expressions from the general species-based design equation, (3.2-3) or (3.2-4), we should know the formation rate of species j, (r_j), at any point in the reactor. To express (r_j), the local concentrations of all species as well as the local temperatures throughout the reactor should be provided. To obtain these quantities, we should solve the overall continuity equation, the individual species continuity equations, and the energy balance equation. This is a formidable task, and, in most situations, we cannot reduce these equations to useful forms and have to resort to numerical or approximate solution methods that are beyond the scope of this book. Therefore, in most instances, we apply the design equations to simplified reactor configurations (or mathematical models) that approximate the operations of actual reactors. In this monograph, we discuss the application of the reactor design equations to the following reactor configurations:

- well-mixed batch (closed) reactor (ideal batch reactor),
- steady, well-mixed continuous reactor (continuous stirred tank reactor, CSTR),
- steady plug flow reactor (PFR), and
- certain other special reactor configurations.

In Section 3.3, we derive the design equations for the first three ideal reactor models. The other reactor configurations are discussed in Chapter 8.

3.3 SPECIES-BASED DESIGN EQUATIONS OF IDEAL REACTORS

3.3.1 Ideal Batch Reactor

For a batch (closed) reactor, $F_{jout} = F_{jin} = 0$, and (3.2-3) becomes

$$\frac{dN_j}{dt} = \int_{V_R} (r_j) \cdot dV.$$

For a well-mixed (ideal) batch reactor, the same conditions (concentration and temperature) exist everywhere; hence, at any instant, (r_j) is the same throughout the reactor. Thus, (3.2-3) becomes

$$\frac{dN_j}{dt} = (r_j) \cdot V_R(t), \tag{3.3-1}$$

which is the species-based design equation of an ideal batch reactor, written for species

j. To obtain the integral form of the design, we separate the variables and integrate (3.3-1),

$$t = \int_{N_j(0)}^{N_j(t)} \frac{dN_j}{(r_j) \cdot V_R} . \tag{3.3-2}$$

Eq. (3.3-2) is the integral, species-based design equation for an ideal batch reactor, written for any species j. It provides a relation between the operating time, t, the amount of the species in the reactor, $N_j(t)$ and $N_j(0)$, the species formation rate, (r_j), and the reactor volume, V_R.

For reactors with **single** chemical reactions, it has been customary to write the species-based design equation for the limiting reactant A, and (3.3-1) reduces to

$$-\frac{dN_A}{dt} = (-r_A) \cdot V_R(t). \tag{3.3-3}$$

Using the conversion definition (1.5-7), $dN_A/dt = -N_A(0) \cdot df_A/dt$, and (3.3-3) becomes

$$N_A(0) \cdot \frac{df_A}{dt} = (-r_A) \cdot V_R(t). \tag{3.3-4}$$

To obtain the integral form of the design, we separate the variables and integrate (3.3-4),

$$t = N_A(0) \int_{0}^{f_A(t)} \frac{df_A}{(-r_A) \cdot V_R} . \tag{3.3-5}$$

3.3.2 Continuous Stirred Tank Reactor (CSTR)

A continuous stirred tank reactor (CSTR) is a reactor model based on two assumptions: (i) steady state operation, and (ii) the same conditions exist everywhere inside the reactor (due to good mixing). For steady operations, the accumulation term in the design equation vanishes. For a well-mixed reactor, the same conditions exist everywhere, and the rate (r_j) is the same throughout the reactor and is equal to the rate at the reactor effluent, $(r_j)_{out}$. Hence, the general, species-based design equation (3.2-3) reduces to

$$F_{j_{out}} - F_{j_{in}} = (r_j)_{out} \cdot V_R. \tag{3.3-6}$$

We can rearrange (3.3-6) as

$$V_R = \frac{F_{j_{out}} - F_{j_{in}}}{(r_j)_{out}} . \tag{3.3-7}$$

Eq. (3.3-7) is the species-based design equations for a CSTR, written for any species j.

It provides a relation between the species flow rate at the inlet and outlet of the reactor, F_{jin} and F_{jout}, the species formation rate (r_j), and the reactor volume, V_R.

For a CSTR with **single** chemical reactions, it has been customary to write the species-based design equation for the limiting reactant A, and (3.3-7) reduces to

$$V_R = \frac{F_{A_{in}} - F_{A_{out}}}{(-r_A)_{out}}.$$ (3.3-8)

Using the conversion definition (1.5-8), $F_{Ain} = F_{A0}(1 - f_{Ain})$, $F_{Aout} = F_{A0}(1 - f_{Aout})$, and (3.3-8) becomes

$$\frac{V_R}{F_{A0}} = \frac{f_{A_{out}} - f_{A_{in}}}{(-r_A)_{out}}.$$ (3.3-9)

Eq. (3.3-9) is the species-based design equation of a CSTR, expressed in terms of the conversion of reactant A.

3.3.3 Plug-Flow Reactor (PFR)

A plug-flow reactor is a reactor model based on two assumptions: (i) steady state operation, and (ii) a flat velocity profile (plug flow) with the same conditions at any given cross section area (no concentration or temperature gradients in the direction perpendicular to the flow). Since the species compositions change along the reactor and the species formation rate, (r_j), varies along the reactor, we consider a differential

$$F_j \longrightarrow \boxed{dV_R} \longrightarrow F_j + dF_j$$

reactor of volume dV_R, $F_{jin} = F_j$, and $F_{jout} = F_j + dF_j$. For steady operations, the accumulation term vanishes, and (3.2-4) reduces to

$$dV_R = \frac{dF_j}{(r_j)}.$$ (3.3-10)

Eq. (3.3-10) is the species-based differential design equations for a plug flow reactor, written for any species j. To obtain the integral form of the design equation for a plug-flow reactor, we integrate (3.3-10),

$$V_R = \int_{F_{jin}}^{F_{jout}} \frac{dF_j}{(r_j)}.$$ (3.3-11)

Eq. (3.3-11) provides a relation between the species flow rate at the inlet and outlet of the reactor, F_{jin} and F_{jout}, the species formation rate, (r_j), and the volume of the reactor,

V_R, for a plug flow reactor.

For plug-flow reactors with **single** chemical reactions, it has been customary to write the species-based design equation for the limiting reactant A, and (3.3-10) reduces to

$$dV_R = -\frac{dF_A}{(-r_A)}. \tag{3.3-12}$$

Differentiating the conversion definition (1.5-8), $dF_A = -F_{A0} \cdot df_A$, and (3.3-12) reduces to

$$dV_R = F_{A0} \frac{df_A}{(-r_A)}. \tag{3.3-13}$$

Eq. (3.3-13) is the species-based differential design equation of a plug-flow reactor, expressed in terms of the conversion of reactant A. To obtain the integral form of the design equation, we separate the variables and integrate (3.3-13),

$$\frac{V_R}{F_{A0}} = \int_{f_{A\,in}}^{f_{A\,out}} \frac{df_A}{(-r_A)}. \tag{3.3-14}$$

Eq. (3.3-14) is the species-based integral design equation of a plug-flow reactor, expressed in terms of the conversion of reactant A.

3.4 REACTION-BASED DESIGN EQUATIONS

The species-based design equations derived in Section 3.3 are the cornerstones of the conventional reactor design methodology. However, as discussed in the Introduction, to analyze and design chemical reactors more effectively and to obtain insight into the operation, reaction-based design formulation should be used. In this section, we derive the reaction-based design equations for the three ideal reactor models. Reaction-based design equations of other reactor configurations are derived in Chapter 8.

3.4.1 Ideal Batch Reactor

For an ideal batch reactor, the species-based design equation, written for species j, is given by (3.3-1),

$$\frac{dN_j}{dt} = (r_j) \cdot V_R(t). \tag{3.4-1}$$

Using stoichiometric relation (1.5-3), $N_j(t)$ is expressed in terms of the extents of the independent reactions by

$$N_j(t) = N_j(0) + \sum_m^{n_{ind}} (s_j)_m \cdot X_m(t),$$

which, upon differentiation, becomes

$$\frac{dN_j}{dt} = \sum_m^{n_{ind}} (s_j)_m \frac{dX_m}{dt}. \tag{3.4-2}$$

Using (2.2-6) to express the formation rate of species j in terms of the rates of the chemical reactions, we obtain

$$(r_j) = \sum_i^{n_{all}} (s_j)_i \cdot r_i. \tag{3.4-3}$$

Substituting (3.4-2) and (3.4-3) into (3.4-1), the design equation becomes

$$\sum_m^{n_{ind}} (s_j)_m \frac{dX_m}{dt} = \sum_i^{n_{all}} (s_j)_i \cdot r_i \cdot V_R(t). \tag{3.4-4}$$

Note that the summation on the left hand side of (3.4-4) is over **independent** reactions only, whereas the summation on the right hand side is over **all** reactions that take place in the reactor. Using (1.6-8), we write the right hand side of (3.4-4) as a sum of dependent reactions and independent reactions,

$$\sum_m^{n_{ind}} (s_j)_m \frac{dX_m}{dt} = \left(\sum_m^{n_{ind}} (s_j)_m \cdot r_m + \sum_k^{n_{dep}} (s_j)_k \cdot r_k \right) \cdot V_R(t). \tag{3.4-5}$$

The stoichiometric coefficient of species j in the k-th dependent reaction, $(s_j)_k$, relates to the stoichiometric coefficients of species j in the independent reactions, $(s_j)_m$, by (1.6-9),

$$\sum_m^{n_{ind}} \alpha_{km} \cdot (s_j)_m = (s_j)_k,$$

where α_{km} is the multiplier of the m-th independent reaction to obtain the k-th dependent reaction. Substituting this relation into (3.4-5), we obtain

$$\sum_m^{n_{ind}} (s_j)_m \frac{dX_m}{dt} = \left(\sum_m^{n_{ind}} (s_j)_m \cdot r_m + \sum_k^{n_{dep}} \sum_m^{n_{ind}} \alpha_{km} \cdot (s_j)_m \cdot r_k \right) \cdot V_R(t). \tag{3.4-6}$$

Since all multipliers α_{km} are constants, we can switch the order of the two summations in the second term in the parentheses and obtain,

$$\sum_{m}^{n_{ind}} (s_j)_m \frac{dX_m}{dt} = \left(\sum_{m}^{n_{ind}} (s_j)_m \cdot r_m + \sum_{m}^{n_{ind}} (s_j)_m \cdot \sum_{k}^{n_{dep}} \alpha_{km} \cdot r_k \right) \cdot V_R(t). \quad (3.4\text{-}7)$$

Since the coefficients and summations of all the terms are identical, (3.4-7) reduces to,

$$\frac{dX_m}{dt} = \left(r_m + \sum_{k}^{n_{dep}} \alpha_{km} \cdot r_k \right) \cdot V_R(t). \quad (3.4\text{-}8)$$

Eq. (3.4-8) is the **reaction-based**, differential design equation of an ideal batch reactor, written for the m-th independent reaction. As will be discussed below, to describe the operation of a reactor with multiple chemical reactions, we have to write (3.4-8) for each of the independent reactions. Note that the reaction-based design equation is invariant of the specific species used in the derivation.

For an ideal batch reactor with a **single** chemical reaction, (3.4-8) reduces to

$$\frac{dX}{dt} = r \cdot V_R(t). \quad (3.4\text{-}9)$$

When a single chemical reaction takes place, the extent of the reaction is proportional to the conversion of reactant A, f_A, given by (1.5-8),

$$f_A = -\frac{s_A}{N_A(0)} X(t),$$

where $N_A(0)$ is the mole content of reactant A initially in the reactor. Hence, we can readily express the design equation in terms of the conversion of the limiting reactant. We differentiate (1.5-8) and obtain

$$\frac{dX}{dt} = -\frac{N_A(0)}{s_A} \frac{df_A}{dt}.$$

Also, using kinetic relation (2,2-5), the depletion rate of reactant A, $(-r_A)$, relates to the reaction rate by

$$(-r_A) = -s_A \cdot r. \quad (3.4\text{-}10)$$

Substituting these two relations into (3.4-9), we obtain

$$N_A(0)\frac{df_A}{dt} = (-r_A) \cdot V_R(t),$$

which is identical to (3.3-4), the differential, species-based design equations of an ideal batch reactor with a single chemical reaction, expressed in terms of the conversion of reactant A.

3.4.2 Plug-Flow Reactor

For a differential steady flow reactor, the species-based design equation, written for species j, is given by (3.3-10),

$$dF_j = (r_j) \cdot dV_R. \tag{3.4-11}$$

We use stoichiometric relation (1.4-6) to express the local molar flow rate of species j, F_j, in terms of extents of the independent reactions,

$$F_j = F_{j_0} + \sum_{m}^{n_{ind}} (s_j)_m \cdot \dot{X}_m. \tag{3.4-12}$$

Differentiating (3.4-12),

$$dF_j = \sum_{m}^{n_{ind}} (s_j)_m \cdot d\dot{X}_m,$$

and substituting this and (3.4-3) into (3.4-11), the design equation becomes

$$\sum_{m}^{n_{ind}} (s_j)_m \frac{d\dot{X}_m}{dV_R} = \sum_{i}^{n_{all}} (s_j)_i \cdot r_i. \tag{3.4-13}$$

Note that here too, the summation on the left-hand side is over the **independent** reactions only, whereas the summation on the right-hand side is over **all** the reactions that take place in the reactor. We follow the same procedure as in the case of the ideal batch reactor. We first write the summation on the right as two sums over dependent reactions and independent reactions. Next, we express the stoichiometric coefficient of species j in the k-th dependent reaction, $(s_j)_k$, in terms of the stoichiometric coefficients of species j in the independent reactions, $(s_j)_m$, using (1.6-9), and then switch the order of the summations to obtain

$$\frac{d\dot{X}_m}{dV_R} = r_m + \sum_{k}^{n_{dep}} \alpha_{km} \cdot r_k. \tag{3.4-14}$$

Eq. (3.4-14) is the **reaction-based**, differential design equation for steady flow reactors, written for the m-th independent reaction. As will be discussed below, to describe the operation of the reactor with multiple reactions, we have to write (3.4-14) for each of the independent reactions.

For steady flow reactors with a **single** chemical reaction, (3.4-14) reduces to

$$\frac{d\dot{X}}{dV_R} = r. \tag{3.4-15}$$

When a single chemical reaction takes place, we can readily express the design equation in terms of the conversion of the limiting reactant using (1.5-11),

$$f_A = -\frac{s_A}{F_{A0}}\dot{X}, \tag{3.4-16}$$

where F_{A0} is the molar flow rate of reactant A in the reactor inlet. We differentiate (3.4-16), $d\dot{X} = -(F_{A0}/s_A)\, df_A$, use (3.4-10), and substitute them into (3.4-15) to obtain

$$\frac{df_A}{dV_R} = \frac{1}{F_{A0}}(-r_A),$$

which is identical to (3.3-13), the differential, species-based design equation of a plug-flow reactor with a single chemical reaction, expressed in terms of the conversion of reactant A.

3.4.3 Continuous Stirred Tank Reactor (CSTR)

For steady, continuous stirred-tank reactors (CSTRs), the species-based design equation is given by (3-15),

$$F_{j_{out}} - F_{j_{in}} = (r_j)_{out} \cdot V_R. \tag{3.4-17}$$

Using stoichiometric relation (3.4-12), the species molar flow rate at the inlet and the outlet are

$$F_{j_{in}} = F_{j_0} + \sum_{m}^{n_{ind}} (s_j)_m \cdot \dot{X}_{m_{in}}$$

$$F_{j_{out}} = F_{j_0} + \sum_{m}^{n_{ind}} (s_j)_m \cdot \dot{X}_{m_{out}},$$

where F_{j_0} is the molar flow rate of species j in a reference stream, and $F_{j_{in}}$ and $F_{j_{out}}$ are, respectively, the extents per time of the m-th independent reaction between the reference stream and the reactor inlet and outlet. Substituting these and (3.4-3) into (3.4-17), the design equation becomes

$$\sum_{m}^{n_{ind}} (s_j)_m \left(\dot{X}_{m_{out}} - \dot{X}_{m_{in}} \right) = \sum_{i}^{n_{all}} (s_j)_i \cdot r_{i_{out}} \cdot V_R. \tag{3.4-18}$$

Here too, the summation on the left-hand side is over the **independent** reactions only, whereas the summation on the right-hand side is over **all** the reactions that take place in the reactor. We follow the same procedure as in the case of the ideal batch and plug-flow reactors. We first write the summation on the right as two sums over dependent reactions and independent reactions. Next, we express the stoichiometric coefficient of species j in the k-th dependent reaction, $(s_j)_k$, in terms of the stoichiometric coefficients of species j in the independent reactions, $(s_j)_m$, using (1.6-9), and then switch the order of the summations to obtain

$$\dot{X}_{m_{out}} - \dot{X}_{m_{in}} = \left(r_{m_{out}} + \sum_{k}^{n_{dep}} \alpha_{km} \cdot r_{k_{out}} \right) \cdot V_R . \qquad (3.4\text{-}19)$$

Eq. (3.4-19) is the **reaction-based** design equation for CSTRs, written for the m-th independent reaction. To describe the operation of the reactor with multiple reactions, we have to write (3.4-19) for each independent reaction.

For a CSTR with a **single** chemical reaction, (3.4-19) reduces to

$$\dot{X}_{out} - \dot{X}_{in} = r \cdot V_R . \qquad (3.4\text{-}20)$$

For a single chemical reaction, the extent of the reaction is proportional to the conversion of reactant A, f_A, given by (3.4-16). Hence,

$$\dot{X}_{in} = -\frac{F_{A0}}{s_A} f_{A_{in}} \quad \text{and} \quad \dot{X}_{out} = -\frac{F_{A0}}{s_A} f_{A_{out}} ,$$

and, using (3.4-10), the depletion rate of reactant A is $(-r_A) = -s_A \cdot r$. Substituting these into (3.4-20), we obtain

$$V_R = F_{A0} \frac{f_{A_{out}} - f_{A_{in}}}{(-r_A)_{out}} ,$$

which is identical to (3.3-9), the species-based design equation of a CSTR with a single chemical reaction, expressed in terms of the conversion of reactant A.

3.4.4 Formulation Procedure

Reaction-based design equations (3.4-8), (3.4-14), and (3.4-19) are written for the m-th independent reaction. Since all the state variables of the reactor depend on the extents of the **independent** reactions, to design a chemical reactor with multiple reactions, we have to write a design equation for each of the independent chemical reactions. Since there are always more chemical species than independent reactions, by formulating the design in terms of reaction-based design equations, we express the design by the **smallest** number of design equations.

As indicated in Chapter 1, we can select different sets of independent reactions. The question then arises as to what is the most appropriate set of independent reactions for the design formulation. Since the design equations include the rates of **all** chemical reactions that actually take place in the reactor, by selecting a set of independent reactions among them, we minimize the number of terms in each design equation. Hence, we adopt the following heuristic rule:

> **Select a set of independent reactions among the chemical reactions whose rate expressions are provided,**
>
> or
>
> **do not select a set of independent reactions that includes a chemical reaction whose rate expression is not provided.**

By adopting this heuristic rule, the design equations consist of the **least** number of terms. Considering that each term is a rate expression, which is a function of temperature, and in many instances, a stiff function, we formulate the design in terms of the **most robust** set of algebraic or differential equations for numerical solutions (see Example 3-3 below).

3.5 DIMENSIONLESS DESIGN EQUATIONS AND OPERATING CURVES

Reaction-based design equations (3.4-8), (3.4-14), and (3.4-19) are expressed in terms of extensive quantities such as reaction extents, reactor volume, molar flow rates, etc. To describe the generic behavior of chemical reactors, we would like to express the design equations in terms of intensive, dimensionless variables. This is done in two steps:
- selecting a convenient framework (or "basis") for the calculation and
- selecting a characteristic time constant.

Below, we reduce the design equations of the three ideal reactors to dimensionless forms. Dimensionless design equations for other reactor configurations are derived in Chapter 8.

To reduce the design equation of an ideal batch reactor, (3.4-9), to dimensionless form, we first select a reference state of the reactor (usually, the initial state) and use the dimensionless extent, Z_m, of the m-th independent reaction, defined by (1.5-1),

$$ Z_m \equiv \frac{X_m}{(N_{tot})_0}, \qquad \left[\frac{\text{mole}}{\text{mole}} \right] \qquad (3.5\text{-}1) $$

where $(N_{tot})_0$ is the total number of moles of the reference state, which is usually the initial state of the reactor. For liquid-phase reactions, we usually define a system that consists of all the species that participate in the chemical reactions and leave out inert species (e.g., solvents). Hence, $(N_{tot})_0$ is the sum of the reactants and products that are initially in the reactor. On the other hand, for gas-phase reactions, we define a system that consists of all species in the reactor, including inert species. Hence, $(N_{tot})_0$ is the sum of all species that are initially in the reactor. We also take the initial reactor volume as the reference volume and define the reference concentration, C_0, by

$$C_0 \equiv \frac{(N_{tot})_0}{V_R(0)}.$$
(3.5-2)

Next, we define the dimensionless operating time by

$$\tau \equiv \frac{\text{operating time}}{\text{characteristic reaction time}} = \frac{t}{t_{cr}},$$
(3.5-3)

where t_{cr} is the characteristic reaction time, defined by (2.5-1). To reduce the reaction-based design equation to dimensionless form, we differentiate (3.5-1) and (3.5-2),

$$dX_m = (N_{tot})_0 \cdot dZ_m,$$

$$dt = t_{cr} \cdot d\tau,$$

and substitute these into (3.4-9) and obtain

$$\frac{dZ_m}{d\tau} = \left(r_m + \sum_{k}^{n_{dep}} \alpha_{km} \cdot r_k \right) \cdot \left(\frac{V_R}{V_R(0)} \right) \cdot \left(\frac{t_{cr}}{C_0} \right).$$
(3.5-4)

Eq. (3.5-4) is the dimensionless, reaction-based design equation of an ideal batch reactor, written for the m-th independent reaction. The factor (t_{cr}/C_0) is a scaling factor that converts the design equation to dimensionless form. Its physical significance is discussed below after Eq. (3.5-12).

To reduce the design equations of flow reactors to dimensionless forms, we select a convenient reference stream as a basis for the calculation. In most cases, it is convenient to select the inlet stream into the reactor as the reference stream, but, in some cases, it is more convenient to select another stream, even an imaginary stream. There is no restriction on the selection of the reference stream, except that we should be able to relate the reactor composition to it in terms of the reaction extents. Once we select the reference stream, we use the dimensionless extent, Z_m, of the m-th independent reaction, defined by (1.5-2),

$$Z_m = \frac{\dot{X}_m}{(F_{tot})_0}, \qquad \left[\frac{\text{mole/time}}{\text{mole/time}} \right]$$
(3.5-5)

where $(F_{tot})_0$ is the total molar flow rate of a reference stream, usually, the inlet stream. For liquid-phase reactions, we usually define a reference stream that consists only of species that participate in the chemical reactions and leave out inert species (e.g., solvents). For gas-phase reactions, we define a system that consists of all species in the reactor, including inert species.

The difficulty in analyzing flow reactors is that the transformations of chemical reactions take place over space (volume), whereas the reaction operation is a rate process. To relate the reactor volume to a time domain, we select a reference volumetric flow rate for the reference stream, v_0, and define the reactor space time, t_{sp}, by

$$t_{sp} \equiv \frac{V_R}{v_0}.$$ (3.5-6)

Note that the space time is an artificial quantity that depends on the selection of the reference stream and v_0. Once we select v_0, we also define the reference concentration, C_0, by

$$C_0 \equiv \frac{(F_{tot})_0}{v_0}.$$ (3.5-7)

Next, we define the dimensionless space time by

$$\tau \equiv \frac{\text{reactor space time}}{\text{characteristic reaction time}} = \frac{V_R}{v_0 \cdot t_{cr}},$$ (3.5-8)

where t_{cr} is the characteristic reaction time, defined by (2.5-1).

To reduce the reaction-based design equation of a plug-flow reactor to dimensionless form, we differentiate (3.5-5) and (3.5-8),

$$d\dot{X}_m = (F_{tot})_0 \cdot dZ_m,$$ (3.5-9)

$$dV_R = t_{cr} \cdot v_0 \cdot d\tau,$$ (3-5-10)

and substitute these into (3.4-14). Using (3.5-7), we obtain

$$\frac{dZ_m}{d\tau} = \left(r_m + \sum_k^{n_{dep}} \alpha_{km} \cdot r_k \right) \left(\frac{t_{cr}}{C_0} \right).$$ (3.5-11)

Eq. (3.5-11) is the dimensionless, reaction-based design equation of a plug-flow reactor, written for the m-th independent reaction. To reduce the reaction-based design equation of a CSTR to dimensionless form, we substitute (3.5-5) and (3.5-8) into (3.4-19) and, using (3.5-7),

$$Z_{m_{out}} - Z_{m_{in}} = \left(r_{m_{out}} + \sum_k^{n_{dep}} \alpha_{km} \cdot r_{k_{out}} \right) \cdot \tau \cdot \left(\frac{t_{cr}}{C_0} \right).$$ (3.5-12)

Eq. (3.5-12) is the dimensionless, reaction-based design equation of a CSTR, written for the m-th independent reaction. The factor (t_{cr}/C_0) in (3.5-11) and (3.5-12) is a dimensional scaling factor that converts the design equations to dimensionless forms.

To understand the structure of the dimensionless reaction-based design equations, recall the definition of the characteristic reaction time, (2.5-1), $t_{cr} = C_0/r_0$. It follows that the scaling factor is $(t_{cr}/C_0) = 1/r_0$, where r_0 is the reference rate of a selected chemical reaction. Substituting this into design equations (3.5-4), (3.5-11), and (3.5-12) reduce, respectively, to

$$\frac{dZ_m}{d\tau} = \left[\left(\frac{r_m}{r_0}\right) + \sum_{k}^{n_{dep}} \alpha_{km} \cdot \left(\frac{r_k}{r_0}\right)\right] \cdot \left(\frac{V_R}{V_R(0)}\right) \tag{3.5-13}$$

$$\frac{dZ_m}{d\tau} = \left(\frac{r_m}{r_0}\right) + \sum_{k}^{n_{dep}} \alpha_{km} \cdot \left(\frac{r_k}{r_0}\right) \tag{3.5-14}$$

$$Z_{m_{out}} - Z_{m_{in}} = \left[\left(\frac{r_{m_{out}}}{r_0}\right) + \sum_{k}^{n_{dep}} \alpha_{km} \cdot \left(\frac{r_{k_{out}}}{r_0}\right)\right] \cdot \tau. \tag{3.5-15}$$

Hence, the dimensionless design equations are expressed in terms of relative reaction rates, (r_m/r_0)'s and (r_k/r_0)'s, with respect to a conveniently-selected reaction rate.

To solve the dimensionless, reaction-based design equations, we have to express the rates of the chemical reactions, r_m's and r_k's, in terms of the dimensionless extents of the independent reactions. Since the reaction rates depend on species concentrations, we have to express the species concentrations in terms of the extents of the independent reactions. These relationships depend on the reactor configuration and the way in which reactants are injected but can be readily derived by using the stoichiometric relations of Chapter 1. This procedure will be discussed in detail in Part Two of the text, where we cover the design of different chemical reactor configurations. The reaction rates also depend on the temperature, θ, whose variation is obtained by solving simultaneously the energy balance equation. By substituting the concentration relations into the rate expressions and the latter into the design equations, we obtain equations in r, Z_m's, and θ.

For ideal batch and plug-flow reactors, we obtain a set of nonlinear, first-order, differential equations of the form

$$\frac{dZ_{m_1}}{d\tau} = G_1(\tau, Z_{m_1}, Z_{m_2}, \dots, Z_{n_{ind}}, \theta),$$

$$\frac{dZ_{m_2}}{d\tau} = G_2(\tau, Z_{m_1}, Z_{m_2}, \dots, Z_{n_{ind}}, \theta),$$

$$\vdots \tag{3.5-16}$$

$$\frac{dZ_{m_{ind}}}{d\tau} = G_{n_{ind}}(\tau, Z_{m_1}, Z_{m_2}, \dots, Z_{n_{ind}}, \theta),$$

where τ, the dimensionless operating time (or space time), is the independent variable. We have here a set of n_{ind} equations with $n_{ind}+1$ dependent variables. The additional equation is the energy balance equation (see Chapter Four) that reduces to the general form

$$\frac{d\theta}{d\tau} = G_{n_{ind}+1}(\tau, Z_{m_1}, Z_{m_2}, \ldots, Z_{n_{ind}}, \theta). \tag{3.5-17}$$

We solve these equations for $Z_{m1}, \ldots, Z_{nind}$ and θ as functions of τ, subject to the initial condition that at $\tau = 0$, the dimensionless extents and the dimensionless temperature are specified. For isothermal operations, (3.5-17) reduces to $d\theta/d\tau = 0$; thus, only the design equations should be solved.

For ideal stirred-tank reactors, the design formulation is a set of nonlinear, homogeneous, algebraic equations of the form

$$G_1(\tau, Z_{m_1}, Z_{m_2}, \ldots, Z_{n_{ind}}, \theta) = 0$$

$$G_2(\tau, Z_{m_1}, Z_{m_2}, \ldots, Z_{n_{ind}}, \theta) = 0$$

$$\vdots \tag{3.5-18}$$

$$G_{n_{ind}}(\tau, Z_{m_1}, Z_{m_2}, \ldots, Z_{n_{ind}}, \theta) = 0$$

and the energy balance equation reduces to the form

$$G_{n_{ind}+1}(\tau, Z_{m_1}, Z_{m_2}, \ldots, Z_{n_{ind}}, \theta) = 0. \tag{3.5-19}$$

We solve (3.5-18) and (3.5-19) simultaneously for $Z_{m1}, \ldots, Z_{nind}$ and θ for different values of dimensionless reactor space times, τ.

The solutions of the design equations (Z_m's versus τ) and the energy balance equation (θ versus τ) provide, respectively, the dimensionless reaction operating curves and the dimensionless temperature curve of the reactor operation. The dimensionless reaction operating curves describe generically the progress of the independent chemical reactions without regard to a specific operating time or reactor volume. The dimensionless temperature curve describes the generic temperature variation during the reactor operation. Using stoichiometric relations derived in Chapter 1, we can readily use the reaction operating curves to obtain the dimensionless operating curves for each species in the reactor, $N_j(\tau)/(N_{tot})_0$ or $F_j/(F_{tot})_0$, versus τ. These curves are discussed in detail in Part Two (Applications) of the book.

The formulation of the reaction-based design equations is illustrated in the three examples below. Example 3-1 shows how a case that is conventionally formulated in terms of four species-based design equations with sixteen terms is formulated here in terms of three reaction-based design equations with six terms. Example 3-3 illustrates how, by adopting the heuristic rule on selecting a set of independent reactions, we formulate the design by the most robust set of equations.

Example 3-1: The following three reversible chemical reactions represent gas-phase cracking of hydrocarbons:

$$A \leftrightarrow 2B$$

$$A + B \leftrightarrow C$$

$$A + C \leftrightarrow D.$$

Species B is the desired product. Formulate the reaction-based design equations, expressed in terms of the reaction rates, for

a. ideal batch reactor,

b. plug-flow reactor and,

c. CSTR.

Solution First, we determine the number of independent reactions and select a set of independent reactions. We represent the given chemical reactions by the following six reactions:

$$\text{Reaction 1:} \quad A \rightarrow 2\,B$$

$$\text{Reaction 2:} \quad 2\,B \rightarrow A$$

$$\text{Reaction 3:} \quad A + B \rightarrow C$$

$$\text{Reaction 4:} \quad C \rightarrow A + B$$

$$\text{Reaction 5:} \quad A + C \rightarrow D$$

$$\text{Reaction 6:} \quad D \rightarrow A + C.$$

We can construct the matrix of stoichiometric coefficients and reduce it to a diagonal form to determine the number of independent reactions. However, in this case, we have three reversible reactions, and, since each of the three forward reactions has a species that does not appear in the other two, we have three independent reactions and three dependent reactions. We select the three forward reactions as the set of independent reactions. Hence, the indices of the independent reactions are m = 1, 3, 5, and we describe the reactor operation in terms of their dimensionless extents, Z_1, Z_3, and Z_5. The indices of the dependent reactions are k = 2, 4, 6. Since this set of independent reactions consists of chemical reactions whose rate expressions are known, the heuristic rule on selecting independent reactions is satisfied. The stoichiometric coefficients of the selected three independent reactions are:

$$s_{A1} = \text{-}1; \;\; s_{B1} = 2; \;\; s_{C1} = 0; \;\; s_{D1} = 0; \;\; \Delta_1 = 1;$$

$$s_{A3} = \text{-}1; \;\; s_{B3} = \text{-}1; \;\; s_{C3} = 1; \;\; s_{D3} = 0; \;\; \Delta_3 = \text{-}1;$$

$$s_{A5} = \text{-}1; \;\; s_{B5} = 0; \;\; s_{C5} = \text{-}1; \;\; s_{D5} = 1; \;\; \Delta_5 = \text{-}1.$$

To determine the α_{km} multipliers for each of the dependent reactions, we use stoichiometric relation (1.6-9). The values are:

$$\alpha_{21} = \text{-}1; \;\; \alpha_{23} = 0; \;\; \alpha_{25} = 0;$$

$$\alpha_{41} = 0; \;\; \alpha_{43} = \text{-}1; \;\; \alpha_{45} = 0; \tag{a}$$

$$\alpha_{61} = 0; \;\; \alpha_{63} = 0; \;\; \alpha_{65} = \text{-}1.$$

a. For an ideal batch reactor, we write design equation (3.5-4) for each of the three independent reactions. Hence, for this case, the design equations are:

$$\frac{dZ_1}{d\tau} = (r_1 - r_2) \cdot \left(\frac{V_R}{V_R(0)} \right) \cdot \left(\frac{t_{cr}}{C_0} \right) \qquad \text{(b)}$$

$$\frac{dZ_3}{d\tau} = (r_3 - r_4) \cdot \left(\frac{V_R}{V_R(0)} \right) \cdot \left(\frac{t_{cr}}{C_0} \right) \qquad \text{(c)}$$

$$\frac{dZ_5}{d\tau} = (r_5 - r_6) \cdot \left(\frac{V_R}{V_R(0)} \right) \cdot \left(\frac{t_{cr}}{C_0} \right). \qquad \text{(d)}$$

Note that $(r_1 - r_2)$, $(r_3 - r_4)$, and $(r_5 - r_6)$ are the net forward rates of the three independent reactions. To solve these equations, we have to select a reference state, define reference concentration, C_0, and characteristic reaction time t_{cr}, and then express the individual reaction rates and $V_R/V_R(0)$ in terms of τ, Z_1, Z_3, and Z_5.

b. For a steady plug-flow reactor, we write design equation (3.5-11) for each of the three independent reactions. Hence, for this case, the design equations are:

$$\frac{dZ_1}{d\tau} = (r_1 - r_2) \cdot \left(\frac{t_{cr}}{C_0} \right) \qquad \text{(e)}$$

$$\frac{dZ_3}{d\tau} = (r_3 - r_4) \cdot \left(\frac{t_{cr}}{C_0} \right) \qquad \text{(f)}$$

$$\frac{dZ_5}{d\tau} = (r_5 - r_6) \cdot \left(\frac{t_{cr}}{C_0} \right). \qquad \text{(g)}$$

To solve these equations, we have to select a reference stream, define reference concentration, C_0, and characteristic reaction time t_{cr}, and then express the individual reaction rates in terms of τ, Z_1, Z_3, and Z_5.

c. For a CSTR, we write design equation (3.5-12) for each of the three independent reactions. Hence, for this case, the design equations are:

$$Z_{1_{out}} - Z_{1_{in}} - (r_{1_{out}} - r_{2_{out}}) \cdot \tau \cdot \left(\frac{t_{cr}}{C_0} \right) = 0 \qquad \text{(h)}$$

$$Z_{3_{out}} - Z_{3_{in}} - (r_{3_{out}} - r_{4_{out}}) \cdot \tau \cdot \left(\frac{t_{cr}}{C_0} \right) = 0 \qquad \text{(i)}$$

$$Z_{5_{out}} - Z_{5_{in}} - (r_{5_{out}} - r_{6_{out}}) \cdot \tau \cdot \left(\frac{t_{cr}}{C_0} \right) = 0. \qquad \text{(j)}$$

To solve these design equations, we have to express the individual reaction rates in terms of $Z_{1_{out}}$, $Z_{3_{out}}$, and $Z_{5_{out}}$, and then, for different values of dimensionless space time τ, we can solve (h), (i), and (j) simultaneously for Z_{1out}, Z_{3out}, and Z_{5out}.

Example 3-2 Acrolein is being produced by catalytic oxidation of propylene on bismuth molybdate in a packed-bed (plug-flow) reactor. The following reactions are taking place in the reactor:

$$\text{Reaction 1:} \quad C_3H_6 + O_2 \rightarrow C_3H_4O + H_2O$$

$$\text{Reaction 2:} \quad C_3H_4O + 3.5\ O_2 \rightarrow 3\ CO_2 + 2\ H_2O$$

$$\text{Reaction 3:} \quad C_3H_6 + 4.5\ O_2 \rightarrow 3\ CO_2 + 3\ H_2O.$$

Adams et al. (J. of Catalysis **3**, 379, 1964) investigated these reactions and expressed the rate of each as second-order (first-order with respect to each reactant). Formulate the dimensionless, reaction-based design equations for an ideal batch reactor, plug-flow reactor, and a CSTR.

Solution The purpose of this example is to illustrate the design formulations for chemical reactions where the dependency between dependent reactions and independent reactions is not due to reversible reactions. First, we determine the number of independent reactions and select a set of independent reactions. We construct a matrix of stoichiometric coefficients for the given reactions

$$\begin{matrix} C_3H_6 & O_2 & C_3H_4O & H_2O & CO_2 \end{matrix}$$
$$\begin{bmatrix} -1 & -1 & 1 & 1 & 0 \\ 0 & -3.5 & -1 & 2 & 3 \\ -1 & -4.5 & 0 & 3 & 3 \end{bmatrix}. \tag{a}$$

Applying Gaussian elimination procedure, matrix (a) reduces to

$$\begin{bmatrix} -1 & -1 & 1 & 1 & 0 \\ 0 & -3.5 & -1 & 2 & 3 \\ 0 & 0 & 0 & 0 & 0 \end{bmatrix}. \tag{b}$$

Matrix (b) is a reduced matrix since all the elements below the diagonal are zeros, and, because it has two nonzero rows, we have here two independent chemical reactions. We select the first two reactions as a set of independent reactions; hence, for this case, $m = 1$, 2, and $k = 3$. To determine the multipliers α_{km}'s of the dependent reaction (Reaction 3) and the two independent reactions (Reactions 1 and 2), we write (1.6-9) for propylene and CO_2,

$$\alpha_{31} \cdot (-1) + \alpha_{32} \cdot (0) = -1 \tag{c}$$

$$\alpha_{31} \cdot (0) + \alpha_{32} \cdot (3) = 3. \tag{d}$$

Solving (c) and (d), we obtain $\alpha_{31} = 1$ and $\alpha_{32} = 1$. (Indeed, Reaction 3 is the sum of Reaction 1 and Reaction 2.) Now that the α_{km} factors are known, we can formulate the design equations. For an ideal batch reactor, we write design equation (3.5-4) for each of the two independent reactions. Hence, for this case, the design equations are:

$$\frac{dZ_1}{d\tau} = (r_1 + r_3) \cdot \left(\frac{V_R}{V_R(0)}\right) \cdot \left(\frac{t_{cr}}{C_0}\right) \tag{e}$$

$$\frac{dZ_2}{d\tau} = (r_2 + r_3) \cdot \left(\frac{V_R}{V_R(0)}\right) \cdot \left(\frac{t_{cr}}{C_0}\right). \tag{f}$$

To solve these design equations, we have to express r_1, r_2, r_3, and $V_R/V_R(0)$ in terms of Z_1 and Z_2. For a plug-flow reactor, we write design equation (3.5-11) for each of the three independent reactions. Hence, for this case, the design equations are:

$$\frac{dZ_1}{d\tau} = (r_1 + r_3) \cdot \left(\frac{t_{cr}}{C_0}\right) \tag{g}$$

$$\frac{dZ_2}{d\tau} = (r_2 + r_3) \cdot \left(\frac{t_{cr}}{C_0}\right). \tag{h}$$

To solve these design equations, we have to express r_1, r_2, and r_3 in terms of Z_1 and Z_2. For a CSTR, we write design equation (3.5-12) for each of the three independent reactions. Hence, for this case, the design equations are:

$$Z_{1out} - Z_{1in} - (r_{1out} + r_{3out}) \cdot \tau \cdot \left(\frac{t_{cr}}{C_0}\right) = 0 \tag{i}$$

$$Z_{2out} - Z_{2in} - (r_{2out} + r_{3out}) \cdot \tau \cdot \left(\frac{t_{cr}}{C_0}\right) = 0. \tag{j}$$

To solve these design equations, we have to express r_{1out}, r_{2out}, and r_{3out} in terms of Z_{1out} and Z_{2out} and then, for different values of τ, we solve for Z_{1out} and Z_{2out}.

Example 3-3 a. Formulate the dimensionless, reaction-based design equations for an ideal batch reactor, plug-flow reactor, and a CSTR if the following gaseous chemical reactions take place in the reactor:

Reaction 1: $4\,NH_3 + 5\,O_2 \rightarrow 4\,NO + 6\,H_2O$

Reaction 2: $4\,NH_3 + 3\,O_2 \rightarrow 2\,N_2 + 6\,H_2O$

Reaction 3: $2\,NO + O_2 \rightarrow 2\,NO_2$

Reaction 4: $4\,NH_3 + 6\,NO \rightarrow 5\,N_2 + 6\,H_2O.$

b. Formulate the dimensionless, reaction-based design equations for an ideal batch reactor using a set of independent reactions obtained from the reduced matrix.

Solution First, we determine the number of independent reactions and select a set of independent reactions. The matrix of stoichiometric coefficients for these reactions was constructed and reduced to a diagonal matrix in Example 1-11. Since there are three nonzero rows in the reduced matrix, there are three independent reactions. The nonzero rows in the reduced matrix represent the following set of independent reactions:

$$\text{Reaction 1:} \quad 4\,NH_3 + 5\,O_2 \rightarrow 4\,NO + 6\,H_2O$$
$$\text{Reaction 5:} \quad 2\,NO \rightarrow O_2 + N_2$$
$$\text{Reaction 6:} \quad 4\,NO \rightarrow N_2 + 2\,NO_2.$$

Note that this set of independent reactions includes two reactions (Reaction 5 and Reaction 6) that are not among the given reactions.

a. Applying the heuristic rule of selecting a set of independent reactions whose rate expressions are known, we should select a set of three independent reactions from the given chemical reactions. Reactions 1, 2, and 3, represent such a set; hence, m = 1, 2, 3 and k = 4, and we express the design equations in terms of Z_1, Z_2, and Z_3. The stoichiometric coefficients of the selected independent reactions are:

$(s_{NH3})_1 = -4;\ (s_{O2})_1 = -5;\ (s_{NO})_1 = 4;\ (s_{H2O})_1 = 6;\quad (s_{N2})_1 = 0;\ (s_{NO2})_1 = 0;\ \Delta_1 = 1;$

$(s_{NH3})_2 = -4;\ (s_{O2})_2 = -3;\ (s_{NO})_2 = 0;\ (s_{H2O})_2 = 6;\quad (s_{N2})_2 = 2;\ (s_{NO2})_2 = 0;\ \Delta_2 = 1;$

$(s_{NH3})_3 = 0;\ (s_{O2})_3 = -1;\ (s_{NO})_3 = -2;\ (s_{H2O})_3 = 0;\quad (s_{N2})_3 = 0;\ (s_{NO2})_3 = 2;\ \Delta_3 = -1.$

To determine the multipliers α_{km}'s of the dependent reaction (Reaction 4) and the three independent reactions, we write stoichiometric relation (1.6-9) for NO_2,

$$\alpha_{41} \cdot (0) + \alpha_{42} \cdot (0) + \alpha_{43} \cdot (2) = 0, \tag{a}$$

and obtain $\alpha_{43} = 0$. We write (1.6-9) for N_2,

$$\alpha_{41} \cdot (0) + \alpha_{42} \cdot (2) + \alpha_{43} \cdot (0) = 5, \tag{b}$$

and obtain $\alpha_{42} = 2.5$. We write (1.6-9) for NH_3,

$$\alpha_{41} \cdot (-4) + \alpha_{42} \cdot (-4) + \alpha_{43} \cdot (0) = -4, \tag{c}$$

and obtain $\alpha_{41} = -1.5$. Now that the α_{km} factors are known, we can formulate the design equations. For an ideal batch reactor, we write design equation (3.5-4) for each of the three independent reactions. Hence, the design equations are:

$$\frac{dZ_1}{d\tau} = (r_1 - 1.5 \cdot r_4) \cdot \left(\frac{V_R}{V_R(0)}\right) \cdot \left(\frac{t_{cr}}{C_0}\right) \tag{d}$$

$$\frac{dZ_2}{d\tau} = (r_2 + 2.5 \cdot r_4) \cdot \left(\frac{V_R}{V_R(0)}\right) \cdot \left(\frac{t_{cr}}{C_0}\right) \tag{e}$$

$$\frac{dZ_3}{d\tau} = r_3 \cdot \left(\frac{V_R}{V_R(0)}\right) \cdot \left(\frac{t_{cr}}{C_0}\right). \tag{f}$$

For a plug-flow reactor, we write design equation (3.5-11) for each of the three independent reactions. Hence, the design equations are:

$$\frac{dZ_1}{d\tau} = (r_1 - 1.5 \cdot r_4) \cdot \left(\frac{t_{cr}}{C_0}\right) \tag{g}$$

$$\frac{dZ_2}{d\tau} = (r_2 + 2.5 \cdot r_4) \cdot \left(\frac{t_{cr}}{C_0}\right) \tag{h}$$

$$\frac{dZ_3}{d\tau} = r_3 \cdot \left(\frac{t_{cr}}{C_0}\right). \tag{i}$$

For a CSTR, we write design equation (3.5-12) for each of the three independent reactions. Hence, for this case, the design equations are:

$$Z_{1_{out}} - Z_{1_{in}} - (r_{1_{out}} - 1.5 \cdot r_{4_{out}}) \cdot \tau \cdot \left(\frac{t_{cr}}{C_0}\right) = 0 \tag{j}$$

$$Z_{2_{out}} - Z_{2_{in}} - (r_{2_{out}} + 2.5 \cdot r_{4_{out}}) \cdot \tau \cdot \left(\frac{t_{cr}}{C_0}\right) = 0 \tag{k}$$

$$Z_{3_{out}} - Z_{3_{in}} - r_{3_{out}} \cdot \tau \cdot \left(\frac{t_{cr}}{C_0}\right) = 0. \tag{l}$$

To solve the design equations, we have to express the rates of the individual reactions in terms of the dimensionless extents of the independent reactions, Z_1, Z_2, and Z_3.
b. Taking the reactions of the reduced matrix as a set of independent reactions, m = 1, 5, 6, and k = 2, 3, 4. Hence, we express the design equations in terms of Z_1, Z_5, and Z_6. The stoichiometric coefficients of the three independent reactions are:

$(s_{NH3})_1 = -4$; $(s_{O2})_1 = -5$; $(s_{NO})_1 = 4$; $(s_{H2O})_1 = 6$; $(s_{N2})_1 = 0$; $(s_{NO2})_1 = 0$; $\Delta_1 = 1$;

$(s_{NH3})_5 = 0$; $(s_{O2})_5 = 1$; $(s_{NO})_5 = -2$; $(s_{H2O})_5 = 0$; $(s_{N2})_5 = 1$; $(s_{NO2})_5 = 0$; $\Delta_5 = 0$;

$(s_{NH3})_6 = 0$; $(s_{O2})_6 = 0$; $(s_{NO})_6 = -4$; $(s_{H2O})_6 = 0$; $(s_{N2})_6 = 1$; $(s_{NO2})_6 = 2$; $\Delta_6 = -1$.

To determine the multipliers α_{km}'s of the three dependent reactions (Reactions 2, 3 and 4) and the three independent reactions, we write stoichiometric relation (1.6-9). For k = 2, we write (1.6-9) for NO_2,

$$\alpha_{21} \cdot (0) + \alpha_{25} \cdot (0) + \alpha_{26} \cdot (2) = 0, \tag{m}$$

and obtain $\alpha_{26} = 0$. We write (1.6-9) for N_2,

$$\alpha_{21} \cdot (0) + \alpha_{25} \cdot (1) + \alpha_{26} \cdot (1) = 2, \tag{n}$$

and obtain $\alpha_{25} = 2$. We write (1.6-9) for NH_3,

$$\alpha_{21} \cdot (-4) + \alpha_{25} \cdot (0) + \alpha_{26} \cdot (0) = -4, \tag{o}$$

and obtain $\alpha_{21} = 1$. For $k = 3$, we write (1.6-9) for NO_2,

$$\alpha_{31} \cdot (-4) + \alpha_{35} \cdot (0) + \alpha_{36} \cdot (2) = 2, \tag{p}$$

and obtain $\alpha_{36} = 1$. We write (1.6-9) for N_2,

$$\alpha_{31} \cdot (0) + \alpha_{35} \cdot (1) + \alpha_{36} \cdot (1) = 0, \tag{q}$$

and obtain $\alpha_{35} = -1$. We write (1.6-9) for NH_3,

$$\alpha_{31} \cdot (-4) + \alpha_{35} \cdot (0) + \alpha_{36} \cdot (0) = -4, \tag{r}$$

and obtain $\alpha_{31} = 0$. For $k = 4$, we write (1.6-9) for NO_2,

$$\alpha_{41} \cdot (0) + \alpha_{45} \cdot (0) + \alpha_{46} \cdot (2) = 0, \tag{s}$$

and obtain $\alpha_{46} = 0$. We write (1.6-9) for N_2,

$$\alpha_{41} \cdot (0) + \alpha_{45} \cdot (1) + \alpha_{46} \cdot (1) = 5, \tag{t}$$

and obtain $\alpha_{45} = 5$. We write (1.6-9) for NH_3,

$$\alpha_{41} \cdot (-4) + \alpha_{45} \cdot (0) + \alpha_{46} \cdot (0) = -4, \tag{u}$$

and obtain $\alpha_{41} = 1$. Hence, the α_{km} factors for the dependent reactions are

$$\alpha_{21} = 1; \quad \alpha_{25} = 2; \quad \alpha_{26} = 0;$$

$$\alpha_{31} = 0; \quad \alpha_{35} = -1; \quad \alpha_{36} = 1;$$

$$\alpha_{41} = 1; \quad \alpha_{45} = 5; \quad \alpha_{46} = 0.$$

Now that the α_{km} factors are known, we can formulate the design equations. For an ideal batch reactor, we write design equation (3.5-4) for each of the three independent reactions. Noting that independent reactions 5 and 6 do not actually take place $r_5 = r_6 = 0$, and the design equations are:

$$\frac{dZ_1}{d\tau} = (r_1 + r_2 + r_4) \cdot \left(\frac{V_R}{V_R(0)} \right) \cdot \left(\frac{t_{cr}}{C_0} \right) \tag{v}$$

$$\frac{dZ_5}{d\tau} = (2 \cdot r_2 - r_3 + 5 \cdot r_4) \cdot \left(\frac{V_R}{V_R(0)} \right) \cdot \left(\frac{t_{cr}}{C_0} \right) \tag{w}$$

$$\frac{dZ_6}{d\tau} = r_3 \cdot \left(\frac{V_R}{V_R(0)}\right) \cdot \left(\frac{t_{cr}}{C_0}\right). \tag{x}$$

Note that, in this case, the design equations have seven terms (seven rate expressions), whereas in the formulation in part (a), design equations (d), (e), and (f), have only five terms. This illustrates that, by adopting the heuristic rule, we minimize the number of terms in the design equations. To solve the design equations, we have to express the rates of the individual reactions in terms of the dimensionless extents of the independent reactions, Z_1, Z_5, and Z_6.

3.6 SUMMARY

In this chapter, we discussed the application of the conservation of mass principle to derive design equations for chemical reactors. The following topics were covered:
- We derived the species continuity equations, describing the variations in species concentrations at any point in the reactor.
- We integrated the continuity equation over the reactor volume and obtained the general, species-based design equation.
- We derived the species-based design equations for three ideal reactor models: ideal batch reactor, plug-flow reactor, and CSTR.
- We derived the reaction-based design equations for three ideal reactors.
- We converted the reaction-based design equations to dimensionless forms that, upon solution, provide the dimensionless operating curves.

Tables 3-2 and 3.3 summarize the dimensionless reaction-based design equations for ideal batch reactor and flow reactors, respectively, and provide the definitions of key parameters.

BIBLIOGRAPHY

More comprehensive treatments of the species continuity equations and the molar fluxes can be found in

Bird, R. B., W. E. Stewart and E. N. Lightfoot, *Transport Phenomena*. New York: John Wiley and Sons, 1960.

Welty, J. R., C. E. Wicks and R. E. Wilson, *Fundamentals of Momentum Heat and Mass Transfer*, 2nd Edition: New York, John Wiley and Sons, 1984.

Bird, R. B., *The Basic Concepts in Transport Phenomena*, **Chem. Eng. Education**, p. 102, Winter 1994.

Applications of the continuity equations for turbulent flows can be found in

Hinze, J. O., *Turbulence*, Second Edition. New York: McGraw-Hill, 1975.

Tennekes, H. and J. L. Lumley, *A First Course in Turbulence*. Cambridge, MA: MIT press, 1972.

Applications of the species continuity equation in reacting systems is found in Rosner, D. E., *Transport Processes in Chemically Reacting Flow Systems*. Boston: Butterworth-Heineman, 1986.

Table 3.2: Dimensionless Reaction-Based Design Equations for Ideal Batch Reactor

	For the m-th Independent Reaction
Design Equation	$$\frac{dZ_m}{d\tau} = \left(r_m + \sum_{k}^{n_{dep}} \alpha_{km} \cdot r_k \right) \cdot \left(\frac{V_R}{V_R(0)} \right) \cdot \left(\frac{t_{cr}}{C_0} \right) \quad \text{(A)}$$
Definitions	Dimensionless Operating Time $$\tau \equiv \frac{t}{t_{cr}} \quad \text{(B)}$$ Dimensionless Extent of the m-th Independent Reaction $$Z_m(\tau) \equiv \frac{X_m(\tau)}{(N_{tot})_0} \quad \text{(C)}$$ Reference Concentration $$C_0 \equiv \frac{(N_{tot})_0}{V_R(0)} \quad \text{(D)}$$ Characteristic Reaction Time (n-th order Reaction) $$t_{cr} \equiv \frac{\text{Characteristic concentration}}{\text{Characteristic reaction rate}} = \frac{C_0}{r_0} \quad \text{(E)}$$

PROBLEMS

3.1$_2$ A 200 liter/min stream of fresh water is fed to and withdrawn from a well-mixed tank with a volume of 1000 liter. Initially, the tank contains pure water. At time $t = 0$, the feed is switched to a brine stream with a concentration of 180 g/liter that is also fed at a rate of 200 liter/min.

a. Derive a differential equation for the salt concentration in the tank.

b. Separate the variables and integrate to obtain an expression for the salt concentra-

Figure 3.3: Dimensionless Reaction-Based Design Equations for Steady Flow Reactors

	Plug-Flow Reactor	CSTR
Design Equation	For the m-th Independent Reaction $$\frac{dZ_m}{d\tau} = \left(r_m + \sum_{k}^{n_{dep}} \alpha_{km} \cdot r_k \right) \cdot \left(\frac{t_{cr}}{C_0} \right) \quad (A)$$	For the m-th Independent Reaction $$Z_{m_{out}} - Z_{m_{in}} = \left(r_{m_{out}} + \sum_{k}^{n_{dep}} \alpha_{km} \cdot r_{k_{out}} \right) \cdot \tau \cdot \left(\frac{t_{cr}}{C_0} \right) \quad (A)$$
Definitions	Dimensionless Space Time Dimensionless Extent Reference Concentration Characteristic Reaction Time	$\tau \equiv \dfrac{V_R}{v_0 \cdot t_{cr}}$ (B) $Z_m = \dfrac{\dot{X}_m}{(F_{tot})_0}$ (C) $C_0 = \dfrac{(F_{tot})_0}{v_0}$ (D) $t_{cr} = \dfrac{\text{Characteristic Concentration}}{\text{Characteristic Reaction Rate}} = \dfrac{C_0}{r_0}$ (E)

tion as a function of time.

c. At what time will the salt concentration in the tank reach a level of 100 g/liter?

3.2₂ Many specialty chemicals are produced in semi-batch reactors where a reactant is added gradually into a batch reactor. This problem concerns the governing equations of such operations without considering chemical reactions. A well-mixed batch reactor initially contains 200 liter of pure water. At time t = 0, we start feeding a brine stream with a salt concentration of 180 g/liter into the tank at a constant rate of 50 liter/min. Calculate:

a. the time the salt concentration in the tank is 60 g/liter,

b. the volume of the tank at that time, and

c. the time and salt concentration in the tank when the volume of the tank is 600 liter.

Assume the density of the brine is the same as the density of the water.

3.3₃ Solve Problem 3.2 when the feed rate of the brine is not constant. Solve for the case where the feed rate is a function of time given by $v_{in}(t) = 100 - 10t$ liter/min (t is in minutes), and the exit flow rate is constant at 20 lit/min.

a. Plot a graph of the salt concentration as a function of time.

b. What is the salt concentration when the amount of solution in the tank is maximum?

c. What is the salt concentration of the last drop in the tank?

Hint: You may resort to numerical solution, but you have to show the equations you solve and the initial conditions.

3-4₁ Ethylene oxide is produced by a catalytic reaction of ethylene and oxygen at controlled conditions. However, side reactions cannot be completely prevented, and it is believed that the following reactions take place:

$$2\ C_2H_4 + O_2 \rightarrow 2\ C_2H_4O$$

$$C_2H_4 + 3\ O_2 \rightarrow 2\ CO_2 + 2\ H_2O$$

$$C_2H_4 + 2\ O_2 \rightarrow 2\ CO + 2\ H_2O$$

$$2\ CO + O_2 \rightarrow 2\ CO_2$$

$$CO + H_2O \rightarrow CO_2 + H_2.$$

Formulate the dimensionless, reaction-based design equations for an ideal batch reactor, plug-flow reactor, and a CSTR.

3-5₁ Cracking of naphtha to produce olefins is a common process in the petrochemical industry. The cracking reactions are represented by the simplified elementary gas-phase reactions:

$$C_{10}H_{22} \rightarrow C_4H_{10} + C_6H_{12}$$

$$C_4H_{10} \rightarrow C_3H_6 + CH_4$$

$$C_6H_{12} \rightarrow C_2H_4 + C_4H_8$$

$$C_4H_8 \rightarrow 2\,C_2H_4.$$

Formulate the dimensionless, reaction-based design equations for a plug-flow reactor.

3-6₁ The cracking of propane to produce ethylene is represented by the simplified kinetic model:

$$C_3H_8 \rightarrow C_2H_4 + CH_4$$

$$C_3H_8 \rightarrow C_3H_6 + H_2$$

$$C_3H_8 + C_2H_4 \rightarrow C_2H_6 + C_3H_6$$

$$2\,C_3H_6 \rightarrow 3\,C_2H_4.$$

$$C_3H_6 \rightarrow C_2H_2 + CH_4$$

$$C_2H_4 + C_2H_2 \rightarrow C_4H_6.$$

Formulate the dimensionless, reaction-based design equations for a plug-flow reactor.

3-7₁ The following reactions are believed to take place during the direct oxidation of methane:

$$2\,CH_4 + 3\,O_2 \rightarrow 2\,CO + 4\,H_2O$$

$$CH_4 + 2\,O_2 \rightarrow CO_2 + 2\,H_2O$$

$$2\,CH_4 + O_2 \rightarrow 2\,CH_3OH$$

$$CH_4 + O_2 \rightarrow HCHO + H_2$$

$$CO + H_2O \rightarrow CO_2 + H_2$$

$$CO + 2\,H_2 \rightarrow CH_3OH$$

$$CO + H_2 \rightarrow HCHO$$

$$2\,CH_4 + O_2 \rightarrow 2\,CO + 4\,H_2$$

$$2\,CO + O_2 \rightarrow 2\,CO_2.$$

Formulate the dimensionless, reaction-based design equations for a plug-flow reactor.

3-8₁ In the reforming of methane, the following chemical reactions occur:

$$CH_4 + 2\,H_2O \rightarrow CO_2 + 4\,H_2$$

$$CH_4 + 2\,O_2 \rightarrow CO_2 + 2\,H_2O$$

$$2\ CH_4 + O_2 \rightarrow 2\ CH_3OH$$

$$2\ CH_4 + 3\ O_2 \rightarrow 2\ CO + 4\ H_2O$$

$$CO_2 + 3\ H_2 \rightarrow CH_3OH + H_2O$$

$$2\ CH_3OH + 3\ O_2 \rightarrow 2\ CO_2 + 4\ H_2O$$

$$2\ CO + O_2 \rightarrow 2\ CO_2$$

$$CO + H_2O \rightarrow CO_2 + H_2.$$

Formulate the dimensionless, reaction-based design equations for a plug-flow reactor.

CHAPTER FOUR

ENERGY BALANCES

This chapter deals with the fourth fundamental concept used in the analysis and design of chemical reactors — conservation of energy (or the first law of thermodynamics). We use energy balances to express the temperature variations during reactor operations. Section 4.1 provides a brief review of basic thermodynamic quantities and relations used in reactor design. We define the "heat of reaction" and equilibrium constant of a chemical reaction and discuss how they vary with temperature and pressure. In Section 4-2, we apply the first law of thermodynamics to derive the energy balance equation for closed and open systems. We reduce these equations to dimensionless forms and derive the energy balance equations for the three ideal reactor configurations: ideal batch reactor, plug-flow reactor, and continuous stirred tank reactor (CSTR).

4.1 REVIEW OF THERMODYNAMIC RELATIONS

4.1.1 Heat of Reaction

The heat of reaction, or more accurately, the enthalpy change during a chemical reaction, ΔH_R, indicates the amount of energy being absorbed or released when a chemical transformation takes place at given operating conditions. The common standard conditions for which heat of reactions are reported are pressure of one atmosphere and temperature of 298°K. The standard heat of reaction of a chemical reaction, denoted by $\Delta H_{R_{298}}^0$, is calculated by

$$\Delta H_{R_{298}}^0 = \sum_{j}^{all} s_j \cdot H_{f_j}^0 , \qquad (4.1\text{-}1)$$

where $H_{f_j}^0$ is the standard heat of formation of species j at 298°K, and s_j is the stoichiometric coefficient of species j, defined by (1.1-1). The values of $H_{f_j}^0$'s of many species are tabulated in thermodynamic textbooks and in handbooks. Note that the dimensions of ΔH_R are energy per mole extent. Also, we adopt the common convention in applying the first law of thermodynamics — heat added to a system is positive and heat removed from a system is negative. Therefore, for exothermic reactions, ΔH_R is negative, and for endothermic reactions, ΔH_R is positive.

The heat of reaction is a function of temperature and pressure. Since the heat of reaction is a state quantity, we can readily express ΔH_R at any T and P. We relate the heat of reaction at any temperature, T, and the standard pressure of one atmosphere, to the standard heat of reaction by

$$\Delta H_R^0(T) = \Delta H_{R\,298}^0 + \int_{298}^{T} \sum_{j}^{all} s_j \cdot \hat{c}_{pj} \cdot dT, \tag{4.1-2}$$

where $\hat{c}_{pj}$ is the molar heat capacity of species j. Note that (4.1-2) represents the calculation of DH_R by cooling the reactants from T to 298°K, carrying out the reaction at 298°K, and then heating up the products from 298°K to T, as shown schematically in

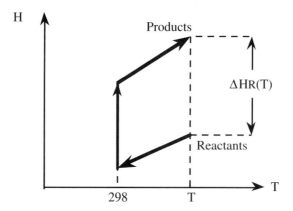

Figure 4-1: Calculation of $\Delta H_R(T)$

Fig. 4-1. Once the heat of reaction at temperature T and one atmosphere is known, we can calculate the heat of reaction at T and pressure P by

$$\Delta H_R(T,P) = \Delta H_R^0(T,1) + \sum_{j}^{all} s_j \cdot \int_{1}^{P} \left[\hat{V}_j - T \cdot \left(\frac{\partial \hat{V}_j}{\partial T} \right)_P \right] \cdot dP, \tag{4.1-3}$$

where $\hat{V}_j$ is the molar volume of species j. The term inside the integral indicates the change in the enthalpy of species j due to the change in pressure at temperature T. In most applications, this term is small, and, unless the reactor is operated at a very high pressure (say, several hundreds bars or higher), we can approximate the heat of reaction by $\Delta H_R(T,1)$, the heat of reaction at one atmosphere.

Example 4-1 Determine the heat of combustion of n-octane at 620°C and 1 atm. The heats of formations of the various species and their heat capacities are given below.
Data: The standard heats of formation of the species (at 298°K) are:

$C_8H_{18}(g) = -208,750$ J/mole
$CO_2(g) = -393,509$ J/mole
$H_2O(g) = -241,818$ J/mole.

The heat capacities (assumed constant at this temperature range):

$C_8H_{18}(g) = 254$ J/mole°K
$O_2(g) = 30.2$ J/mole°K
$CO_2(g) = 40.3$ J/mole°K
$H_2O(g) = 34.0$ J/mole°K.

Solution: The chemical reaction is:

$$C_8H_{18}(g) + 12.5\ O_2(g) \rightarrow 8\ CO_2(g) + 9\ H_2O(g),$$

and the stoichiometric coefficients are:

$$s_{C8H18} = -1;\quad s_{O2} = -12.5;\quad s_{CO2} = 8;\quad s_{H2O} = 9;\quad \Delta = 3.5.$$

We determine the standard heat of reaction by (4.1-1), noting that the heat of formation of an element is, by definition, zero:

$$\Delta H^0_{R\,298} = \sum_j^{all} s_j \cdot H^0_{f_j} = (-1) \cdot H^0_{fC8H18} + (-12.5) \cdot H^0_{fO2} + (8) \cdot H^0_{fCO2} + (9) \cdot H^0_{fH2O} =$$

$$= (-1)\,(-208,750) + (-12.5)\,(0) + (8)\,(-393,509) + (9)\,(-241,818) =$$

$$= -5,115.68 \text{ kJ/mole.} \tag{a}$$

Since ΔH_R is negative, the reaction is exothermic. We determine the heat of reaction at 620°C by using (a) and (4.1-2). For this case, the individual species heat capacities are constant; therefore, the term inside the integral is independent of temperature,

$$\sum_j^{all} s_j \cdot \hat{c}_{p_j} = (-1)\,(254) + (-12.5)\,(30.2) + (8)\,(40.3) + (9)\,(34.0) =$$

$$= -3.10 \text{ J/mole°K.} \tag{b}$$

Substituting the values from (a) and (c) into (4.1-2),

$$\Delta H^0_R(893°\,K) = -5,115.68 \cdot 10^3 + \int_{298}^{893} (-3.10) \cdot dT$$

$$= -5,117.52 \text{ kJ/mole.} \tag{c}$$

Note that, in this case, because the difference between the heat capacity of the reactants and the heat capacity of the products is small, the heat of reaction changes very slightly with temperature.

4.1.2 Effect of Temperature on Equilibrium Constant

The equilibrium constant of a chemical reaction relates to the heat of reaction by

$$\frac{d(\ln K)}{dt} = -\frac{\Delta H_R^0}{R \cdot T^2}. \tag{4.1-4}$$

To express how the reaction equilibrium constant varies with temperature, we separate the variables and integrate,

$$\ln \frac{K(T)}{K(T_0)} = -\int_{T_0}^{T} \frac{\Delta H_R(T)}{R \cdot T^2} dT. \tag{4.1-5}$$

In many instances, the heat of reaction is essentially constant, and (4.1-5) reduces

$$\ln \frac{K(T)}{K(T_0)} = -\frac{\Delta H_R(T_0)}{R}\left(\frac{1}{T} - \frac{1}{T_0}\right). \tag{4.1-6}$$

The equilibrium constant, K, relates to the composition of the individual species at equilibrium. For a general chemical reaction of the form $b\,B + c\,C \rightarrow r\,R + s\,S$,

$$K = \frac{a_R^r \cdot a_S^s}{a_B^b \cdot a_C^c}, \tag{4.1-7}$$

where a_j is the activity coefficient of species j.

4.2 ENERGY BALANCES

The first law of thermodynamics is concerned with the conservation of energy. In its most general form, we can write it as the following statement:

$$\left\{\begin{array}{c}\text{rate}\\\text{energy}\\\text{enters}\\\text{system by}\\\text{streams}\end{array}\right\} + \left\{\begin{array}{c}\text{rate heat}\\\text{energy}\\\text{enters}\\\text{system}\end{array}\right\} = \left\{\begin{array}{c}\text{rate}\\\text{energy}\\\text{leaves}\\\text{system by}\\\text{streams}\end{array}\right\} + \left\{\begin{array}{c}\text{rate}\\\text{work}\\\text{done by}\\\text{system}\end{array}\right\} + \left\{\begin{array}{c}\text{rate}\\\text{energy}\\\text{accumulates}\\\text{in system}\end{array}\right\}. \tag{4.2-1}$$

When applying (4.2-1), we have to consider **all** types of energy and account for **all** forms of work. We usually derive simplified and useful relations by imposing certain assumptions and incorporating other thermodynamic relations (equation of state, Maxwell's equations, etc.) into the energy balance equation. Therefore, before applying any of the relations derived below, it is essential to identify all the assumptions made and examine whether they are valid.

Let E denote the total amount of energy, in all its forms, contained in a system, and let "e" denote the specific energy (energy per unit mass). In most chemical processes, electric, magnetic, and nuclear energies are negligible. Therefore, we restrict the treatment of the first law of thermodynamics to applications where energy is present in three forms: internal energy, U, kinetic energy, KE, and potential (gravitational) energy, PE; hence,

$$E = U + KE + PE. \tag{4.2-2}$$

The specific energy is

$$e = u + \frac{1}{2}v^2 + g \cdot z, \tag{4.2-3}$$

where u is the specific internal energy, v is the linear velocity, and z is the vertical elevation.

Below, we apply the energy balances for macroscopic systems. First, we derive the energy balance equation for closed systems (batch reactors) and then for open systems (flow reactors). Microscopic energy balances, used to describe point-to-point temperature variations inside a chemical reactor, are outside the scope of this book.

4.2.1 Closed Systems (Batch Reactors)

For closed systems (no material flows into or out of the system), the energy balance is written in the common form of the first law of thermodynamics

$$\Delta E(t) = Q(t) - W(t), \tag{4.2-4}$$

where $\Delta E(t)$ is the change in the energy of the system, $Q(t)$ is the heat added to the system, and $W(t)$ is the work done by the system on the surroundings during operating time t. For a stationary systems, the only energy changed is internal energy; hence, $\Delta E(t) = \Delta U(t)$, and (4.2-4) reduces to

$$\Delta U(t) = Q(t) - W(t). \tag{4.2-5}$$

For batch reactors, the work term consists of two components: shaft work (work done by a mechanical device such as a stirrer) and work done by expanding the boundaries of the system against the surroundings,

$$W = W_{sh} + \int P \cdot dV. \tag{4.2-6}$$

Expressing the internal energy in terms of the enthalpy,

$$\Delta U = \Delta H - \int P \cdot dV - \int V \cdot dP,$$

and, substituting into (4.2-6), the energy balance becomes

$$\Delta H(t) = Q(t) - W_{sh}(t) + \int V \cdot dP. \tag{4.2-7}$$

Most reactor operations take place at isobaric or near isobaric conditions, and, in general, the enthalpy is a weak function of the pressure. Hence, the last term is comparatively small, and the energy balance equation reduces to

$$\Delta H(t) = Q(t) - W_{sh}(t). \tag{4.2-8}$$

In many instances, the reacting fluid is not viscous, and the shaft work is small in comparison with the heat added to the system; hence

$$\Delta H(t) \approx Q(t). \tag{4.2-9}$$

For batch systems with chemical reactions, enthalpy variations are due to changes in both composition and temperature. We usually select a reference state at some convenient composition and temperature T_0 (usually the initial temperature). Assuming no phase change, the change in the enthalpy in operating time t is

$$\Delta H(t) = \sum_m^{n_{ind}} \Delta H_{R_m}(T_0) \cdot X_m(t) + \int_{T_0}^{T(t)} (\Sigma N_j \cdot \hat{c}_{p_j})_t dT - \int_{T_0}^{T(0)} (\Sigma N_j \cdot \hat{c}_{p_j})_0 dT, \tag{4.2-10}$$

where $(\Sigma N_j \cdot \hat{c}_{p_j})_0$ and $(\Sigma N_j \cdot \hat{c}_{p_j})_t$ indicate the heat capacity of the reacting fluid initially and at time t, respectively. Note that since enthalpy is a state quantity, the calculation of the energy difference is as shown schematically in Figure 4-2 and that the summation in the first term on the right is over the independent reactions. The first term on the right hand side of (4.2-10) represents the change in enthalpy due to composition changes (by chemical reactions) at the reference temperature, T_0, whereas the other two terms indicate the change in the "sensible heat" (enthalpy change due to variation in temperature). Substituting (4.2-10) into (4.2-8), the general energy balance equation of closed systems becomes

$$Q(t) = \sum_m^{n_{ind}} \Delta H_{R_m}(T_0) \cdot X_m(t) +$$

$$+ \int_{T_0}^{T(t)} (\Sigma N_j \cdot \hat{c}_{p_j})_t dT - \int_{T_0}^{T(0)} (\Sigma N_j \cdot \hat{c}_{p_j})_0 dT + W_{sh}(t). \tag{4.2-11}$$

It is convenient to select the initial temperature as the reference temperature, $T_0 = T(0)$, and (4.2-11) reduces to

$$Q(t) = \sum_m^{n_{ind}} \Delta H_{R_m}(T_0) \cdot X_m(t) + \int_{T_0}^{T(t)} (\Sigma N_j \cdot \hat{c}_{p_j})_t dT + W_{sh}(t). \tag{4.2-12}$$

Eq. (4.2-12) is the integral form of the energy balance equation for batch reactors, relating the reactor temperature, $T(t)$, to the extents of the independent reactions, $X_m(t)$, the heat added to the reactor, $Q(t)$, and the mechanical work done by the system, $W_{sh}(t)$, during operating time t.

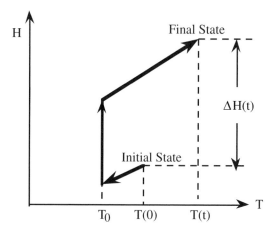

Figure 4-2: Calculation of Enthalpy Difference

The differential energy balance equation is obtained either by formal differentiation of (4.2-12) or by writing (4.2-12) for $t + \Delta t$ and taking the limit of $Q(t + \Delta t) - Q(t)$ as $\Delta t \to 0$,

$$\frac{dQ}{dt} = \dot{Q}(t) = \sum_{m}^{n_{ind}} \Delta H_{R_m}(T_0) \cdot \frac{dX_m}{dt} + (\Sigma N_j \cdot \hat{c}_{p_j}) \frac{dT}{dt} + \dot{W}_{sh}(t), \quad (4.2\text{-}13)$$

where $\dot{Q}(t)$ is the rate heat is added to the reactor, and $\dot{W}_{sh}(t)$ is the rate of mechanical work done by the reactor at time t. For an ideal batch reactor, the temperature is uniform throughout, and

$$\dot{Q}(t) = U \cdot S(t) \cdot [T_F - T(t)], \quad (4.2\text{-}14)$$

where U is the overall heat transfer coefficient, S(t) is the heat transfer area at time t, and T_F is the temperature of the heating (or cooling) fluid. The heat transfer area S(t) relates to the reactor volume by

$$S(t) = \left(\frac{S}{V}\right) \cdot V_R(t), \quad (4.2\text{-}15)$$

where (S/V) is the heat transfer area per unit volume, a system characteristic that depends on the geometry of the reactor. Substituting (4.2-14) and (4.2-15) into (4.2-13) and rearranging, we obtain,

$$\frac{dT}{dt} = \frac{U \cdot V_R(t)}{(\Sigma N_j \cdot \hat{c}_{p_j})} \left(\frac{S}{V}\right) \cdot [T_F - T(t)] -$$

$$- \sum_{m}^{n_{ind}} \frac{\Delta H_{R_m}(T_0)}{(\Sigma N_j \cdot \hat{c}_{p_j})} \frac{dX_m}{dt} - \frac{1}{(\Sigma N_j \cdot \hat{c}_{p_j})} \overset{\circ}{W}_{sh}(t). \quad (4.2\text{-}16)$$

Equation (4.2-16) is the differential form of the energy balance equation, expressing the changes in the reactor temperature during the reactor operation. Note that the first term on the right hand side represents the rate heat is added to the reactor at time t divided by the heat capacity of the fluid,

$$\frac{1}{(\Sigma N_j \cdot \hat{c}_{p_j})} \dot{Q}(t) = \frac{U \cdot V_R(t)}{(\Sigma N_j \cdot \hat{c}_{p_j})} \left(\frac{S}{V}\right) \cdot \left[T_F - T(t)\right]. \qquad (4.2\text{-}17)$$

To reduce (4.2-16) to a dimensionless form, we use dimensionless temperature, defined by (2.3-4), $\theta = T/T_0$, the dimensionless extent, defined by (1.5-1), $Z_m = X_m/(N_{tot})_0$, and the dimensionless operating time, defined by (3.5-3), $\tau = t/t_{cr}$. Dividing both sides of (4.2-16) by T_0 and $(N_{tot})_0$ and multiplying both sides by the characteristic reaction time, t_{cr}, we obtain

$$\frac{d\theta}{d\tau} = \frac{U \cdot V_R \cdot t_{cr}}{(\Sigma N_j \cdot \hat{c}_{p_j})} \left(\frac{S}{V}\right) \cdot (\theta_F - \theta) -$$

$$- \frac{(N_{tot})_0}{T_0 \cdot (\Sigma N_j \cdot \hat{c}_{p_j})} \sum_m^{n_{ind}} \Delta H_{R_m}(T_0) \frac{dZ_m}{d\tau} - \frac{(N_{tot})_0}{T_0 \cdot (\Sigma N_j \cdot \hat{c}_{p_j})} \frac{dW_{sh}}{d\tau}. \quad (4.2\text{-}18)$$

Eq. (4.2-18) is the **dimensionless**, differential energy balance equation of ideal batch reactors, relating the reactor temperature, $\theta(\tau)$, to the extents of the independent reactions, $Z_m(\tau)$, at dimensionless operating time τ. Note that individual $dZ_m/d\tau$'s are expressed in terms of $\theta(\tau)$ and $Z_m(\tau)$ by the reaction-based design equations as described in Chapter Three.

Dimensionless energy balance equation (4.2-18) is not conveniently used because the heat capacity term, $(\Sigma N_j \cdot \hat{c}_{p_j})$, is a function of the temperature and reaction extents; hence, it varies during the operation. To simplify the equation and obtain dimensionless quantities for heat-transfer, we define heat capacity of the reference state and relate $(\Sigma N_j \cdot \hat{c}_{p_j})$ at any instant to it by

$$(\Sigma N_j \cdot \hat{c}_{p_j}) = (N_{tot})_0 \cdot \hat{c}_{p0} \cdot CF(Z_m, \theta), \qquad (4.2\text{-}19)$$

where $\hat{c}_{p0}$ is the specific molar heat capacity of the reference state and $CF(Z_m, \theta)$ is a correction factor that adjusts the value of the heat capacity as Z_m and θ vary. Substituting (4.2-19) into (4.2-18) and noting that $(N_{tot})_0 = C_0 \cdot V_R(0)$, we obtain

$$\frac{d\theta}{d\tau} = \frac{1}{CF(Z_m, \theta)} \left[\frac{U \cdot t_{cr}}{C_0 \cdot \hat{c}_{p0}} \left(\frac{S}{V}\right) \cdot \left(\frac{V_R}{V_R(0)}\right) \cdot (\theta_F - \theta) \right] -$$

$$- \frac{1}{CF(Z_m, \theta)} \left[\sum_m^{n_{ind}} \frac{\Delta H_{R_m}(T_0)}{T_0 \cdot \hat{c}_{p0}} \frac{dZ_m}{d\tau} + \frac{d}{d\tau} \left(\frac{W_{sh}}{T_0 \cdot (N_{tot})_0 \cdot \hat{c}_{p0}} \right) \right]. \quad (4.2\text{-}20)$$

Eq. (4.2-20) is a simplified, dimensionless differential energy balance equation of batch reactors, where each term is divided by $(T_0 \cdot (N_{tot})_0 \cdot \hat{c}_{p0})$, the "reference thermal energy" of the system.

The first term in the bracket of (4.2-20) represents the dimensionless heat-transfer rate with a dimensionless driving force, $(\theta_F - \theta)$,

$$\frac{d}{d\tau}\left(\frac{Q}{T_0 \cdot N_{tot}(0) \cdot \hat{c}_{p0}}\right) = \left(\frac{U \cdot t_{cr}}{C_0 \cdot \hat{c}_{p0}}\left(\frac{S}{V}\right) \cdot \frac{V_R(t)}{V_R(0)}\right) \cdot (\theta_F - \theta), \qquad (4.2\text{-}21)$$

where

$$\frac{Q}{T_0 \cdot (N_{tot})_0 \cdot \hat{c}_{p0}} \qquad (4.2\text{-}22)$$

is the dimensionless heat added to the reactor. Using (4.2-14) and (4.2-15), we define the dimensionless heat-transfer number of the reactor, HTN, by

$$\text{HTN} = \frac{U \cdot t_{cr}}{C_0 \cdot \hat{c}_{p0}}\left(\frac{S}{V}\right). \qquad (4.2\text{-}23)$$

The dimensionless heat-transfer number lumps together all the effects related to the heat transfer in relation to the time scale of the chemical reaction; it is analogous to Nusselt number. Note that the HTN is proportional to the heat transfer coefficient, U, which depends on the flow conditions, the properties of the fluid, and the heat-transfer area per unit volume, (S/V).

The second term in the bracket of (4.2-20) represents the dimensionless heat of reactions. The coefficient inside the summation is the dimensionless heat of reaction of the m-th independent chemical reaction,

$$\text{DHR}_m = \frac{\Delta H_{R_m}(T_0)}{T_0 \cdot \hat{c}_{p0}}. \qquad (4.2\text{-}24)$$

It is a characteristic of the chemical reaction, the reference temperature, and the composition of the reference state. The third term in the bracket of (4.2-20) represents the dimensionless rate of shaft work, where

$$\frac{W_{sh}}{T_0 \cdot (N_{tot})_0 \cdot \hat{c}_{p0}} \qquad (4.2\text{-}25)$$

is the dimensionless shaft work. Using these dimensionless quantities, (4.2-20) reduces to

$$\frac{d\theta}{d\tau} = \frac{1}{CF(Z_m,\theta)}\left[\text{HTN} \cdot \left(\frac{V_R}{V_R(0)}\right) \cdot (\theta_F - \theta)\right] -$$

$$- \frac{1}{CF(Z_m,\theta)}\left[\sum_m^{n_{ind}}\text{DHR}_m \frac{dZ_m}{d\tau} + \frac{d}{d\tau}\left(\frac{W_{sh}}{T_0 \cdot (N_{tot})_0 \cdot \hat{c}_{p0}}\right)\right]. \qquad (4.2\text{-}26)$$

The definition of the specific molar heat capacity of the reference state, $\hat{c}_{p0}$, and the determination of the correction factor, $CF(Z_m, \theta)$, deserves a closer examination. Note that the summation in $(\Sigma N_j \cdot \hat{c}_{pj})$ is over all species present in the reactor, including inert species. Also note that, for liquid-phase reactions, the "system" in the reactor design formulation is defined as the species participating in the chemical reactions with the exclusion of solvents (see Section 3.5). However, in the energy balance equation, the heat capacity of the solvent should be considered.

Using stoichiometric relation (1.5-3) to express the mole content of species j in terms of the extents of the independent reactions,

$$(\Sigma N_j \cdot \hat{c}_{pj}) = (N_{tot})_0 \left(\sum_j^{all} y_j(0) \cdot \hat{c}_{p_j}(\theta) + \sum_j^{all} \hat{c}_{p_j}(\theta) \cdot \sum_m^{n_{ind}} (s_j)_m \cdot Z_m \right), \quad (4.2\text{-}27)$$

where $\hat{c}_{p_j}(\theta)$ is the specific molar heat capacity of species j at dimensionless temperature θ. For the reference state, $Z_m = 0$ for all reactions, and (4.2-27) reduces to

$$(\Sigma N_j \cdot \hat{c}_{p_j})_0 = (N_{tot})_0 \sum_j^{all} y_j(0) \cdot \hat{c}_{p_j}(\theta). \quad (4.2\text{-}28)$$

We define the average specific molar heat capacity of the reference state, $\hat{c}_{p0}$, by

$$(\Sigma N_j \cdot \hat{c}_{p_j})_0 = (N_{tot})_0 \cdot \hat{c}_{p0}. \quad (4.2\text{-}29)$$

Substituting (4.2-28) into (4.2-29),

$$\hat{c}_{p0} = \sum_j^{all} y_j(0) \cdot \hat{c}_{p_j}(1). \quad (4.2\text{-}30)$$

Eq. (4.2-30) is used to determine the specific molar heat capacity of the reference state, $\hat{c}_{p0}$, for gas-phase reactions. For liquid-phase reactions, the heat capacity of the reacting system is given in terms of the mass-based specific heat capacity of the reacting fluid, $\bar{c}_p$. Hence, we define the specific molar heat capacity of the reference state, $\hat{c}_{p0}$, by

$$\hat{c}_{p0} = \frac{M}{(N_{tot})_0} \bar{c}_p, \quad (4.2\text{-}31)$$

where M is the mass of the reference state, $\bar{c}_p$ is the mass-based heat capacity, and $(N_{tot})_0$ is the total number of moles of the reference state.

To determine the correction factor of the heat capacity, $CF(Z_m, \theta)$, we use (4.2-19) and the appropriate definition of the specific molar heat capacity of the reference state, $\hat{c}_{p0}$. For gas-phase reactions, using (4.2-30),

$$CF(Z_m,\theta) = \frac{\sum\limits_{j}^{all} y_j(0) \cdot \hat{c}_{p_j}(\theta)}{\sum\limits_{j}^{all} y_j(0) \cdot \hat{c}_{p_j}(1)} + \frac{\sum\limits_{j}^{all} \hat{c}_{p_j}(\theta) \cdot \sum\limits_{m}^{n_{ind}} (s_j)_m \cdot Z_m}{\sum\limits_{j}^{all} y_j(0) \cdot \hat{c}_{p_j}(1)}. \tag{4.2-32}$$

For liquid-phase reactions, using (4.2-31),

$$(\Sigma N_j \cdot \hat{c}_{p_j}) = \frac{(N_{tot})_0}{M \cdot \overline{c}_p} \left(\sum\limits_{j}^{all} y_j(0) \cdot \hat{c}_{p_j}(\theta) + \sum\limits_{j}^{all} \hat{c}_{p_j}(\theta) \cdot \sum\limits_{m}^{n_{ind}} (s_j)_m \cdot Z_m \right). \tag{4.2-33}$$

In practice, we encounter three levels of difficulties in solving the dimensionless energy balance equation (4.2-26):

a. The heat capacity of the reacting fluid is **independent** of **both** the composition and the temperature. In this case, $CF(Z_m,\theta) = 1$. This situation commonly occurs in liquid-phase reactions as well as in gas-phase reactions where the heat capacities of the products are close to that of the reactants and do not vary with temperature.

b. The heat capacity of the reacting fluid **depends** on the composition of the stream but is **independent** of temperature. In this case, $\hat{c}_{p_j}(\theta) = \hat{c}_{p_j}(1)$ for all species, and (4.2-32) reduces to

$$CF(Z_m,\theta) = 1 + \frac{\sum\limits_{j}^{all} \hat{c}_{p_j}(1) \sum\limits_{m}^{n_{ind}} (s_j)_m \cdot Z_m}{\sum\limits_{j}^{all} y_j(0) \cdot \hat{c}_{p_j}(1)}. \tag{4.2-34}$$

This situation occurs when the specific heat capacities of the individual species do not depend on the temperature.

c. The heat capacity of the reacting fluid **depends** on **both** the composition and the temperature. In this case, the specific heat capacities of the individual species depend on the temperature and are usually expressed in the form

$$\frac{\hat{c}_{p_j}}{R} = A_j + B_j \cdot T + C_j \cdot T^2 + D_j \cdot T^{-2}, \tag{4.2-35}$$

where A_j, B_j, C_j, and D_j are tabulated constants, characteristic to each species, and R is the universal gas constant. For this case, the individual specific molar heat capacity of species j at dimensionless temperature θ is

$$\hat{c}_{p_j}(\theta) = R\left[A_j + (B_j \cdot T_0) \cdot \theta + (C_j \cdot T_0^2) \cdot \theta^2 + \left(\frac{D_j}{T_0^2} \right) \cdot \theta^{-2} \right], \tag{4.2-36}$$

and these functions should be substituted in (4.2-32) to determine the values of $CF(Z_m,\theta)$.

This is readily done numerically to obtain solutions of the dimensionless design and energy balance equations. Once $CF(Z_m, \theta)$ is known, the energy balance equation, (4.2-26), can be solved simultaneously with the dimensionless design equations.

For isothermal operations, $d\theta/d\tau = 0$, and using (4.2-14), for batch reactors with negligible shaft work, (4.2-26) reduces to

$$\frac{d}{d\tau}\left(\frac{Q}{T_0 \cdot (N_{tot})_0 \cdot \hat{c}_{p0}}\right) = \sum_{m}^{n_{ind}} DHR_m \frac{dZ_m}{d\tau}, \qquad (4.2\text{-}37)$$

which, using (4.2-24), can be further simplified to

$$dQ = (N_{tot})_0 \sum_{m}^{n_{ind}} \Delta H_{R_m}(T_0) \cdot dZ_m. \qquad (4.2\text{-}38)$$

Eq. (4.2-38) provides the heating (or cooling) load needed to maintain isothermal conditions. For adiabatic operations $(S/V) = 0$ (no heat-transfer area), and for batch reactors with negligible shaft work (4.2-26) reduces to

$$\frac{d\theta}{d\tau} = \frac{-1}{CF(Z_m, \theta)} \sum_{m}^{n_{ind}} DHR_m \frac{dZ_m}{d\tau}, \qquad (4.2\text{-}39)$$

which can be further simplified to

$$d\theta = \frac{-1}{CF(Z_m, \theta)} \sum_{m}^{n_{ind}} DHR_m \, dZ_m. \qquad (4.2\text{-}40)$$

Eq. (4.2-40) relates changes in the temperature to changes in the extents of the independent reactions during adiabatic operations.

Example 4-2 The following gas-phase chemical reactions take place in a batch reactor:

$$A + B \rightarrow C$$

$$C + B \rightarrow D.$$

Initially, the reactor is at $420°K$ and 2 atm and contains a mixture of 45% A, 45% B, and 10% I (by mole). Based on the data below, determine
a. the average specific molar heat capacity of the reference state,
b. the dimensionless heat of reaction of each chemical reaction, and
c. the correction factor of heat capacity.
d. Formulate the dimensionless energy balance equation for adiabatic operation.
Data: At $420°K$, $\Delta H_{R1} = -11,000$ cal/mole extent; $\Delta H_{R2} = -8,000$ cal/mole extent;
$\hat{c}_{pA} = 20$ cal/mole°K; $\hat{c}_{pB} = 7$ cal/mole°K; $\hat{c}_{pC} = 22$ cal/mole°K;
$\hat{c}_{pD} = 25$ cal/mole°K; $\hat{c}_{pI} = 14$ cal/mole°K.

Solution The two chemical reactions are independent, and their stoichiometric coefficients are:

$$s_{A1} = -1; \quad s_{B1} = -1; \quad s_{C1} = 1; \quad s_{D1} = 0; \quad \Delta_1 = -1;$$

$$s_{A2} = 0; \quad s_{B2} = -1; \quad s_{C2} = -1; \quad s_{D2} = 1; \quad \Delta_2 = -1.$$

We select the initial temperature as the reference temperature; hence, $T_0 = T(0) = 420°K$.
a. For gas-phase reactions, the average specific molar heat capacity of the reference
state is determined using (4.2-30):

$$\hat{c}_{p0} = y_A(0) \, \hat{c}_{pA}(1) + y_B(0) \, \hat{c}_{pB}(1) + y_C(0) \, \hat{c}_{pC}(1) + y_I(0) \, \hat{c}_{pI}(1) =$$

$$= (0.45) \, 20 + (0.45) \, 7 + (0.0) \, 22 + (0.1) \, 14 = 13.55 \text{ cal/mole}°K.$$

b. Using (4.2-24), the dimensionless heat of reactions of the two reactions are:

$$DHR_1 = \frac{\Delta H_{R_1}(T_0)}{T_0 \cdot \hat{c}_{p0}} = \frac{(-11 \cdot 10^3)}{(420) \cdot (13.55)} = -1.933$$

$$DHR_2 = \frac{\Delta H_{R_2}(T_0)}{T_0 \cdot \hat{c}_{p0}} = \frac{(-8 \cdot 10^3)}{(420) \cdot (13.55)} = -1.406.$$

c. For gas-phase reactions, the correction factor of heat capacity, $CF(Z_m, \theta)$, is determined by (4.2-32). In this case, the species heat capacities are independent of the
temperature; hence

$$CF(Z_m, \theta) = 1 + \frac{\sum\limits_{j}^{all} \hat{c}_{p_j}(\theta) \cdot \sum\limits_{m}^{n_{ind}} (s_j)_m \cdot Z_m}{\sum\limits_{j}^{all} y_j(0) \cdot \hat{c}_{p_j}(1)} =$$

$$= 1 + \frac{1}{13.55} \left\{ \begin{array}{l} \hat{c}_{pA} \cdot s_{A1} \cdot Z_1 + \hat{c}_{pB} \cdot (s_{B1} \cdot Z_1 + s_{B2} \cdot Z_2) + \\ + \hat{c}_{pC} \cdot (s_{C1} \cdot Z_1 + s_{C2} \cdot Z_2) + \hat{c}_{pD} \cdot s_{D2} \cdot Z_2 \end{array} \right\} =$$

$$= 1 + \frac{1}{13.55} \left\{ \begin{array}{l} 20 \cdot (-1) \cdot Z_1 + 7 \cdot [(-1) \cdot Z_1 + (-1) \cdot Z_2] + \\ + 22 \cdot [(1) \cdot Z_1 + (-1) \cdot Z_2] + 25 \cdot (1) \cdot Z_2 \end{array} \right\} =$$

$$= \frac{13.55 - 5 \cdot Z_1 - 4 \cdot Z_2}{13.55}.$$

d. For adiabatic operations, $(S/V) = 0$. Substituting the values calculated above in
(4.2-26), the energy balance equation is

$$\frac{d\theta}{d\tau} = \frac{13.55}{13.55 - 5 \cdot Z_1 - 4 \cdot Z_2} \left[(1.933) \cdot \frac{dZ_1}{d\tau} + (1.406) \cdot \frac{dZ_2}{d\tau} \right],$$

where $dZ_1/d\tau$ and $dZ_2/d\tau$ are formulated by the design equations.

Example 4-3 The following simultaneous chemical reactions take place in an aqueous solution in a batch reactor:

$$2 A \rightarrow B$$

$$2 B \rightarrow C.$$

Two hundreds liters of a 4 mole/liter solution of reactant A are charged into a batch reactor. The density of the solution is 1.05 kg/liter, and its initial temperature is 310°K. Based on the data below, determine

a. the average specific molar heat capacity of the reference state,
b. the dimensionless heat of reaction of each chemical reaction, and
c. the correction factor of heat capacity.
d. Formulate the dimensionless energy balance equation for adiabatic operation.
Data: At 310°K, $\Delta H_{R1} = -9,000$ cal/mole A; $\Delta H_{R2} = -8,000$ cal/mole B;
The heat capacity of the solution is that of water (1 kcal/kg) and does not vary with the solution composition or the temperature.

Solution The two chemical reactions are independent, and their stoichiometric coefficients are:

$$s_{A1} = -2; \quad s_{B1} = 1; \quad s_{C1} = 0;$$

$$s_{A2} = 0; \quad s_{B2} = -2; \quad s_{C2} = 1.$$

We select the initial reactor content as the reference state; hence $T_0 = T(0) = 310°K$, $V_R(0) = 200$ liter, and the mass of the reacting fluid is $M = (1.05$ kg/lit$)(200$ liter$) = 210$ kg. Since only reactant A is initially present in the reactor, the reference concentration is $C_0 = C_A(0) = 4$ mole/liter. Hence, $(N_{tot})_0 = C_0 V_R(0) = 800$ mole.

a. For liquid-phase reactions, the average specific molar heat capacity of the reference state is defined by (4.2-31):

$$\hat{c}_{p0} = \frac{M}{(N_{tot})_0}\overline{\hat{c}}_p = \frac{(210 \text{ kg})}{(800 \text{ mole})}(1000 \text{ cal/kg °K}) = 262.5 \text{ cal/mole°K}. \quad (a)$$

b. The given heats of reactions are given per mole of A and B, respectively, but the selected reactions contain 2 moles of A and B, respectively. Hence, for the selected reactions, $\Delta H_{R1} = -18,000$ cal/mole extent and $\Delta H_{R2} = -16,000$ cal/mole extent. Using (4.2-24), the dimensionless heats of reactions of the two chemical reactions are:

$$DHR_1 = \frac{\Delta H_{R_1}(T_0)}{T_0 \cdot \hat{c}_{p0}} = \frac{(-18 \cdot 10^3)}{(310) \cdot (262.5)} = -0.221$$

$$DHR_2 = \frac{\Delta H_{R_2}(T_0)}{T_0 \cdot \hat{c}_{p0}} = \frac{(-16 \cdot 10^3)}{(310) \cdot (262.5)} = -0.197.$$

c. For liquid-phase reactions with fluids whose heat capacities are independent of composition and the temperature, $CF(Z_m,\theta) = 1$.

d. For adiabatic operations, $(S/V) = 0$ or $HTN = 0$. Substituting the values calculated above in (4.2-26), the energy balance equation is

$$\frac{d\theta}{d\tau} = (0.221)\cdot\frac{dZ_1}{d\tau} + (0.197)\cdot\frac{dZ_2}{d\tau},$$

where $dZ_1/d\tau$ and $dZ_2/d\tau$ are formulated by the design equations.

4.2.2 Open Systems (Flow Reactors)

Consider a general flow system with one inlet and one outlet as shown schematically in Figure 4-3. We write (4.2-1) for this system,

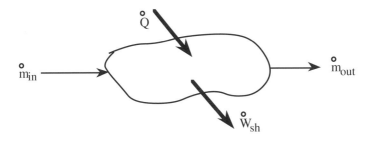

Figure 4-3: Schematic Diagram of a General Flow System

$$\frac{dE_{sys}}{dt} = \dot{Q} - \dot{W} + \left(u + \frac{1}{2}v^2 + g\cdot z\right)_{in}\dot{m}_{in} - \left(u + \frac{1}{2}v^2 + g\cdot z\right)_{out}\dot{m}_{out}, \quad (4.2\text{-}41)$$

where E_{sys} is the total energy of the system, and $\dot{m}_{in}$ and $\dot{m}_{out}$ are, respectively, the mass flow rates into and out of the system. The work term, $\dot{W}$, in (4.2-41), consists of three components: rate of work, $\dot{W}_{sh}$, the system is doing on the surroundings by a mechanical device (shaft work), rate of viscous work, $\dot{W}_{vis}$, and work done by pushing the streams into and out of the system,

$$\dot{W} = \dot{W}_{sh} + \dot{W}_{vis} + \left(\frac{P}{\rho}\right)_{out}\dot{m}_{out} - \left(\frac{P}{\rho}\right)_{in}\dot{m}_{in}. \quad (4.2\text{-}42)$$

Using the definition of the specific enthalpy, h, (in energy/mass)

$$h = u + \frac{P}{\rho}, \quad (4.2\text{-}43)$$

and substituting (4.2-42) and (4.2-43) into (4.2-41), the energy balance equation becomes

$$\frac{dE_{sys}}{dt} = \dot{Q} - \dot{W}_{sh} - \dot{W}_{vis} +$$

$$+\left(h + \frac{1}{2}v^2 + g\cdot z\right)_{in}\dot{m}_{in} - \left(h + \frac{1}{2}v^2 + g\cdot z\right)_{out}\dot{m}_{out}. \qquad (4.2\text{-}44)$$

Eq. (4.2-44) is the general energy balance equation for flow systems. For steady operations, $dE_{sys}/dt = 0$, $\dot{m}_{in} = \dot{m}_{out} = \dot{m}$, and (4.2-44) reduces to

$$\dot{Q} - \dot{W}_{sh} - \dot{W}_{vis} = \left[\left(h + \frac{1}{2}v^2 + g\cdot z\right)_{out} - \left(h + \frac{1}{2}v^2 + g\cdot z\right)_{in}\right]\cdot\dot{m}. \qquad (4.2\text{-}45)$$

For most chemical processes, the kinetic and potential energy of the streams are negligible in comparison with the enthalpy, and the viscous work is usually small; hence, (4.2-45) reduces to

$$\dot{Q} - \dot{W}_{sh} = (h_{out} - h_{in})\cdot\dot{m} = \Delta\dot{H}. \qquad (4.2\text{-}46)$$

For flow systems with chemical reactions, the enthalpy varies due to changes in composition and temperature. We select the reference state as one atmosphere and temperature T_0. Assuming no phase change, the enthalpy difference between the outlet and the inlet is

$$\Delta\dot{H} =$$

$$\sum_m^{n_{ind}} \Delta H_{R_m}(T_0)\cdot(\dot{X}_{m_{out}} - \dot{X}_{m_{in}}) + \int_{T_0}^{T_{out}}(\Sigma F_j\cdot\hat{c}_{p_j})_{out}dT - \int_{T_0}^{T_{in}}(\Sigma F_j\cdot\hat{c}_{p_j})_{in}dT, (4.2\text{-}47)$$

where $(\dot{X}_{m_{out}} - \dot{X}_{m_{in}})$ is the extent per unit time of the m-th independent reaction in the reactor. Hence, the general energy balance equation for steady flow reactors is

$$\dot{Q} - \dot{W}_{sh} = \sum_m^{n_{ind}} \Delta H_{R_m}(T_0)\cdot(\dot{X}_{m_{out}} - \dot{X}_{m_{in}}) +$$

$$+ \int_{T_0}^{T_{out}}(\Sigma F_j\cdot\hat{c}_{p_j})_{out}dT - \int_{T_0}^{T_{in}}(\Sigma F_j\cdot\hat{c}_{p_j})_{in}dT. \qquad (4.2\text{-}48)$$

Eq. (4.2-48) is the integral form of the general energy balance equation for steady flow reactors.

The differential energy balance equation for steady flow reactors with no mechanical work is obtained either by formal differentiation of (4.2-48) or by writing (4.2-48) for $T_{out} = T + \Delta T$, $T_{in} = T$, $(\dot{X}_{m_{out}} - \dot{X}_{m_{in}}) = \Delta\dot{X}_m$ and taking the limit as $\Delta T \to 0$ and $\Delta\dot{X}_m \to 0$,

$$d\dot{Q} = \sum_m^{n_{ind}} \Delta H_{R_m}(T_0)d\dot{X}_m + (\Sigma F_j\cdot\hat{c}_{p_j})\cdot dT. \qquad (4.2\text{-}49)$$

The rate heat is added to the differential reactor, $d\dot{Q}$, is

$$d\dot{Q} = U \cdot dS \cdot (T_F - T), \tag{4.2-50}$$

where U is the heat transfer coefficient, dS is the heat-transfer area, and T_F is the temperature of the heating (or cooling) fluid. The heat transfer area, dS, relates to the reactor volume by

$$dS = \left(\frac{S}{V}\right) \cdot dV_R, \tag{4.2-51}$$

where (S/V) is the heat transfer area per unit volume, a characteristic that depends on the geometry of the reactor. For cylindrical reactors of diameter D, (S/V) = 4/D. Substituting (4.2-50) and (4.2-51) into (4.2-49) and rearranging, we obtain,

$$\frac{dT}{dV_R} = \frac{U}{(\Sigma F_j \cdot \hat{c}_{p_j})} \left(\frac{S}{V}\right) \cdot (T_F - T) - \sum_m^{n_{ind}} \left[\frac{\Delta H_{R_m}(T_0)}{(\Sigma F_j \cdot \hat{c}_{p_j})}\right] \frac{d\dot{X}_m}{dV_R}. \tag{4.2-52}$$

Equation (4.2-52) relates the changes in temperature to the reactor volume.

   To reduce (4.2-52) to dimensionless form, we use dimensionless temperature, defined by (2.3-4), $\theta = T/T_0$, the dimensionless extent, defined by (1.5-2), $Z_m = \dot{X}_m/(F_{tot})_0$, and the dimensionless space time, defined by (3.5-8), $\tau = V_R/v_0 \cdot t_{cr}$. Dividing both sides of (4.2-52) by T_0 and $(F_{tot})_0$ and multiplying both sides by the characteristic reaction time, t_{cr}, we obtain

$$\frac{d\theta}{d\tau} = \frac{U \cdot v_0 \cdot t_{cr}}{(\Sigma F_j \cdot \hat{c}_{p_j})} \left(\frac{S}{V}\right) \cdot (\theta_F - \theta) - \frac{(F_{tot})_0}{T_0} \sum_m^{n_{ind}} \left[\frac{\Delta H_{R_m}(T_0)}{(\Sigma F_j \cdot \hat{c}_{p_j})}\right] \frac{dZ_m}{d\tau}. \tag{4.2-53}$$

Eq. (4.2-52) is the **dimensionless**, differential energy balance equation for steady flow reactors, relating the temperature, θ, to the extents of the independent reactions, Z_m, as functions of space time τ. Note that individual $dZ_m/d\tau$'s are expressed in terms of $\theta(t)$ and $Z_m(\tau)$ by the reaction-based design equations, as described in Chapter 3.

   As in the case of batch reactors, dimensionless energy balance equation (4.2-53) is not conveniently used because the heat capacity term, $(\Sigma F_j \cdot \hat{c}_{p_j})$, is a function of the temperature and reaction extents and, consequently, varies along the reactor. To simplify the equation and obtain dimensionless quantities for heat-transfer, we define the heat capacity of the reference stream and relate $(\Sigma F_j \cdot \hat{c}_{p_j})$ at any point in the reactor to it by

$$(\Sigma F_j \cdot \hat{c}_{p_j}) = (F_{tot})_0 \cdot \hat{c}_{p0} \cdot CF(Z_m, \theta), \tag{4.2-54}$$

where $\hat{c}_{p0}$ is the specific molar heat capacity of the reference stream, and $CF(Z_m, \theta)$ is a correction factor that adjusts the value of the heat capacity as Z_m and θ vary. Substituting (4.2-54) into (4.2-53) and noting that $(F_{tot})_0 = C_0 \cdot v_0$, we obtain

$$\frac{d\theta}{d\tau} = \frac{1}{CF(Z_m,\theta)}\left[\frac{U \cdot t_{cr}}{C_0 \cdot \hat{c}_{p0}}\left(\frac{S}{V}\right) \cdot (\theta_F - \theta) - \sum_{m}^{n_{ind}}\left(\frac{\Delta H_{R_m}(T_0)}{T_0 \cdot \hat{c}_{p0}}\right)\frac{dZ_m}{d\tau}\right]. \quad (4.2\text{-}55)$$

Eq. (4.2-55) is a simplified dimensionless differential energy balance equation of steady flow reactors, where each term is divided by $(T_0 \cdot (F_{tot})_0 \cdot \hat{c}_{p0})$, the "reference thermal energy rate." The first term in the bracket of (4.2-55) represents the dimensionless heat-transfer rate with a dimensionless driving force $(\theta_F - \theta)$, where

$$\frac{\dot{Q}}{T_0 \cdot (F_{tot})_0 \cdot \hat{c}_{p0}} \quad (4.2\text{-}56)$$

is the dimensionless heat rate added to the reactor. Using the heat-transfer number, HTN, defined by (4.2-23), and the dimensionless heat of reaction of the m-th independent chemical reaction, DHR_m, defined by (4.2-24), (4.2-55) reduces to

$$\frac{d\theta}{d\tau} = \frac{1}{CF(Z_m,\theta)}\left[HTN \cdot (\theta_F - \theta) - \sum_{m}^{n_{ind}}DHR_m \cdot \frac{dZ_m}{d\tau}\right]. \quad (4.2\text{-}57)$$

As in the case of batch reactors, the definition of the specific molar heat capacity of the reference state, $\hat{c}_{p0}$, and the determination of the correction factor, $CF(Z_m,\theta)$, deserve a closer examination. Using stoichiometric relation (1.5-5) to express the molar flow rate of species j in terms of the extents of the independent reactions,

$$(\Sigma F_j \cdot \hat{c}_{p_j}) = (F_{tot})_0\left(\sum_{j}^{all} y_{j0} \cdot \hat{c}_{p_j}(\theta) + \sum_{j}^{all}\hat{c}_{p_j}(\theta) \cdot \sum_{m}^{n_{ind}}(s_j)_m \cdot Z_m\right). \quad (4.2\text{-}58)$$

For gas-phase reactions, we define the average specific molar heat capacity of the reference stream, $\hat{c}_{p0}$, by

$$\hat{c}_{p0} = \sum_{j}^{all} y_{j0} \cdot \hat{c}_{p_j}(1). \quad (4.2\text{-}59)$$

For liquid-phase reactions, we define the average specific molar heat capacity of the reference stream, $\hat{c}_{p0}$,

$$\hat{c}_{p0} = \frac{\dot{m}}{(F_{tot})_0}\overline{c}_p, \quad (4.2\text{-}60)$$

where $\dot{m}$ is the mass flow rate of the reference stream, $\overline{c}_p$ is the mass-based heat capacity, and $(F_{tot})_0$ is the total molar flow rate of the reference stream.

To determine the correction factor of the heat capacity, $CF(Z_m,\theta)$, we use (4.2-54) and the appropriate definition of the specific molar heat capacity of the reference stream, $\hat{c}_{p0}$. For gas-phase reactions, using (4.2-30),

$$CF(Z_m,\theta) = \frac{\sum\limits_j^{all} y_{j_0} \cdot \hat{c}_{p_j}(\theta)}{\sum\limits_j^{all} y_{j_0} \cdot \hat{c}_{p_j}(1)} + \frac{\sum\limits_j^{all} \hat{c}_{p_j}(\theta) \cdot \sum\limits_m^{n_{ind}} (s_j)_m \cdot Z_m}{\sum\limits_j^{all} y_{j_0} \cdot \hat{c}_{p_j}(1)}. \qquad (4.2\text{-}61)$$

For liquid-phase reactions, using (4.2-31),

$$CF(Z_m,\theta) = \frac{(F_{tot})_0}{\dot{m} \cdot \bar{c}_p} \left(\sum\limits_j^{all} y_{j_0} \cdot \hat{c}_{p_j}(\theta) + \sum\limits_j^{all} \hat{c}_{p_j}(\theta) \cdot \sum\limits_m^{n_{ind}} (s_j)_m \cdot Z_m \right). \qquad (4.2\text{-}62)$$

For isothermal operations, $d\theta/dt = 0$, and, using (4.2-56), for plug-flow reactors (4.2-57) reduces to

$$\frac{d}{d\tau}\left(\frac{\dot{Q}}{T_0 \cdot (F_{tot})_0 \cdot \hat{c}_{p0}} \right) = \sum\limits_m^{n_{ind}} \left(\frac{\Delta H_{R_m}(T_0)}{T_0 \cdot \hat{c}_{p0}} \right) \frac{dZ_m}{d\tau}, \qquad (4.2\text{-}63)$$

which can be further simplified to

$$d\dot{Q} = (F_{tot})_0 \sum\limits_m^{n_{ind}} \Delta H_{R_m}(T_0) \cdot dZ_m. \qquad (4.2\text{-}64)$$

Eq. (4.2-64) provides the heating (or cooling) rate along the reactor needed to maintain isothermal conditions. For adiabatic operations, $(S/V) = 0$ (no heat-transfer area); hence, HTN = 0, and for plug-flow reactors (4.2-55) reduces to

$$\frac{d\theta}{d\tau} = \frac{-1}{CF(Z_m,\theta)} \sum\limits_m^{n_{ind}} DHR_m \frac{dZ_m}{d\tau}, \qquad (4.2\text{-}65)$$

which can be further simplified to

$$d\theta = \frac{-1}{CF(Z_m,\theta)} \sum\limits_m^{n_{ind}} DHR_m \cdot dZ_m. \qquad (4.2\text{-}66)$$

Eq. (4.2-66) relates changes in the temperature to changes in the extents of the independent reactions along the reactor.

To derive the dimensionless energy balance equation for a CSTR, we use the integral form of the energy balance equation, (4.2-48). Since the temperature in the reactor is uniform, $T = T_{out}$, and the rate heat is transferred to the reactor is

$$\dot{Q} = U \cdot \left(\frac{S}{V} \right) \cdot V_R \cdot (T_F - T_{out}), \qquad (4.2\text{-}67)$$

where (S/V) is the heat transfer area per unit volume of reactor. We substitute (4.2-67) into (4.2-48), divide both sides of the equation by T_0 and $(F_{tot})_0$, and multiply both sides by t_{cr}. We then use (4.2-54) and rearrange the energy balance equation to obtain

$$HTN \cdot \tau \cdot (\theta_F - \theta_{out}) - \frac{\dot{W}_{sh}}{T_0 \cdot (F_{tot})_0 \cdot \hat{c}_{p0}} = \sum_m^{n_{ind}} DHR_m \cdot (Z_{m_{out}} - Z_{m_{in}}) +$$

$$+ \int_1^{\theta_{out}} CF(Z_m, \theta)_{out} \cdot d\theta - \int_1^{\theta_{in}} CF(Z_m, \theta)_{in} \cdot d\theta. \qquad (4.2\text{-}68)$$

This is the dimensionless energy balance equation for a CSTR, relating the outlet temperature, θ_{out}, to the inlet temperature, θ_{in}, and the extents of the independent reactions in the reactor, $(Z_{m_{out}} - Z_{m_{in}})$'s.

For isothermal CSTRs with negligible shaft work, $\theta_{out} = \theta_{in} = 1$, and, using (4.2-56), (4.2-68) reduces to

$$\frac{\dot{Q}}{T_0 \cdot (F_{tot})_0 \cdot \hat{c}_{p0}} =$$

$$= \sum_m^{n_{ind}} DHR_m \cdot (Z_{m_{out}} - Z_{m_{in}}) + \int_1^{\theta_{in}} \left[CF(Z_m, \theta)_{out} - CF(Z_m, \theta)_{in} \right] \cdot d\theta. \quad (4.2\text{-}69)$$

When the correction factors at the inlet and outlet are the same, (4.2-69) can be further simplified to

$$\dot{Q} = (F_{tot})_0 \cdot \sum_m^{n_{ind}} \Delta H_{R_m}(T_0) \cdot (Z_{m_{out}} - Z_{m_{in}}). \qquad (4.2\text{-}70)$$

Eq. (4.2-70) provides the heating (or cooling) rate needed to maintain isothermal conditions. For adiabatic CSTRs, $(S/V) = 0$ (no heat-transfer area), and assuming negligible shaft work, (4.2-68) reduces to

$$-\sum_m^{n_{ind}} DHR_m \cdot (Z_{m_{out}} - Z_{m_{in}}) = \int_1^{\theta_{out}} CF(Z_m, \theta)_{out} \cdot d\theta - \int_1^{\theta_{in}} CF(Z_m, \theta)_{in} \cdot d\theta. \quad (4.2\text{-}71)$$

Eq. (4.2-71) relates the temperatures at the reactor inlet and outlet to the extents of the independent reactions.

Example 4-4 The following gas-phase chemical reactions take place in a catalytic reactor producing ethylene oxide:

$$C_2H_4 + 0.5\, O_2 \rightarrow C_2H_4O$$

$$C_2H_4 + 3\, O_2 \rightarrow 2\, CO_2 + 2\, H_2O.$$

A feed stream at $600°K$ and 2 atm, consisting of 67.67% C_2H_4 and 33.33% O_2 (by mole), is introduced into the reactor. Based on the data below, determine $CF(Z_m, \theta)$. Data: The specific molar heat capacities of the species (in cal/mole°K) are:

$\hat{c}_{pC2H4} = 2.83 + 28.6 \cdot 10^{-3} \, T + 8.73 \cdot 10^{-3} \, T^2$;

$\hat{c}_{pO2} = 7.23 + 1.005 \cdot 10^{-3} \, T - 0.451 \cdot 10^5 \, T^{-2}$;

$\hat{c}_{pC2H4O} = -0.765 + 46.62 \cdot 10^{-3} \, T + 18.47 \cdot 10^{-3} \, T^2$

$\hat{c}_{pCO2} = 10.84 + 2.076 \cdot 10^{-3} \, T - 2.30 \cdot 10^5 \, T^{-2}$;

$\hat{c}_{pH2O} = 6.895 + 2.881 \cdot 10^{-3} \, T + 0.240 \cdot 10^5 \, T^{-2}$.

Solution This problem illustrates how to determine $CF(Z_m, \theta)$ that depends on both Z_m and θ. The two chemical reactions are independent, and their stoichiometric coefficients are:

$$(s_{C2H4})_1 = -1; \quad (s_{O2})_1 = -0.5; \quad (s_{C2H4O})_1 = 1; \quad (s_{CO2})_1 = 0; \quad (s_{H2O})_1 = 0;$$

$$(s_{C2H4})_2 = -1; \quad (s_{O2})_2 = -3; \quad (s_{C2H4O})_2 = 0; \quad (s_{CO2})_2 = 2; \quad (s_{H2O})_2 = 2.$$

We select the inlet stream as the reference stream; hence $T_0 = 600°K$, and $(y_{C2H4})_0 = 0.667$, $(y_{O2})_0 = 0.333$, $(y_{C2H4O})_0 = 0$, $(y_{CO2})_0 = 0$, and $(y_{H2O})_0 = 0$. Using $T = T_0 \cdot \theta$, we express the specific molar heat capacities of the species in terms of θ:

$\hat{c}_{pC2H4}(\theta) = 2.83 + 17.16 \cdot \theta + 3.143 \cdot \theta^2$;

$\hat{c}_{pO2}(\theta) = 7.23 + 0.603 \cdot \theta - 0.125 \cdot \theta^{-2}$;

$\hat{c}_{pC2H4O}(\theta) = -0.765 + 27.97 \cdot \theta + 6.65 \cdot \theta^2$;

$\hat{c}_{pCO2}(\theta) = 10.84 + 1.246 \cdot \theta - 0.639 \cdot \theta^{-2}$;

$\hat{c}_{pH2O}(\theta) = 6.895 + 1.729 \cdot \theta + 0.0667 \cdot \theta^{-2}$.

For gas-phase reactions, the reference specific molar heat capacity is determined using (4.2-59):

$$\hat{c}_{pO} = \sum_j^{all} y_{j_0} \cdot \hat{c}_{p_j}(1) = (0.667) \cdot 23.13 + (0.333) \cdot 7.71 = 18.0 \text{ cal/mole°K}. \quad (a)$$

For gas-phase reactions, the correction factor is expressed by (4.2-61),

$$CF(Z_m, \theta) = \frac{\sum_j^{all} y_{j_0} \cdot \hat{c}_{p_j}(\theta)}{\sum_j^{all} y_{j_0} \cdot \hat{c}_{p_j}(1)} + \frac{\sum_j^{all} \hat{c}_{p_j}(\theta) \cdot \sum_m^{n_{ind}} (s_j)_m \cdot Z_m}{\sum_j^{all} y_{j_0} \cdot \hat{c}_{p_j}(1)}. \quad (b)$$

Substituting the $\hat{c}_{p_j}(\theta)$'s, the first term on the right reduces to

$$\frac{1}{18} \left((0.667) \cdot \hat{c}_{pC2H4}(\theta) + (0.333) \cdot \hat{c}_{pO2}(\theta) \right) =$$

$$= \frac{1}{18} \left\{ \begin{array}{l} (0.667) \cdot (2.83 + 17.16 \cdot \theta + 3.243 \cdot \theta^2) + \\ + (0.333) \cdot (7.23 + 0.608 \cdot \theta - 0.125 \cdot \theta^{-2}) \end{array} \right\} =$$

$$= \frac{4.295 + 11.65 \cdot \theta + 2.096 \cdot \theta^2 - 0.0416 \cdot \theta^{-2}}{18}.$$

Expanding the second term on the right of (b),

$$= \frac{1}{18} \left\{ \begin{array}{l} (-Z_1 - Z_2) \cdot \hat{c}_{pC2H4}(\theta) + Z_1 \cdot \hat{c}_{pC2H4O}(\theta) + (-0.5 \cdot Z_1 - 3 \cdot Z_2) \cdot \hat{c}_{pO2}(\theta) \\ + 2 \cdot Z_2 \cdot \hat{c}_{pCO2}(\theta) + 2 \cdot Z_2 \cdot \hat{c}_{pH2O}(\theta) \end{array} \right\} =$$

$$= \frac{1}{18} \left\{ \begin{array}{l} (-Z_1 - Z_2) \cdot (2.83 + 17.16 \cdot \theta + 3.143 \cdot \theta^2) + \\ + Z_1 \cdot (-0.765 + 27.97 \cdot \theta + 6.65 \cdot \theta^2) + \\ + (-0.5 \cdot Z_1 - 3 \cdot Z_2) \cdot (7.233 + 0.603 \cdot \theta - 0.125 \cdot \theta^2) + \\ + 2 \cdot Z_2 \cdot (10.84 + 1.246 \cdot \theta - 0.639 \cdot \theta^{-2}) + \\ + 2 \cdot Z_2 \cdot (6.895 + 1.729 \cdot \theta - 0.067 \cdot \theta^{-2})) \end{array} \right\} =$$

$$= \frac{1}{18} \left\{ \begin{array}{l} (-7.21 \cdot Z_1 + 10.95 \cdot Z_2) - (10.51 \cdot Z_1 + 13.02 \cdot Z_2) \cdot \theta + \\ + (3.08 \cdot Z_1 - 3.143 \cdot Z_2) \cdot \theta^2 + (0.0625 \cdot Z_1 - 0.770 \cdot Z_2) \cdot \theta^{-2} \end{array} \right\}.$$

Combining the two terms, the correction factor is

$$CF(Z_m, \theta) = \frac{1}{18} \left\{ \begin{array}{l} (4,295 - 7.21 \cdot Z_1 + 10.95 \cdot Z_2) + \\ + (11.65 + 10.51 \cdot Z_1 + 13.02 \cdot Z_2) \cdot \theta + \\ + (2.096 + 3.507 \cdot Z_1 - 3.143 \cdot Z_2) \cdot \theta^2 + \\ + (0.0416 - 0.0625 \cdot Z_1 - 0.770 \cdot Z_2) \cdot \theta^{-2} \end{array} \right\}.$$

4.3 SUMMARY

In this chapter, we reviewed the thermodynamic relations related to reactor operations and derived the energy balance equations for batch and flow reactors. Main topics covered are:

a. Heat of reaction of a chemical reaction, and its dependence on temperature and pressure.

b. Equilibrium constant of a chemical reaction, and its dependence on temperature.

c. Definition of the specific molar heat capacity of the reference state, $\hat{c}_{p0}$, for gas-phase and liquid-phase reactions.
d. Definition of the correction factor of heat capacity, $CF(Z_m,\theta)$, for gas-phase and liquid-phase reactions.
e. Definition of the dimensionless heat of reaction, DHR_m.
f. Definition of the heat-transfer number, HTN.
h. Derivation of the macroscopic energy balance equation of batch reactors and reducing it to a dimensionless form.
i. Derivation of the integral and differential energy balance equations for flow reactors and reducing these equations to dimensionless forms.

Tables 4-1 and 4-2 provide, respectively, the dimensionless energy balance equations and auxiliary relations for ideal batch reactors and steady flow reactors (plug-flow and CSTR)

BIBLIOGRAPHY

More detailed treatment of the energy balance equations and deriving them for microscopic systems can be found in:

Bird, R. B., W. E. Stewart and E. N. Lightfoot, *Transport Phenomena*. New York: John Wiley and Sons, 1960.
Welty, J. R., C. E. Wicks and R. E. Wilson, *Fundamentals of Momentum Heat and Mass* Transfer, 2nd Edition. New York: John Wiley and Sons, 1984.

More detailed treatment of chemical equilibrium and thermodynamic properties can be found in:

Smith, J. M. and H. C. Van Ness, *Introduction to Chemical Engineering Thermodynamics*, Fourth Edition, New York, McGraw-Hill, 1987.

PROBLEMS

4.1₁ The following liquid-phase chemical reactions take place in a CSTR:

$$A \rightarrow 2\,B$$

$$B \rightarrow C + D.$$

Both reactions are first-order. An aqueous solution whose concentration is $C_{A0} = 3$ mole/liter is fed at a rate of 200 liter/min into a 500 liter CSTR with a heat transfer area of 20 m^2/m^3. The inlet temperature is 60°C, and the temperature of the jacket is 100°C. The density and heat capacity of the solution are essentially the same as those of water ($\rho = 1$ g/cm^3; $\bar{c}_p = 1$ cal/g°C). The heat transfer coefficient is estimated as 5 cal/sec·cm^2·°C. Based on the data below, determine
a. the average specific molar heat capacity of the reference stream,
b. the dimensionless heat of reaction of each chemical reaction,

Table 4-1: Energy Balance Equation for Ideal Batch Reactor

Energy Balance Equation

$$\frac{d\theta}{d\tau} = \frac{1}{CF(Z_m, \theta)} \cdot \left[HTN \cdot (\theta_F - \theta) - \sum_m^{n_{ind}} DHR_m \cdot \frac{dZ_m}{d\tau} - \frac{d}{d\tau} \left(\frac{W_{sh}}{(N_{tot})_0 \cdot T_0 \cdot \hat{c}_{p0}} \right) \right]$$

(A)

Definitions and Auxiliary Relations

Dimensionless Temperature

$$\theta \equiv \frac{T}{T_0} \tag{B}$$

Specific Molar Heat Capacity of Reference State

$$\hat{c}_{p0} = \sum_j^{all} y_j(0) \cdot \hat{c}_{p_j}(T_0) \quad \text{or} \quad \hat{c}_{p0} = \frac{M}{(N_{tot})_0} \overline{c}_p \tag{C}$$

Dimensionless Heat of Reaction of the m-th Independent Reaction

$$DHR_m = \frac{\Delta H_{R_m}(T_0)}{T_0 \cdot \hat{c}_{p0}} \tag{D}$$

Dimensionless Heat Transfer Number

$$HTN = \frac{U \cdot t_{cr}}{C_0 \cdot \hat{c}_{p0}} \left(\frac{S}{V} \right) \tag{E}$$

Dimensionless Heat

$$\frac{Q}{(N_{tot})_0 \cdot T_0 \cdot \hat{c}_{p0}} \tag{F}$$

Correction Factor of Heat Capacity (Gas-Phase Reactions)

$$CF(Z_m, \theta) = \frac{\sum_j^{all} y_j(0) \cdot \hat{c}_{p_j}(\theta)}{\sum_j^{all} y_j(0) \cdot \hat{c}_{p_j}(1)} + \frac{\sum_j^{all} \hat{c}_{p_j}(\theta) \cdot \sum_m^{n_{ind}} (s_j)_m \cdot Z_m}{\sum_j^{all} y_j(0) \cdot \hat{c}_{p_j}(1)} \tag{G}$$

Table 4-2: Energy Balance Equations for Steady Flow Reactors

For Plug-Flow Reactor

$$\frac{d\theta}{d\tau} = \frac{1}{CF(Z_m, \theta)} \cdot \left[HTN \cdot (\theta_F - \theta) - \sum_m^{n_{ind}} DHR_m \cdot \frac{dZ_m}{d\tau} \right] \tag{A}$$

For Continuous Stirred Tank Reactors

$$HTN \cdot \tau \cdot (\theta_F - \theta_{out}) - \frac{\dot{W}_{sh}}{T_0 \cdot (F_{tot})_0 \cdot \hat{c}_{p0}} = \sum_m^{n_{ind}} DHR_m \cdot (Z_{m_{out}} - Z_{m_{in}}) +$$

$$+ \int_1^{\theta_{out}} CF(Z_m, \theta)_{out} \cdot d\theta - \int_1^{\theta_{in}} CF(Z_m, \theta)_{in} \cdot d\theta. \tag{B}$$

Definitions and Auxiliary Relations

Specific Molar Heat Capacity of Reference State

$$\hat{c}_{P0} = \sum_j^{all} y_{j0} \cdot \hat{c}_{p_j}(T_0) \quad \text{or} \quad \hat{c}_{P0} = \frac{\dot{m}}{(F_{tot})_0} \bar{c}_p, \tag{C}$$

Dimensionless Heat of Reaction of the m-th Independent Reaction

$$DHR_m = \frac{\Delta H_{R_m}(T_0)}{T_0 \cdot \hat{c}_{P0}} \tag{D}$$

Dimensionless Heat Transfer Number

$$HTN = \frac{U \cdot t_{cr}}{C_0 \cdot \hat{c}_{P0}} \left(\frac{S}{V} \right) \tag{E}$$

Dimensionless Heat-Transfer Rate

$$\frac{\dot{Q}}{(F_{tot})_0 \cdot T_0 \cdot \hat{c}_{P0}} \tag{F}$$

Correction Factor of Heat Capacity (Gas-Phase Reactions)

$$CF(Z_m, \theta) = \frac{\sum_j^{all} y_{j0} \cdot \hat{c}_{p_j}(\theta)}{\sum_j^{all} y_{j0} \cdot \hat{c}_{p_j}(1)} + \frac{\sum_j^{all} \hat{c}_{p_j}(\theta) \cdot \sum_m^{n_{ind}} (s_j)_m \cdot Z_m}{\sum_j^{all} y_{j0} \cdot \hat{c}_{p_j}(1)} \tag{G}$$

c. the dimensionless heat-transfer number, and
d. the correction factor of heat capacity.
e. Formulate the dimensionless energy balance equation for the operation.

Data: At 60°C, $\Delta H_{R1} = 20{,}000$ cal/mole extent; $\Delta H_{R2} = 12{,}000$ cal/mole extent; $k_1 = 0.8$ min^{-1}; $k_2 = 0.2$ min^{-1}.

4.2$_2$ A stream consisting of 40% O_2 and 60% CH_4 is fed into a plug-flow reactor where the following reactions take place:

$$2\ CH_4 + O_2 \rightarrow 2\ CH_3OH$$

$$CH_4 + 2\ O_2 \rightarrow CO_2 + 2\ H_2O.$$

The feed temperature is 600°K. Based on the data below, determine
a. the dimensionless heat of reaction of each chemical reaction at 600°K,
b. the specific molar heat capacity of the feed, and
c. the correction factor of heat capacity, $CF(Z_m, \theta)$.

Data: The standard heat of formations of the species at 298°K (in Joule/mole):

Methane: -74,520
Methanol (gas): -200,600
CO_2: -393,509
Water (gas): -241,818

The species heat capacities in Joule/mole °K (T in °K)

Methane $= 1.702 + 9.081\ 10^{-3}\ T - 2.164\ 10^{-6}\ T^2$
Methanol $= 2.211 + 12.216\ 10^{-3}\ T - 3.450\ 10^{-6}\ T^2$
Oxygen $= 3.639 + 0.506\ 10^{-3}\ T - 0.227\ 10^5\ T^{-2}$
CO_2 $= 5.457 + 1.045\ 10^{-3}\ T - 1.157\ 10^5\ T^{-2}$
Water $= 3.470 + 1.450\ 10^{-3}\ T - 0.121\ 10^5\ T^{-2}$

PART TWO
APPLICATIONS

Part One covered the four fundamental concepts used to formulate the design equations of chemical reactors (stoichiometry, chemical kinetics, species balances, and heat balance). Dimensionless, reaction-based design equations were derived for different reactor configurations. Part Two deals with solving the design equations and obtaining the dimensionless operating curves that describe the reactor operations. It describes how to apply the design equations and energy balance equation to different reactor configurations.

In order to solve the design equations (obtain the dimensionless extents of the independent reactions in terms of either the dimensionless operating time or space time), it is necessary to express the rates of the individual chemical reactions in terms of these variables. The general design formulation for chemical reactors with multiple reactions goes as follows:

1. Identify all chemical reactions and select a set of independent reactions, applying the heuristic rule of Chapter 3.
2. Select a reference state (or stream) and determine $(N_{tot})_0$ (or $(F_{tot})_0$), C_0, T_0, and the specific molar-based heat capacity $\hat{c}_{p0}$.
3. Write the dimensionless reaction-based design equations for each independent reaction, as described in Chapter 3.
4. Express the concentrations of the individual species and the correction factor of the heat capacity, $CF(Z_m,\theta)$ (see Chapter 4), in terms of the extents of the independent reactions, taking into consideration the geometry of the reactor and the way the reactants are fed.
5. Select one chemical reaction and determine the characteristic reaction time as described in Chapter 2.

6. Solve the design equations simultaneously with the energy balance equation, and obtain the dimensionless reaction and temperature operating curves (Z_m and θ) as functions of the dimensionless operating time (or space time), τ.

7. Use the stoichiometric relations of Chapter 1 to obtain the dimensionless operating curves of the individual species.

Part Two describes how the general design formulation is applied to different reactor configurations and various operation modes. Chapter 5 deals with ideal batch reactors, Chapter 6 with plug-flow reactors, Chapter 7 with continuous stirred tank reactors, and Chapter 8 with other reactor configurations. In each case, we first express the concentrations of the individual species in terms of the extents of the independent reactions. In order to acquaint the reader with the new reactor design methodology, first isothermal operations with single chemical reactions are considered, followed by isothermal operations with multiple chemical reactions. Each chapter is then concluded with the general case of non-isothermal operations with multiple chemical reactions. Chapter 9 provides a brief discussion of applying the reaction-based design formulation to economic-based optimization of reactor operations.

CHAPTER FIVE

IDEAL BATCH REACTOR

A batch reactor is a reactor without an input or an output. Reactants are charged into the reactor before the commencement of the operation, and the content of the reactor is discharged at the end of the operation. An **ideal** batch reactor, schematically described in Figure 5-1, is a simplified mathematical model of batch reactors. The model is based on the assumption that, due to good agitation, the same conditions (species compositions and temperature) exist **everywhere** inside the reactor. Consequently, the same reaction rates prevail throughout the entire reactor volume.

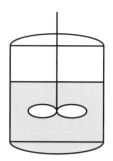

Figure 5-1: Schematic Description of an Ideal Batch Reactor

In practice, batch reactors vary in size, ranging from a few cubic centimeters in a test tube, through a beaker agitated by a magnetic stirrer in a chemical laboratory, to a few-liters, pilot-scale reactor, to several thousand gallon production reactors. Achieving good mixing in small reactors is relatively simple, but it is a difficult task in large reactors. Mixing in very large reactors is an important practical issue that requires careful attention and proper mechanical design of the reactor and agitator. In general, the ideal batch reactor model provides a good mathematical representation of the actual operation when the mixing time is short in comparison to the reaction time. For very large reactors and reactors filled with viscous liquids, the ideal batch reactor model serves only as a first approximation of the actual reactor operation.

In this chapter, we analyze the operation of ideal batch reactors. In Section 5.1, we review how the dimensionless design equations are utilized and discuss the auxiliary relations that should be incorporated to solve the design equations. In the rest of the chapter, we cover numerous applications of batch reactors of increased level of complexity. Sections 5.2 and 5.3 concern isothermal operations, where only the design equations should be solved. Section 5.2 covers operations with single reactions, where single design equations should be solved. Section 5.3 covers isothermal operations with multiple chemical reactions, where several design equations should be solved simultaneously. In Section 5.4, we discuss the general case, involving any mode of operation (isothermal or non-isothermal) with any number of chemical reactions. Hence, multiple design equations should be solved simultaneously with the energy balance equation. Since Sections 5.2 and 5.3 cover special, simplified cases, readers familiar with the operation of ideal batch reactors can proceed directly from Section 5.1 to Section 5.4. Readers who do not have previous exposure to batch reactor operations are encouraged to cover Section 5.2 and 5.3 first to familiarize themselves with key operational issues.

5.1 DESIGN EQUATIONS AND AUXILIARY RELATIONS

The differential reaction-based, dimensionless design equation of an ideal batch reactor, written for the m-th independent reaction, was derived in Chapter 3,

$$\frac{dZ_m}{d\tau} = \left(r_m + \sum_{k}^{n_{dep}} \alpha_{km} \cdot r_k \right) \cdot \left(\frac{V_R}{V_R(0)} \right) \cdot \left(\frac{t_{cr}}{C_0} \right). \tag{5.1-1}$$

Recall that Z_m is the dimensionless extent of the m-th independent chemical reaction defined by

$$Z_m = \frac{X_m}{(N_{tot})_0}, \tag{5.1-2}$$

and τ is the dimensionless operating time,

$$\tau = \frac{t}{t_{cr}}, \tag{5.1-3}$$

where t_{cr} is a conveniently-selected characteristic reaction time. Also, C_0 is the reference concentration defined by

$$C_0 = \frac{(N_{tot})_0}{V_R(0)}, \tag{5.1-4}$$

where $(N_{tot})_0$ is the reference total molar content, and $V_R(0)$ is the initial reactor volume. Also, recall that the dimensionless design equation is expressed in terms of relative reaction rates given by (3.5-13).

As discussed in Chapter 3, to describe the operation of a reactor with multiple chemical reactions, we have to write (5.1-1) for each of the independent chemical reactions. To solve the design equations (to obtain relationships between Z_m's and τ), we have to express $V_R(\tau)$ and the rates of the individual chemical reactions, r_m's and r_k's, in terms of Z_m's and τ. Below, we discuss the auxiliary relations needed to express the design equations explicitly in terms of Z_m's and τ.

The volume-based rate of the i-th chemical reaction (see Section 2.3) is

$$r_i = k_i(T_0) \cdot e^{\gamma_i \frac{\theta-1}{\theta}} \cdot h_i(C_j's), \tag{5.1-5}$$

where $k_i(T_0)$ is the reaction rate constant at reference temperature T_0, γ_i is the dimensionless activation energy ($\gamma_i = Ea_i/R \cdot T_0$), θ is the dimensionless temperature, and $h_i(C_j's)$ is a function of the species concentrations, given by the rate expression. To express the rates of the chemical reactions in terms of Z_m's and τ, we have to relate the species concentrations to Z_m's and τ. For ideal batch reactors, the concentration of species j at dimensionless operating time τ is

$$C_j(\tau) = \frac{N_j(\tau)}{V_R(\tau)}. \tag{5.1-6}$$

Using stoichiometric relation (1.5-3),

$$N_j(\tau) = (N_{tot})_0 \cdot \left(y_j(0) + \sum_m^{n_{ind}} (s_j)_m \cdot Z_m(\tau) \right), \tag{5.1-7}$$

where

$$y_j(0) = \frac{N_j(0)}{(N_{tot})_0} \tag{5.1-8}$$

is the mole fraction of species j in the reference state. Substituting (5.1-7) into (5.1-6), the concentration of species j at time τ is

$$C_j(\tau) = \frac{(N_{tot})_0}{V_R(\tau)} \cdot \left(y_j(0) + \sum_m^{n_{ind}} (s_j)_m \cdot Z_m(\tau) \right). \tag{5.1-9}$$

For constant-volume batch reactors, $V_R(\tau) = V_R(0)$, and (5.1-9) reduces to

$$C_j(\tau) = C_0 \cdot \left(y_j(0) + \sum_m^{n_{ind}} (s_j)_m \cdot Z_m(\tau) \right). \tag{5.1-10}$$

Eq. (5.1-10) provides the species concentrations in terms of the extents of the independent reactions for constant-volume batch reactors.

For gaseous, variable-volume batch reactors, like the one shown schematically in Figure 5-2, $V_R(\tau)$ varies during the operation. Assuming ideal gas behavior, the reactor volume at dimensionless operating time τ is

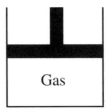

Figure 5-2: Variable-Volume Batch Reactor

$$\frac{V_R(\tau)}{V_R(0)} = \left(\frac{N_{tot}(\tau)}{(N_{tot})_0}\right) \cdot \left(\frac{T(\tau)}{T(0)}\right) \cdot \left(\frac{P(0)}{P(\tau)}\right). \qquad (5.1\text{-}11)$$

Using stoichiometric relation (1.5-4) to express the total number of moles in terms of the extents of the chemical reactions, (5.1-11) becomes

$$\frac{V_R(\tau)}{V_R(0)} = \left(1 + \sum_m^{n_{ind}} \Delta_m \cdot Z_m(\tau)\right) \cdot \theta(\tau) \cdot \left(\frac{P(0)}{P(\tau)}\right), \qquad (5.1\text{-}12)$$

where θ is the dimensionless temperature, T/T_0. Substituting (5.1-12) into (5.1-9), for **isobaric** operations of variable volume batch reactors, we obtain

$$C_j(\tau) = C_0 \cdot \frac{y_j(0) + \sum_m^{n_{ind}} (s_j)_m \cdot Z_m(\tau)}{\left(1 + \sum_m^{n_{ind}} \Delta_m \cdot Z_m(\tau)\right) \cdot \theta(\tau)}. \qquad (5.1\text{-}13)$$

Eq. (5.1-13) provides species concentrations in terms of the extents of the independent reactions for gaseous, variable-volume batch reactors. Table 5.1 provides a summary of the design equations and auxiliary relations used in the design formulation of ideal batch reactors.

Note that (5.1-5) and (5.1-13) contain another dependent variable, θ, the dimensionless temperature, whose variation during the reactor operation is expressed by the energy balance equation. For ideal batch reactors with negligible mechanical shaft work, the dimensionless energy balance equation, derived in Chapter 4, is

Table 5.1: Design Equations and Related Quantities

Design Equation	For the m-th Independent Reaction $$\frac{dZ_m}{d\tau} = \left(r_m + \sum_{k}^{n_{dep}} \alpha_{km} \cdot r_k \right) \cdot \left(\frac{V_R}{V_R(0)} \right) \cdot \left(\frac{t_{cr}}{C_0} \right) \quad \text{(A)}$$
Definitions	Dimensionless Operating Time $$\tau \equiv \frac{t}{t_{cr}} \quad \text{(B)}$$ Dimensionless Extent of the m-th Independent Reaction $$Z_m(\tau) \equiv \frac{X_m(\tau)}{(N_{tot})_0} \quad \text{(C)}$$ Reference Concentration $$C_0 \equiv \frac{(N_{tot})_0}{V_R(0)} \quad \text{(D)}$$ Characteristic Reaction Time $$t_{cr} \equiv \frac{\text{Characteristic concentration}}{\text{Characteristic reaction rate}} = \frac{C_0}{r_0} \quad \text{(E)}$$
Species Concentrations	For Constant-Volume Reactor, $V_R = V_R(0)$ $$C_j = C_0 \cdot \left(y_j(0) + \sum_{m}^{n_{ind}} (s_j)_m \cdot Z_m \right) \quad \text{(F)}$$ For Gaseous, Variable-Volume Reactor $$C_j = C_0 \frac{y_j(0) + \sum_{m}^{n_{ind}} (s_j)_m \cdot Z_m}{\left(1 + \sum_{m}^{n_{ind}} \Delta_m \cdot Z_m \right) \cdot \theta} \cdot \left(\frac{P}{P_0} \right) \quad \text{(G)}$$

$$\frac{d\theta}{d\tau} = \frac{1}{CF(Z_m, \theta)} \cdot \left[HTN \cdot \left(\frac{V_R}{V_R(0)} \right) \cdot (\theta_F - \theta) - \sum_{m}^{n_{ind}} DHR_m \cdot \frac{dZ_m}{d\tau} \right], \quad (5.1\text{-}14)$$

where HTN is the dimensionless heat-transfer number of the reactor, defined by

$$HTN = \frac{U \cdot t_{cr}}{C_0 \cdot \hat{c}_{p0}} \cdot \left(\frac{S}{V} \right), \quad (5.1\text{-}15)$$

DHR_m is of the dimensionless heat of reaction of the m-th independent chemical reaction, defined by

$$DHR_m = \frac{\Delta H_{Rm}(T_0)}{T_0 \cdot \hat{c}_{p0}}, \quad (5.1\text{-}16)$$

and $CF(Z_m, \theta)$ is the correction factor of the heat capacity, defined by (4.2-19). The first term inside the bracket of (5.1-14) represents the dimensionless heat-transfer rate,

$$\frac{d}{d\tau} \left(\frac{Q}{T_0 \cdot (N_{tot})_0 \cdot \hat{c}_{p0}} \right) = \left[\frac{U \cdot t_{cr}}{C_0 \cdot \hat{c}_{p0}} \cdot \left(\frac{S}{V} \right) \cdot \left(\frac{V_R}{V_R(0)} \right) \right] \cdot (\theta_F - \theta), \quad (5.1\text{-}17)$$

where

$$\frac{Q}{T_0 \cdot (N_{tot})_0 \cdot \hat{c}_{p0}} \quad (5.1\text{-}18)$$

is the dimensionless heat added to the reactor. Note that the dimensionless heat-transfer number, HTN, is proportional to the heat transfer coefficient, U, which depends on the flow conditions, the properties of the fluid, and the heat-transfer area per unit volume, (S/V). The second term inside the bracket of (5.1-14) represents the heat generated (or absorbed) by the chemical reactions. Note that the dimensionless heat of reaction of the m-th independent chemical reaction, DHR_m, is a characteristic of the chemical reaction, the reference temperature, and the composition of the reference state. Also, note that the specific molar heat capacity of the reference state, $\hat{c}_{p0}$, is defined differently for gas-phase and liquid-phase reactions: for gas-phase it is defined by (4.2-59) and for liquid-phase by (4.2-60). To design non-isothermal batch reactors, we have to solve the design equations simultaneously with the energy balance equation, (5.1-14). Table 5.2 provides a summary of the energy balance equation and auxiliary relations used in the design formulation of ideal batch reactors.

Below, we discuss how to apply the design equations and the energy balance equation to determine various quantities concerning the operations of ideal batch reactors. We first examine isothermal operations with single reactions to illustrate how the rate expressions are incorporated into the design equation. Then, we will discuss how the reaction rate is determined. Next, we will expand the analysis to isothermal opera-

Table 5.2: Energy Balance Equation and Related Quantities

Energy Balance Equation

$$\frac{d\theta}{d\tau} = \frac{1}{CF(Z_m,\theta)} \cdot \left[HTN \cdot (\theta_F - \theta) - \sum_{m}^{n_{ind}} DHR_m \cdot \frac{dZ_m}{d\tau} - \frac{d}{d\tau}\left(\frac{W_{sh}}{(N_{tot})_0 \cdot T_0 \cdot \hat{c}_{p0}} \right) \right]$$

(A)

Definitions and Auxiliary Relations

Dimensionless Temperature

$$\theta \equiv \frac{T}{T_0}$$

(B)

Specific Molar Heat Capacity of Reference State

$$\hat{c}_{p0} = \sum_{j}^{all} y_j(0) \cdot \hat{c}_{p_j}(T_0) \quad or \quad \hat{c}_{p0} = \frac{M}{(N_{tot})_0}\bar{c}_p$$

(C)

Dimensionless Heat of Reaction of the m-th Independent Reaction

$$DHR_m \equiv \frac{\Delta H_{R_m}(T_0)}{T_0 \cdot \hat{c}_{p0}}$$

(D)

Dimensionless Heat of Reaction

$$HTN = \frac{U \cdot t_{cr}}{C_0 \cdot \hat{c}_{p0}}\left(\frac{S}{V}\right)$$

(E)

Dimensionless Heat

$$\frac{Q}{(N_{tot})_0 \cdot T_0 \cdot \hat{c}_{p0}}$$

(F)

Correction Factor of Heat Capacity (Gas-Phase Reactions)

$$CF(Z_m,\theta) = \frac{\sum_{j}^{all} y_j(0) \cdot \hat{c}_{p_j}(\theta)}{\sum_{j}^{all} y_j(0) \cdot \hat{c}_{p_j}(1)} + \frac{\sum_{j}^{all}\hat{c}_{p_j}(\theta)\cdot \sum_{m}^{n_{ind}}(s_j)_m \cdot Z_m}{\sum_{j}^{all} y_j(0) \cdot \hat{c}_{p_j}(1)}$$

(G)

tions with multiple reactions. In the final section, we consider non-isothermal operations with multiple reactions.

5.2 ISOTHERMAL OPERATIONS WITH SINGLE REACTIONS

We start the analysis of ideal batch reactors by considering isothermal operations with **single** reactions. Note that isothermal operation is a mathematical condition imposed on the design equation and the energy balance equation. In practice, it rarely occurs because it requires that the heat generated (or consumed) by chemical reactions is identical to the heat removal rate.

When a single chemical reaction takes place in the reactor, the operation is described by a single design equation, and (5.1-1) reduces to

$$\frac{dZ}{d\tau} = \left(\frac{t_{cr}}{C_0}\right) \cdot \left(\frac{V_R(\tau)}{V_R(0)}\right) \cdot r, \qquad (5.2\text{-}1a)$$

whereas (3.5-13) reduces to

$$\frac{dZ}{d\tau} = \left(\frac{r}{r_0}\right) \cdot \left(\frac{V_R(\tau)}{V_R(0)}\right), \qquad (5.2\text{-}1b)$$

where Z is the dimensionless extent of the reaction and r is its rate. For isothermal operations, $d\theta/d\tau = 0$, and we have to solve only the design equation. The energy balance equation provides the heating (or cooling) load necessary to maintain isothermal conditions. Combining (5.1-14) and (5.1-17), the energy balance equation becomes

$$dQ = (N_{tot})_0 \cdot \Delta H_{R_m}(T_0) \cdot dZ. \qquad (5.2\text{-}2)$$

Furthermore, for isothermal operations, the individual reaction rates depend only on the species concentrations, and (5.1-5) reduces to

$$r_i = k_i(T_0) \cdot h_i(C_j's). \qquad (5.2\text{-}3)$$

The solution of the design equation, $Z(\tau)$ versus τ, is the dimensionless operating curve of the reactor. It describes the progression of the chemical reaction with time. Furthermore, once $Z(\tau)$ is known, we can apply stoichiometric relation (5.1-7) to obtain the composition of the individual species at any time τ. Also, note that if one prefers to express the design equation in terms of the actual operating time t, rather than the dimensionless operating time τ, by dividing both sides of (5.2-1a) by t_{cr}, we obtain

$$\frac{dZ}{dt} = \left(\frac{1}{C_0}\right) \cdot \left(\frac{V_R(t)}{V_R(0)}\right) \cdot r. \qquad (5.2\text{-}4)$$

For reactors with single chemical reactions, the common practice has been to express the design equation in terms of the conversion of the limiting reactant A, f_A. We can easily express design equation (5.2-1) in terms of f_A since, for single reactions, the conversion is proportional to the extent. Using stoichiometric relation (1.5-8),

$$Z(t) = -\frac{y_A(0)}{s_A} f_A(t), \qquad (5.2-5)$$

and (5.2-1) becomes

$$\frac{df_A}{d\tau} = \left(\frac{t_{cr}}{C_A(0)}\right) \cdot \left(\frac{V_R(\tau)}{V_R(0)}\right) \cdot (-r_A), \qquad (5.2-6)$$

where $(-r_A) = -s_A \cdot r$, is the depletion rate of reactant A. In this text, we formulate the design of reactors with single reactions in terms of dimensionless extent rather than the conversion because it is a special case of the general formulation. Those who prefer to use the conversion can use (5.2-5) to substitute $f_A(\tau)$ for $Z(\tau)$ in the final formulation.

Note that design equation (5.2-1) has three variables: the operating time, τ, the reaction extent, Z, and the reaction rate, r. The design equation is applied to determine any one of these variables when the other two are provided. A typical design problem is to determine the reactor operating time necessary to obtain a specified extent (or conversion) for a given reaction rate. The second application is to determine the extent (or conversion) achieved in a specified operating time τ for a given reaction rate. The third application is to determine the reaction rate when the extent (or conversion) is provided as a function of operating time.

5.2.1 Constant-Volume Reactors

First, we consider constant-volume batch reactors — reactors whose volumes do not change during the operation, $V_R(\tau) = V_R(0)$. In practice, this condition is satisfied either when the walls of the reactor are stationary (constant-volume reactor) or when the reaction takes place in a liquid phase. In the latter case, the assumption is that the density difference between the reactants and the products is small. For most liquid-phase reactions, the density variations are small.

For constant-volume batch reactors with single reactions, the dimensionless re-action-based design equation, (5.2-1a) and (5.2-1b), reduces to

$$\frac{dZ}{d\tau} = r \cdot \left(\frac{t_{cr}}{C_0}\right) = \frac{r}{r_0}. \qquad (5.2-7)$$

The equation can be solved when the reaction rate, r, is expressed in terms of τ and Z. Eq. (5.2-7) is an initial value problem to be solved for initial value Z(0). In some instances, we can solve it by separating the variables and integrating to obtain

$$\tau = \left(\frac{C_0}{t_{cr}}\right) \int_{Z(0)}^{Z(\tau)} \frac{dZ}{r} = \int_{Z(0)}^{Z(\tau)} \left(\frac{r_0}{r}\right) dZ . \tag{5.2-8}$$

Eq. (5.2-8) is the integral form of the dimensionless, reaction-based design equation for an ideal, constant-volume batch reactor. Note that the value of $Z(0)$ depends on the selection of the reference state, and, when the initial state is taken as the reference state, $Z(0) = 0$. Below, we analyze the operation of constant-volume reactors with single reactions for different types of chemical reactions.

To solve the design equations (5.2-7) or (5.2-8), we have to express the reaction rate, r, in terms of the dimensionless extent, Z. In general, the reaction rate is provided either in the form of experimental data or as an algebraic expression. Consider first a case where the reaction rate is provided in the form of experimental data without a rate expression. Examining the structure of (5.2-8), we see that if we know how r varies with Z, we can then plot (r_0/r) versus Z, and the area under the curve between $Z(0)$ and $Z(\tau)$ is equal to the dimensionless operating time, τ, (see Figure 5-3). Hence, we can solve the design equation numerically by obtaining a relationship between $Z(\tau)$ and τ and plotting the dimensionless operating curve. From this curve, we can determine the necessary τ for any specified extent and calculate the actual operating time by $t = t_{cr} \cdot \tau$.

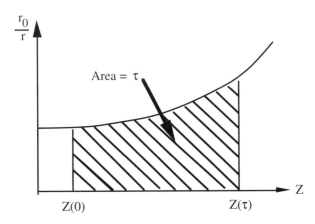

Figure 5-3: Determination of Operating Time from Experimental Rate Data

In many applications, the experimental rate data of a chemical reaction are provided in terms of the depletion rate of reactant A, $(-r_A)$, for different values of the concentration of A, C_A. To obtain a relation between r and Z, we have to relate r to $(-r_A)$ and C_A to Z. Using (2.2-5),

$$r = -\frac{(-r_A)}{s_A} . \tag{5.2-9}$$

Using (5.1-9) for a single reaction,

$$C_A(\tau) = C_0[y_A(0) + s_A \cdot Z(\tau)], \qquad (5.2\text{-}10)$$

and after rearrangement, we obtain

$$Z(\tau) = \frac{1}{s_A} \cdot \left(\frac{C_A(\tau)}{C_0} - y_A(0) \right). \qquad (5.2\text{-}11)$$

Hence, for each given value of C_A, we can calculate the values of r and Z.

Example 5-1 The liquid-phase reaction $2\,A \rightarrow B$ is carried out in an isothermal batch reactor. A solution with initial concentration of $C_A(0) = 7$ mole/liter is charged into the reactor. Based on the experimental rate data below:
a. plot the dimensionless operating curve of the reactor,
b. plot the dimensionless operating curves of each species, and
c. determine the operating time needed to reduce the concentration of A to 1 mole/liter.

C_A (mole/lit)	0.0	1.0	2.0	3.0	4.0	5.0	6.0	7.0	8.0
$(-r_A)$ (mole/lit. min.)	0.0	0.118	0.222	0.316	0.400	0.476	0.546	0.609	0.667

Solution This example illustrates how to construct the dimensionless operating curves of an ideal batch reactor when the reaction rate is provided in the form of data instead of an expression. The stoichiometric coefficients of the reaction are: $s_A = -2$, $s_B = 1$. To utilize all the given data, we select the reference state for the first data point, $C_0 = 8$ mole/liter, and, using (5.1-8), $y_A(0) = C_A(0)/C_0 = 0.875$, $y_B(0) = 0$. Applying (5.2-9), the reaction rate is

$$r = -\frac{(-r_A)}{s_A} = \frac{(-r_A)}{2}, \qquad (a)$$

and we calculate r for each respective C_A. At the reference state, $r_0 = 0.333$ mole/lit. min. Using (2.5-1), the characteristic reaction time is

$$t_{cr} = \frac{C_0}{r_0} = \frac{8.0}{0.333} = 24 \text{ min}. \qquad (b)$$

We use (5.2-11) to calculate Z for each value of C_A,

$$Z = \frac{1}{2} \cdot \left(y_A(0) - \frac{C_A}{8.0} \right). \qquad (c)$$

The calculated values of r and Z are shown in the table below:

C_A (mole/lit)	0.0	1.0	2.0	3.0	4.0	5.0	6.0	7.0	8.0
Z (dimensionless)	.4375	.375	.3125	.250	.1875	.125	.0625	0.0	—
r (mole/lit. min.)	0.0	.059	.111	.158	.200	.238	.273	.305	.333
1/r (lit. min./mole)	∞	17.00	9.00	6.64	5.00	4.20	3.66	3.28	3.00
(r_0/r) (dimensionless)	∞	5.65	3.00	2.11	1.67	1.40	1.22	1.09	1.00

The figure below shows the plot of (r_0/r) versus Z for this reaction.
a. Now that the curve is known, by calculating the area under the curve, we can determine the needed dimensionless operating time τ for any value of $Z(\tau)$ and construct the reaction operating curve. For this case, $Z(0) = 0$ and $Z(\infty) = 0.4375$. We integrate numerically using the trapezoidal method, and the area of the i-th increment is

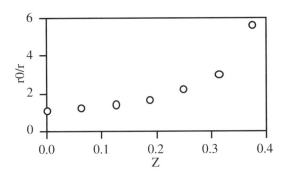

$$\Delta\tau_i = \frac{1}{2}\cdot\left[\left(\frac{r_0}{r}\right)_i + \left(\frac{r_0}{r}\right)_{i-1}\right]\cdot(Z_i - Z_{i-1}).$$

The calculated values of Z and dimensionless operating time τ are shown in the table below:

Z (dimensionless)	0.4375	0.375	0.3125	0.2500	0.187	0.125	0.0625	0.0
τ (dimensionless)	∞	0.798	0.528	0.368	0.154	0.258	0.0722	0.0

The figure below shows the plot of $Z(\tau)$ versus τ — the dimensionless reaction operating curve.

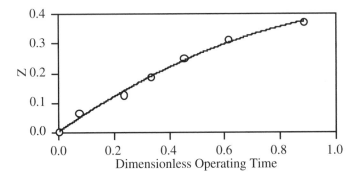

b. To obtain the species operating curves, we use (5.1-7) that reduces to

$$N_j(\tau) = (N_{tot})_0\left[y_j(0) + s_j\cdot Z(\tau)\right].\qquad(d)$$

For the given reaction, we obtain

$$\frac{N_A(\tau)}{(N_{tot})_0} = 0.875 - 2\cdot Z(\tau) \quad\text{and}\quad \frac{N_B(\tau)}{(N_{tot})_0} = Z(\tau).\qquad(e)$$

The figure below shows the species operating curves.

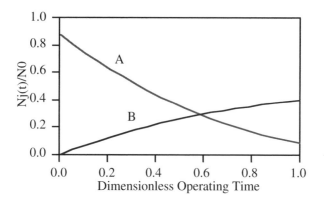

c. From the reaction operating curve (or the corresponding area under the curve in first figure), $Z = 0.375$ is achieved at $\tau = 0.798$. From (b), the reaction characteristic time is 24 min., and the actual operating time is

$$t = t_{cr} \cdot \tau = (24 \text{ min.}) \cdot (0.798) = 19.15 \text{ min.}$$

Next, we consider the application of the design equation when the reaction rate is provided in the form of algebraic expressions. We start with chemical reactions involving a single reactant of the general form A → Products, whose rate expression is of the form $r = k \cdot C_A^{\alpha}$. For these reactions $s_A = -1$, and, if only reactant A is charged into the reactor, we select the initial state as the reference state; hence, $C_0 = C_A(0)$ and $y_A(0) = 1$. Using (5.1-9), the concentration of A is

$$C_A(t) = C_0\left[1 - Z(t)\right]. \tag{5.2-12}$$

Hence, the reaction rate is

$$r = k \cdot C_0^{\alpha}(1 - Z)^{\alpha}. \tag{5.2-13}$$

Using (2.5-1), for α-th order reactions, the characteristic reaction time, t_{cr}, is

$$t_{cr} = \frac{C_0}{k \cdot C_0^{\alpha}} = \frac{1}{k \cdot C_0^{\alpha-1}}, \tag{5.2-14}$$

and the dimensionless operating time, τ, is

$$\tau = \frac{t}{t_{cr}} = k \cdot C_0^{\alpha-1} \cdot t. \tag{5.2-15}$$

Substituting (5.2-13) and (5.2-14) into (5.2-7), the design equation reduces to

$$\frac{dZ}{d\tau} = (1-Z)^{\alpha}. \qquad (5.2\text{-}16)$$

We have to solve differential equation (5.2-16), subject to the initial condition that at τ = 0, $Z(0) = 0$. To obtain an analytical solution, we separate the variables and integrate,

$$\tau = \int_{0}^{Z(\tau)} \frac{dZ}{(1-Z)^{\alpha}}.$$

The solution is

$$Z(\tau) = 1 - e^{-\tau} \qquad \text{for } \alpha = 1 \qquad (5.2\text{-}17a)$$

and

$$Z(\tau) = 1 - \left\{ \frac{1}{1+(\alpha-1)\tau} \right\}^{1/\alpha-1} \qquad \text{for } \alpha \neq 1 \qquad (5.2\text{-}17b)$$

Figure 5-4 shows the reaction operating curve for different values of α. Note that each reaction order is represented by a single curve, independent of the numerical value of the rate constant and the initial reactant concentration. We determine the reaction time needed to achieve a certain extent by reading from the chart the value of τ for the respective order and then calculating the actual operating time t from (5.2-15).

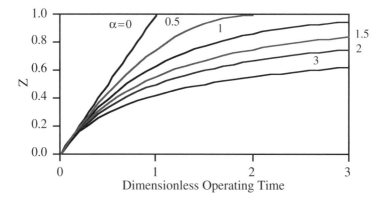

Figure 5-4: Operating Curves of an Ideal Constant-Volume Batch Reactor with a Chemical reaction A → products.

Example 5-2 At 800°K, dimethyl ether (CH_3OCH_3) decomposes according to the first-order, irreversible, gas-phase reaction

$$CH_3OCH_3 \rightarrow CH_4 + CO + H_2.$$

Dimethyl ether is introduced into an evacuated, constant-volume batch reactor, and the initial pressure is 310 mm Hg (absolute). After 780 minutes, the reactor pressure is 490 mm Hg.

a. Derive an expression for the reactor pressure as a function of operating time.

b. Determine the value of the reaction rate constant at 800°K.

Solution The chemical reaction is $A \rightarrow B + C + D$, and its stoichiometric coefficients are:

$$s_A = -1; \quad s_B = 1; \quad s_C = 1; \quad s_D = 1; \quad \text{and} \quad \Delta = 2.$$

We select the initial state as the reference state, and, since only ether is charged into the reactor, $C_0 = C_A(0)$, $y_A(0) = 1$, $y_B(0) = 0$, $y_C(0) = 0$, and $y_D(0) = 0$. The reaction is first-order; hence, $r = k \cdot C_A$, and, for constant-volume reactor, using (5.2-10), $C_A = C_0 \cdot (1 - Z)$, and

$$r = k \cdot C_0 \cdot (1 - Z). \tag{a}$$

Using (5.2-14), for a first-order reaction, the characteristic reaction time is

$$t_{cr} = \frac{1}{k}, \tag{b}$$

and $\tau = k \cdot t$. Substituting (a) and (b) into the dimensionless design equation of constant-volume batch reactor, (5.2-7), we obtain

$$\frac{dZ}{d\tau} = 1 - Z. \tag{c}$$

We solve (c), subject to the initial condition $Z(0) = 0$, and the solution is given by (5.2-17),

$$Z(\tau) = 1 - e^{-\tau}. \tag{d}$$

In this case, we have data for $P(t)$, so we should derive a relation between $P(\tau)$ and $Z(\tau)$. At high temperature and low pressure, we can assume ideal gas behavior,

$$P \cdot V_R = N_{tot} \cdot R \cdot T,$$

and, for a constant-volume isothermal reactor,

$$\frac{P}{P(0)} = \frac{N_{tot}}{N_{tot}(0)}. \tag{e}$$

Using the relation between the total number of moles and the extent, (1.5-4),

$$\frac{P(\tau)}{P(0)} = 1 + \Delta \cdot Z(\tau), \tag{f}$$

which, for the given reaction, can be written as,

$$Z(\tau) = \frac{1}{2}\left(\frac{P(\tau)}{P(0)} - 1\right). \tag{g}$$

a. Substituting (g) into (d), we obtain

$$P(\tau) = P(0) \cdot \left(3 - 2 \cdot e^{-\tau}\right). \tag{h}$$

b. At t = 780 min.,

$$\frac{P(\tau)}{P(0)} = \frac{490}{310} = 1.58,$$

and, solving (h), $\tau = 0.343$. But, $\tau = k \cdot t$; thus,

$$k = \frac{\tau}{t} = \frac{0.343}{(780 \text{ min})} = 4.4 \ 10^{-4} \text{ min}^{-1}.$$

Example 5-3 A biological waste, A, is decomposed by an enzymatic reaction A → B + C in aqueous solution. The rate expression of the reaction is

$$r = \frac{k \cdot C_A}{K_m + C_A}.$$

An aqueous solution with a concentration of 2 g-mole A/liter is charged into a batch reactor. For the enzyme type and concentration used, $k = 0.1$ mole/lit min and $K_m = 4$ mole/lit.
a. Derive and plot the dimensionless operating curve.
b. How long should we operate the reactor to achieve 80% conversion?
Solution This example illustrates how to derive dimensionless operating curves for ideal batch reactors with reactions whose rate expressions are not power functions of the concentrations.
a. The reaction is A → B + C, and the stoichiometric coefficients are $s_A = -1$, $s_B = 1$, and $s_C = 1$. We select the initial state as the reference state, and, since A is the only species charged into the reactor, $C_0 = C_A(0)$ and $y_A(0) = 1$, $y_B(0) = 0$, $y_C(0) = 0$. The dimensionless design equation of a constant-volume batch reactor, (5.2-7), is

$$\frac{dZ}{d\tau} = r \cdot \left(\frac{t_{cr}}{C_0}\right). \tag{a}$$

In this case, the rate expression is

$$r = \frac{k \cdot C_A}{K_m + C_A}, \tag{b}$$

and, using (5.2-10) for a constant-volume batch reactor, $C_A = C_0 (1 - Z)$, and (b) becomes

$$r = k \cdot \frac{1 - Z}{\dfrac{K_m}{C_0} + (1 - Z)} \cdot \tag{c}$$

For rate expressions that are not power functions of the concentrations, we follow the procedure described in Section 3.1: First, we express the concentrations in terms of the dimensionless extent. Then, we arrange the rate expression, $h(C_j's)$, such that it has a dimensional part and a dimensionless part. The dimensional term is then taken as the characteristic rate to be substituted in (2.5-1). In this case, from (c) the dimensional term is k, so the characteristic reaction time is

$$t_{cr} = \frac{C_0}{k} = \frac{(2 \text{ mole/liter})}{(0.1 \text{ mole/lit min})} = 20 \text{ min}, \tag{d}$$

and the dimensionless operating time is

$$\tau = \frac{k}{C_0} \cdot t = \frac{t}{20} \cdot \tag{e}$$

Substituting (c) and (d) into (a), the dimensionless design equation becomes

$$\frac{dZ}{d\tau} = \frac{1 - Z}{\dfrac{K_m}{C_0} + (1 - Z)} \cdot \tag{f}$$

Separating the variables and integrating, noting that $Z(0) = 0$, we obtain

$$\tau = Z(\tau) - \left(\frac{K_m}{C_0} \right) \cdot \ln[1 - Z(\tau)]. \tag{g}$$

The figure below shows the dimensionless reaction operating curve for different values of K_m/C_0.

b. Using stoichiometric relation (5.2-5), the dimensionless extent for $f_A = 0.8$ is

$$Z = -\frac{y_A(0)}{s_A} f_A = -\frac{1}{(-1)} 0.80 = 0.80.$$

We can determine the dimensionless operating time needed for $Z = 0.80$ in two ways: by using the operating curve or by using the design equation. In this case, $K_m/C_0 = 2$, and, from the chart, $Z = 0.80$ is reached at $\tau = 4.0$. Alternatively, we substitute $Z(\tau) = 0.8$ in (g) and find that the exact value is $\tau = 4.019$. We calculate the operating time t from the definition of the dimensionless time, (e),

$$t = \tau \cdot t_{cr} = (4.019) \cdot (20 \text{ min}) = 80.4 \text{ min}.$$

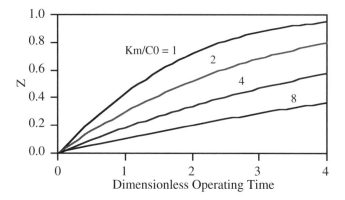

We continue the analysis of ideal, constant-volume batch reactors with single reactions and consider now chemical reactions involving more than one reactant of the general form

$$A + b\,B \;\rightarrow\; \text{Products}. \qquad (5.2\text{-}18)$$

Note that this is a general presentation of reactions between two reactants where $b = s_B / s_A$. The stoichiometric coefficients of the reaction are: $s_A = -1$ and $s_B = -b$. We consider here chemical reactions whose rate expressions are of the form $r = k \cdot C_A{}^{\alpha} \cdot C_B{}^{\beta}$. We have to express C_A and C_B in terms of the dimensionless extent, Z, for constant-volume batch reactor, using (5.2-10)

$$C_A(\tau) = C_0\big[y_A(0) - Z(\tau)\big], \qquad (5.2\text{-}19)$$

$$C_B(\tau) = C_0\big[y_B(0) - b \cdot Z(\tau)\big], \qquad (5.2\text{-}20)$$

where $y_A(0) = C_A(0)/C_0$ and $y_B(0) = C_B(0)/C_0$. Substituting these into the rate expression, we obtain

$$r = k \cdot C_0{}^{\alpha+\beta}\big[y_A(0) - Z\big]^{\alpha} \cdot \big[y_B(0) - b \cdot Z\big]^{\beta}. \qquad (5.2\text{-}21)$$

Using (2.5-1), the characteristic reaction time is

$$t_{cr} = \frac{1}{k \cdot C_0{}^{\alpha+\beta-1}}, \qquad (5.2\text{-}22)$$

and the dimensionless reactor operating time is

$$\tau = k \cdot C_0{}^{\alpha+\beta-1} \cdot t. \qquad (5.2\text{-}23)$$

Substituting (5.2-21) and (5.2-22) into the design equation for a constant-volume reactor, (5.2-7), the dimensionless design equation is

$$\frac{dZ}{d\tau} = [y_A(0) - Z]^{\alpha} \cdot [y_B(0) - b \cdot Z]^{\beta}. \tag{5.2-24}$$

We solve (5.2-24) numerically, subject to the initial condition that at $\tau = 0$, $Z(0) = 0$. Figure 5-5 shows the dimensionless reaction operating curves with reaction (5.2-18) with $b = 1$ and $\alpha = \beta = 1$ for different proportions of the reactants. Note that when the reactants are in stoichiometric proportion, $y_B(0) = b \cdot y_A(0)$, and (5.2-24) reduces to

$$\frac{dZ}{d\tau} = b^{\beta} [y_A(0) - Z]^{\alpha + \beta}. \tag{5.2-25}$$

Note that for all other cases $b \cdot y_A(0)/y_B(0) \leq 1$, since species A is the limiting reactant.

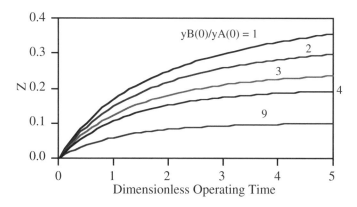

Figure 5-5: Operating Curves for Constant-Volume Batch Reactors with the Reaction A + B → Products and the Rate Expression r = k·C_A·C_B.

Example 5-4 The liquid-phase reaction A + B → C is investigated in an isothermal batch reactor. The reaction is first-order with respect to each reactant. A solution with initial concentrations $C_B(0) = 2 \cdot C_A(0)$ is charged into the reactor, and the concentration of the product, $C_C(t)$, is measured. Based on the data below:
a. derive an expression for the dimensionless reactor operating curve, and
b. determine the reaction rate constant at the operating temperature.

Data:

t (min)	0	0.5	1.0	2.0	4.0	6.0	10	∞
C_C (mole/lit)	0	0.490	0.820	1.24	1.64	1.82	1.95	2.0

Solution The chemical reaction is A + B → C, and its stoichiometric coefficients are: $s_A = -1$; $s_B = -1$; $s_C = 1$; and $\Delta = -1$. We select the initial state as the reference state, and $C_0 = C_A(0) + C_B(0) + C_C(0)$. Since reactants A and B are charged in proportion of 1:2,

$y_A(0) = 0.333$, $y_B(0) = 0.667$, and $y_C(0) = 0$. The rate expression is

$$r = k \cdot C_A \cdot C_B. \tag{a}$$

For a constant-volume reactor, the species concentrations are calculated by (5.2-10),

$$C_A(\tau) = C_0 [0.333 - Z(\tau)] \tag{b}$$

$$C_B(\tau) = C_0 [0.667 - Z(\tau)] \tag{c}$$

$$C_C(\tau) = C_0 \cdot Z(\tau). \tag{d}$$

Substituting (b) and (c) into (a), the reaction rate is

$$r = k \cdot C_0^2 (0.333 - Z) \cdot (0.667 - Z). \tag{e}$$

Using (2.5-1), the characteristic reaction time is

$$t_{cr} = \frac{1}{k \cdot C_0}, \tag{f}$$

and the dimensionless operating time is

$$\tau = k \cdot C_0 \cdot t. \tag{g}$$

Note that C_0 is not known (only the initial proportion of the reactants is given). However, the concentration of species C at completion is provided, $C_C(\infty) = 2$ mole/liter. At completion, $r = 0$ and, from (e), $Z(\infty) = 0.333$. Substituting these values into (d), we obtain $C_0 = 6$ mole/liter. Substituting (e) and (f) into (5-18), the dimensionless design equation is

$$\frac{dZ}{d\tau} = (0.333 - Z) \cdot (0.667 - Z), \tag{h}$$

which should be solved subject to the initial condition that $Z(0) = 0$.
a. We solve (h) by separating the variables and integrating,

$$\tau = 3 \cdot \ln \left(0.5 \frac{0.667 - Z(\tau)}{0.333 - Z(\tau)} \right). \tag{i}$$

To express $Z(\tau)$ explicitly, we rearrange (i)

$$Z(\tau) = \frac{2}{3} \left(\frac{1 - e^{-\tau/3}}{2 - e^{-\tau/3}} \right). \tag{j}$$

b. To determine the value of k from the measure values of C(t), we use design equation (i). We denote the right hand side of (i) by $G[Z(\tau)]$. Noting that $\tau = t/t_{cr}$,

$$t = \frac{1}{t_{cr}} 3 \cdot \ln\left(0.5 \frac{0.667 - Z(\tau)}{0.333 - Z(\tau)}\right), \tag{k}$$

and, by plotting $G[Z(\tau)]$ versus t, we obtain a straight line whose slope is $1/t_{cr}$. Using (d), we determine the values of the dimensionless extent at the given values of t by

$$Z(t) = \frac{C_C(t)}{C_0} = \frac{C_C(t)}{6.0}. \tag{l}$$

The calculated values of $Z(t)$ and $G[Z(t)]$ are provided in the table below:

t (min)	0	0.5	1.0	2.0	4.0	6.0	10.0
C_C (mole/lit)	0	0.490	0.820	1.24	1.65	1.82	1.95
$Z(t)$		0.0	0.0817	0.137	0.207	0.275	0.303	0.325
$G[Z(t)]$		0.0	0.452	0.905	1.81	3.65	5.41	9.19

The plot of $G[Z(t)]$ versus t is shown in the figure below. The slope of the line is 0.916, and, from (e), $k\,C_0 = 0.916$ min^{-1}. Hence, $k = 0.916/6 = 0.153$ lit mole^{-1} min^{-1}.

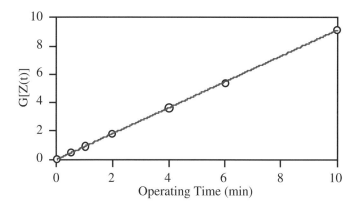

5.2.2 Gaseous, Variable-Volume Reactors

In this section, we consider ideal batch reactors whose volume changes during the operation, like the one shown schematically in Figure 5-2. In general, when a gas-phase reaction results in a net change of the total number of moles, the volume of the reactor changes during the operation if the reactor is maintained at a constant pressure and temperature. For example, the combustion stage of a car engine cycle (while the valves are closed) is carried out in a variable-volume reactor; in fact, the change in the volume is utilized to generate mechanical work. For such operations, we should incorporate the changes in the reactor volume into the design equation.

The variation of the reactor volume with the reaction extent is given by (5.1-11), which for isothermal, isobaric operations with single reactions, reduces to

$$\frac{V_R(t)}{V_R(0)} = 1 + \Delta \cdot Z(t). \tag{5.2-26}$$

Note that (5.2-26) is also readily expressed in terms of the conversion of the limiting reactant, f_A, using (5.2-5),

$$\frac{V_R(t)}{V_R(0)} = 1 + \varepsilon_A \cdot f_A(t), \tag{5.2-27}$$

where ε_A is an expansion factor defined by

$$\varepsilon_A = -\frac{\Delta}{s_A} \cdot \frac{N_A(0)}{(N_{tot})_0}. \tag{5.2-28}$$

The factor ε_A is extensively used in the literature in design formulations for reactors with single chemical reactions, expressed in terms of reactant conversion.

Substituting (5.2-26) into design equation (5.1-1), we obtain

$$\frac{dZ}{d\tau} = r \cdot (1 + \Delta \cdot Z) \cdot \left(\frac{t_{cr}}{C_0}\right) = \left(\frac{r}{r_0}\right) \cdot (1 + \Delta \cdot Z). \tag{5.2-29}$$

Eq. (5.2-29) is the differential dimensionless design equation for isothermal-isobaric, gaseous variable-volume batch reactors. It is solved subject to the initial condition $Z(0) = 0$. In some cases, we can obtain analytical solutions by separating the variables and integrating,

$$\tau = \left(\frac{C_0}{t_{cr}}\right) \int_0^{Z(\tau)} \frac{dZ}{r \cdot (1 + \Delta \cdot Z)} = \int_0^{Z(\tau)} \frac{r_0}{r \cdot (1 + \Delta \cdot Z)} \cdot dZ. \tag{5.2-30}$$

Eqs. (5.2-30) is the integral form of the dimensionless design equation. Note that for the special case when $\Delta = 0$ (no change in the number of moles), (5.2-29) and (5.2-30) reduce, respectively, to (5.2-7) and (5.2-8) — the dimensionless design equations for a constant-volume batch reactor. Note that by multiplying both sides of (5.2-30) by t_{cr}, we express the design equation in terms of the actual operating time (rather than the dimensionless time),

$$t = C_0 \int_0^{Z(t)} \frac{dZ}{r \cdot (1 + \Delta \cdot Z)}. \tag{5.2-31}$$

To solve the design equation, we have to express the reaction rate, r, in terms of Z and, to do so, we relate the species concentrations to the dimensionless extent. From (5.1-12), for isothermal operations with single reactions,

$$C_j(t) = C_0 \frac{y_j(0) + s_j \cdot Z(t)}{1 + \Delta \cdot Z(t)} . \tag{5.2-32}$$

Note that, when $\Delta \neq 0$, the species concentrations are not proportional to the extent. To calculate $Z(t)$ when $C_j(t)$ is given, we rearrange (5.2-32),

$$Z(t) = \frac{C_j(t) - y_j(0) \cdot C_0}{s_j \cdot C_0 - \Delta \cdot C_j(t)} . \tag{5.2-33}$$

Below, we analyze the operation of variable-volume, gaseous batch reactors and describe how to apply the design equation for different cases. We start by considering the design when the reaction rate is given in the form of experimental data. Examining the structure of the design equation, (5.2-30), we see that if we can determine how the reaction rate, r, relates to extent, Z, we can then plot $r_0/r \cdot (1 + \Delta \cdot Z)$ versus Z, and the area under the curve is equal to τ.

Example 5-5 The gas-phase reaction $A \rightarrow B + C$ is being carried out in an isothermal, isobaric, variable-volume batch reactor. The reactor is charged with reactant A, and the initial concentration of $C_A(0) = 0.02$ mole/liter. Based on the experimental rate data provided:
a. plot the dimensionless operating curve, and
b. determine the reactor operating time needed to obtain 40% conversion.
Data: $C_A \cdot 10^2$ (mole/lit)   0.2   0.4   0.6   0.8   1.0   1.2   1.4   1.6   1.8   2.0
      $(-r_A)$ (mole/lit. min.) .143   .222   .273   .308   .333   .353   .368   .381   .391   .400
Solution The reaction is $A \rightarrow B + C$, and the stoichiometric coefficients are:

$$s_A = -1; \quad s_B = 1; \quad s_C = 1; \quad \text{and} \quad \Delta = 1.$$

We select the initial condition as the reference state, and, since only reactant A is charged into the reactor, $C_0 = C_A(0) = 2 \cdot 10^{-2}$ mole/lit, $y_A(0) = 1$, and $y_B(0) = y_C(0) = 0$. Using (5.2-9),

$$r = -\frac{(-r_A)}{(-1)} = (-r_A), \tag{a}$$

we calculate r for each respective $(-r_A)$, and $r_0 = 0.400$ mole/lit. min. Using (2.5-1), the characteristic reaction time is

$$t_{cr} = \frac{C_0}{r_0} = \frac{0.02}{0.400} = 0.05 \text{ min} = 3 \text{ sec}. \tag{b}$$

a. We use (5.2-33) to calculate the corresponding dimensionless extent Z for each value of the C_A,

$$Z = \frac{C_0 - C_A}{C_0 + C_A} . \tag{c}$$

The calculated values are shown in the table below:

$C_A \cdot 10^2$ (mole/lit)	0.2	0.4	0.6	0.8	1.0	1.2	1.4	1.6	1.8	2.0
Z (dimensionless)	.474	.444	.412	.375	.333	.286	.231	.167	0.09	0.00
r (mole/lit. min.)	.143	.222	.273	.300	.333	.353	.368	.381	.391	.400
$r_0/r\,(1+\Delta \cdot Z)$		1.90	1.25	1.04	.944	.900	.881	.883	.900	.939	1.00

The figure below shows the plot of $r_0/r \cdot (1+\Delta \cdot Z)$ versus Z for this reaction.

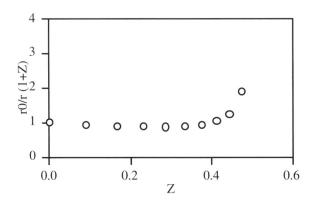

b. Now that the curve is known, we can determine the needed dimensionless operating time τ for any value of $Z(\tau)$ by calculating the area under the curve. We integrate numerically using the trapezoidal method; thus, the area of the i-th increment is

$$\Delta\tau_i = \frac{1}{2}\left\{\left(\frac{r_0}{[r\cdot(1+\Delta\cdot Z)]}\right)_i + \left(\frac{r_0}{[r\cdot(1+\Delta\cdot Z)]}\right)_{i-1}\right\}\cdot(Z_i - Z_{i-1}).$$

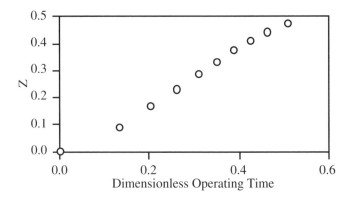

The calculated values of Z and τ are shown in the table below:

Z (dimensionless) 0.474 0.444 0.412 0.375 0.333 0.286 0.231 0.167 0.009 0.0

τ (dimensionless) 0.510 0.463 0.426 0.389 0.351 0.309 0.260 0.203 0.132 0.0

The figure above shows the reaction operating curve (Z versus τ) for the operation.

b. Using (5.2-5), the dimensionless extent for conversion of 40% is Z = 0.40. From the reaction operating curve (or the corresponding area under the curve in the first figure) a dimensionless extent of 0.40 is achieved at $\tau = 0.408$. From (b), the reaction characteristic time is 3 sec; hence

$$t = t_{cr} \cdot \tau = (0.408) \cdot (3 \text{ sec}) = 1.22 \text{ sec}.$$

Next, we discuss the application of the design equation for a variable-volume batch reactor when the reaction rate is provided in the form of an algebraic expression. We start by considering chemical reactions of the form A $\rightarrow$ Products, whose rate expression is $r = k \cdot C_A^\alpha$. For this case, $s_A = -1$, and $y_A(0) = 1$. Using (5.2-32), the rate expression becomes

$$r = k \cdot C_A^\alpha = k \cdot C_0^\alpha \cdot \left\{ \frac{1-Z}{1+\Delta \cdot Z} \right\}^\alpha. \tag{5.2-34}$$

For α-th order reactions, the characteristic reaction time is

$$t_{cr} = \frac{1}{k \cdot C_0^{\alpha-1}}. \tag{5.2-35}$$

Substituting (5.2-34) and (5.2-35) into (5.2-29), we obtain

$$\frac{dZ}{d\tau} = \frac{(1-Z)^\alpha}{(1+\Delta \cdot Z)^{\alpha-1}}. \tag{5.2-36}$$

Eq. (5.2-36) is the differential dimensionless design equation and should be solved subject to the specified initial condition, Z(0). Separating the variables and integrating,

$$\tau = \int_{Z(0)}^{Z(\tau)} \frac{(1+\Delta \cdot Z)^{\alpha-1}}{(1-Z)^\alpha} \cdot dZ. \tag{5.2-37}$$

Eq. (5.2-37) is the integral dimensionless design equation for this case. Figure 5-6 shows the performance curves of a second-order reaction ($\alpha = 2$) for different values of Δ. Note that for larger values of Δ, longer operating times are required for achieving a given conversion. Also note that the curve for $\Delta = 0$ represents the performance of a constant-volume batch reactor.

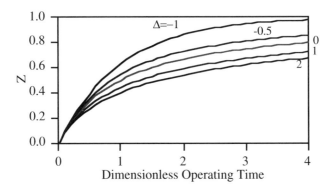

Figure 5-6: Operating Curves of Isothermal-Isobaric Batch Reactor with Second-Order Gaseous Reaction A → Products for Different Δ Values

We continue the analysis of gaseous, variable-volume batch reactors and consider now chemical reactions with two reactants of the general form,

$$A + b\,B \;\rightarrow\; Products, \qquad (5.2\text{-}38)$$

whose rate expression is of the form $r = k \cdot C_A{}^\alpha \cdot C_B{}^\beta$. For this case, $s_A = -1$, $s_B = -b$, and the initial composition, $y_A(0)$ and $y_B(0)$, is specified. Using (5.2-32), the rate expression becomes

$$r = k \cdot C_0{}^{\alpha+\beta} \cdot \frac{[y_A(0) - Z]^\alpha \cdot [y_B(0) - b \cdot Z]^\beta}{(1 + \Delta \cdot Z)^{\alpha+\beta}}. \qquad (5.2\text{-}39)$$

Using (2.5-1), for reactions whose overall order is $\alpha + \beta$, the characteristic reaction time is

$$t_{cr} = \frac{1}{k \cdot C_0{}^{\alpha+\beta-1}}, \qquad (5.2\text{-}40)$$

and the dimensionless operating time is

$$\tau = k \cdot C_0{}^{\alpha+\beta-1} \cdot t.$$

Substituting (5.2-39) and (5.2-40) into (5.2-36), the design equation reduces to

$$\frac{dZ}{d\tau} = \frac{[y_A(0) - Z]^\alpha \cdot [y_B(0) - b \cdot Z]^\beta}{(1 + \Delta \cdot Z)^{\alpha+\beta-1}}. \qquad (5.2\text{-}41)$$

Eq. (5.2-41) is the differential dimensionless design equation for this case and should be solved subject to the initial condition that at $\tau = 0$, $Z = 0$. Separating the variables and integrating, we obtain

$$\tau = \int\limits_{0}^{Z(\tau)} \frac{(1+\Delta \cdot Z)^{\alpha+\beta-1}}{[y_A(0)-Z]^{\alpha} \cdot [y_B(0)-b \cdot Z]^{\beta}} \cdot dZ. \qquad (5.2\text{-}42)$$

Eq. (5.2-42) is the integral dimensionless design equation for this case.

Example 5-6 The gas-phase reaction $A + B \rightarrow C$ is carried out in an isothermal batch reactor operated at 250°C and constant pressure of 2 atm. The reaction is first-order with respect to A and B. A gas mixture consisting of 25% A and 75% B (by mole) is charged into a reactor whose initial volume is 100 liters. At 250°C, the reaction rate constant is k = 0.8 liter/mole min.

a. Plot the dimensionless reaction operating curve.
b. Determine the operating time needed for 80% conversion.
c. Calculate the volume of the reactor at the end of the operation.
d. Calculate the concentration of C at the end of the operation.

Solution The chemical reaction is $A + B \rightarrow C$, and its stoichiometric coefficients are:

$$s_A = -1; \quad s_B = -1; \quad s_C = 1; \quad \Delta = -1.$$

We select the initial state as the reference state; hence, $C_0 = C_A(0) + C_B(0) + C_C(0)$, and $y_A(0) = 0.25$, $y_B(0) = 0.75$, and $y_C(0) = 0$. The rate expression is $r = k \cdot C_A \cdot C_B$, and, using (5.2-32), the species concentrations are

$$C_A = C_0 \cdot \frac{0.25 - Z}{1 - Z} \qquad (a)$$

$$C_B = C_0 \cdot \frac{0.75 - Z}{1 - Z} \qquad (b)$$

$$C_C = C_0 \cdot \frac{Z}{1 - Z}. \qquad (c)$$

Using (a) and (b), the reaction rate expression is

$$r = k \cdot C_0^2 \cdot \frac{(0.25 - Z) \cdot (0.75 - Z)}{(1 - Z)^2}. \qquad (d)$$

Using (2.5-1), for a second-order reaction, the characteristic reaction time is

$$t_{cr} = \frac{1}{k \cdot C_0}, \qquad (f)$$

and the dimensionless operating time is

$$\tau = k \cdot C_0 \cdot t. \qquad (g)$$

Substituting (e) and (f) into (5.2-29), the design equation for this case is

$$\frac{dZ}{d\tau} = \frac{(0.25 - Z) \cdot (0.75 - Z)}{(1 - Z)}. \qquad (h)$$

We solve (h) numerically, subject to the initial condition that at $\tau = 0$, $Z = 0$, and plot the reaction operating curve shown in the figure below.

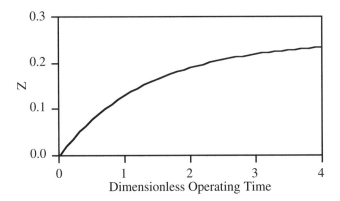

b. To determine the operating time required for 80% conversion, we have to calculate first the corresponding extent. Using (5.2-5),

$$Z = \frac{y_A(0)}{s_A} \cdot f_A = \frac{0.25}{(-1)} \cdot (0.80) = 0.20 .$$

From the operating curve (or the numerical solution of the design equation), $Z = 0.20$ is reached at $\tau = 2.25$. To calculate the actual operating time, we first have to determine the characteristic reaction time. We calculate C_0, which for an ideal gas is

$$C_0 = \frac{P_0}{R \cdot T} = \frac{(2 \text{ atm})}{(82.05 \ 10^{-3} \text{ lit - atm/mole}^\circ\text{K}) \cdot (523^\circ\text{K})} = 4.68 \ 10^{-2} \text{ mole/liter} .$$

Using the definition of the dimensionless time, (f),

$$t = \frac{\tau}{k \cdot C_0} = \frac{(2.25)}{(0.8 \text{ liter/mole min}) \cdot (4.68 \ 10^{-2} \text{ mole/liter})} = 60.4 \text{ min} .$$

c. Using (5.1-11), the reactor volume at the end of the operation is

$$V_R(t) = V_R(0) \cdot [1 + \Delta \cdot Z] = (100 \text{ liter}) \cdot [1 - (0.2)] = 80 \text{ liter} .$$

d. Using (c), the concentration of species C is

$$C_C = C_0 \cdot \frac{Z}{1 - Z} = (4.68 \ 10^{-2}) \cdot \frac{0.20}{1 - 0.20} = 1.17 \ 10^{-2} \text{ mole/liter} .$$

We derived the design equations for gaseous variable-volume batch reactors under two assumptions: (i) all the species are gaseous, and (ii) the mixture behaves as an ideal gas. In some operations, one or more of the species in the reactor may be vapors at saturated conditions. While the ideal gas relation provides a reasonable approxima-tion for the volume of species in the vapor phase, it cannot be applied for their volume in the liquid phase. Below, we modify the design equations for a variable-volume batch reactor with saturated vapors.

Recognizing that, with the exception of operations at very high pressures, the specific volume of a species in the liquid-phase is two to three orders of magnitude smaller than its specific volume in the vapor-phase, the volume of the liquid present in a gaseous reaction can be neglected. Hence, when considering variations in the reactor volume, we should account only for changes in the number of moles in the gas phase, Δ_{gas}. Thus, the factor Δ used in the design equation should be calculated on the basis of the change in the number of moles in the gas phase only. The calculation of the reac-tion factor Δ needs modification, since Δ_{gas} depends on the extent when the two phases are in equilibrium as illustrated in the example below.

Example 5-7 The elementary gas-phase reaction $A + B \rightarrow C$ is carried out in an iso-thermal-isobaric 234.5 liter batch reactor operated at 1.2 atm and 70°C. At this temperature, k = 3.2 lit/(mole min), and the vapor pressure of the product C is 0.3 atm. The reactor is filled with a stoichiometric mixture of A and B; determine:

a. the extent when C starts to condense,
b. the operating time when C starts to condense,
c. the operating time needed for 80% conversion, and
d. plot the performance curve.

Solution At the beginning of the operation, all the species are gaseous, the reaction is

$$A(g) + B(g) \rightarrow C(g),$$

and the stoichiometric coefficients are

$$s_A = -1; \quad s_B = -1; \quad s_C = 1; \quad \text{and} \quad \Delta_{gas} = \Delta = -1.$$

We select the initial state as the reference state, and the initial composition is $y_A(0) = y_B(0) = 0.5$ and $y_C(0) = 0$. Using (5.2-32), the concentration of the species are:

$$C_A = C_B = C_0 \cdot \frac{0.5 - Z}{1 - Z} \tag{a}$$

$$C_C = C_0 \cdot \frac{Z}{1 - Z}. \tag{b}$$

Applying the equation of state for an ideal gas, the initial total number of moles is

$$(N_{tot})_0 = \frac{P_{tot}(0) \cdot V_R}{R \cdot T} = \frac{(1.2 \text{ atm}) \cdot (234.5 \text{ lit})}{(0.08205 \text{ lit - atm/mole}°K) \cdot (343.13°K)} = 10 \text{ mole},$$

and the reference concentration is

$$C_0 = \frac{(N_{tot})_0}{V_R(0)} = \frac{(10 \text{ mole})}{(234.5 \text{ lit})} = 0.0426 \text{ mole/liter}.$$

a. First, we determine the extent of the reaction when C starts to condense. Condensation of C commences when its partial pressure in the reactor is 0.3 atm. But,

$$P_C(t) = y_C(t) \cdot P_{tot}(t) = \frac{N_C(t)}{N_{tot}(t)} \cdot P_{tot}(t).$$

Using the stoichiometric relations (1.5-3) and (1.5-4),

$$P_C(t) = \frac{Z(t)}{1 - Z(t)} \cdot P_{tot}(t) = \frac{Z}{1 - Z} \cdot (1.2) = 0.3. \tag{c}$$

Solving (c), we obtain $Z(t) = 0.20$.
b. To calculate the operating time needed to reach $Z = 0.2$, we use the design equation for an isothermal-isobaric, variable-volume batch reactor, (5.2-29),

$$\frac{dZ}{d\tau} = r \cdot (1 - Z) \cdot \left(\frac{t_{cr}}{C_0}\right). \tag{d}$$

Using (a), the rate expression is

$$r = k \cdot C_A \cdot C_B = k \cdot C_0^2 \cdot \frac{(0.5 - Z)^2}{(1 - Z)^2}. \tag{e}$$

Using (2.5-1), for a second-order reaction, the characteristic reaction time is

$$t_{cr} = \frac{1}{k \cdot C_0} = \frac{1}{(3.2 \text{ lit/mole min}) \cdot (0.0426 \text{ mole/liter})} = 7.34 \text{ min}, \tag{f}$$

and $\tau = k \cdot C_0 \cdot t$. Substituting (e) and (f) into (d), the dimensionless design equation reduces to

$$\frac{dZ}{d\tau} = \frac{(0.5 - Z)^2}{(1 - Z)}. \tag{g}$$

The dimensionless operating time for $Z = 0.20$ is

$$\tau = \int_0^{0.2} \frac{1 - Z}{(0.5 - Z)^2} \cdot dZ = 1.178, \tag{h}$$

and, using (f), the actual operating time is

$$t = \frac{\tau}{k \cdot C_0} = \frac{1.178}{(3.2 \text{ lit/mole min}) \cdot (0.0426 \text{ mole/liter})} = 8.64 \text{ min}.$$

c. After $Z = 0.20$ (40% conversion of A) is reached, product C is formed in a liquid phase, and the following reaction takes place in the reactor

$$A(g) + B(g) \rightarrow C(\text{liq}). \tag{i}$$

For this reaction, Δ_{gas} = -2. Hence, for $Z > 0.20$, a portion of C is formed in the gas phase and a portion in liquid phase (with negligible volume). The total number of moles in the gas phase is now (for $Z > 0.20$)

$$(N_{tot})_{gas} = (N_{tot})_0 \cdot [1 + (-1) \cdot 0.2 + (-2) \cdot (Z - 0.2)] = (N_{tot})_0 \cdot (1.2 - 2 \cdot Z),$$

and

$$\frac{V_R}{V_R(0)} = 1.2 - 2 \cdot Z. \tag{j}$$

Modifying (5.2-29), the design equation is

$$\frac{dZ}{d\tau} = r \cdot (1.2 - 2 \cdot Z) \cdot \left(\frac{t_{cr}}{C_0} \right). \tag{k}$$

Modifying (5.2-32), the concentrations of the two reactants are now (for $Z > 0.2$)

$$C_A = C_B = C_0 \cdot \frac{0.5 - Z}{1.2 - 2 \cdot Z}. \tag{l}$$

Substituting (l) and (f) into (k), the dimensionless design equation for this stage is

$$\frac{dZ}{d\tau} = \frac{(0.5 - Z)^2}{(1.2 - 2 \cdot Z)}. \tag{m}$$

To determine the dimensionless operating time for 80% conversion ($Z = 0.4$), we separate the variables and integrate (numerically) between 0.2 and 0.4,

$$\tau = \int_{0.2}^{0.4} \frac{1.2 - 2 \cdot Z}{(0.5 - Z)^2} \cdot dZ = 3.5305. \tag{n}$$

Using (f), the actual operating time for this stage is

$$t = \frac{\tau}{k \cdot C_0} = \frac{3.5305}{(3.2 \text{ lit/mole min}) \cdot (0.0426 \text{ mole/liter})} = 25.90 \text{ min}.$$

The total operating time needed for 80% conversion is $t = 8.64 + 25.90 = 34.54$ min.

d. The reactor operation is described by two design equations: (g) for $0 \leq Z \leq 0.2$ and

(m) for $Z > 0.2$. The figure below shows the reaction operating curve. Note that in the second stage of the operation, the Δ factor is smaller than in the first stage; consequently, the total operating time is smaller than that calculated by ignoring the condensation of species C. If we integrated (h) between 0 and 0.4, we would have obtained $\tau = 5.61$ which corresponds to a total operating time of 41.15 min.

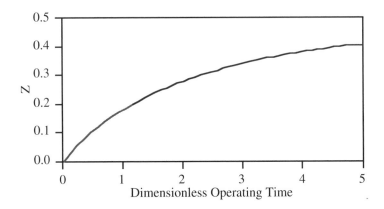

We conclude the discussion on the applications of the design equation of ideal batch reactors with two important comments. First, note that in all the cases discussed above, the design equation provides only the reactor operating time (either dimensionless or dimensional) needed to obtain a given extent level. The design equation does not indicate what reactor size (volume) we should use. Second, what is the best conversion (or extent) we should specify for a given operation? These two issues are determined by other considerations. The reactor size is determined by the required production rate and by the down time between batches (for loading, downloading, and cleaning). The optimal conversion level is determined by the cost of the reactants, the value of the products, the cost of the equipment, and the operating expenses. These issues are discussed in Chapter 9.

5.2.3 Determination of the Reaction Rate Expression

In the preceding two sections, we discussed how to apply the design equation when the rate expression is known and when the rate is provided experimentally. We derived the design equations, plotted the reaction operating curves for constant-volume and variable-volume ideal batch reactors, and determined the operating time needed to reach a specified conversion. In this section, we will describe methods to determine the rate expression from operating data obtained on batch reactors.

The determination of the rate expression is usually carried out in three steps:
* determining the form of the rate expression (function $h(C_j\text{'s})$ such as (2.3-7)), (for

rate expressions in the form of powers of the concentrations, it entails determining the orders of the individual species),
* determining the value of the rate constant, k(T), at a given operating temperature, and
* determining the activation energy.

The determination of the rate constant's parameters (activation energy and the frequency factor) is carried out by determining the rate constant at a series of different temperatures and then applying the procedure described in Section 2-3. Below, we describe methods to determine the rate constant and the orders of individual species from data collected on batch reactors. The determination of other rate expressions and their parameters is covered later.

Determination of the Reaction Rate Constant

The method of determining the reaction rate constant, k, is an extension of applying the design equation for isothermal operations, described in Sections 5.2.1 and 5.2.2. Example 5-4 illustrated how to determine the rate constant from experimental data. Below, we describe the generic method to determine the value of the rate constant, k, at a given temperature.

Consider the general design equation of an ideal batch reactor with a single reaction, (5.2-4). When the rate expression is known, we express r and V_R in terms of Z, separate the variables, and obtain

$$\tau = \left(\frac{C_0}{t_{cr}}\right) \int_0^{Z(\tau)} \left(\frac{1}{r}\right) \left(\frac{V_R(0)}{V_R}\right) \cdot dZ = G[Z(\tau)]. \tag{5.2-43}$$

The right hand side of (5.2-43) is a function of Z(t), denoted by $G[Z(\tau)]$, that depends on the extent, the rate expression, the stoichiometry of the reaction, and the initial proportion of reactants. For rate expressions that are powers of the concentrations, we can readily calculate the values of the function $G[Z(\tau)]$ at different values of τ. For example, for constant-volume reactors with reactions whose stoichiometry is A $\rightarrow$ Products and whose rate expression is of the form $r = k\,C_A^\alpha$, we derive G[Z(t)] from (5.2-17). For $\alpha = 1$,

$$G[Z(t)] = -\ln[1 - Z(t)], \tag{5.2-44a}$$

and for $\alpha \neq 1$,

$$G[Z(t)] = \frac{1}{\alpha - 1}\left\{\left(\frac{Z(t)}{1 - Z(t)}\right)^{\alpha - 1} - 1\right\}. \tag{5.2-44b}$$

Now, recalling the definition of the dimensionless operating time, $\tau = t/t_{cr}$, and, using (5.2-43), by plotting G[Z(t)] versus the operating time, t, we obtain a straight line that passes through the origin, whose slope is $1/t_{cr}$. Once we determine the value of t_{cr} from the slope, we calculate the value of the rate constant, k, using (5.2-15).

Two comments on the method of determining the rate constant may be in order.

First, the main drawback of the method is that the extent, $Z(t)$, is not a measurable quantity. Therefore, to obtain $G[Z(t)]$, we have to express $Z(t)$ in terms of a measurable quantity (species concentration, total pressure, etc.) using the stoichiometric relations. Second, in many cases, when we express $Z(t)$ in terms of another quantity, we may obtain a linear relation of operating time, t, that does not pass through the origin. This is illustrated in Example 5-8 below.

Example 5-8 Di-tert-butyl peroxide decomposes according to the first-order chemical reaction

$$(CH_3)_3COOC(CH_3)_3 \rightarrow C_2H_6 + 2\ CH_3COCH_3.$$

Pure di-tert-butyl peroxide was charged into an isothermal, constant-volume batch reactor operated at 480°C. The progress of the reaction was monitored by measuring the total pressure of the reactor. Based on the data below, determine:
a. the order of the reaction, and
b. the reaction rate constant at 480°C.

Data: time (min.)   0.0   2.5   5.0   10.0   15.0   20.0
     Pressure (mm Hg) 7.5 10.5 12.5 15.8   17.9   19.4

Solution We can describe the reaction as $A \rightarrow B + 2\ C$, whose stoichiometric coefficients are:

$$s_A = -1;\quad s_B = 1;\quad s_C = 2;\quad \text{and}\quad \Delta = 2.$$

We select the initial state as the reference state, and, since only reactant A is charged into the reactor, $y_A(0) = 1$ and $y_B(0) = y_C(0) = 0$.
a. Using the design equation for a constant-volume batch reactor, (5.2-7), and the definition of the dimensionless operating time, $\tau = t/t_{cr}$,

$$\frac{dZ}{d\tau} = \left(\frac{t_{cr}}{C_0}\right) \cdot r. \tag{a}$$

In this case, the rate expression is $r = k \cdot C_A$, and, from (2.5-1), for first-order reactions,

$$t_{cr} = \frac{1}{k}. \tag{b}$$

For a constant-volume batch reactor with a single reaction, using (5.1-9), the concentration is

$$C_A(\tau) = C_0 \cdot [1 - Z(\tau)]. \tag{c}$$

Substituting (c) into the rate expression and the latter into (a) and using (b), the design equation becomes

$$\frac{dZ}{d\tau} = (1 - Z). \tag{d}$$

Separating the variables and integrating, using the initial condition that at $Z(0) = 0$,

$$\tau = \frac{t}{t_{cr}} = -\ln(1-Z). \tag{e}$$

So, to determine k, we have to plot $\ln(1 - Z)$ versus t and obtain a straight line whose slope is -k. In this case, we have data in terms of P(t), so we have to derive a relation between P(t) and Z(t). For an ideal gas, using the stoichiometric relations (1.5-3) and (1.5-4), for a constant-volume reactor

$$\frac{P(t)}{P(0)} = \frac{N_{tot}(t)}{N_{tot}(0)} = 1 + 2 \cdot Z(t); \tag{f}$$

hence,

$$Z(t) = \frac{1}{2}\left(\frac{P(t)}{P(0)} - 1\right) \tag{g}$$

$$1 - Z(t) = \frac{1}{2}\left(3 - \frac{P(t)}{P(0)}\right). \tag{h}$$

Substituting (h) into (e) and using (b), we obtain

$$\ln\left(3 - \frac{P(t)}{P(0)}\right) = \ln 2 - k \cdot t. \tag{i}$$

Hence, by plotting $\ln [3 - P(t)/P(0)]$ versus operating time, t, we obtain a straight line whose slope is -k. We calculate the needed quantities and show the plot in the figure below. From the plot, the slope is:

$$-k = \frac{\ln\ (2.0) - \ln\ (0.5)}{0 - 17.5} = -7.92\ 10^{-2}\,\text{min}^{-1}.$$

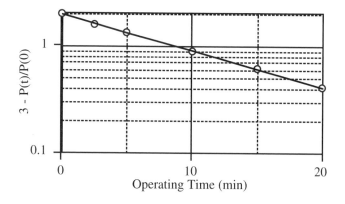

Therefore, the reaction rate constant at 480°C is $k = 7.92 \ 10^{-2} \ \text{min}^{-1}$.

Determination of Species' Orders

Next, we discuss how to determine the orders of individual species for chemical reactions whose rate expressions are powers of the concentrations,

$$r = k \cdot C_A{}^\alpha \cdot C_B{}^\beta. \qquad (5.2\text{-}45)$$

When the reaction rate is determined at different reactant concentrations, we rewrite (5.2-45) as

$$\ln r = \ln k + \alpha \cdot \ln C_A + \beta \cdot \ln C_B, \qquad (5.2\text{-}46)$$

and, using multivariable linear regression, we can calculate the orders of the individual species (α, β, ...). To obtain reliable values for the orders, we need many isothermal data of the reaction rate at different reactant concentrations. In many cases, we want a quick estimate of the orders. To obtain these, we conduct a few experiments while maintaining the concentration of some reactants constant. We write (5.2-45) for two runs and take the ratio between them,

$$\frac{r_1}{r_2} = \frac{k_1}{k_2} \left(\frac{C_{A1}}{C_{A2}} \right)^\alpha \cdot \left(\frac{C_{B1}}{C_{B2}} \right)^\beta. \qquad (5.2\text{-}47)$$

Thus if, for example, the two runs are conducted at the same temperature, ($k_1 = k_2$), and in both of them, C_B is the same ($C_{B1} = C_{B2}$), we can calculate the order of A, α, from

$$\alpha = \frac{\ln\left(\dfrac{r_1}{r_2}\right)}{\ln\left(\dfrac{C_{A1}}{C_{A2}}\right)}. \qquad (5.2\text{-}48)$$

Example 5-9 Kinetic measurements of the reaction $A + B \rightarrow C$ were carried out in an isothermal, constant-volume batch reactor at 80°C. Based on the data below, determine the rate expression.

Data:

Run	C_A	C_B	Rate (mole/min lit.)
		Concentration (mole/lit.)	
1	0.5	0.5	1.30
2	1.0	0.5	2.60
3	0.5	1.0	1.84

Solution Noting that $C_{B1} = C_{B2}$, and, since the temperature is constant, $k_1 = k_2$, we write (5.2-47) for Runs 1 and 2,

$$\frac{1.30}{2.60} = \left(\frac{0.5}{1.0} \right)^\alpha,$$

and find that $\alpha = 1$. Similarly, noting that $C_{A1} = C_{A3}$ and $k_1 = k_3$, we write (5.2-47) for Runs 1 and 3,

$$\frac{1.30}{1.84} = \left(\frac{0.5}{1.0}\right)^\beta,$$

and find that $\beta = 0.5$. The rate expression is $r = k \cdot C_A \cdot C_B^{0.5}$. Now that we know the orders, we can determine the value of the rate constant at 80°C. For each run,

$$k = \frac{r}{C_A^2 \cdot C_B^{0.5}}.$$

The average value of k is 6.13 lit.$^{-1.5}$ mole.$^{-1.5}$ min.$^{-1}$.

The main difficulty in using (5.2-45) and (5.2-46) is that the reaction rate, r, is not a measurable quantity since it is defined in terms of the extent, which itself is not a measurable quantity. Therefore, we have to derive a relationship between the reaction rate, r, and the appropriate measurable quantity. We do so by using the design equation and stoichiometric relations. Assume that we measure the concentration of species j, $C_j(t)$, as a function of time in an isothermal, constant-volume batch reactor. To derive a relation between the reaction rate, r, and $C_j(t)$, we divide both sides of the differential design equation for a constant-volume batch reactor, (5.2-7), by t_{cr} and obtain

$$\frac{dZ}{dt} = \frac{1}{C_0} \cdot r. \tag{5.2-49}$$

From (5.2-12), for single reactions in constant-volume batch reactors,

$$C_j(t) = C_0\left[y_j(0) + s_j \cdot Z(t)\right]. \tag{5.2-50}$$

Differentiating (5.2-50),

$$\frac{dZ}{dt} = \frac{1}{C_0 \cdot s_j} \cdot \frac{dC_j}{dt}, \tag{5.2-51}$$

and substituting it into (5.2-49), we obtain

$$r = \frac{1}{s_j} \cdot \frac{dC_j}{dt}. \tag{5.2-52}$$

Note that, to obtain the reaction rate, r, we differentiate the data with respect to time; hence, this method for determining the orders of the reaction is commonly called the "differential method." Now, combining (5.2-52) with (5.2-46),

$$\ln\left(\frac{1}{s_j}\cdot\frac{dC_j}{dt}\right) = \ln k + \alpha\cdot\ln C_A + \beta\cdot\ln C_B. \tag{5.2-53}$$

Also, note that when species j is a reactant, both dC_j/dt and s_j are negative, but the argument inside the log on the left hand side is positive.

Before we illustrate how to apply (5.2-53), a comment on numerical differentiation is in order. To achieve higher accuracy of the derivatives, we use second-order differentiation relations (see Appendix A). Therefore, when the data points are equally spaced, we calculate the slope of each point, except the two endpoints, using the central differentiation equation. For the first point, we use the backward differentiation equation, and, for the last point, we use the forward differentiation equation. When the points are not equally spaced, we use the central differentiation equation for the midpoint between any two adjacent data points. Hence, for n data points on species concentrations, we obtain (n-1) derivative values for the midpoints concentrations.

Next, we consider the application of (5.2-53) to determine the orders of the various species. We start with an irreversible chemical reaction of the form A → Products, whose rate expression is $r = k\cdot C_A{}^\alpha$. Example 5-10 illustrates the determination of the reaction order, a, when the progress of the reaction is monitored by measuring the reactor pressure.

Example 5-10 Consider the decomposition of di-tert-butyl peroxide in a constant-volume batch reactor, discussed in Example 5-8. Here, we want to use the pressure data to determine the order of the reaction. The chemical reaction is

$$(CH_3)_3COOC(CH_3)_3 \rightarrow C_2H_6 + 2\ CH_3COCH_3,$$

and di-tert-butyl peroxide is charged into an isothermal, constant-volume batch reactor operated at 480°C. Based on the data below, determine the order of the reaction.

Data: Time (min.)   0.0   2.5   5.0   10.0   15.0   20.0
       Pressure (mm Hg)   7.5   10.5   12.5   15.8   17.9   19.4

Solution We can describe the reaction as A → B + 2 C, whose stoichiometric coefficients are:

$$s_A = -1; \quad s_B = 1; \quad s_C = 2; \quad \Delta = 2.$$

Since only reactant A is charged into the reactor, $y_A(0) = 1$ and $y_B(0) = y_C(0) = 0$.
a. From (5.2-49),

$$r = C_0\frac{dZ}{dt}, \tag{a}$$

where the rate expression is $r = k\cdot C_A{}^\alpha$. Using (5.2-12), for a constant-volume batch reactor with a single reaction,

$$C_A(t) = C_0(1 - Z). \tag{b}$$

Substituting (b) into the rate expression and the latter in the design equation,

$$\frac{dZ}{dt} = k \cdot C_0{}^{\alpha-1} \cdot (1-Z)^{\alpha}. \tag{c}$$

Taking the log of both sides,

$$\ln \left(\frac{dZ}{dt} \right) = \ln \left(k \cdot C_0{}^{\alpha-1} \right) + \alpha \cdot \ln (1-Z). \tag{d}$$

So, to determine α, we have to plot $\ln(dZ/dt)$ versus $\ln(1 - Z)$ and obtain α from the slope. In this case, we have data in terms of $P(t)$, so we have to derive a relation between $P(t)$ and $Z(t)$. In Example 5-8, we obtained for this case,

$$\frac{P(t)}{P(0)} = \frac{N_{tot}(t)}{N_{tot}(0)} = 1 + 2 \cdot Z(t), \tag{e}$$

which reduces to

$$Z(t) = \frac{1}{2} \left(\frac{P(t)}{P(0)} - 1 \right). \tag{f}$$

Differentiating (f),

$$\frac{dZ}{dt} = \frac{1}{2 \cdot P(0)} \cdot \frac{dP}{dt}. \tag{g}$$

We substitute (g) and (f) into (d) and obtain

$$\ln \left(\frac{dP}{dt} \right) = (\text{constant}) + \alpha \cdot \ln \left(3 - \frac{P(t)}{P(0)} \right). \tag{h}$$

Thus, we plot $\ln[dP/dt]$ versus $\ln[3 - P(t)/P(0)]$ to determine the value of α. We calculate the derivatives numerically, using second-order forward, center, and backward numerical differentiation formulas, and obtain

	t (min)	P (mmHg)	Δt	dP/dt	3 - P(t)/P(0)
1.	0	7.5		1.40	2.000
2.	2.5	10.5	2.5	1.00	1.600
3.	5.0	12.5	2.5	0.69	1.333
4.	10.0	15.8	5.0	0.54	0.893
5.	15.0	17.9	5.0	0.42	0.613
6.	20.0	19.4	5.0	0.30	0.413

The plot is shown in the figure below. The slope of the line is

$$\alpha = \frac{\ln (1.0) - \ln (0.2)}{\ln (1.6) - \ln (0.31)} = 0.98 \approx 1.0.$$

Note that the points are spread around the line. This is typical in plots used for the determination of reaction orders, and it is due to the inaccuracy introduced by the numerical differentiation. Consequently, it is common to round the value on the order to

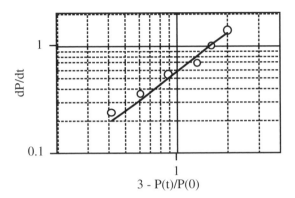

$$\text{dP/dt}$$

$$1$$

$$0.1$$

$$1$$

$$3 - P(t)/P(0)$$

the nearest multiple of 0.5. Once the order of the reaction is known, the value of the reaction rate constant is determined by using the integral form of the design equation, as described in Example 5-8.

We continue the discussion of methods to determine the reaction orders and consider now chemical reactions with more than one reactant, whose rate expression is of the form,

$$r = k \cdot C_A{}^\alpha \cdot C_B{}^\beta.$$

We have to determine now the values of both α and β. Depending which quantity is being measured, we substitute (5.2-52) or an equivalent relation into (5.2-46) and obtain the orders. When many values of the rate, r, at different values of C_A and C_B are available, we use multi-linear regression of (5.2-46) to determine the orders of the individual species. However, when only few measurements can be taken, we have to set up the experiments such that they yield the desired quantities. Usually, we mix the reactants in proportions that are convenient for the determination of the individual orders or the overall orders. Commonly, one of the following mixtures is used:

a. **Stoichiometric Proportion**   When the two reactants are in stoichiometric proportion,

$$C_B(0) = \frac{s_B}{s_A} \cdot C_A(0),$$

it follows from stoichiometric relation (1.2-3) that for constant-volume reactors,

$$C_B(t) = \frac{s_B}{s_A} \cdot C_A(t). \tag{5.2-54}$$

Substituting (5.2-54), (5.2-46) becomes

$$r = k' \cdot C_A^{\alpha+\beta}, \qquad (5.2\text{-}55)$$

where $k' = k \cdot (s_B/s_A)$ is a new constant. When the concentration of species j, $C_j(t)$, is the measured quantity, from (5.2-53),

$$\ln \left(\frac{1}{s_j} \cdot \frac{dC_j}{dt} \right) = \ln \ k' + (\alpha + \beta) \cdot \ln \ C_A. \qquad (5.2\text{-}56)$$

By plotting $\ln[(1/s_A) \cdot (dC_A/dt)]$ versus $\ln C_A$, we obtain a straight line whose slope is $\alpha + \beta$. Hence, by conducting experiments with the reactants in stoichiometric proportion, we obtain the overall order of the reaction. To determine the orders of the individual species, we use one of the other methods below.

    b. **Large excess of one reactant** When $C_B(0) >> (s_B/s_A) \cdot C_A(0)$, say for $s_A = s_B = -1$ $C_B(0) = 10$ $C_A(0)$, the relative change in C_B is small. Note that when limiting reactant A is completely depleted, the concentration of reactant B does not change by more than 10%. Therefore, we can assume that $C_B(t) \approx C_B(0)$, and (5.2-45) becomes

$$r = k'' \cdot C_A^{\alpha}, \qquad (5.2\text{-}57)$$

where $k'' = k \cdot C_B(0)$. When $C_A(t)$ is the measured quantity, we use (5.2-49) and write (5.2-57) as

$$\ln \left(\frac{1}{s_A} \cdot \frac{dC_A}{dt} \right) = \ln \ k'' + \alpha \cdot \ln \ C_A. \qquad (5.2\text{-}58)$$

By plotting $\ln[1/s_A \cdot (dC_A/dt)]$ versus $\ln C_A$, we obtain α. Hence, by conducting experiments with a large excess of one reactant, we obtain the order of the limiting reactant. We determine the individual species orders by repeating the procedure with different mixtures, each time using a different limiting reactant.

    c. **Keeping the concentration of one reactant constant.** Assume we conduct an experiment while maintaining C_B constant. Eq. (5.2-45) becomes

$$r = k''' \cdot C_A^{\alpha} \qquad (5.2\text{-}59)$$

where $k''' = k \cdot C_B$. The method is similar to the one using the excess amount, except that experimentally, this one is more difficult to perform.

    The method for determining the reaction rate constant, described above, can also be used to determine the form of the rate expression (the species orders). The determination of the rate expression goes as follows: first, we assume (hypothesize) the form of the rate expression (first-order, second-order, etc.) and use (5.2-8) to derive the function $G[Z(t)]$. Then, we plot $G[Z(t)]$ versus the operating time, t, to check whether the experimental data fit the function $G[Z(t)]$. When the experimental data fit the derived straight line, we accept the hypothesized rate expression. Then, we use the slope of the line to determine the reaction rate constant. If the experimental data do not fit the straight line for the derived $G[Z(t)]$, we reject the hypothesized rate expression and test a different rate expression. Because the procedure involves integrating the rate expression, it is commonly referred to as the "integral method." Note that each trial is rather long since it involves the integration of the hypothesized rate expression. The procedure is tedious and time-consuming and, therefore, is rarely applied. Hence, we usually apply the "differential method" to determine the orders of the reaction and then use the integral method to determine the value of the reaction rate constant.

Special Methods

a. Initial Rate Method The "differential method" described above is based on calculating the reaction rate from data obtained during the entire operation; therefore, it applies only to irreversible reactions. For reversible reactions, we use a modified differential method — the initial rate method. In this case, a series of experiments are conducted at selected initial reactant compositions, and each run is terminated at low conversion. From the collected data, we calculate (by numerical differentiation) the reaction rate at the initial conditions. Since the reaction extent is low, the reverse reaction is negligible, and we can readily determine the orders of the forward reaction from the known initial compositions. The rate of the reversible reaction is determined by conducting a series of experiments when the reactor is charged with selected initial product compositions. The initial rate method is also used to determine the rates for complex reactions, since it enables us to isolate the effect of different reactants.

b. Fraction-Life Time Method The fraction-life time method is a modification of the integral method. It is useful when we have a means to detect when a <u>certain</u> extent (conversion) level is reached (e.g., Geiger counter). The method is based on measuring the operating time, t_f, needed to reach a specified conversion level for different initial concentrations of the reactant. The time t_f is commonly referred to as the "fraction-life time." The relation between t_f and the initial concentration provides the order of the reaction.

To derive the relation between t_f, the initial concentration, and the reaction order, consider the integral form of the dimensionless design equations for constant-volume and variable-volume batch reactors with an α-th order reaction, (5.2-17) and (5.2-42), respectively. For a given conversion level, $f_A(\tau) = f$, the extent is the same, the right hand side of these equations is constant, and, consequently

$$t_f = t_{cr} \cdot (\text{constant}). \tag{5.2-60}$$

The value of the constant in (5.2-60) depends on the selected value of the extent, Z_f, and, in the case of variable-volume reactors, also on the stoichiometric factor Δ. Substituting the characteristic reaction time, t_{cr}, for α-th order reaction, (2.5-3), into (5.2-60), we obtain

$$t_f = \frac{(\text{constant})}{k \cdot C_0^{\alpha-1}}. \tag{5.2-61}$$

Note that for all orders, except when $\alpha = 1$, the fraction life time depends on C_0. For isothermal operations, k is constant, and (5.2-61) reduces to

$$t_f = \frac{(\text{constant})}{C_0^{\alpha-1}}. \tag{5.2-62}$$

To determine the order of a reaction, we conduct a series of experiments, each with a different C_0, and measure t_f. Modifying (5.2-62),

$$\ln(t_f) = \ln(\text{constant}) - (\alpha - 1) \cdot \ln(C_0). \tag{5.2-63}$$

Hence, by plotting $\ln(t_f)$ versus $\ln[C_0]$, we can readily determine the order of the reaction from the slope of the line. If we conduct only two runs, we can write (5.2-63) for each of them and obtain

$$\alpha - 1 = \frac{\ln\left(\dfrac{t_{f1}}{t_{f2}}\right)}{\ln\left(\dfrac{C_{02}}{C_{01}}\right)}. \tag{5.2-64}$$

For the special case that the conversion is 0.5 (f = 0.5), $t_f = t_{1/2}$, it is called the half-life time. For first-order reactions of the form A $\rightarrow$ Products, (5.2-61) reduces to

$$t_{1/2} = \frac{\ln 2}{k}. \tag{5.2-65}$$

Hence, by indicating the value of the half-life time, $t_{1/2}$, we actually specify the value of the reaction rate constant, k. This is the common way the rate constants of radioactive reactions are reported in the literature.

We can also use the fraction-life time method to determine the activation energy of the chemical reaction. It is done by conducting a series of experiments at different temperatures while maintaining C_0 constant. For this case, (5.2-61) reduces to

$$t_f = \frac{(\text{constant})}{k(T)}, \tag{5.2-66}$$

and we can write it as

$$\ln(t_f) = \ln(\text{constant}) - \ln[k(T)]. \tag{5.2-67}$$

Using the Arrhenius equation, (2.3-3), (5.2-67) becomes

$$\ln(t_f) = \ln(\text{constant}) - \ln(k_0) + \frac{E_a}{R} \cdot \frac{1}{T}. \tag{5.2-68}$$

Hence, by plotting $\ln(t_f)$ versus $1/T$, we obtain a straight line whose slope is E_a/R.

Example 5-11 The gas-phase reaction A $\rightarrow$ 3 R is investigated in a constant-volume batch reactor. An infrared (IR) analyzer is set to detect the time when conversion of

0.201 is reached. Based on the data below, determine:
a. the order of the reaction, and
b. the activation energy.

Data:	Initial pressure	T	f_A	t
Run	(mm Hg)	(°C)		(sec)
1	800	300	0.201	150
2	400	300	0.201	212
3	800	320	0.201	78

Solution The reaction is $A \rightarrow 3\,R$, and the stoichiometric coefficients are: $s_A = -1$ and $s_R = 3$. The rate expression is $r = k \cdot C_A$, and we have to determine the value of n. Since the conversion is the same for all runs, the constant in (5.2-61) is the same. For Runs 1 and 2, $T_1 = T_2$; therefore, $k_1 = k_2$. For an ideal gas,

$$C_0 = \frac{P(0)}{R \cdot T}, \tag{a}$$

and, substituting in (5.2-64), we obtain

$$n - 1 = \frac{\ln\left(\dfrac{t_{f1}}{t_{f2}}\right)}{\ln\left(\dfrac{P(0)_2}{P(0)_1}\right)} = \frac{\ln\left(\dfrac{150}{212}\right)}{\ln\left(\dfrac{400}{800}\right)} = 0.5. \tag{b}$$

Hence, the order of the reaction is n = 1.5.
b. Now that we know the reaction order, we can write (5.2-52) for Runs 1 and 3. Note that since $T_1 \neq T_3$, $C_A(0)$ is not the same in both runs. Using (b), we obtain

$$\frac{k_1}{k_3} = \left(\frac{t_{f3}}{t_{f1}}\right) \cdot \left(\frac{P(0)_3}{P(0)_1} \cdot \frac{T_1}{T_3}\right)^{n-1} = \left(\frac{78}{150}\right) \cdot \left(\frac{300 + 273}{320 + 273}\right)^{0.5} = 0.511. \tag{c}$$

From (2.3-3),

$$\ln\left(\frac{k_1}{k_3}\right) = -\frac{E_a}{R} \cdot \left(\frac{1}{T_1} - \frac{1}{T_3}\right).$$

Hence,

$$\ln\,(0.511) = -\frac{E_a}{R} \cdot \left(\frac{1}{300 + 273} - \frac{1}{320 + 273}\right), \tag{d}$$

and the activation energy is $E_a = 22.66$ kcal/mole.

5.3 ISOTHERMAL OPERATIONS WITH MULTIPLE REACTIONS

When more than one chemical reaction takes place in the reactor, we have to address several issues before we start the design procedure. We have to determine how many independent reactions take place in the reactor and select a set of independent reactions for the design formulation. Next, we have to identify all the reactions that actually take place (including the dependent reactions) and express their rates. To determine the reactor compositions and all other state quantities, we have to write (5.1-1) for each of the independent chemical reactions. To solve the design equations (obtain relationships between Z_m's and τ), we have to express the rates of the individual chemical reactions, r_m's and r_k's, in terms of Z_m's and τ. The procedure for designing batch reactors with multiple reactions goes as follows:

a. Identify all the chemical reactions taking place in the reactor, and define the stoichiometric coefficients of each species in each reaction.
b. Determine the number of independent chemical reactions.
c. Select a set of independent reactions among the reactions whose rate expressions are given.
d. For each dependent reaction, determine its α_{km} multipliers with each of the independent reactions using (1.6-9).
e. Select a reference state (determine T_0, C_0, $V_R(0)$ and the reference species compositions, $y_j(0)$'s.
f. Write design equation (5.1-1) for each independent chemical reaction.
g. Select a leading (or desirable) reaction, and determine the expression form of the characteristic reaction time, t_{cr}, and its numerical value.
h. Express the reaction rates of all chemical reactions in terms of the dimensionless extents of the independent reactions, Z_m's, and the dimensionless temperature, θ.
i. Solve the design equations and the energy balance equation simultaneously for Z_m's and θ as functions of the dimensionless operating time, τ, and obtain the dimensionless reaction operating curves.
j. Calculate the dimensionless species operating curves of all species, using (1.5-3).
k. Determine the actual reactor operating time based on the most desirable value of τ obtained from the dimensionless operating curves.

Below, we describe the design formulation of isothermal batch reactors with multiple reactions for various types of chemical reactions (reversible, series, parallel, etc.). In most cases, we solve the equations numerically by applying a numerical technique such as the Runge-Kutta method, but, in some simple cases, we can obtain analytical solutions. Note that, for isothermal operations, we do not have to consider the effect of temperature variation; therefore, we do not have to solve the design equations simultaneously with the energy balance equation.

We start the analysis with single reversible reactions. When a single reversible reaction takes place, there is only one independent reaction, but we have to consider the rates of both reactions, the forward reaction and the backward reaction. Hence, we have to solve only one design equation to describe the reactor operation, but we have to account for the effect of the reversible reaction on the rate. The design procedure is similar to the one discussed in Section 5.2. To illustrate the effect of the reverse reaction, consider the reversible elementary isomerization reaction $A \leftrightarrow B$ in a constant-volume batch reactor. We treat a reversible reaction as two chemical reactions:

$$\text{Reaction 1} \qquad A \to B$$

$$\text{Reaction 2} \qquad B \to A$$

whose stoichiometric coefficients are

$$s_{A1} = -1, \; s_{B1} = 1, \; \Delta_1 = 0 \quad \text{and} \quad s_{A2} = 1, \; s_{B2} = -1, \; \Delta_2 = 0.$$

We select the forward reaction as the independent reaction and the reverse reaction as the dependent reaction. Hence, the index of the independent reaction is $m = 1$, the index of the dependent reaction is $k = 2$, and, since Reaction 2 is the reverse of Reaction 1, $\alpha_{21} = -1$. Using the design equation (5.1-1), for a constant-volume reactor, it reduces to

$$\frac{dZ_1}{d\tau} = (r_1 - r_2) \cdot \left(\frac{t_{cr}}{C_0} \right). \tag{5.3-1}$$

We define the characteristic reaction time on the basis of the forward reaction (Reaction 1),

$$t_{cr} = \frac{1}{k_1}.$$

For constant-volume reactors, the species concentrations are calculated by (5.1-10), and the respective reaction rates are

$$r_1 = k_1 \cdot C_0 \left[y_A(0) - Z_1 \right]$$

$$r_2 = k_2 \cdot C_0 \left[y_B(0) + Z_1 \right].$$

Substituting these relations into (5.3-1), the design equation reduces to

$$\frac{dZ_1}{d\tau} = \left[y_A(0) - Z_1 \right] - \left(\frac{k_2}{k_1} \right) \cdot \left[y_B(0) + Z_1 \right]. \tag{5.3-2}$$

To simplify the design equation, we rearrange (5.3-2) and notice that, at equilibrium, $dZ_1/d\tau = 0$; hence,

$$Z_{1_{eq}} = \frac{y_A(0) - \left(\dfrac{k_2}{k_1} \right) \cdot y_B(0)}{1 + \left(\dfrac{k_2}{k_1} \right)}, \tag{5.3-3}$$

where $Z_{1_{eq}}$ is the dimensionless extent at equilibrium. Using (5.3-3), the design equation becomes

$$\frac{dZ_1}{d\tau} = Z_{1_{eq}} - \left(\frac{k_1 + k_2}{k_1}\right) \cdot Z_1. \tag{5.3-4}$$

Separating the variables and integrating, we obtain

$$Z(\tau) = Z_{1_{eq}} \cdot (1 - e^{-\left(\frac{k_1 + k_2}{k_1}\right)\tau}). \tag{5.3-5}$$

Figure 5-8 shows the reaction operating curve for different values of k_2/k_1. Note that

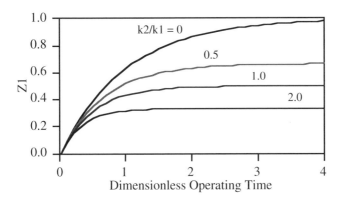

Figure 5-8: Operating Curve of a Batch Reactor with a Single Reversible Reaction.

the design equation for batch reactors with single reversible reactions has two param-eters (k_1 and k_2), whereas the design equation for reactors with an irreversible reaction has only one parameter. Also note that for an irreversible reaction, $k_2 = 0$, and, from (5.3-3), $Z_{1eq} = y_A(0)$; hence, (5.3-4) reduces to the design equation for the irreversible reaction. When the reactor is charged with species B, ($y_A(0) = 0$, $y_B(0) = 1$), $Z_{1eq} = -k_2/(k_1 + k_2)$, and, from (5.3-3), the extent is negative. This indicates that the independent reaction (Reaction 1) proceeds in the opposite direction.

Example 5-12 The reversible gas-phase chemical reaction A $\leftrightarrow$ 2 B takes place in an isothermal, isobaric, ideal batch reactor. The forward reaction is first-order, and the backward reaction is second-order. We want to process a 1,000 mole/min stream of pure A at 120°C and 2 atm and achieve 40% conversion. At the operating conditions (120°C), $k_1 = 0.1$ min^{-1} and $k_2 = 0.322$ lit mole^{-1} min^{-1}.
a. Derive the design equation, and plot the dimensionless reaction operating for a constant-volume batch reactor.
b. What is the equilibrium composition at 120°C in a constant-volume reactor?

c. What is the operating time to obtain 90% of the equilibrium conversion?
d. Derive the design equation, and plot the dimensionless reaction operating for an isobaric, variable-volume batch reactor.
e. What is the equilibrium composition at 120°C in an isobaric variable-volume reactor?
f. What is the operating time to obtain 90% of the equilibrium conversion?

Solution We write the reversible reaction as two separate reactions:

$$\text{Reaction 1} \qquad \text{A} \rightarrow 2\,\text{B} \qquad r_1 = k_1 \cdot C_A$$

$$\text{Reaction 1} \qquad 2\,\text{B} \rightarrow \text{A} \qquad r_2 = k_2 \cdot C_B^2$$

whose stoichiometric coefficients are

$$s_{A1} = -1; \quad s_{B1} = 2; \quad \Delta_1 = 1;$$

$$s_{A2} = 1; \quad s_{B2} = -2; \quad \Delta_2 = -1.$$

We select Reaction 1 as the independent reaction and express the reactor composition in terms of Z_1; hence, the reaction indices are $m = 1$ and $k = 2$, and, for a reversible reaction, $\alpha_{21} = -1$. We select the initial state as the reference state; hence, $Z_1(0) = 0$,

$$C_0 = \frac{P(0)}{R \cdot T(0)} = \frac{(2\ \text{atm})}{(82.06 \cdot 10^{-3}\,\text{lit} - \text{atm/mole}^\circ\text{K})} = 6.20 \cdot 10^{-2}\ \text{mole/lit}, \qquad \text{(a)}$$

and, since only reactant A is charged to the reactor, $y_A(0) = 1$ and $y_B(0) = 0$.

a. For a constant-volume batch reactor, (5.1-1) reduces to

$$\frac{dZ_1}{d\tau} = (r_1 - r_2) \cdot \left(\frac{t_{cr}}{C_0} \right). \qquad \text{(b)}$$

We define the characteristic reaction time on the basis of Reaction 1,

$$t_{cr} = \frac{C_0}{k \cdot C_0} = \frac{1}{k_1} = 10\ \text{min}. \qquad \text{(c)}$$

The amounts of species in the reactor at any time are:

$$N_A = (N_{tot})_0 \cdot \left[y_A(0) + s_{A_1} \cdot Z_1 \right] = (N_{tot})_0 \cdot (1 - Z_1) \qquad \text{(d)}$$

$$N_B = (N_{tot})_0 \cdot \left[y_B(0) + s_{B_1} \cdot Z_1 \right] = (N_{tot})_0 \cdot (2 \cdot Z_1). \qquad \text{(e)}$$

For a constant-volume reactor, the species concentrations are expressed by (5.2-10), and the respective reaction rates are

$$r_1 = k_1 \cdot C_0 \cdot (1 - Z_1) \quad \text{and} \quad r_2 = k_2 \cdot C_0^2 \cdot (2 \cdot Z_1)^2. \qquad \text{(f)}$$

Substituting the rates and (c) into (b), the design equation becomes

$$\frac{dZ_1}{d\tau} = 1 - Z_1 - \left(\frac{k_2 \cdot C_0}{k_1}\right) \cdot 4 \cdot Z_1^2. \tag{g}$$

For the given data, $k_2 \cdot C_0 / k_1 = 0.2$, and we solve (g) subject to the initial condition that $Z_1(0) = 0$. The figure below shows the reaction operating curve.

b. At equilibrium, $dZ_1/d\tau = 0$, and, from (g), the equilibrium extent is $Z_{1eq} = 0.656$. Since there is only one independent reaction, we use (1.5-8) to determine the equilibrium conversion,

$$f_{A_{eq}} = -\frac{s_{A_1}}{y_A(0)} \cdot Z_{1eq} = 0.656.$$

Using (d) and (e), the species molar fractions at equilibrium are

$$y_{A_{eq}} = \frac{1 - Z_{1eq}}{1 + Z_{1eq}} = 0.208 \quad \text{and} \quad y_{B_{eq}} = \frac{2 \cdot Z_{1eq}}{1 + Z_{1eq}} = 0.792.$$

c. A level of 90% equilibrium conversion corresponds to $Z = 0.590$, and, from the operating curve, it is reached at $\tau = 1.26$. Using (c), the operating time is 12.56 min.

d. For an isobaric, variable-volume batch reactor, the design equation (5.1-1) is

$$\frac{dZ_1}{d\tau} = (r_1 - r_2) \cdot (1 + \Delta_1 \cdot Z_1) \cdot \left(\frac{t_{cr}}{C_0}\right). \tag{h}$$

The species concentrations are expressed by (5.1-13), and the reaction rates are

$$r_1 = k_1 \cdot C_0 \left(\frac{1 - Z_1}{1 + Z_1}\right) \quad \text{and} \quad r_2 = k_2 \cdot C_0^2 \left(\frac{2 \cdot Z_1}{1 + Z_1}\right)^2. \tag{i}$$

Substituting the rates and (c) into (h), the design equation becomes

$$\frac{dZ_1}{d\tau} = 1 - Z_1 - \left(\frac{k_2 \cdot C_0}{k_1}\right) \cdot \frac{4 \cdot Z_1^2}{1 + Z_1}. \tag{j}$$

We solve (j) subject to the initial condition that $Z_1(0) = 0$. The figure shows the reaction operating curve.

e. At equilibrium, $dZ_1/d\tau = 0$, and, from (j), the equilibrium extent is $Z_{1eq} = 0.745$. Therefore, the equilibrium extent is $f_{A_{eq}} = 0.745$. Using (d) and (e), the composition at equilibrium is

$$y_{A_{eq}} = \frac{1 - Z_{1eq}}{1 + Z_{1eq}} = 0.147 \quad \text{and} \quad y_{B_{eq}} = \frac{2 \cdot Z_{1eq}}{1 + Z_{1eq}} = 0.853.$$

f. A level of 90% equilibrium conversion corresponds to $Z = 0.671$ and, from the operating curve, it is reached at $\tau = 1.55$; using (c), the operating time is 15.5 min.

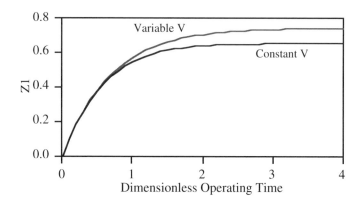

Next, we consider series (consecutive) chemical reactions. These are reactions where the product of one reaction reacts to form an undesirable species. In such cases, it is important to consider the amount of desirable and undesirable products formed in addition to the conversion of the reactant. In many instances, the yield of the desirable product provides a measure of the reactor performance.

Example 5-13 A valuable product B is produced in a batch reactor by the liquid-phase, first-order chemical reaction

$$A \rightarrow 2\,B.$$

At the reactor operating conditions, B decomposes according to the first-order reaction

$$B \rightarrow C + D.$$

The reactor is charged with 200 liter of an organic solution ($C_A(0) = 4$ mole/lit) and is operated at 120°C. At the reactor temperature, $k_1 = 0.05$ min^{-1} and $k_2 = 0.025$ min^{-1}.
a. Derive the design equations, and plot the reaction and species operating curves.
b. Calculate the yield of product B, and plot it as a function of the operating time.
c. Determine the operating time when the production of B is maximized.
d. Determine the conversion of A and the amount of B produced at optimal operating time.
Solution This is an example of series (sequential) chemical reactions. The chemical reactions taking place in the reactor are:

$$\text{Reaction 1:} \quad A \rightarrow 2\,B$$

$$\text{Reaction 2:} \quad B \rightarrow C + D$$

and their stoichiometric coefficients are:

$$s_{A1} = -1; \quad s_{B1} = 2; \quad s_{C1} = 0; \quad s_{D1} = 0; \quad \Delta_1 = 1;$$

$$s_{A2} = 0; \quad s_{B2} = -1; \quad s_{C2} = 1; \quad s_{D2} = 1; \quad \Delta_2 = 1.$$

a. Since each reaction has a species that does not participate in the other, the two reactions are independent, and there is no dependent reaction. We write the dimensionless design equation, (5.1-1), for each of the independent reactions (m = 1, 2),

$$\frac{dZ_1}{d\tau} = r_1 \cdot \left(\frac{t_{cr}}{C_0} \right) \tag{a}$$

$$\frac{dZ_2}{d\tau} = r_2 \cdot \left(\frac{t_{cr}}{C_0} \right). \tag{b}$$

We select the initial state as the reference sate; hence, $Z_1(0) = Z_2(0) = 0$. Since only reactant A is charged into the reactor, $C_0 = C_A(0)$, and $y_A(0) = 1$, and $y_B(0) = y_C(0) = y_D(0) = 0$. Also,

$$(N_{tot})_0 = V_R(0) \cdot C_0 = (200 \text{ lit}) \cdot (4 \text{ mole/lit} = 800 \text{ mole}.$$

Using stoichiometric relation (1.5-8), we express the molar content of each species in the reactor in terms of Z_1 and Z_2,

$$N_A = (N_{tot})_0 \cdot \left(y_A(0) + s_{A_1} \cdot Z_1 + s_{A_2} \cdot Z_2 \right) = (N_{tot})_0 \cdot (1 - Z_1) \tag{c}$$

$$N_B = (N_{tot})_0 \cdot \left(y_B(0) + s_{B_1} \cdot Z_1 + s_{B_2} \cdot Z_2 \right) = (N_{tot})_0 \cdot (2 \cdot Z_1 - Z_2) \tag{d}$$

$$N_C = (N_{tot})_0 \cdot \left(y_C(0) + s_{C_1} \cdot Z_1 + s_{C_2} \cdot Z_2 \right) = (N_{tot})_0 \cdot Z_2 \tag{e}$$

$$N_D = (N_{tot})_0 \cdot \left(y_D(0) + s_{D_1} \cdot Z_1 + s_{D_2} \cdot Z_2 \right) = (N_{tot})_0 \cdot Z_2. \tag{f}$$

For constant-volume batch reactors, the concentration of each species is expressed by (5.1-6) and the rates of the two reactions are

$$r_1 = k_1 \cdot C_0 \cdot (1 - Z_1) \tag{g}$$
$$r_2 = k_2 \cdot C_0 \cdot (2 \cdot Z_1 - Z_2). \tag{h}$$

We select Reaction 1 and define the characteristic reaction time by

$$t_{cr} = \frac{1}{k_1} = 20 \text{ min}. \tag{i}$$

Substituting (g), (h), and (i) into (a) and (b), the two design equations reduce to

$$\frac{dZ_1}{d\tau} = (1 - Z_1) \tag{j}$$

$$\frac{dZ_2}{d\tau} = \left(\frac{k_2}{k_1}\right) \cdot (2 \cdot Z_1 - Z_2). \tag{k}$$

We have to solve (j) and (k) subject to the initial condition that at $\tau = 0$, $Z_1(0) = Z_2(0) = 0$. In this case, the two design equations are not coupled, and we can obtain analytical solutions for (j) and (k). We first solve (j) and obtain,

$$Z_1(\tau) = 1 - e^{-\tau}. \tag{l}$$

We substitute (l) into (k), and the second design equation reduces to

$$\frac{dZ_2}{d\tau} + \left(\frac{k_2}{k_1}\right) \cdot Z_2 = 2 \cdot \left(\frac{k_2}{k_1}\right) \cdot (1 - e^{-\tau}). \tag{m}$$

This is a first-order linear differential equation in Z_2 that can be solved by using an integrating factor. Using the initial condition that at $\tau = 0$, $Z_2(0) = 0$, we obtain

$$Z_2(\tau) = 2 \cdot \left[1 - \left(\frac{k_1}{k_1 - k_2}\right) \cdot e^{-\frac{k_2}{k_1} \cdot \tau} + \left(\frac{k_2}{k_1 - k_2}\right) \cdot e^{-\tau} \right]. \tag{n}$$

a. For the given data, $k_2/k_1 = 0.5$, we can plot $Z_1(\tau)$ and $Z_2(\tau)$, the operating curves of the two independent reactions (see figure below). Once we have $Z_1(\tau)$ and $Z_2(\tau)$, we use (c) through (f) to calculate the dimensionless species contents, $N_j(\tau)/(N_{tot})_0$, as a function of the dimensionless space time. The figures below show these curves.

b. The yield is defined by (1.8-3). In this case, the desirable chemical reaction is Reaction 1, and the desirable product is B; hence

$$\eta_B(\tau) = -\left(\frac{s_{A_1}}{s_{V_1}}\right) \cdot \left(\frac{N_B(\tau) - N_B(0)}{N_A(0)}\right),$$

and, using (d),

$$\eta_B(\tau) = \frac{1}{2} \cdot [2 \cdot Z_1(\tau) - Z_2(\tau)] = Z_1(\tau) - 0.5 \cdot Z_2(\tau). \tag{o}$$

The plot of the yield is shown in the figure below.

c. To determine the operating time that provides the highest production of product B, we use either the species' operating curves or the analytical solutions. From the graph above, we determine that maximum N_B is reached at $\tau = 1.4$. Alternatively, we use (c), (l), and (n) to obtain

$$\frac{N_B(\tau)}{N_{tot}(0)} = 2 \cdot (1 - e^{-\tau}) - 2 \cdot \left[1 - \left(\frac{k_1}{k_1 - k_2}\right) \cdot e^{-\frac{k_2}{k_1} \cdot \tau} + \left(\frac{k_2}{k_1 - k_2}\right) \cdot e^{-\tau} \right]. \tag{p}$$

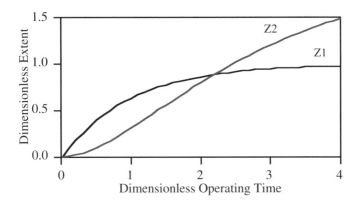

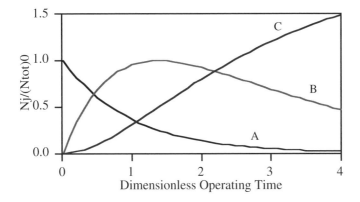

To find the maximum, we take the derivative of (p) with respect to τ and equate it to zero and obtain

$$\tau_{maxB} = \left(\frac{k_1}{k_1 - k_2} \right) \cdot \ln \left(\frac{k_1}{k_1 - k_2} \right). \qquad (q)$$

For $k_2/k_1 = 0.5$, $\tau_{maxB} = 1.38$. Using (i), the optimal operating time is

$$t = \tau \cdot t_{cr} = (1.38)\,(20\ min) = 27.6\ min.$$

d. For $k_2/k_1 = 0.5$ and $\tau = 1.38$, the solutions of (l) and (n) are $Z_1 = 0.748$ and $Z_2 = 0.497$. Using (c) through (f), the species amounts at $\tau = 1.39$ are:

$$N_A = (800) \cdot (1 - 0.748) = 201.3\ mole$$

$$N_B = (400) \cdot (2 \cdot 0.753 - 0.497) = 799.9\ mole$$

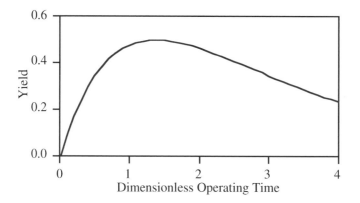

$$N_C = (800) \cdot 0.507 = 397.5 \text{ mole}$$

$$N_D = (800) \cdot 0.507 = 397.5 \text{ mole.}$$

The conversion of reactant A is

$$f_A = \frac{400 - 98.8}{400} = 0.748.$$

Next, we consider parallel chemical reactions. These are reactions in which a reactant of the desirable reaction also reacts in another reaction to form an undesirable species. In such cases, it is important to consider not only the conversion of the reactant but also the amount of desirable and undesirable products formed. The yield of the desirable product provides a measure of the reactor performance.

Example 5-14 The catalytic oxidation of propylene on bismuth molybdate to produce acrolein is investigated in a constant-volume (spinning basket) batch reactor. The following chemical reactions take place in the reactor

$$C_3H_6 + O_2 \rightarrow C_3H_4O + H_2O$$

$$C_3H_4O + 3.5 O_2 \rightarrow 3 CO_2 + 2 H_2O$$

$$C_3H_6 + 4.5 O_2 \rightarrow 3 CO_2 + 3 H_2O.$$

These rate of each reaction is second-order (first-order with respect to each reactant), and their constants at 460°C are: $k_1 = 0.5$ liter mole^{-1} sec^{-1}, $k_2/k_1 = 0.25$, $k_3/k_1 = 0.10$. A gas mixture consisting of 60% propylene and 40% oxygen is charged into a 4 liter isothermal reactor operated at 460°C. The initial pressure is 1.2 atm.
a. Derive the design equations, and plot the reaction and species operating curves.
b. Determine the operating time needed for 75% conversion of oxygen.

c. What is the reactor composition at 75% conversion of oxygen?

Solution The reactor design formulation of these chemical reactions was discussed in Example 3-2. Here, we complete the design for an isothermal batch reactor and obtain the dimensionless reaction and species operating curves. Recall that there are two independent reactions and one dependent reaction, and, following the heuristic rule, we select a set of three reactions from the given reactions. We select Reactions 1 and 2 as a set of independent reactions; hence, $m = 1, 2$, and $k = 3$, and we express the design equations in terms of Z_1 and Z_2. The stoichiometric coefficients of the independent reactions are:

$$(s_{C3H6})_1 = -1; \ (s_{O2})_1 = -1; \ (s_{C3H4O})_1 = 1; \ (s_{H2O})_1 = 1; \ (s_{CO2})_1 = 0; \ \text{and} \ \Delta_1 = 0;$$

$$(s_{C3H6})_2 = 0; \ (s_{O2})_2 = -3.5; \ (s_{C3H4O})_2 = -1; \ (s_{H2O})_2 = 2; \ (s_{CO2})_2 = 3; \ \text{and} \ \Delta_2 = 0.5.$$

Recall from Example 3-2 that the multipliers a_{km}'s of the dependent reaction (Reaction 3) and the two independent reactions (Reactions 1 and 2) are $\alpha_{31} = 1$ and $\alpha_{32} = 1$ (Reaction 3 is the sum of Reaction 1 and Reaction 2). We select the initial state as the reference state; hence, $Z_1(0) = Z_2(0) = 0$ and $y_{C3H6}(0) = 0.6$, $y_{O2}(0) = 0.40$, $y_{C3H4O}(0) = y_{H2O}(0) = y_{CO2}(0) = 0$. Also,

$$C_0 = \frac{P(0)}{R \cdot T_0} = \frac{(1.2 \ \text{atm})}{(0.08205 \ \text{lit - atm/mole}^\circ\text{K}) \cdot (733^\circ\text{K})} = 2.00 \cdot 10^{-2} \ \text{mole/lit} \quad \text{(a)}$$

$$(N_{tot})_0 = V_R(0) \cdot C_0 = (4 \ \text{liter}) \cdot (2.00 \cdot 10^{-2} \ \text{mole/lit.}) = 0.08 \ \text{mole.}$$

a. To design a constant-volume batch reactor, we write design equation (5.1-1) with $V_R(\tau) = V_R(0)$ for each of the independent reactions,

$$\frac{dZ_1}{d\tau} = (r_1 + r_3) \cdot \left(\frac{t_{cr}}{C_0}\right) \quad \text{(b)}$$

$$\frac{dZ_2}{d\tau} = (r_2 + r_3) \cdot \left(\frac{t_{cr}}{C_0}\right). \quad \text{(c)}$$

Using stoichiometric relation (1.5-3), the amount of each species in the reactor at any instant is

$$N_{C3H6}(\tau) = (N_{tot})_0 \cdot \left[y_{C3H6}(0) - Z_1(\tau)\right] = (N_{tot})_0 \cdot \left[0.6 - Z_1(\tau)\right] \quad \text{(d)}$$

$$N_{O2}(\tau) = (N_{tot})_0 \cdot [y_{O2}(0) - Z_1(\tau) - 3.5 \cdot Z_2(\tau)] = (N_{tot})_0 [0.4 - Z_1(\tau) - 3.5 \cdot Z_2(\tau)] \quad \text{(e)}$$

$$N_{C3H4O}(\tau) = (N_{tot})_0 [y_{C3H4O}(0) + Z_1(\tau) - Z_2(\tau)] = (N_{tot})_0 [Z_1(\tau) - Z_2(\tau)] \quad \text{(f)}$$

$$N_{H2O}(\tau) = (N_{tot})_0 [y_{H2O}(0) + Z_1(\tau) + 2 \cdot Z_2(\tau)] = (N_{tot})_0 [Z_1(\tau) + 2 \cdot Z_2(\tau)] \quad \text{(g)}$$

$$N_{CO2}(\tau) = (N_{tot})_0 [y_{CO2}(0) + 3 \cdot Z_2(\tau)] = (N_{tot})_0 \cdot 3 \cdot Z_2(\tau). \quad \text{(h)}$$

Using (1.5-4), the total molar content in the reactor is

$$N_{tot}(\tau) = (N_{tot})_0 [1 + 0.5 \cdot Z_2(\tau)]. \tag{i}$$

For constant-volume batch reactors, we express the species concentrations by (5.1-9), and the rates of the three chemical reactions are:

$$r_1 = k_1 \cdot C_0^2 (0.6 - Z_1) \cdot (0.4 - Z_1 - 3.5 \cdot Z_2) \tag{j}$$

$$r_2 = k_2 \cdot C_0^2 (Z_1 - Z_2) \cdot (0.4 - Z_1 - 3.5 \cdot Z_2) \tag{k}$$

$$r_3 = k_3 \cdot C_0^2 (0.6 - Z_1) \cdot (0.4 - Z_1 - 3.5 \cdot Z_2). \tag{l}$$

We select Reaction 1 as the leading reaction, and, using (2.5-1), the characteristic reaction time is

$$t_{cr} = \frac{C_0}{k_1 \cdot C_0^2} = \frac{1}{k_1 \cdot C_0} = 100 \text{ sec}. \tag{m}$$

Substituting (j), (k), (l), and (m) into (b) and (c), the design equations become

$$\frac{dZ_1}{d\tau} = \left(1 + \frac{k_3}{k_1}\right) \cdot (0.6 - Z_1) \cdot (0.4 - Z_1 - 3.5 \cdot Z_2) \tag{n}$$

$$\frac{dZ_2}{d\tau} = \left(\frac{k_2}{k_1}\right) \cdot (Z_1 - Z_2) \cdot (0.4 - Z_1 - 3.5 \cdot Z_2) +$$

$$+ \left(\frac{k_3}{k_1}\right) \cdot (0.6 - Z_1) \cdot (0.4 - Z_1 - 3.5 \cdot Z_2). \tag{r}$$

Substituting the numerical values $k_2/k_1 = 0.25$ and $k_3/k_1 = 0.1$ into (n) and (r), we solve them numerically for Z_1 and Z_2, subject to the initial conditions that at $\tau = 0$, $Z_1 = Z_2 = 0$. The first figure below shows the operating curves of the two independent reactions. Using the solutions of Z_1 and Z_2, we apply (d) through (h) to calculate the dimension-

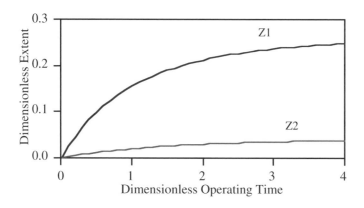

less molar amounts of the species as a function of τ. The second figure shows the operating curves of all the species in the reactor.

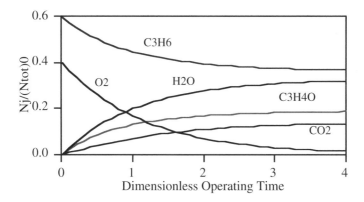

b. Using the definition of the conversion, at 75% conversion of oxygen,

$$N_{O2} = (1 - 0.75)\, N_{O2}(0) = (1 - 0.75)\, 0.4\, (N_{tot})_0;$$

hence, $N_{O2}/(N_{tot})_0 = 0.1$. From the species operating curve (or tabulated values of the numerical solution), $N_{O2}/(N_{tot})_0 = 0.1$ is reached at $\tau = 1.78$. Using the definition of the dimensionless operating time and (m), the operating time is

$$t = \tau \cdot t_{cr} = (1.78)\,(100\ \text{sec}) = 178\ \text{sec}.$$

c. From the reaction operating curves at $\tau = 1.78$, $Z_1 = 0.204$ and $Z_2 = 0.0275$. Using (d) through (h) and (i), the mole fraction of the individual species are: $y_{C3H6} = 0.391$, $y_{O2} = 0.099$, $y_{C3H4O} = 0.174$, $y_{H2O} = 0.255$, and $y_{CO2} = 0.081$.

Example 5-15 Selective oxidation of ammonia is investigated in an isothermal constant-volume batch reactor. The following gaseous chemical reactions take place in the reactor:

Reaction 1	$4\,NH_3 + 5\,O_2 \rightarrow 4\,NO + 6\,H_2O$	$r_1 = k_1 \cdot C_{NH3} \cdot C_{O2}$
Reaction 2	$4\,NH_3 + 3\,O_2 \rightarrow 2\,N_2 + 6\,H_2O$	$r_2 = k_2 \cdot C_{NH3} \cdot C_{O2}$
Reaction 3	$2\,NO + O_2 \rightarrow 2\,NO_2$	$r_3 = k_3 \cdot C_{O2} \cdot C_{NO}$
Reaction 4	$4\,NH_3 + 6\,NO \rightarrow 5\,N_2 + 6\,H_2O$	$r_4 = k_4 \cdot C_{NH3} \cdot C_{NO}$

NO is the desired product. A gas mixture consisting of 50% NH_3 and 50% O_2 (mole %) is charged into a 4 liter reactor, and the initial pressure is 2 atm The reactor operates at 609°K. Based on the rate data below,
a. derive the design equations, and plot the reaction and species operating curves,
b. determine the operating time at which the maximum amount of NO is reached, and
c. calculate the reactor composition at that time.

Data: At 609°K, $k_1 = 20$ (lit/mole)^2 min^{-1}; $k_2 = 0.04$ (lit/mole) min^{-1};
$k_3 = 40$ (lit/mole)^2 min^{-1}; $k_4 = 0.0274$ (lit/mole)$^{0.667}$ min^{-1}.

Solution The reactor design formulation of these chemical reactions was discussed in Example 3-3. Here, we complete the design for an isothermal ideal batch reactor and obtain the dimensionless reaction and species operating curves. Recall that there are three independent reactions and one dependent reaction, and, following the heuristic rule, we select a set of three reactions from the given reactions. We select Reactions 1, 2, and 3 as a set of independent reactions; hence, $m = 1, 2, 3, k = 4$, and we express the design equations in terms of Z_1, Z_2, and Z_3. The stoichiometric coefficients of the independent reactions are:

$(s_{NH3})_1 = -4$; $(s_{O2})_1 = -5$; $(s_{NO})_1 = 4$; $(s_{H2O})_1 = 6$; $(s_{N2})_1 = 0$; $(s_{NO2})_1 = 0$; $\Delta_1 = 1$;

$(s_{NH3})_2 = -4$; $(s_{O2})_2 = -3$; $(s_{NO})_2 = 0$; $(s_{H2O})_2 = 6$; $s_{N2})_2 = 2$; $(s_{NO2})_2 = 0$; $\Delta_2 = 1$;

$(s_{NH3})_3 = 0$; $(s_{O2})_3 = -1$; $(s_{NO})_3 = -2$; $(s_{H2O})_3 = 0$; $(s_{N2})_3 = 0$; $(s_{NO2})_3 = 2$; $\Delta_3 = -1$.

Recall from Example 3-2 that the multipliers α_{km}'s of the dependent reaction (Reaction 4) and the three independent reactions are: $\alpha_{43} = 0$, $\alpha_{42} = 2.5$, and $\alpha_{41} = -1.5$. We select the initial mixture as the reference state; hence, $Z_1(0) = Z_2(0) = Z_3(0) = 0$, and the reference concentration is

$$C_0 = \frac{P(0)}{R \cdot T_0} = \frac{(2 \text{ atm})}{(0.08205 \text{ lit - atm/mole}°K) \cdot (609°K)} = 0.04 \text{ mole/liter} . \qquad (a)$$

For the selected reference state, $y_{NH3}(0) = 0.5$, $y_{O2}(0) = 0.5$, $y_{NO}(0) = y_{H2O}(0) = y_{N2}(0) = y_{NO2}(0) = 0$. The total mole content of the reference state is

$$(N_{tot})_0 = V_R(0) \cdot C_0 = (4 \text{ lit}) \cdot (0.04 \text{ mole/liter}) = 0.16 \text{ mole.}$$

Using stoichiometric relation (1.5-3), we express the molar content of each species at any instant in the reactor in terms of Z_1, Z_2, and Z_3,

$$N_{NH3} = (N_{tot})_0 (y_{NH3}(0) - 4 \cdot Z_1 - 4 \cdot Z_2) \qquad (b)$$

$$N_{O2} = (N_{tot})_0 (y_{O2}(0) - 5 \cdot Z_1 - 3 \cdot Z_2 - Z_3) \qquad (c)$$

$$N_{NO} = (N_{tot})_0 (y_{NO}(0) + 4 \cdot Z_1 - 2 \cdot Z_3) \qquad (d)$$

$$N_{H2O} = (N_{tot})_0 (y_{H2O}(0) + 6 \cdot Z_1 + 6 \cdot Z_2) \qquad (e)$$

$$N_{N2} = (N_{tot})_0 (y_{N2}(0) + 2 \cdot Z_2) \qquad (f)$$

$$N_{NO2} = (N_{tot})_0 (y_{NO2}(0) + 2 \cdot Z_3). \qquad (g)$$

Using (1.5-4), the total molar content in the reactor is

$$N_{tot} = (N_{tot})_0 (1 + Z_1 + Z_2 - Z_3). \qquad (h)$$

a. To design a constant-volume batch reactor, we write the design equation (5.1-1) with $V_R(0) = V_R(0)$, for each of the independent chemical reactions,

$$\frac{dZ_1}{d\tau} = (r_1 - 1.5 \cdot r_4) \cdot \left(\frac{t_{cr}}{C_0} \right) \tag{i}$$

$$\frac{dZ_2}{d\tau} = (r_2 + 2.5 \cdot r_4) \cdot \left(\frac{t_{cr}}{C_0} \right) \tag{j}$$

$$\frac{dZ_3}{d\tau} = r_3 \cdot \left(\frac{t_{cr}}{C_0} \right). \tag{k}$$

For constant-volume batch reactors, we express the species concentrations by (5.1-9), and the rates of the four chemical reactions, expressed in terms of dimensionless extents, are:

$$r_1 = k_1 \cdot C_0^3 \cdot [y_{NH_3}(0) - 4 \cdot Z_1 - 4 \cdot Z_2] \cdot [y_{O_2}(0) - 5 \cdot Z_1 - 3 \cdot Z_2 - Z_3]^2 \tag{l}$$

$$r_2 = k_2 \cdot C_0^2 \cdot [y_{NH_3}(0) - 4 \cdot Z_1 - 4 \cdot Z_2] \cdot [y_{O_2}(0) - 5 \cdot Z_1 - 3 \cdot Z_2 - Z_3] \tag{m}$$

$$r_3 = k_3 \cdot C_0^3 \cdot [y_{O_2}(0) - 5 \cdot Z_1 - 3 \cdot Z_2 - Z_3] \cdot [y_{NO}(0) + 4 \cdot Z_1 - 2 \cdot Z_3]^2 \tag{n}$$

$$r_4 = k_4 \cdot C_0^{5/3} \cdot [y_{NH_3}(0) - 4 \cdot Z_1 - 4 \cdot Z_2]^{2/3} \cdot [y_{NO}(0) + 4 \cdot Z_1 - 2 \cdot Z_3]. \tag{o}$$

We select Reaction 1, the leading reaction, as the reference reaction and define the characteristic reaction time by

$$t_{cr} = \frac{C_0}{k_1 \cdot C_0^3} = \frac{1}{k_1 \cdot C_0^2} = \frac{1}{(20) \cdot (0.04)^2} = 31.25 \text{ min}. \tag{p}$$

Substituting (l) through (m) and (p) into (a), (b), and (c), the design equations reduce to

$$\frac{dZ_1}{d\tau} = [y_{NH_3}(0) - 4 \cdot Z_1 - 4 \cdot Z_2] \cdot [y_{O_2}(0) - 5 \cdot Z_1 - 3 \cdot Z_2 - Z_3]^2 -$$
$$-1.5 \cdot \left(\frac{k_4}{k_1 \cdot C_0^{4/3}} \right) \cdot [y_{NH_3}(0) - 4 \cdot Z_1 - 4 \cdot Z_2]^{2/3} \cdot [y_{NO}(0) + 4 \cdot Z_1 - 2 \cdot Z_3] \tag{q}$$

$$\frac{dZ_2}{d\tau} = \left(\frac{k_2}{k_1 \cdot C_0} \right) \cdot [y_{NH_3}(0) - 4 \cdot Z_1 - 4 \cdot Z_2] \cdot [y_{O_2}(0) - 5 \cdot Z_1 - 3 \cdot Z_2 - Z_3] +$$
$$+2.5 \cdot \left(\frac{k_4}{k_1 \cdot C_0^{4/3}} \right) \cdot [y_{NH_3}(0) - 4 \cdot Z_1 - 4 \cdot Z_2]^{2/3} \cdot [y_{NO}(0) + 4 \cdot Z_1 - 2 \cdot Z_3] \tag{r}$$

$$\frac{dZ_3}{d\tau} = \left(\frac{k_3}{k_1}\right) \cdot [y_{O_2}(0) - 5 \cdot Z_1 - 3 \cdot Z_2 - Z_3] \cdot [y_{NO}(0) + 4 \cdot Z_1 - 2 \cdot Z_3]^2. \qquad (s)$$

The parameters of the dimensionless design equations are

$$\left(\frac{k_2}{k_1 \cdot C_0}\right) = 0.05; \quad \left(\frac{k_3}{k_1}\right) = 2; \quad \left(\frac{k_4}{k_1 \cdot C_0^{4/3}}\right) = 0.1.$$

We substitute these values into (q), (r), and (s), and solve them numerically for Z_1, Z_2, and Z_3, subject to the initial conditions that at $\tau = 0$, $Z_1 = Z_2 = Z_3 = 0$, using a mathematical software. The first figure below shows the solutions of the design equations — the reaction operating curves. Based on the solutions of Z_1, Z_2, and Z_3, we use (c) through (g) to calculate the dimensionless species content as functions of τ. The second figure below shows the species operating curves.

b. From the species operating curves (or the table of calculated data), maximum amount

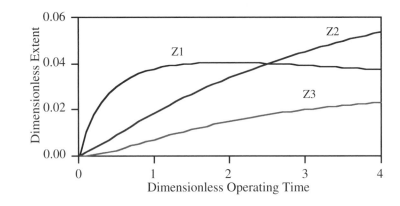

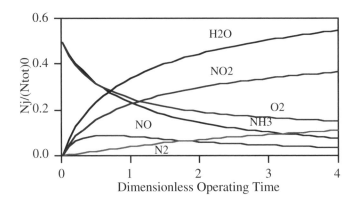

of NO is achieved at $\tau = 1.0$, and, using the definition of the dimensionless operating time and (p), the required operating time is

$$t = \tau t_{cr} = (1.0)(31.25 \text{ min}) = 31.25 \text{ min}.$$

At $\tau = 1.0$, $Z_1 = 0.0296$, $Z_2 = 0.0306$, and $Z_3 = 0.0045$. Substituting these values into (c) through (g) and using (h), the species molar fractions are: $y_{NH3} = 2.488$, $y_{O2} = 2.455$, $y_{NO} = 1.050$, $y_{H2O} = 3.468$, $y_{N2} = 0.588$, and $y_{NO2} = 0.086$.

5.4 NON-ISOTHERMAL OPERATIONS

The design formulation of non-isothermal batch reactors with multiple reactions follows the same procedure outlined in the previous section — we write design equation (5.1-1) for each independent reaction. However, since the reactor temperature may vary during the operation, the design equations have an additional variable, θ, and we should solve them simultaneously with the energy balance equation, (5.1-14). Note that the energy balance equation contains another variable — the temperature of the heating (or cooling) fluid, θ_F, that may also vary during the operation (θ_F is constant only when the fluid either evaporates or condenses). In general, we have to write an energy balance equation for the heating/cooling fluid to express changes in θ_F, but, in most applications, it is assumed constant. Also, because of the complex geometry of the heat-transfer surface (shell or a coil), an average of the inlet temperature and the outlet temperature is usually used.

The procedure for setting up the dimensionless energy balance equation goes as follows:

a. Define the **reference state**, and identify T_0, C_0, $(N_{tot})_0$, and y_{j0}'s.
b. Determine the specific molar heat capacity of the reference state, $\hat{c}_{p0}$.
c. Determine the dimensionless activation energies, γ_i's, of **all** chemical reactions.
d. Determine the dimensionless heat of reactions, DHR_m's, of the **independent** reactions.
e. Determine the correction factor of the heat capacity, $CF(Z_m, \theta)$.
f. Specify the dimensionless heat-transfer number, HTN.
g. Determine (or specify) the dimensionless temperature of the heating/cooling fluid, θ_F.
h. Determine (or specify) the initial dimensionless temperature, $\theta(0)$.
i. Solve the energy balance equation simultaneously with the design equations to obtain Z_m's and θ as functions of the dimensionless operating time, τ.

Hence, the design formulation of non-isothermal batch reactors consists of ($n_{ind} + 1$) nonlinear first-order differential equations,

$$\frac{dZ_1}{d\tau} = G_1(\tau, Z_1, Z_2, ..., Z_{n_{ind}}, \theta),$$

$$\frac{dZ_2}{d\tau} = G_1(\tau, Z_1, Z_2, ..., Z_{n_{ind}}, \theta),$$

$$\vdots \tag{5.5-1}$$

$$\frac{dZ_{n_{ind}}}{d\tau} = G_{n_{ind}}(\tau, Z_1, Z_2, ..., Z_{n_{ind}}, \theta),$$

$$\frac{d\theta}{d\tau} = G_{n_{ind}+1}(\tau, Z_1, Z_2, ..., Z_{n_{ind}}, \theta),$$

where τ is the dimensionless operating time defined by (5.1-3). The solutions of these equations provide Z_m's and θ as functions of τ.

The examples below illustrate the design of non-isothermal ideal batch reactors.

Example 5-16 The first-order, liquid-phase reaction A $\rightarrow$ 2 B takes place in an aqueous solution. A 200 liter solution with a concentration of $C_A(0) = 0.8$ mole/liter is charged into a batch reactor. Based on the data below:
a. Derive the operating curve when the reactor is operated isothermally, and determine the operating time needed to achieve 80% conversion.
b. Determine the heating/cooling load during the isothermal operation.
c. Derive the operating curve when the reactor is operated adiabatically, and determine the operating time needed to achieve 80% conversion.
Data: $T_0 = 47°C = 320°K$; $\Delta H_R(T_0) = -40$ kcal/mole B;
           At 47°C, $k = 0.04$ min^{-1}; $E_a = 9,000$ cal/mole;
           $\rho = 1.0$ kg/liter; $\bar{c}_p = 1.0$ kcal/kg °K.
Solution The chemical reaction is A $\rightarrow$ 2 B, and its stoichiometric coefficients are: $s_A = -1$ and $s_B = 2$. The rate expression is $r = k \cdot C_A$, and the heat of reaction is

$$\Delta H_R(320°K) = \left(-40 \frac{kcal}{mole\ B}\right) \cdot \left(2 \frac{mole\ B}{mole\ extent}\right) = -80 \frac{kcal}{mole\ extent}.$$

Since only one reaction takes place, only one design equation is needed to describe the operation, and, using (5.1-1), for constant-volume batch reactors, the design equation is

$$\frac{dZ}{d\tau} = r \cdot \left(\frac{t_{cr}}{C_0}\right). \tag{a}$$

We select the initial sate as the reference sate; hence, $C_0 = C_A(0)$, $y_A(0) = 1$, $y_B(0) = 0$, $Z(0) = 0$,

$$(N_{tot})_0 = V_R(0) \cdot C_0 = (200\ liter) \cdot (0.8\ mole/liter) = 160\ mole. \tag{b}$$

Using the stoichiometric relations,

$$N_A(t) = (N_{tot})_0 \cdot [y_A(0) - Z(t)] = (N_{tot})_0 \cdot [1 - Z(t)] \tag{c}$$

$$N_B(t) = (N_{tot})_0 \cdot [y_B(0) + 2 \cdot Z(t)] = (N_{tot})_0\ 2 \cdot Z(t). \tag{d}$$

For a liquid-phase reaction, using (5.1-10), $C_A(t) = C_0 \cdot [1 - Z(t)]$, and

$$r = k(T_0) \cdot C_0 \cdot (1 - Z) \cdot e^{\gamma \frac{\theta - 1}{\theta}} , \tag{e}$$

where γ is the dimensionless activation energy,

$$\gamma = \frac{E_a}{R \cdot T_0} = \frac{9000}{(1.987) \cdot (320)} = 14.154 .$$

We define the characteristic reaction time,

$$t_{cr} = \frac{1}{k(T_0)} = \frac{1}{0.04} = 25 \text{ min} . \tag{f}$$

Substituting (e) and (f) into (a), the design equation reduces to

$$\frac{dZ}{d\tau} = (1 - Z) \cdot e^{\gamma \frac{\theta - 1}{\theta}} . \tag{g}$$

a. For isothermal operation, $\theta = 1$, and design equation (g) reduces to

$$\frac{dZ}{d\tau} = (1 - Z), \tag{h}$$

and the dimensionless reaction operating curve is given by

$$Z(\tau) = 1 - e^{-\tau}. \tag{i}$$

Using (1.5-8), for 80% conversion, the extent is $Z = 0.80$, and, from (i), $\tau = 1.61$. Using (f), the operating time for 80% conversion is

$$t = t_{cr} \cdot \tau = (25 \text{ min}) \cdot (1.61) = 40.25 \text{ min}.$$

b. For isothermal operations, $d\theta/d\tau = 0$, and the energy balance equation is (5.1-15),

$$\frac{d}{d\tau} \left(\frac{Q}{T_0 \cdot N_{tot}(0) \cdot \hat{c}_{p0}} \right) = \sum_{m}^{n_{ind}} \left(\frac{\Delta H_{R_m}(T_0)}{T_0 \cdot \hat{c}_{p0}} \right) \cdot \frac{dZ_m}{d\tau} ,$$

which, upon integration, becomes

$$Q(t) = (N_{tot})_0 \cdot \Delta H_R(T_0) \cdot Z(t) = (160 \text{ mole}) \cdot \left(-80 \frac{\text{kcal}}{\text{mole}} \right) \cdot (0.8) = -10.24 \cdot 10^3 \text{ kcal}.$$

The negative sign indicates that heat is removed from the reactor.

c. For adiabatic operation, $(S/V) = 0$; hence, $HTN = 0$, and (5.1-14) reduces

$$\frac{d\theta}{d\tau} = \frac{-1}{CF(Z_m, \theta)} DHR \frac{dZ}{d\tau} . \tag{j}$$

Using (4.2-55), for liquid-phase reactions, the specific molar heat capacity of the reference state is

$$\hat{c}_{p0} = \frac{M}{(N_{tot})_0}\overline{c}_p = \frac{200 \text{ kg}}{160 \text{ mole}}\left(1\frac{\text{kcal}}{\text{kg}^\circ\text{K}}\right) = 1.25\frac{\text{kcal}}{\text{mole}^\circ\text{K}}, \tag{k}$$

and the dimensionless heat of reaction is

$$\text{DHR} = \left(\frac{\Delta H_R(T_0)}{T_0\cdot\hat{c}_{p0}}\right) = \frac{-80}{(320)\cdot(1.25)} = -0.20.$$

Assuming constant heat capacity, $CF(Z_m,\theta) = 1$, and (j) reduces to

$$\frac{d\theta}{d\tau} = -\left[\frac{-80}{(320)\cdot(1.25)}\right]\cdot(1-Z)\cdot e^{14.15\frac{\theta-1}{\theta}}. \tag{l}$$

We solve (l) simultaneously with (g), subject to the initial condition that at $\tau = 0$, $Z = 0$ and $\theta = 1$. The solutions are shown in the figure below. From the tabulated solution, $Z = 0.8$ is reached at $\tau = 0.53$. Using (g), the operating time is

$$t = t_{cr}\cdot\tau = (25 \text{ min})\cdot(0.53) = 13.25 \text{ min}.$$

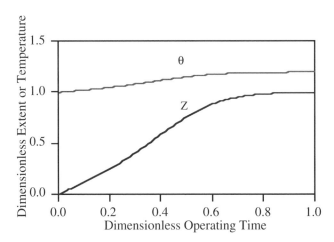

Example 5-17 The second-order gas-phase reaction $2 A \rightarrow B$ is carried out in a constant-volume batch reactor. The reactor volume is 250 ft³, $C_A(0) = 0.008$ lbmole/ft3, $C_B(0) = 0$, and $T(0) = T_0 = 600°R$. The heat of reaction $\Delta H_R(T_0) = -4,000$ Btu/lbmole A and the species heat capacities are: $\hat{c}_{p_A} = 20$ Btu/lbmole°R, and $\hat{c}_{p_B} = 30$ Btu/lbmole°R. At 600°R, the reaction rate constant is $k(T_0) = 100$ ft³/lbmole hr, and the activation energy is $E_a/R = 8,400°R$. Derive the reaction operating curve, and deter-

mine the reaction time for 70% conversion at the following operating conditions:

a. The reactor is operated isothermally at 600°R.

b. Calculate the overall heating (or cooling) load in (a).

c. The reactor is operated adiabatically.

d. Heat is removed with U = 30 Btu/ft² hr °R, S = 2 ft², and T_F = 580°R.

Solution The chemical reaction is $2 A \rightarrow B$, and its stoichiometric coefficients are: s_A = -2 and s_B = 1. The rate expression is $r = k \cdot C_A^2$, and, for this reaction, the heat of reaction is

$$\Delta H_R (600°R) = \left(\frac{-4,000 \text{ Btu}}{\text{lbmole A}} \right) \cdot \left(\frac{2 \text{ lbmole A}}{\text{lbmole extent}} \right) = -8,000 \frac{\text{Btu}}{\text{lbmole extent}}.$$

From (5-1), the dimensionless design equation for a constant-volume batch reactor with a single reaction is

$$\frac{dZ}{d\tau} = r \cdot \left(\frac{t_{cr}}{C_0} \right). \tag{a}$$

We select the initial state as the reference state; hence, $C_0 = C_A(0)$, $y_A(0) = 1$, $y_B(0) = 0$, and $Z(0) = 0$. Also,

$$(N_{tot})_0 = V_R(0) \cdot C_0 = (250 \text{ ft}^3) \cdot (0.008 \text{ lbmole/ft3}, = 2 \text{ lbmole.} \tag{b}$$

Using the stoichiometric relations,

$$N_A(t) = (N_{tot})_0 \cdot [y_A(0) - 2 \cdot Z(t)] = (N_{tot})_0 \cdot [1 - 2 \cdot Z(t)] \tag{c}$$

$$N_B(t) = (N_{tot})_0 \cdot [y_B(0) + Z(t)] = (N_{tot})_0 \cdot Z(t). \tag{d}$$

For a constant-volume batch reactor, using (5.1-10), the concentration at any time is

$$C_A(t) = C_0 [1 - 2 \cdot Z(t)], \tag{e}$$

and the reaction rate is

$$r = k(T_0) \cdot C_0^2 \cdot (1 - 2 \cdot Z)^2 \cdot e^{\gamma \frac{\theta - 1}{\theta}}, \tag{f}$$

where γ is the dimensionless activation energy,

$$\gamma = \frac{E_a}{R \cdot T_0} = \frac{8,400}{600} = 14.$$

We define the characteristic reaction time,

$$t_{cr} = \frac{1}{k(T_0) \cdot C_0} = \frac{1}{(100) \cdot (0.008)} = 1.25 \text{ hr.} \tag{g}$$

Substituting (f) and (g) into (a), the design equation reduces to

$$\frac{dZ}{d\tau} = (1 - 2 \cdot Z)^2 \cdot e^{\gamma \frac{\theta - 1}{\theta}}.$$ (h)

a. For isothermal operation, $\theta = 1$, and, separating the variables, (h) becomes

$$\frac{dZ}{d\tau} = (1 - 2 \cdot Z)^2.$$ (i)

We solve (i) subject to the initial condition that, at $\tau = 0$, $Z = 0$ and obtain

$$Z(\tau) = \frac{\tau}{1 + 2 \cdot \tau}.$$ (j)

Using (1.5-11), for 70% conversion the extent is $Z = 0.35$, and, from (j), for $Z = 0.35$, $\tau = 1.167$. Using (g), the operating time is

$$t = \tau \cdot t_{cr} = (1.167) \cdot (1.25 \text{ hr}) = 1.46 \text{ hr}.$$

b. For isothermal operation, $d\theta/d\tau = 0$, and the energy balance equation reduces to

$$\frac{d}{d\tau} \cdot \left(\frac{Q}{T_0 \cdot (N_{tot})_0 \cdot \hat{c}_{p0}} \right) = \left(\frac{\Delta H_R(T_0)}{T_0 \cdot \hat{c}_{p0}} \right) \cdot \frac{dZ}{d\tau},$$

which, upon integration, becomes

$$Q(\tau) = (N_{tot})_0 \cdot \Delta H_R(T_0) \cdot Z(\tau) = (2 \text{ lbmole}) \cdot \left(\frac{-8,000 \text{ Btu}}{\text{lbmole extent}} \right) = -5,600 \text{ Btu}.$$

The negative sign indicates that heat is removed from the reactor.

c. For adiabatic operations, $(S/V) = 0$; hence, $HTN = 0$, and (5.1-14) reduces to

$$\frac{d\theta}{d\tau} = \frac{-1}{CF(Z_m, \theta)} DHR \frac{dZ}{d\tau}.$$ (k)

In this case, the specific heat capacity of the reference state, defined by (4.2-59), is

$$\hat{c}_{p0} = \sum_j^{all} y_j(0) \cdot \hat{c}_{p_j}(1) = y_A(0) \cdot \hat{c}_{pA}(1) = 20 \text{ Btu/lbmole}^{\circ}R,$$ (l)

and the dimensionless heat of reaction is

$$DHR = \left(\frac{\Delta H_R(T_0)}{T_0 \cdot \hat{c}_{p0}} \right) = \frac{-8,000}{(600) \cdot (20)} = -0.667.$$

Using (4.2-61), the correction factor of the heat capacity is

$$CF(Z_m, \theta) = 1 + \frac{1}{\hat{c}_{P0}} \left[\hat{c}_{PA}(\theta) \cdot (-2) \cdot Z + \hat{c}_{PB}(\theta) \cdot Z \right] = 1 - 0.5 \cdot Z. \tag{m}$$

Substituting (l), (m), and (h) into (k), the energy balance equation reduces to

$$\frac{d\theta}{d\tau} = \frac{-0.333}{1 - 0.5 \cdot Z} \cdot (1 - 2 \cdot Z)^2 \cdot e^{\gamma \frac{\theta - 1}{\theta}}. \tag{n}$$

For this case, $\gamma = 14$. We solve (h) and (n) numerically, subject to the initial conditions that, at $\tau = 0$, $Z = 0$, and $\theta = 1$. The figure below shows the reaction operating curve and the temperature variation. From the calculated data, a conversion of $Z = 0.35$ is reached at $\tau = 0.211$, and, at that time, the dimensionless reactor temperature is $\theta = 1.261$. The reactor temperature at the end of the operation is therefore

$$T = \theta \cdot T_0 = (1.261) \cdot (600°R) = 756.6°R.$$

Using (g), the operating time in this case is

$$t = \tau \cdot t_{cr} = (0.211) \cdot (1.25 \text{ hr}) = 0.264 \text{ hr}.$$

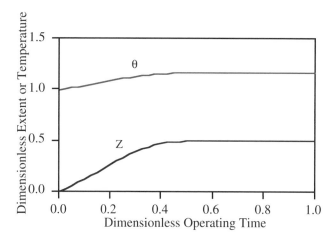

Note that the operating time for adiabatic operation is about one-third that of an isothermal operation.

d. For operations with heat transfer, design equation (h) applies, but the energy balance equation should include the heat transfer term. Using (5.1-14),

$$\frac{d\theta}{d\tau} = \frac{1}{CF(Z, \theta)} \left[HTN \cdot (\theta_F - \theta) - DHR \frac{dZ}{d\tau} \right]. \tag{o}$$

From the given data, the heat transfer number is

$$HTN = \frac{U \cdot t_{cr}}{C_0 \cdot \hat{c}_{P0}} \cdot \left(\frac{S}{V}\right) = \frac{(30) \cdot (1.25)}{(0.008) \cdot (20)} \cdot \left(\frac{2}{250}\right) = 1.875,$$

and $\theta_F = 580/600 = 0.9667$. Substituting the numerical values and (h),

$$\frac{d\theta}{d\tau} = \frac{1}{1 - 0.5 \cdot Z}\left[1.875 \cdot (0.9667 - \theta) + 0.333 \cdot (1 - 2 \cdot Z)^2 \cdot e^{\gamma \frac{\theta - 1}{\theta}}\right]. \qquad (p)$$

We solve (h) and (p) numerically, subject to the initial conditions that at $\tau = 0$, $Z = 0$ and $\theta = 1$. The figure below shows the reaction operating curve and temperature as a function of operating time. From the calculated data, $Z = 0.35$ is reached at $\tau = 0.720$ and, at that time, the dimensionless reactor temperature is $\theta = 1.15$. The reactor temperature at the end of the operation is therefore

$$T = \theta \cdot T_0 = (1.15) \cdot (600^\circ R) = 690^\circ R.$$

Using (e), the operating time in this case is

$$t = \tau \cdot t_{cr} = (0.720) \cdot (1.25 \text{ hr}) = 0.90 \text{ hr.}$$

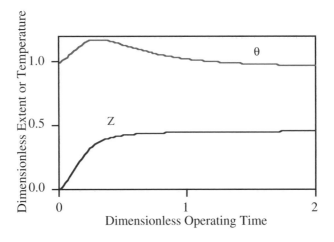

Note that the operating time in this case is longer than the time in the adiabatic case but shorter than that of the isothermal case.

Example 5-18 The gas-phase elementary chemical reactions

$$A + B \rightarrow C$$

$$C + B \rightarrow D$$

are carried out in an isobaric batch reactor. A gas mixture of 50% A and 50% B is charged into the reactor, whose initial volume is 10 liter. At the beginning of the operation, the temperature is 150°C, and a pressure of 2 atm is maintained during operation. Based on the data below, derive the design equations, and plot the operating curves for the operations indicated. In each case, determine the operating time required to obtain 75% conversion of B, the temperature, and the reactor content at the end of the operation. Consider the following operations:

a. Isothermal operation.
b. Determine the heating/cooling load in (a).
c. Adiabatic operation.
d. Operation with heat-transfer. The heat transfer area is 0.25 m², the heat transfer coefficient is 143 cal/m² hr °K, and $T_F = 120°C$.

Data: At 150°C, $k_1 = 0.02$ liter mole⁻¹sec⁻¹; $k_2 = 0.01$ liter mole⁻¹sec⁻¹;

At 150°C, $\Delta H_{R1} = -4,500$ cal/mole extent; $\Delta H_{R2} = -6,000$ cal/mole extent;
$E_{a1} = 10,090$ cal/mole; $E_{a2} = 12,607$ cal/mole;
$\hat{c}_{PA} = 16$ cal/mole°K; $\hat{c}_{PB} = 8$ cal/mole°K; $\hat{c}_{PC} = 20$ cal/mole°K;
$\hat{c}_{PD} = 26$ cal/mole°K.

Solution The reactions are:

$$\text{Reaction 1:} \qquad A + B \rightarrow C$$

$$\text{Reaction 2:} \qquad C + B \rightarrow D$$

and the stoichiometric coefficients are:

$$s_{A1} = -1; \ s_{B1} = -1; \ s_{C1} = 1; \ s_{D1} = 0; \ \Delta_1 = -1;$$

$$s_{A2} = 0; \ s_{B1} = -1; \ s_{C1} = -1; \ s_{D1} = 1; \ \Delta_2 = -1.$$

Since each reaction has a species that does not appear in the other reaction, the two reactions are independent, and there is no dependent reaction. We select the initial state as the reference state,

$$C_0 = \frac{P}{R \cdot T_0} = \frac{2}{(0.08205) \cdot (423)} = 0.0576 \text{ mole/lit},$$

$$(N_{tot})_0 = V_R(0) \cdot C_0 = (10 \text{ lit}) \cdot (0.0576 \text{ mole/lit}) = 0.576 \text{ mole},$$

$y_A(0) = y_B(0) = 0.5$, and $y_C(0) = y_D(0) = 0$. We write the dimensionless design equation for a variable-volume batch reactor, combining (5.1-1) and (5.1-11), for the two independent reactions

$$\frac{dZ_1}{d\tau} = r_1 \cdot \left(1 + \sum_m^{n_{ind}} \Delta_m \cdot Z_m\right) \cdot \theta \cdot \left(\frac{t_{cr}}{C_0}\right) \qquad (a)$$

$$\frac{dZ_2}{d\tau} = r_2 \cdot \left(1 + \sum_m^{n_{ind}} \Delta_m \cdot Z_m\right) \cdot \theta \cdot \left(\frac{t_{cr}}{C_0}\right) \qquad (b)$$

Using the stoichiometric relations, we express the molar content of the species:

$$N_A(t) = (N_{tot})_0 \cdot [y_A(0) - Z_1(t)] \tag{c}$$

$$N_B(t) = (N_{tot})_0 \cdot [y_B(0) - Z_1(t) - Z_2(t)] \tag{d}$$

$$N_C(t) = (N_{tot})_0 \cdot [y_C(0) + Z_1(t) - Z_2(t)] \tag{e}$$

$$N_D(t) = (N_{tot})_0 \cdot [y_D(0) + Z_2(t)]. \tag{f}$$

For variable-volume batch reactors, the species concentrations are expressed by (5.1-13), and the rates of the two reactions are,

$$r_1 = k_1 \cdot C_A \cdot C_B = k_1(T_0) \cdot e^{\gamma_1 \frac{\theta-1}{\theta}} \cdot C_0^2 \frac{[y_A(0) - Z_1] \cdot [y_B(0) - Z_1 - Z_2]}{[(1 - Z_1 - Z_2) \cdot \theta]^2} \tag{g}$$

$$r_2 = k_2 \cdot C_C \cdot C_B = k_2(T_0) \cdot e^{\gamma_2 \frac{\theta-1}{\theta}} \cdot C_0^2 \frac{(Z_1 - Z_2) \cdot [y_B(0) - Z_1 - Z_2]}{[(1 - Z_1 - Z_2) \cdot \theta]^2}. \tag{h}$$

We select a characteristic reaction time on the basis of Reaction 1,

$$t_{cr} = \frac{1}{k_1(T_0) \cdot C_0} = 868 \text{ sec} = 14.46 \text{ min}. \tag{i}$$

Substituting (g) and (h) into (a) and (b), the two design equations become

$$\frac{dZ_1}{d\tau} = \frac{(y_A(0) - Z_1) \cdot (y_B(0) - Z_1 - Z_2)}{(1 - Z_1 - Z_2) \cdot \theta} \cdot e^{\gamma_1 \frac{\theta-1}{\theta}} \tag{j}$$

$$\frac{dZ_2}{d\tau} = \left(\frac{k_2(T_0)}{k_1(T_0)} \right) \cdot \frac{(Z_1 - Z_2) \cdot (y_B(0) - Z_1 - Z_2)}{(1 - Z_1 - Z_2) \cdot \theta} \cdot e^{\gamma_2 \frac{\theta-1}{\theta}}. \tag{k}$$

To write the energy balance equation, we first determine $\hat{c}_{p0}$, using (4.2-59),

$$\hat{c}_{p0} = \sum_{j}^{all} y_j(0) \cdot \hat{c}_{p_j}(1) =$$

$$= y_A(0) \cdot \hat{c}_{p_A}(1) + y_B(0) \cdot \hat{c}_{p_B}(1) + y_C(0) \cdot \hat{c}_{p_C}(1) + y_D(0) \cdot \hat{c}_{p_D}(1) =$$
$$= (0.5) \cdot (16) + (0.5) \cdot (8) = 12.0 \text{ cal/mole}°K.$$

Using (4.2-61), the correction factor of the heat capacity is

$$CF(Z_m, \theta) = 1 + \frac{1}{\hat{c}_{p0}} \sum_{j}^{all} \hat{c}_{p_j}(\theta) \sum_{m}^{n_{ind}} (s_j)_m \cdot Z_m =$$

$$= 1 + \frac{1}{\hat{c}_{P0}} \left[\hat{c}_{PA} \cdot (-Z_1) + \hat{c}_{PB} \cdot (-Z_1 - Z_2) + \hat{c}_{PC} \cdot (Z_1 - Z_2) + \hat{c}_{PD} \cdot Z_2 \right] =$$

$$= 1 + \frac{1}{12} \left[(16) \cdot (-Z_1) + (8) \cdot (-Z_1 - Z_2) + (20) \cdot (Z_1 - Z_2) + (26) \cdot Z_2 \right].$$

Hence,
$$CF(Z_m, \theta) = \frac{6 - 2 \cdot Z_1 - Z_2}{6}. \tag{1}$$

Substituting (1) into (5.1-14), the energy balance equation reduces to

$$\frac{d\theta}{d\tau} = \left(\frac{6}{6 - 2 \cdot Z_1 - Z_2} \right) \cdot \left[HTN \cdot (\theta_F - \theta) - DHR_1 \frac{dZ_1}{d\tau} - DHR_2 \frac{dZ_2}{d\tau} \right]. \tag{m}$$

For the given reactions,

$$\gamma_1 = \frac{E_{a_1}}{R \cdot T_0} = 12; \quad DHR_1 = \left(\frac{\Delta H_{R_1}(T_0)}{T_0 \cdot \hat{c}_{P0}} \right) = -0.886;$$

$$\gamma_2 = \frac{E_{a_2}}{R \cdot T_0} = 15; \quad DHR_2 = \left(\frac{\Delta H_{R_2}(T_0)}{T_0 \cdot \hat{c}_{P0}} \right) = -1.182.$$

a. For isothermal operation ($\theta = 1$), the two design equations, (j) and (k), reduce to

$$\frac{dZ_1}{d\tau} = \frac{(0.5 - Z_1) \cdot (0.5 - Z_1 - Z_2)}{1 - Z_1 - Z_2} \tag{n}$$

$$\frac{dZ_2}{d\tau} = (0.5) \frac{(Z_1 - Z_2) \cdot (0.5 - Z_1 - Z_2)}{1 - Z_1 - Z_2}. \tag{o}$$

We solve (n) and (o) numerically, subject to the initial conditions that, at $\tau = 0$, $Z_1 = Z_2 = 0$. The figures below show, respectively, the reaction operating curves and the species operating curves. Using the conversion definition, for 75% conversion, $N_B/N_B(0) = 0.25$ or $N_B/N_{tot}(0) = 0.125$. Hence, we use the species operating curve of B to determine τ and find that for $N_B/N_{tot}(0) = 0.125$, $\tau = 2.73$. At that time, $Z_1 = 0.318$ and $Z_2 = 0.079$. Hence, using (i), the operating time is

$$t = (2.73) \cdot (14.46 \text{ min}) = 39.5 \text{ min}.$$

Using (c) through (f), the content of the reactor at that time is: $N_A(t) = 0.105$ mole, $N_B(t) = 0.059$ mole, $N_C(t) = 0.138$ mole, and $N_D(t) = 0.045$ mole.

b. For an isothermal operation, $d\theta/d\tau = 0$, and the energy balance equation reduces to

$$\frac{d}{d\tau} \left(\frac{Q}{T_0 \cdot (N_{tot})_0 \cdot \hat{c}_{P0}} \right) = \left(\frac{\Delta H_{R_1}(T_0)}{T_0 \cdot \hat{c}_{P0}} \right) \frac{dZ_1}{d\tau} + \left(\frac{\Delta H_{R_2}(T_0)}{T_0 \cdot \hat{c}_{P0}} \right) \frac{dZ_2}{d\tau},$$

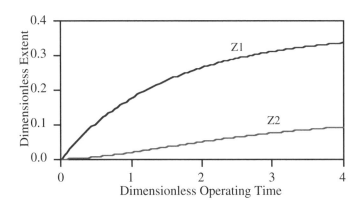

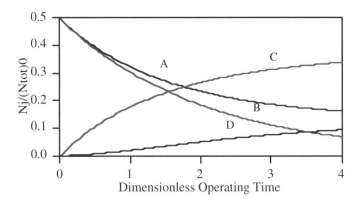

which, upon integration, becomes

$$Q(t) = (N_{tot})_0 \cdot (-4{,}500 \cdot Z_1 - 6{,}000 \cdot Z_2) = -1{,}097 \text{ cal.}$$

The negative sign indicates that heat is removed from the reactor.

c. For adiabatic operation, $(S/V) = 0$; hence, $HTN = 0$, and, with the substitution of (j) and (k), the energy balance equation, (m), reduces to

$$\frac{d\theta}{d\tau} = \left(\frac{6}{6 - 2 \cdot Z_1 - Z_2} \right)(0.886) \frac{(0.5 - Z_1) \cdot (0.5 - Z_1 - Z_2)}{(1 - Z_1 - Z_2) \cdot \theta} \cdot e^{\gamma_1 \frac{\theta - 1}{\theta}} +$$

$$+ \left(\frac{6}{6 - 2 \cdot Z_1 - Z_2} \right)(1.182) \cdot (0.5) \frac{(Z_1 - Z_2) \cdot (0.5 - Z_1 - Z_2)}{(1 - Z_1 - Z_2) \cdot \theta} \cdot e^{\gamma_2 \frac{\theta - 1}{\theta}} \cdot \quad \text{(p)}$$

We solve (j), (k), and (p) numerically, subject to the initial condition that at $\tau = 0$, $Z_1 =$

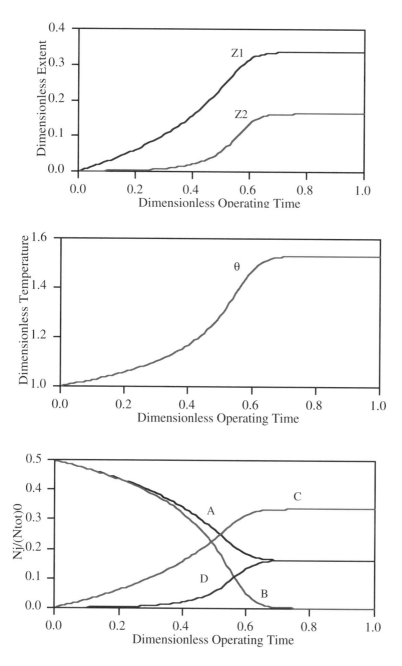

$Z_2 = 0$ and $\theta = 1$. The first graph shows the reaction operating curves, the second shows the temperature variation, and the third the species operating curves. We use the calculated data to determine τ for $N_B/N_{tot}(0) = 0.125$. We find that $\tau = 0.545$, and, at that time, $Z_1 = 0.283$, $Z_2 = 0.096$, and $\theta = 1.39$. Hence, using (i), the operating time is

$$t = (0.545) \cdot (14.46 \text{ min}) = 7.88 \text{ min}.$$

Using (c) through (f), the content of the reactor at that time is: $N_A(t) = 0.125$ mole, $N_B(t) = 0.070$ mole, $N_C(t) = 0.108$ mole, and $N_D(t) = 0.055$ mole.

d. For an operation with heat-transfer, we first determine the heat transfer number,

$$HTN = \frac{U \cdot t_{cr}}{C_0 \cdot \hat{c}_{p0}} \left(\frac{S}{V}\right) = 1.246,$$

and the energy balance equation reduces to

$$\frac{d\theta}{d\tau} = \left(\frac{6}{6 - 2 \cdot Z_1 - Z_2}\right) \left[1.246 \cdot (0.929 - \theta) + (0.886)\frac{dZ_1}{d\tau} + (1.182)\frac{dZ_2}{d\tau}\right]. \quad (q)$$

We solve (j), (k), and (q) numerically, subject to the initial condition that, at $\tau = 0$, $Z_1 = Z_2 = 0$ and $\theta = 1$. The first graph shows the reaction operating curves, the second the

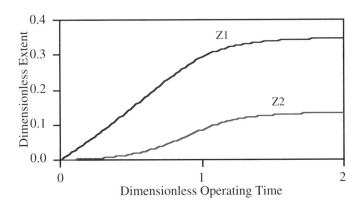

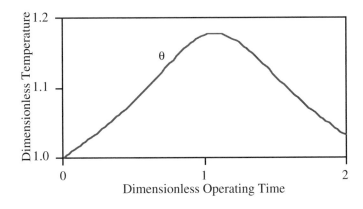

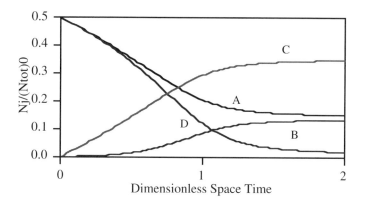

temperature variation, and the third the species operating curves. We use the calculated data to determine τ for $N_B/N_{tot}(0) = 0.125$. We find that $\tau = 0.995$, and, at that time, $Z_1 = 0.292$ and $Z_2 = 0.083$. Hence, using (i), the operating time is

$$t = (0.995) \cdot (14.46 \text{ min}) = 14.4 \text{ min}.$$

Using (c) through (f), the content of the reactor at that time is: $N_A(t) = 0.119$ mole, $N_B(t) = 0.071$ mole, $N_C(t) = 0.121$ mole, and $N_D(t) = 0.0478$ mole.

5.5 SUMMARY

In this chapter, we analyzed the operation of ideal batch reactors. We covered the following topics:

a. The underlying assumptions for ideal batch reactor operations and when they are applied in practice.

b. Derived the dimensionless design equation for isothermal operation with single reactions and obtained the reaction operating curve.

c. Discussed the operation of constant-volume batch reactors.

d. Discussed the operation of gaseous, variable-volume batch reactors.

e. Discussed methods to determine the reaction rate constants for rate expressions.

f. Discussed the design of isothermal operations with multiple chemical reactions.

g. Discussed the design of non-isothermal operations with multiple chemical reactions.

BIBLIOGRAPHY

Information on the mechanical design of well-mixed reactors is available in:
Oldshue, J. Y., *Fluid Mixing Technology*. New York: McGraw Hill, 1983.

Tatterson, G. B., *Fluid Mixing and Gas Dispersion in Agitating Tanks*. New York: McGraw Hill, 1991.

Tatterson, G. B., *Scale-up and Design of Industrial Mixing processes*. New York: McGraw Hill, 1994.

PROBLEMS

5-1$_2$ The gas phase decomposition of ethylene oxide, according to the reaction

$$C_2H_4O(g) \rightarrow CH_4(g) + CO(g),$$

has been investigated in a constant-volume batch reactor at 400°C. The reaction is assumed to be first-order. Based on the data below, determine:
a. Is the first-order assumption valid?
b. What is the reaction rate constant at 400°C?
c. What time is required for 50% conversion?
Data:

Time (min)	0	7	12	18
Pressure (mmHg)	119	130.7	138.2	146.4

(Adapted from Fine and Beall's book)

5-2$_2$ Reactant A decomposes according to the reaction A → 2B + C in an aqueous solution containing a homogeneous catalyst. The measured relation between (-r_A) and C_A is given in the table below. We want to run this reaction in a batch reactor at the same catalyst concentration as used in obtaining the data. Determine:
a. the time needed to lower the concentration of A from $C_A(0) = 10$ mole/lit to $C_A(t) = 2$ mole/lit, and
b. the final concentration after 10 hrs of operation if the reactor is charged with a solution of $C_A(0) = 7$ mole/lit.
Data:

$C_A(t)$ (mole/lit)	1	2	4	6	7	9	12
(-r_A) (mole/lit hr)	0.06	0.1	0.25	1.0	2.0	1.0	0.5

(Adapted from Levenspiel's Omnibook)

5-3$_2$ The reaction A → Products is investigated in a constant-volume batch reactor. After 8 minutes in a batch reactor, 80% of reactant A is converted. After 18 minutes, 90% of A is converted. If $C_A(0) = 1$ mole/liter, determine the rate expression of this reaction. (Adapted from Levenspiel's Omnibook)

5-4$_2$ Dvorko and Shilov (Kinetics and Catalysis, 4, 212, 1964) studied the iodine-catalyzed addition reaction between HI and cyclohexene in benzene solution,

$$HI + C_6H_{10} \rightarrow C_6H_{11}I.$$

The reaction is believed to be first-order with respect to each reactant. An experiment was conducted at 20°C in a benzene solution with an iodine concentration of 4.22 10^{-4}

mole/lit. The initial concentration of cyclohexene was 0.123. mole/lit. Based on the data below, determine:

a. Is the assumed rate expression valid?

b. What is the reaction rate constant at 20°C?

Data:

Time (sec)	C_{HI} (mole/lit)
0	0.106
150	0.090
480	0.087
870	0.076
1500	0.062
2280	0.050

(Adapted from Hill's textbook)

5-5$_2$ Huang and Dauerman (Ind. Eng. Chem. Process. Des. Develop **8**, 227, 1969) studied the reaction between benzyl chloride and sodium acetate in dilute aqueous solution at 102°C

$$NaAc + C_6H_5CH_2Cl \rightarrow C_6H_5CH_2Ac + Na^+ + Cl^-.$$

Initially, the concentration of both sodium acetate and benzyl chloride is 0.757 mole/liter. The fraction of unconverted benzyl as a function of time is given below. Assuming the reaction is first-order with respect to each reactant, determine:

a. Is the assumed rate expression valid?

b. The reaction rate constant at 102°C.

Data:

Time (ksec)	$C_B(t)/C_B(0)$
10.80	0.945
24.48	0.912
46.08	0.846
54.72	0.809
69.48	0.779
88.56	0.730
109.44	0.678
126.72	0.638
133.74	0.619
140.76	0.590

(Adapted from Hill's textbook)

5-6$_2$ The dimerization second-order reaction $2 A \rightarrow R$ takes place in a liquid solution. When the reactor is charged with a solution with $C_A(0) = 1$ mole/liter, 50% conversion is reached after one hour. What will the conversion be after one hour if the initial concentration of A is 10 mole/lit? (Adapted from Levenspiel's Omnibook)

5-7$_2$ Enzyme E catalyzes the chemical reaction $A \rightarrow P + R$ in aqueous solution. The reaction rate expression is

$$(-r_A) = \frac{200 \cdot C_A \cdot C_B}{2 + C_A} \quad \text{(mole/lit min)}.$$

If we charge a solution with $C_E = 0.001$ mole/lit and $C_A(0) = 10$ mole/lit into a batch reactor, find the time it takes for the concentration of A to drop to 0.025 mole/lit. Note that the concentration of the enzyme remains unchanged during the operation. Plot the dimensionless operating curve. (Adapted from Levenspeil's Omnibook)

5-8₂ The reaction of cyclohexanol and acetic acid in dioxane solution as catalyzed by sulfuric acid was studied by McCracken and Dickson (Ind. Eng. Chem. Proc. Des. and Dev., 6, 286, 1967). The esterification reaction can be represented by the following stoichiometric reaction,

$$A + B \rightarrow C + W$$

(acetic acid + cyclohexanol → cyclohexylacetate + water)

The reaction is carried out in a well-stirred batch reactor at 40°C. Under these conditions, the esterification reaction can be considered as irreversible at conversions less than 70%. The following data were obtained using identical sulfuric acid concentrations in both runs.

Run 1: $C_A(0) = C_B(0) = 2.5$ kmoles/m³

$C_A(t)$ (kmoles/m³)	Time, t (ksec)
2.070	7.2
1.980	9.0
1.915	10.8
1.860	12.6
1.800	14.4
1.736	16.2
1.692	18.0
1.635	19.8
1.593	21.6
1.520	25.2
1.460	28.8

Run 2: $C_A(0) = 1$; $C_B(0) = 8$ kmole/m³

$C_A(t)$ (kmoles/m³)	Time, t (ksec)
0.885	1.8
0.847	2.7
0.769	4.5
0.671	7.2
0.625	9.0
0.544	12.6
0.500	15.3
0.463	18.0

Determine the order of the reaction with respect to each reactant and the rate constant for the forward reaction under the conditions of the two runs. The rate constant should differ between runs, but the conditions are such that the order will not differ. The individual and overall orders are integers. (Adapted from Hill's textbook)

5-9₂ The elementary, gas-phase, dimerization reaction $2\,A \rightarrow P$ is carried out in an isobaric, isothermal batch reactor. Pure A is charged into the reactor which operates at 5 atm and 300°C. Based on the data below, determine:
a. the time needed for 80% conversion of A and
b. the volume of the reactor at that time.
c. Plot the performance curve.
Data: The reaction rate constant at 300°C, $k = 1$ min⁻¹ (mole/lit)⁻¹

Initial volume of the reactor: 500 liter.

5-10₂ The following data are typical of the pyrolysis of dimethylether at 504°C,

$$CH_3OCH_3 \rightarrow CH_4 + H_2 + CO.$$

The reaction takes place in the gas phase in an isothermal constant-volume reactor. Determine the order of the reaction and the reaction rate constant. The order may be assumed to be an integer.

Data:

Time (sec)	P(kPa)
0	41.6
390	54.4
777	65.1
1195	74.9
3155	103.9
∞	124.1

(Adapted from Hill's textbook)

5-11₁ For the reaction 2 A + 3 B → 2 C, the following data were obtained:

Run	Concentration (mole/lit)		Rate
	A	B	(mole/lit min)
1	0.50	1.00	0.1
2	0.50	2.00	0.3
3	1.00	1.00	0.4

Determine the form of the rate expression. What is the value of the reaction rate constant?

5-12₂ For the following data:

Concentration (mole/lit)			Rate
A	B	C	(mole/lit min)
0.01	0.20	0.10	2.8
0.01	0.40	0.10	5.6
0.01	0.80	0.05	5.6
0.02	0.10	0.10	2.8

determine:
a. the order of the reaction with respect to A, B, and C, and
b. the value of the reaction rate constant.

5-13₂ At 800°K, di-methyl ether decomposes according to the reaction

$$CH_3OCH_3(g) \rightarrow CH_4(g) + CO(g) + H_2(g).$$

Ether is charged into a constant-volume batch reactor, and the progress of the reaction is monitored measuring the reactor pressure.
Based on the data below, determine
a.   the order of the reaction (use the differential method),
b.   the reaction rate constant at 800°K (use the integral method), and

c.   the half-life time at 800°K.

Data:

Time (min)	0	390	780	1200	3160
Pressure (mmHg)	310	410	490	560	780

5-14$_2$ Hinshelwood and Burk (Proc. Roy. Soc. 106A (284), 1924) have studied the thermal decomposition of nitrous oxide. Consider the following "adjusted" data at 1030°K.

Initial pressure of N_2O (mm Hg)	Half-life (sec)
82.5	860
139	470
296	255
360	212

a.   Determine the order of the reaction and the reaction rate constant.

b.   The following additional data were reported at the temperatures indicated:

Temperature (°K)	Initial pressure (mm Hg)	Half-life (sec)
1085	345	53
1030	360	212
967	294	1520

What is the activation energy of the reaction? (Adapted from from Hill's textbook)

5-15$_2$ The gas phase dimerization of trifluorochloroethylene may be represented by

$$2\ C_2F_3Cl \rightarrow C_4F_6Cl_2.$$

The following data are typical of this reaction at 440°C as it occurs in a constant-volume reactor.

Time, t (sec)	Total pressure (kPa)
0	82.7
100	71.1
200	64.0
300	60.4
400	56.7
500	54.8

Determine the order of the reaction and the reaction rate constant under these conditions. Assume the order is an integer. (Adapted from Hill's textbook)

5-16$_4$ The elementary liquid-phase reactions

$$A + B \rightarrow C$$

$$C + B \rightarrow D$$

take place in a 200 liter batch reactor. A solution ($C_A(0) = 2$ mole/lit, $C_B(0) = 2$ mole/

lit) at 80°C is charged into the reactor. Based on the data below, derive the reaction operating curves and the temperature curve for each of the operations below. For each case, determine the operating time required for maximum production of C and the amount of C and D formed at that time.

a. Isothermal operation at 80°C.

b. The heating/cooling load on the reactor in (a).

c. Adiabatic operation. What is the reactor temperature at the end of the operation?.

d. Heat is removed with U = 1 kcal/m² hr °K, S = 1 m², and $T_F = 80°C$

Data: At 80°C $k_1 = 0.1$ lit mol⁻¹ min⁻¹; $E_{a1} = 6,000$ cal/mole;

At 80°C $k_2 = 0.2$ lit mol⁻¹ min⁻¹; $E_{a2} = 8,000$ cal/mole;

At 80°C $\Delta H_{R1} = -15,000$ cal/mole extent; $\Delta H_{R2} = -10,000$ cal/mole extent;

Density of the solution = 1,000 g/liter; Heat Capacity of the solution = 1 cal/g°C.

5-17₄ The irreversible gas-phase reactions (both are first-order)

$$A \rightarrow 2V$$

$$V \rightarrow 2W$$

are carried out in a constant-volume batch reactor. Species A is charged into a 100 liter reactor, and initially P = 3 atm and T = 731°K ($C_{A0} = 0.05$ mole/liter). Based on the data below, calculate,

a. for isothermal operation, the time needed for 40% conversion of A.

b. The production rate of V in (a).

c. The heating/cooling load in (a).

d. For adiabatic operation, the time needed for 40% conversion of A.

e. The production rate of V in (d).

f. The needed time for 40% conversion if the reactor wall temperature is 750°K and U = 20 cal/cm²min°K.

Data: At 731°K $k_1 = 2$ min⁻¹; $k_2 = 0.5$ min⁻¹;

$E_{a1} = 8,000$ cal/mole; $E_{a2} = 12,000$ cal/mole;

At 731°K $\Delta H_{R1} = 3,000$ cal/mole of V; $\Delta H_{R2} = 4,500$ cal/mole of W;

$\hat{c}_{PA} = 65$ cal/mole °K; $\hat{c}_{PV} = 40$ cal/mole °K; $\hat{c}_{PW} = 25$ cal/mole °K.

5-18₄ The first-order gas-phase reaction A → B + C takes place in a constant-volume batch reactor. Species A is charged into a 200 liter reactor, and the initial conditions are 731°K and 10 atm. Based on the data below, derive the reaction operating curves and the temperature curve for each of the operation below. For each case, determine the operating time required for 80% conversion.

a. Isothermal operation at 731°K.

b. The heating/cooling load in (a).

c. Adiabatic operation.

d. Operation with heat-transfer (U = 10 cal/m² hr °K, S = 1 m², and $T_F = 750°K$).

Data: At 731°K $k = 0.2$ sec⁻¹; $E_a = 12,000$ cal/mole;

At 731°K $\Delta H_R = 10{,}000$ cal/mole extent;

$\hat{c}_{p_A} = 25$ cal/mole °K; $\hat{c}_{p_B} = 15$ cal/mole °K; $\hat{c}_{p_C} = 18$ cal/mole °K.

CHAPTER SIX

PLUG-FLOW REACTOR

The plug-flow reactor (PFR) is a mathematical model that depicts a certain type of continuous reactor operation. The model is based on three assumptions:
- the reactor is operated at steady-state,
- the fluid moves in a flat (piston-like or "plug") velocity profile, and
- there is no spatial (radial) variation in concentrations or temperature at any cross section in the reactor (no radial gradients).

Chemical transformations take place along the reactor and, consequently, species compositions, temperature, and pressure vary from point to point along the reactor.

In practice, the velocity profile is rarely flat, and radial gradients of concentration and temperature do exist, especially in large diameter reactors. Hence, the plug-flow reactor model does not describe exactly the conditions in actual reactors. However, it provides a convenient mathematical means to estimate the performance of some reactors. It also provides a measure of a flow reactor performance when there is no mixing inside. The plug-flow model adequately describes the reactor operation when one of the following two conditions is satisfied:
- Tubular reactors whose length is much larger than their diameter. The acceptable length-to-diameter ratio (L/D) depends on the flow conditions in the reactor. Typically, for turbulent flows, L/D > 20, and, for laminar flows, L/D > 200.
- Packed-bed reactors (the packing disperses the fluid laterally across the bed resulting in a near uniform flow).

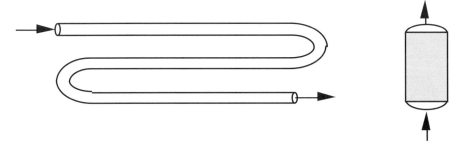

a. Long Tubular Reactor   b. Packed-Bed Reactor
Figure 6-1: Schematic Description of Plug-Flow Reactors

6.1 DESIGN EQUATIONS AND AUXILIARY RELATIONS

The differential, reaction-based design equation of a plug-flow reactor, written for the m-th independent reaction, derived in Chapter 3, is

$$\frac{dZ_m}{d\tau} = \left(r_m + \sum_{k}^{n_{dep}} \alpha_{km} \cdot r_k \right) \cdot \left(\frac{t_{cr}}{C_0} \right), \tag{6.1-1}$$

where Z_m is the dimensionless extent of the m-th independent reaction defined by

$$Z_m = \frac{\dot{X}_m}{(F_{tot})_0}, \tag{6.1-2}$$

τ is the dimensionless space time defined by

$$\tau = \frac{V_R}{v_0 \cdot t_{cr}}, \tag{6.1-3}$$

where t_{cr} is a conveniently-selected characteristic reaction time, and C_0 is a conveniently-selected reference concentration defined by

$$C_0 = \frac{(F_{tot})_0}{v_0}. \tag{6.1-4}$$

As discussed in Chapter 3, to describe the operation of a reactor with multiple reactions, we have to write (6.1-1) for each of the independent chemical reactions. The solutions of the design equations (the relationships between Z_m's and τ) are the operating curves of the reactor and completely describe its operation. To solve the design equations, we have to express the rates of the individual chemical reactions, r_m's and r_k's, in terms of Z_m's and τ. Below, we discuss the auxiliary relations used to express the design equations explicitly in terms of Z_m's and τ.

The volume-based rate of the i-th chemical reaction (see Section 2.3) is

$$r_i = k_i(T_0) \cdot e^{\gamma_i \frac{\theta-1}{\theta}} \cdot h_i(C_j's), \tag{6.1-5}$$

where $k_i(T_0)$ is the reaction rate constant at reference temperature T_0, γ_i is the dimensionless activation energy ($\gamma_i = Ea_i/RT_0$), and $h_i(C_j's)$ is a function of the species concentrations, given by the rate expression. To express the rates of the chemical reactions in terms of the reaction extents, we have to relate the species concentrations to Z_m's and τ. For plug-flow reactors, the concentration of species j at any point in the reactor is

$$C_j = \frac{F_j}{v}, \tag{6.1-6}$$

where F_j is the local molar flow rate of species j, and v is the local volumetric flow rate. Using stoichiometric relation (1.5-5),

$$F_j = (F_{tot})_0 \cdot \left(y_{j_0} + \sum_{m}^{n_{ind}} (s_j)_m \cdot Z_m \right), \qquad (6.1\text{-}7)$$

and the local concentration of species j is

$$C_j = \frac{(F_{tot})_0}{v} \cdot \left(y_{j_0} + \sum_{m}^{n_{ind}} (s_j)_m \cdot Z_m \right). \qquad (6.1\text{-}8)$$

For **liquid-phase** reactions, the density of the reacting fluid is practically constant; hence, $v = v_0$, and (6.1-8) reduces to

$$C_j = C_0 \cdot \left(y_{j_0} + \sum_{m}^{n_{ind}} (s_j)_m \cdot Z_m \right). \qquad (6.1\text{-}9)$$

Hence, (6.1-9) provides the species concentrations in terms of the extents of the independent reactions for liquid-phase reactions.

For **gas-phase** reactions, the volumetric flow rate changes along the reactor. Assuming ideal gas behavior, the local volumetric flow rate is

$$v = v_0 \left(\frac{F_{tot}}{(F_{tot})_0} \right) \cdot \left(\frac{T}{T_0} \right) \cdot \left(\frac{P_0}{P} \right). \qquad (6.1\text{-}10)$$

Using stoichiometric relation (1.5-6) to express the total molar flow rate in terms of the extents of the independent reactions, (6.1-10) becomes

$$v = v_0 \left(1 + \sum_{m}^{n_{ind}} \Delta_m \cdot Z_m \right) \cdot \theta \cdot \left(\frac{P_0}{P} \right), \qquad (6.1\text{-}11)$$

where θ is the dimensionless temperature, T/T_0. Substituting (6.1-11) into (6.1-8), we obtain

$$C_j = C_0 \cdot \frac{y_{j_0} + \sum_{m}^{n_{ind}} (s_j)_m \cdot Z_m}{\left(1 + \sum_{m}^{n_{ind}} \Delta_m \cdot Z_m \right) \cdot \theta} \cdot \left(\frac{P}{P_0} \right). \qquad (6.1\text{-}12)$$

Eq. (6.1-12) provides the species concentrations in terms of the extents of the independent reactions for gas-phase reactions in plug-flow reactors. Table 6.1 provides the design equations and auxiliary relations used in the design formulation of plug-flow reactors.

Note that (6.1-5) and (6.1-12) contain another dependent variable, θ, the dimensionless temperature. Since the temperature may vary along the reactor, we express its variation by applying the energy balance equation. For plug-flow reactors with negligible viscous work, the dimensionless energy balance equation, derived in Section 4.2, is

$$\frac{d\theta}{d\tau} = \frac{1}{CF(Z_m, \theta)} \cdot \left[HTN \cdot (\theta_F - \theta) - \sum_m^{n_{ind}} DHR_m \cdot \frac{dZ_m}{d\tau} \right], \qquad (6.1\text{-}13)$$

where HTN is the dimensionless heat-transfer number defined by (4.2-22),

$$HTN = \frac{U \cdot t_{cr}}{C_0 \cdot \hat{c}_{p0}} \cdot \left(\frac{S}{V} \right), \qquad (6.1\text{-}14)$$

and DHR_m is the dimensionless heat of reaction of the m-th independent chemical reaction, defined by (4.2-23),

$$DHR_m = \frac{\Delta H_{R_m}(T_0)}{T_0 \cdot \hat{c}_{p0}}. \qquad (6.1\text{-}15)$$

The quantity $\hat{c}_{p0}$ is the specific molar heat capacity of the reference stream, defined for gas-phase reactions by (4.2-59) and for liquid-phase reactions by (4.2-60). To design non-isothermal plug-flow reactors, we have to solve the design equations simultaneously with (6.1-13). Table 6.2 provides a summary of the energy balance equation and auxiliary relations used in the design formulation of plug-flow reactors.

Below, we discuss how to apply the design equations and the energy balance equations to determine various quantities concerning the operations of plug-flow reactors. In Section 6.2, we examine isothermal operations with single reactions to illustrate how the rate expressions are incorporated into the design equation and how rate expressions are determined. In Section 6.3, we expand the analysis to isothermal operations with multiple reactions. In Section 6.4, we consider non-isothermal operations with multiple reactions. In all these cases, we assume that the pressure drop along the reactor is negligible. In Section 6.5, we consider the effect of pressure drop on the operations of plug-flow reactors with gas-phase reactions.

6.2 ISOTHERMAL OPERATIONS WITH SINGLE REACTIONS

We start the analysis of plug-flow reactors by considering isothermal operations with single reactions. For isothermal operations, $d\theta/d\tau = 0$, and we have to solve only the design equations, and the energy balance equation provides the heating (or cooling) load necessary to maintain isothermal conditions. Furthermore, for isothermal opera-

Table 6.1: Design Equations and Related Quantities

Design Equation	For the m-th Independent Reaction $$\frac{dZ_m}{d\tau} = \left(r_m + \sum_{k}^{n_{dep}} \alpha_{km} \cdot r_k \right) \cdot \left(\frac{t_{cr}}{C_0} \right) \qquad \text{(A)}$$
Definitions	Dimensionless Space Time $$\tau \equiv \frac{V_R}{v_0 \cdot t_{cr}} \qquad \text{(B)}$$ Dimensionless Extent of the m-th Independent Reaction $$Z_m = \frac{\dot{X}_m}{(F_{tot})_0} \qquad \text{(C)}$$ Reference Concentration $$C_0 = \frac{(F_{tot})_0}{v_0} \qquad \text{(D)}$$ Characteristic Reaction Time $$t_{cr} = \frac{\text{Characteristic Concentration}}{\text{Characteristic Reaction Rate}} = \frac{C_0}{r_0} \qquad \text{(E)}$$
Species Concentrations	For Liquid-Phase Reactions $$C_j = C_0 \cdot \left(y_{j0} + \sum_{m}^{n_{ind}} (s_j)_m \cdot Z_m \right) \qquad \text{(F)}$$ For Gas-Phase Reactions $$C_j = C_0 \frac{y_{j0} + \sum_{m}^{n_{ind}} (s_j)_m \cdot Z_m}{\left(1 + \sum_{m}^{n_{ind}} \Delta_m \cdot Z_m \right) \cdot \theta} \cdot \left(\frac{P}{P_0} \right) \qquad \text{(G)}$$

Table 6.2: Energy Balance Equation and Related Quantities

Energy Balance Equation

$$\frac{d\theta}{d\tau} = \frac{1}{CF(Z_m, \theta)} \cdot \left[HTN \cdot (\theta_F - \theta) - \sum_{m}^{n_{ind}} DHR_m \cdot \frac{dZ_m}{d\tau} \right] \qquad (A)$$

Definitions and Auxiliary Relations

Dimensionless Temperature

$$\theta \equiv \frac{T}{T_0} \qquad (B)$$

Specific Molar Heat Capacity of Reference Stream

$$\hat{c}_{p0} = \sum_{j}^{all} y_{j0} \cdot \hat{c}_{p_j}(T_0) \quad \text{or} \quad \hat{c}_{p0} = \frac{\dot{m}}{(F_{tot})_0} \bar{c}_p \qquad (C)$$

Dimensionless Heat of Reaction of the m-th Independent Reaction

$$DHR_m = \frac{\Delta H_{R_m}(T_0)}{T_0 \cdot \hat{c}_{p0}} \qquad (D)$$

Dimensionless Heat Transfer Number

$$HTN = \frac{U \cdot t_{cr}}{C_0 \cdot \hat{c}_{p0}} \left(\frac{S}{V} \right) \qquad (E)$$

Dimensionless Heat Rate

$$\frac{\dot{Q}}{(F_{tot})_0 \cdot T_0 \cdot \hat{c}_{p0}} \qquad (F)$$

Correction Factor of Heat Capacity (Gas-Phase Reactions)

$$CF(Z_m, \theta) = \frac{\sum_{j}^{all} y_{j0} \cdot \hat{c}_{p_j}(\theta)}{\sum_{j}^{all} y_{j0} \cdot \hat{c}_{p_j}(1)} + \frac{\sum_{j}^{all} \hat{c}_{p_j}(\theta) \cdot \sum_{m}^{n_{ind}} (s_j)_m \cdot Z_m}{\sum_{j}^{all} y_{j0} \cdot \hat{c}_{p_j}(1)} \qquad (G)$$

tions, the individual reaction rates depend only on the species concentrations, and (6.1-5) reduces to

$$r_i = k_i(T_0) \cdot h_i(C_j's). \tag{6.2-1}$$

When single chemical reactions take place in the reactor, the operation is described by a single design equation, and (6.1-1) reduces to

$$\frac{dZ}{d\tau} = r \cdot \left(\frac{t_{cr}}{C_0}\right), \tag{6.2-2}$$

where Z is the dimensionless extent of the reaction, and r is its volume-based rate. We have to solve (6.2-2) subject to the initial condition that at $\tau = 0$, Z is specified. In some instances, we can solve (6.2-2) by separating the variables and integrating to obtain

$$\tau = \left(\frac{C_0}{t_{cr}}\right) \int_{Z_{in}}^{Z_{out}} \frac{dZ}{r} = \int_{Z_{in}}^{Z_{out}} \left(\frac{r_0}{r}\right) dZ, \tag{6.2-3}$$

where Z_{in} and Z_{out} are, respectively, the extent at the reactor inlet and outlet. Eq. (6.2-3) is the integral form of the dimensionless, reaction-based design equation of plug-flow reactors with a single chemical reaction. Note that the limits of the integral depend on the selection of the reference stream. The solution of the design equation, Z_{out} versus τ, is the dimensionless operating curve of the reactor. It describes the progress of the chemical reaction along the reactor. Furthermore, once Z_{out} is known, we can apply stoichiometric relation (6.1-7) to obtain the local molar flow rates of the individual species at any point along the reactor. Note that for given inlet conditions, design equations (6.2-1) and (6.2-3) have three variables: the dimensionless space time, τ, the reaction extent at the reactor outlet, Z_{out}, and the reaction rate, r. The design equation is applied to determine any one of these variables when the other two are provided. In a typical design problem, we determine the reactor volume necessary to obtain a specified extent (or conversion) for a given feed rate and reaction rate. A second application of the design equation is to determine the extent (or conversion) at the reactor outlet for a given reactor volume and reaction rate. The third application is to determine the reaction rate when the extent (or conversion) at the reactor outlet is provided for different feed rates.

Note that if one prefers to express the design equation in terms of the reactor volume, V_R, rather than the dimensionless space time τ, by substituting (6.1-3) and (6.1-4) into (6.2-2),

$$\frac{dZ}{dV_R} = \frac{1}{(F_{tot})_0} \cdot r. \tag{6.2-4}$$

Separating the variables and integrating, (6.2-4) becomes

$$V_R = (F_{tot})_0 \int_{Z_{in}}^{Z_{out}} \frac{dZ}{r}. \qquad (6.2\text{-}5)$$

For reactors with single chemical reactions, the common practice has been to express the design equation in terms of the conversion of the limiting reactant, f_A. We can easily express design equation (6.2-2) in terms of f_A since, for single reactions, the conversion is proportional to the extent. Using stoichiometric relation (1.5-11),

$$Z = -\frac{y_{A0}}{s_A} f_A, \qquad (6.2\text{-}6)$$

and (6.2-2) becomes

$$\frac{df_A}{d\tau} = \left(\frac{t_{cr}}{C_{A0}}\right) \cdot (-r_A), \qquad (6.2\text{-}7)$$

where $(-r_A) = -s_A \cdot r$ is the depletion rate of reactant A.

To solve the design equations (6.2-2) or (6.2-3), we have to express the reaction rate, r, in terms of the dimensionless extent, Z. To do so, we express the species concentrations in terms of Z. For single **liquid-phase** reactions, (6.1-9) reduces to

$$C_j = C_0(y_{j0} + s_j \cdot Z). \qquad (6.2\text{-}8)$$

For single **gaseous** reactions, assuming negligible pressure drop, (6.1-12) reduces to

$$C_j = C_0 \left(\frac{y_{j0} + s_j \cdot Z}{1 + \Delta \cdot Z}\right). \qquad (6.2\text{-}9)$$

Below, we analyze the operation of isothermal plug-flow reactors with single reactions for different types of chemical reactions. For convenience, we divide the analysis into two sections: design and determination of the rate expression. In the former, we determine the size of the reactor for a known reaction rate, specified feed rate, and specified conversion or a species production rate attainable a given reactor. In the second section, we determine the rate expression from reactor operating data.

6.2.1 Design

In general, the reaction rate is provided either in the form of experimental data or as an algebraic expression. First, we discuss the design of a plug-flow reactor when the reaction rate is provided in the form of experimental data without a rate expression. Examining the structure of (6.2-3), we see that if we know how r varies with Z, we can then plot (r_0/r) versus Z, and the area under the curve between Z_{in} and Z_{out} is equal to the dimensionless space time, τ (see Figure 6-2). Hence, we can obtain numerically a relationship between Z_{out} and τ and plot the dimensionless operating curve. From the operating curve, we can determine the necessary τ for any specified extent. The required reactor volume, V_R, is readily determined from the dimensionless space time using (6.1-3) $V_R = v_0 \cdot t_{cr} \cdot \tau$

In many applications, the experimental rate data of a chemical reaction are provided in terms of the formation (or depletion) rate of species j, (r_j), at different concentrations of species j, C_j. To obtain a relation between r and Z, we have to relate r to (r_j) and C_j to Z. Using kinetic relation (2.2-5),

$$r = \frac{(r_j)}{s_j}. \qquad (6.2\text{-}10)$$

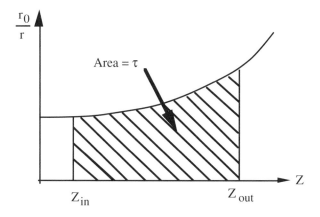

Figure 6-2: Determination of Dimensionless Space Time from Experimental Rate Data

As for the relationship between C_A to Z, we distinguish between liquid-phase and gas-phase reactions. For liquid-phase reactions, we use (6.1-9) and obtain

$$Z = \frac{C_A - y_{A_0} \cdot C_0}{s_A \cdot C_0}. \qquad (6.2\text{-}11)$$

For gas-phase reactions, we use (6.1-12) and obtain

$$Z = \frac{C_A - y_{A_0} \cdot C_0}{s_A \cdot C_0 - \Delta \cdot C_A}. \qquad (6.2\text{-}12)$$

In the two examples below, we illustrate how these relations are used to design plug-flow reactors with liquid-phase and gas-phase reactions.

Example 6-1 The liquid-phase reaction $A \rightarrow B + C$ is to be carried out in an isothermal plug-flow reactor. A solution with a concentration of $C_{A0} = 7$ mole/liter is fed into the reactor at a rate of 50 liter/min. The experimental rate data, obtained in a batch reactor, are provided below:

a. Plot the dimensionless operating curve of a plug-flow reactor, and
b. determine the reactor volume needed to obtain an outlet stream with $C_{Aout} = 1$ mole/liter.
c. If the stream is fed into an existing reactor whose volume is 200 liter, determine the outlet concentration of reactant A.

C_A (mole/lit)	0.0	1.0	2.0	3.0	4.0	5.0	6.0	7.0	8.0
$(-r_A)$ (mole/lit. min.)	0.0	0.118	0.222	0.316	0.400	0.476	0.546	0.609	0.667

Solution This example illustrates how to design a plug-flow reactor when the reaction rate is not provided in the form of an algebraic expression. The stoichiometric coefficients of the reaction are: $s_A = -1$ and $s_B = 1$. To utilize all the given rate data, we select the reference stream as an imaginary stream with $C_0 = 8$ mole/liter and a flow rate of $v_0 = 50$ liter/min. For this basis, $y_{A0} = C_{A0}/C_0 = 0.875$, $y_{B0} = 0$, and $y_{C0} = 0$. We use (6.2-10) to determine the reaction rate,

$$r = -\frac{(-r_A)}{s_A} = (-r_A),$$ (a)

and, using (2.5-1), the characteristic reaction time is

$$t_{cr} = \frac{C_0}{r_0} = \frac{8.0}{0.667} = 12 \text{ min}.$$ (b)

For liquid-phase reactions, we use (6.2-8) to relate the dimensionless extent Z to each corresponding value of the C_A,

$$Z = \frac{C_A - y_{A0} \cdot C_0}{s_A \cdot C_0} = \frac{C_A - 7.0}{(-1) \cdot 8.0} = \frac{7.0 - C_A}{8.0}.$$ (c)

The calculated values of r and Z are shown in the table below:

C_A (mole/lit)	0.0	1.0	2.0	3.0	4.0	5.0	6.0	7.0	8.0
Z (dimensionless)	0.875	0.750	0.625	0.500	0.375	0.250	0.125	0.0	—
r (mole/lit. min.)	0.0	0.118	0.220	0.316	0.400	0.476	0.546	0.609	0.667
(r_0/r) (dimensionless)	∞	5.65	3.00	2.11	1.667	1.40	1.221	1.095	1.00

The first figure below shows the plot of (r_0/r) versus Z for this reaction. Now that the curve is known, we can determine the needed dimensionless operating time τ for any value of Z by calculating the area under the curve. We integrate numerically using the trapezoidal method; thus, the area of the i-th increment is

$$\Delta\tau_i = \frac{1}{2}\left[\left(\frac{r_0}{r}\right)_i + \left(\frac{r_0}{r}\right)_{i-1}\right] \cdot (Z_i - Z_{i-1}).$$

The calculated values of Z and τ are shown in the table below:

Z (dimensionless)	0.875	0.750	0.625	0.500	0.375	0.250	0.125	0.0
τ (dimensionless)	∞	1.596	1.056	0.736	0.500	0.308	0.145	0.0

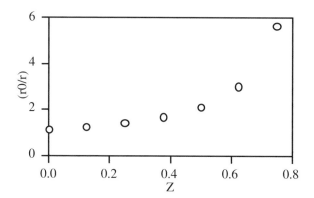

The figure below shows the plot of $Z(\tau)$ versus τ — the dimensionless reaction operating curve.

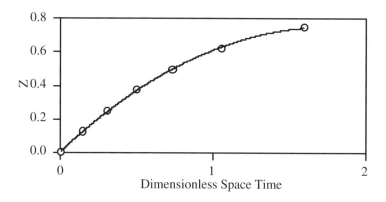

b. For $C_A = 1$ mole/lit, $Z = 0.75$, and, using the dimensionless operating curve, $\tau = 1.596$. Hence, using the definition of the dimensionless space time, (6.1-3), the volume of the reactor is

$$V_R = v_0 \cdot t_{cr} \cdot \tau = (50 \text{ liter/min}) \cdot (12 \text{ min}) \cdot (1.596) = 957.6 \text{ liter.} \qquad \text{(f)}$$

c. Using (6.1-3), for a 200 liter reactor, the dimensionless space time is

$$\tau = \frac{V_R}{v_0 \cdot t_{cr}} = \frac{200}{50 \cdot 12} = 0.333. \qquad \text{(g)}$$

From the reaction operating curve (or interpolation of the values in the table above), for $\tau = 0.333$, $Z = 0.270$. Using (b), the outlet concentration is

$$C_A = 8.0 \cdot (0.875 - Z) = 4.84 \text{ mole/liter.}$$

Example 6-2 The gas-phase reaction $A + B \rightarrow C$ is investigated in a constant-volume batch reactor. The reactor is charged with a mixture of 40% A and 60% B (% mole) at initial pressure of 3 atm. and then operated isothermally at 400°K. The following data was obtained:

$C_A \cdot 10^2$ (mole/lit)   3.656 3.199 2.742 2.285 1.828 1.371 0.914 0.457
$(-r_A) \cdot 10^2$ (mole/lit. min)   3.146 2.700 2.270 1.865 1.483 1.123 0.787 0.463

We now want to design an isothermal plug-flow reactor to process a stream at 3 atm. and 400°K that consists of 40% A and 60% B (% mole). The feed rate to the reactor is 200 mole/min. Based on the experimental rate data,

a. plot the dimensionless operating curve of a plug-flow reactor, and
b. determine the reactor volume needed to obtain 80% conversion.
c. If the stream is fed into an existing reactor whose volume is 200 liter, determine the outlet conversion of reactant A.

Solution The chemical reaction is $A + B \rightarrow C$, and the stoichiometric coefficients are: $s_A = -1$, $s_B = -1$, $s_C = 1$, and $\Delta = -1$. We select the inlet stream as a reference stream; hence, $y_{A0} = 0.4$, $y_{B0} = 0.6$, and $y_{C0} = 0$. The reference concentration is

$$C_0 = \frac{P_0}{R \cdot T_0} = \frac{(3 \text{ atm})}{(0.08206 \text{ lit - atm/mole}°K) \cdot (400°K)} = 9.14 \cdot 10^{-2} \text{ mole/liter}, \quad \text{(a)}$$

and the volumetric flow rate of the reference stream is

$$v_0 = \frac{(F_{tot})_0}{C_0} = \frac{(200 \text{ mole/min})}{(9.14 \cdot 10^{-2} \text{ mole/liter})} = 2,188 \text{ liter/min}. \quad \text{(b)}$$

For gas-phase reactions, we use (6.2-8) to relate the dimensionless extent Z to each corresponding value of the C_A,

$$Z = \frac{C_A - y_{A0} \cdot C_0}{s_A \cdot C_0 - \Delta \cdot C_A} = \frac{0.4 \cdot C_0 - C_A}{C_0 - C_A}. \quad \text{(c)}$$

To obtain the reaction rate at each point, we use (6.2-10),

$$r = -\frac{(-r_A)}{s_A} = (-r_A). \quad \text{(d)}$$

From the given rate data, at the reference stream, $r_0 = 3.146$ and

$$t_{cr} = \frac{C_0}{r_0} = \frac{0.0914 \text{ mole/lit}}{3.146 \text{ mole/lit hr}} = 0.02905 \text{ hr} = 1.74 \text{ min}. \quad \text{(e)}$$

The calculated values of r and Z are shown in the table below:

$C_A \cdot 10^2$ (mole/lit)   3.656   3.199   2.742 2.285 1.828 1.371 0.914 0.457
Z (dimensionless)   0.0   0.0769 0.143 0.200 0.250 0.294 0.333 0.368
(r_0/r) (dimensionless) 1.000   1.165   1.386 1.687 2.121 2.801 3.997 6.795

The figure below shows the plot of $(r_0/r)\cdot 1/r$ versus Z for this operation.

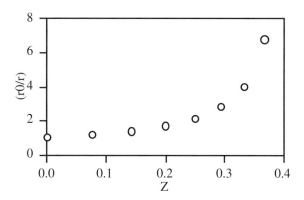

a.   Now that the curve is known, we can construct the dimensionless operating curve by determining the area under the curve (τ) for each value of Z. We integrate numerically using the trapezoidal method (see Appendix); thus, the area of the i-th increment is

$$\Delta\tau_i = \frac{1}{2}\left[\left(\frac{r_0}{r}\right)_i + \left(\frac{r_0}{r}\right)_{i-1}\right]\cdot(Z_i - Z_{i-1}).$$

The calculated values of Z and τ are shown in the table below:

Z	0.0	0.0769	0.143	0.200	0.250	0.294	0.333	0.368
τ	0.0	0.0832	0.1675	0.2551	0.3503	0.4586	0.5912	0.780

The figure below shows the plot of Z versus τ — the reaction operating curve.

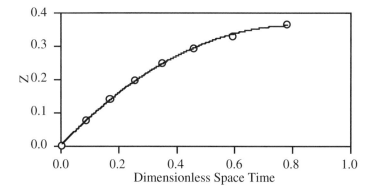

b. We use the operating curve to design the reactor. Applying (6.2-6),

$$Z = -\frac{y_{A0}}{s_A} f_A = -\frac{0.4}{(-1)} 0.8 = 0.32. \tag{f}$$

From the curve (or tabulated values), $Z = 0.32$ is reached at $\tau = 0.55$. Applying the definition of the dimensionless time, (6.1-3),

$$V_R = v_0 t_{cr} \tau = (2{,}188 \text{ liter/min}) \cdot (0.969 \text{ min}) \cdot (0.55) = 1{,}166 \text{ liter}.$$

c. For a 200 liter plug flow reactor, the dimensionless operating time is

$$\tau = \frac{V_R}{v_0 \cdot t_{cr}} = \frac{(200 \text{ liter})}{(2{,}188 \text{ liter/min}) \cdot (0.969 \text{ min})} = 0.0943.$$

From the operating curve, at $\tau = 0.0943$, $Z = 0.110$, and the outlet conversion is

$$f_{A_{out}} = -\frac{s_A}{y_{A0}} Z_{out} = -\frac{-1}{0.4} 0.110 = 0.275. \tag{g}$$

We continue the analysis of isothermal plug-flow reactors with single reactions and consider applications of the design equation when the reaction rate is provided in the form of algebraic expressions. For these cases, we use either (6.2-8) or (6.2-9) to express the species concentrations in terms of dimensionless extent Z, substitute these relations into the rate expression, and substitute the latter into the design equation, (6.2-2). We obtain a first-order differential equation that should be solved, either analytically or numerically, under specified initial (inlet) conditions. The solution of the design equation (Z versus τ) is the dimensionless operating curve of the reactor.

Consider, for example, the first-order gas-phase chemical reaction A $\rightarrow$ Products and a stream with a given composition to be fed to the reactor. For this reaction, $s_A = -1$, Δ is determined by the stoichiometry, $r = k \cdot C_A$, and y_{A0}, y_{B0}, y_{C0}, etc., are specified. Using (6.2-9), the reaction rate expression is

$$r = k \cdot C_0 \cdot \frac{y_{A0} - Z}{1 + \Delta \cdot Z}, \tag{6.2-13}$$

and, substituting (6.2-13) into (6.2-2), the design equation is

$$\frac{dZ}{d\tau} = k \cdot t_{cr} \cdot \frac{y_{A0} - Z}{1 + \Delta \cdot Z}. \tag{6.2-14}$$

Using (2.5-3), for first-order reactions, the characteristic reaction time is $t_{cr} = 1/k$, and, if only species A is fed into the reactor ($y_{A0} = 1$), the design equation reduces to

$$\frac{dZ}{d\tau} = \frac{1 - Z}{1 + \Delta \cdot Z}. \tag{6.2-15}$$

The solution of this design equation for different values of Δ is shown in Figure 6-3.

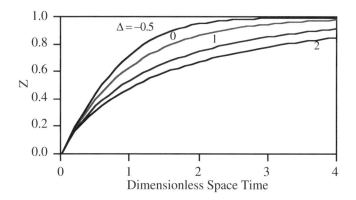

Figure 6-3: Operating Curves for Plug-Flow Reactors with a Single First-Order
Reaction of the Form $A \rightarrow$ Products

The figure illustrates the effect of the change in the volumetric flow rate of the gas
along the reactor. We see that when Δ is larger, a larger reactor volume (V_R is proportional to τ) is required to achieve a given level of extent. Note that the curve for $\Delta = 0$
is also the operating curve for liquid-phase reactions. To grasp what physically happens in the reactor, consider a certain amount of feed near the reactor inlet, and follow
its movement along the reactor, as shown schematically in Figure 6-4. When the reaction increases the total number of moles ($\Delta > 0$), the volume of the selected amount
increases as it moves down the reactor. Since the diameter of the reactor is the same,
the fluid downstream has to move faster than the fluid upstream. Consequently, the
reaction time becomes progressively shorter along the reactor, and a larger reactor volume is needed to achieve a given conversion.

Figure 6-4: Gas Expansion Along a Plug-Flow Reactor

Example 6-3 A gas-phase, decomposition reaction, $A \rightarrow B + C$, is carried out in an
isothermal, plug-flow reactor. The reaction is first-order, and its rate constant at the
operating temperature is $k = 10 \text{ sec}^{-1} = 600 \text{ min}^{-1}$. A gaseous stream of reactant $C_{A0} =$
$4 \cdot 10^{-3}$ lbmole/ft^3 is fed into the reactor at a rate of 100 lbmole/min. Determine:
a. the volume of the reactor needed to achieve 80% conversion and
b. the volume of the reactor needed to achieve 80% conversion if, by mistake, we use
   $\Delta = 0$.

c. What would the actual conversion be if we use a PFR with the volume calculated in (b)?

d. What should be the feed flow rate if we want to maintain 80% on a plug-flow reactor with the volume calculated in (b)?

Solution The chemical reaction is $A \rightarrow B + C$, and the stoichiometric coefficients are:

$$s_A = -1; \quad s_B = 1; \quad s_C = 1; \quad \text{and} \quad \Delta = 1.$$

We select the inlet stream as the reference stream ($Z_{in} = 0$), and, since only species A is fed into the reactor, $y_{A0} = 1$, $y_{B0} = y_{C0} = 0$, $C_0 = C_{A0} = 4 \cdot 10^{-3}$ lbmole/ft^3, and $F_{A0} = (F_{tot})_0 = 100$ lbmole/min. The volumetric feed rate is

$$v_0 = \frac{(F_{tot})_0}{C_0} = \frac{100 \text{ lbmole/min}}{4 \cdot 10^{-3} \text{ lbmole/ft}^3} = 25 \cdot 10^3 \text{ ft}^3/\text{min} = 416.7 \text{ ft}^3/\text{sec}.$$

Using (6.2-6),

$$Z_{out} = -\frac{y_{A0}}{s_A} f_{A_{out}} = 0.80.$$

For plug-flow reactors with a single chemical reaction, the design equation is

$$\frac{dZ}{d\tau} = r \cdot \left(\frac{t_{cr}}{C_0}\right). \tag{a}$$

For gas-phase reactions, we express C_A in terms of Z by (6.2-9), and the reaction rate is

$$r = k \cdot C_0 \cdot \frac{1 - Z}{1 + \Delta \cdot Z}. \tag{b}$$

For a first-order reaction, the characteristic reaction time is

$$t_{cr} = \frac{1}{k} = 10 \text{ sec}. \tag{c}$$

Substituting (b) and (c) into (a), the dimensionless design equation is

$$\frac{dZ}{d\tau} = \frac{1 - Z}{1 + Z}. \tag{d}$$

Separating the variables and integrating, we obtain

$$\tau = \int_0^{0.8} \frac{1 + Z}{1 - Z} dZ = 2.42. \tag{e}$$

Using (6.1-3), the reactor volume needed is

$$V_R = \tau \cdot v_0 \cdot t_{cr} = \frac{\tau \cdot v_0}{k} = \frac{(2.42) \cdot (416.7 \text{ ft}^3/\text{sec})}{(10 \text{ sec}^{-1})} = 100.8 \text{ ft}^3. \qquad (f)$$

b. If we do not account for the change in the volumetric flow rate along the reactor ($\Delta = 0$), the differential design equation is

$$\frac{dZ}{d\tau} = 1 - Z. \qquad (g)$$

We solve (g) by separating the variables and integrating

$$\tau = \int_0^{0.8} \frac{1}{1-Z} dZ = 1.61. \qquad (h)$$

Hence, the calculated reactor volume is

$$V_R = \tau \cdot v_0 \cdot t_{cr} = \frac{\tau \cdot v_0}{k} = \frac{(1.61) \cdot (25 \cdot 10^3 \text{ ft}^3/\text{min})}{(600 \text{ min}^{-1})} = 67.1 \text{ ft}^3. \qquad (i)$$

Note that by using the wrong concentration expression, we specify a reactor volume that is only 67% of the required volume.

c. We now calculate the actual outlet conversion of a plug-flow reactor with volume of 67.1 ft^3. For $V_R = 67.1$ ft^3 and the given feed rate, the dimensionless space time is 1.61. To determine the outlet conversion, we should solve

$$1.61 = \int_0^{Z_{out}} \frac{1+Z}{1-Z} dZ. \qquad (j)$$

The solution is $Z_{out} = f_{Aout} = 0.682$. Hence, if we use the wrongly-specified reactor volume, we will obtain only 68.2% conversion instead of the desired 80%.

d. To attain a conversion of 0.80 on the 67.1 ft^3 reactor, the feed flow rate should be reduced. To determine the feed rate, we solve the dimensionless design equation (e) and then use the wrongly-specified reactor volume. We saw above that for $Z_{out} = 0.80$, $\tau = 2.42$; hence,

$$v_0 = \frac{V_R}{\tau \cdot t_{cr}} = \frac{V_R \cdot k}{\tau} = \frac{(67.1 \text{ ft}^3) \cdot (600 \text{ min}^{-1})}{2.42} = 16.64 \cdot 10^3 \text{ ft}^3/\text{min}. \qquad (k)$$

$$(F_{tot})_0 = v_0 C_0 = (16.64 \cdot 10^3 \text{ ft}^3/\text{min}) \cdot (4 \cdot 10^{-3} \text{ lbmole/ft}^3) = 66.6 \text{ lbmole/min}.$$

If we use the wrongly-specified reactor volume and want to maintain the specified conversion, we can process only 66.6 lbmole/min instead of 100 lbmole/min.

The design of chemical reactors and the values of calculated quantities (reactor volume, production rates, etc.) do not depend on the specific reference stream selected. In fact, we may use a fictitious stream as a reference stream (provided no portion of the stream is diverted from the reactor). We also indicated that, usually, it is convenient to select the inlet stream to the system as the reference stream. This is illustrated in Example 6-4 below. Example 6-5 illustrates the use of a fictitious reference stream. Example 6-6 shows the application of the design equation when the rate expression is not a power function of the concentrations.

Example 6-4 The second-order, gas-phase chemical reaction $2\,A \rightarrow B$ is conducted in two isothermal plug-flow reactors connected in series (see diagram). A feed stream consisting of 80% A and 20% inert (I) is fed into the system at a molar flow-rate of 125 mole/min. The concentration of reactant A in the feed stream is $C_{A0} = 5 \cdot 10^{-2}$ mole/liter. The molar flow rate of A at the outlet of the first reactor is $F_{A1} = 50$ mole/min, and we want to design the second reactor such that the molar flow rate of A at its outlet is $F_{A2} = 20$ mole/min. If the volume of the first reactor is 1,000 liter, determine the volume of Reactor 2. Carry out the calculations in two ways:
a. by selecting the inlet stream to the system as the reference stream and
b. by selecting the stream entering Reactor 2 as the reference stream.

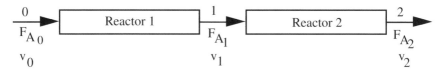

Solution The chemical reaction is $2\,A \rightarrow B$, and the stoichiometric coefficients are:

$$s_A = -2; \quad s_B = 1; \quad s_I = 0; \quad \text{and} \quad \Delta = -1.$$

a. We select the inlet stream to the system (Stream 0) as the reference stream. Hence, $y_{A0} = 0.8$, $y_{I0} = 0.2$, $y_{B0} = 0$, the reference concentration is

$$C_0 = \frac{C_{A0}}{y_{A0}} = \frac{0.05}{0.80} = 0.0625 \text{ mole/liter},$$

and the volumetric flow rate of the reference stream is

$$v_0 = \frac{(F_{tot})_0}{C_0} = \frac{125 \text{ mole/min}}{0.0625 \text{ mole/liter}} = 2{,}000 \text{ liter/min}.$$

For a gas-phase reaction, using (6.2-9), the rate expression is

$$r = k \cdot C_0^2 \cdot \left(\frac{y_{A0} + s_A \cdot Z}{1 + \Delta \cdot Z}\right)^2. \tag{a}$$

For a second-order reaction, the characteristic reaction time, t_{cr}, is

$$t_{cr} = \frac{1}{k \cdot C_0} . \tag{b}$$

Substituting (a) and (b) into (6.2-2), the design equation is

$$\frac{dZ}{d\tau} = \left(\frac{0.8 - 2 \cdot Z}{1 - Z}\right)^2 . \tag{c}$$

Separating the variables, the design equation for Reactor 2 is

$$\tau_2 = \int_{Z_{in}}^{Z_{out}} \left(\frac{1 - Z}{0.8 - 2 \cdot Z}\right)^2 dZ , \tag{d}$$

where $\tau_2 = V_{R2}/v_0 \cdot t_{cr}$ is the dimensionless space time of Reactor 2. To evaluate the integral, we have to specify Z_{in} and Z_{out}. Using stoichiometric relation (1.5-5), (1-18),

$$Z_{in} = Z_1 = \frac{F_{A_1} - (F_{tot})_0 \cdot y_{A_0}}{s_A \cdot (F_{tot})_0} = \frac{50 - 125 \cdot (0.8)}{(-2) \cdot 125} = 0.20 \tag{e}$$

$$Z_{out} = Z_2 = \frac{F_{A_2} - (F_{tot})_0 \cdot y_{A_0}}{s_A \cdot (F_{tot})_0} = \frac{20 - 125 \cdot (0.8)}{(-2) \cdot 125} = 0.32 . \tag{f}$$

Substituting these values into (d), we obtain

$$\tau_2 = \int_{0.20}^{0.32} \left(\frac{1 - Z}{0.8 - 2 \cdot Z}\right)^2 dZ = 0.980 . \tag{g}$$

Using (6.1-3) and (b), the volume of Reactor 2 is

$$V_{R_2} = \tau_2 \cdot v_0 \cdot t_{cr} = (0.980) \cdot (2,000 \text{ liter/min}) \cdot t_{cr} . \tag{h}$$

Since the value of k is not given, we do not know the value of t_{cr}. We determine it by writing the design equation for Reactor 1. Using design equation (6.2-3),

$$\tau_1 = \int_{0}^{0.20} \left(\frac{1 - Z}{0.8 - 2 \cdot Z}\right)^2 dZ = 0.483 , \tag{i}$$

and substituting into (6.1-3), the characteristic reaction time is

$$t_{cr} = \frac{V_{R_1}}{\tau_1 \cdot v_0} = \frac{(1,000 \text{ liter})}{(0.4829) \cdot (2,000 \text{ liter/min})} = 1.035 \text{ min} . \tag{j}$$

Substituting (j) into (h), the volume of Reactor 2 is

$$V_{R_2} = (0.9799) \cdot (2.000 \text{ liter/min}) \cdot (1.035 \text{ min}) = 2,029 \text{ liter}.$$

From (b), the value of the reaction rate constant is

$$k = \frac{1}{t_{cr} \cdot C_0} = \frac{1}{(1.035 \text{ min}) \cdot (6.25 \cdot 10^{-2} \text{ mole/liter})} = 15.46 \frac{\text{liter}}{\text{mole/min}}. \qquad (k)$$

b. Now, we calculate the volume of Reactor 2 when the inlet stream to Reactor 2 (Stream 1) is selected as the reference stream. Using stoichiometric relation (1.5-6), the molar flow rate of this stream is

$$(F_{tot})_1 = (F_{tot})_0 (1 + \Delta \cdot Z_1) = (125 \text{ mole/min}) \cdot (1 - 0.2) = 100 \text{ mole/min}, \qquad (l)$$

and, for isothermal and isobaric operation, the volumetric flow rate of Stream 1 is

$$v_1 = \frac{(F_{tot})_1}{(F_{tot})_0} v_0 = \frac{100}{125} (2,000 \text{ liter/min}) = 1,600 \text{ liter/min},$$

and

$$y_{A_1} = \frac{F_{A_1}}{(F_{tot})_1} = \frac{50}{100} = 0.50. \qquad (m)$$

For this case, $Z_{in} = 0$, and

$$Z_{out} = \frac{F_{A_2} - (F_{tot})_1 \cdot y_{A_1}}{s_A \cdot (F_{tot})_1} = \frac{20 - 100 \cdot (0.50)}{(-2) \cdot 100} = 0.15. \qquad (n)$$

Note that since the temperature and pressure are the same, $C_1 = C_0$ and t_{cr} is the same. Substituting these values in (6.2-3), the dimensionless design equation for Reactor 2 is

$$\tau_2 = \int_0^{0.15} \left(\frac{1 - Z}{0.50 - 2 \cdot Z} \right)^2 dZ = 1.225. \qquad (o)$$

Using (6.1-3) and writing (d) for Stream 1, the volume of Reactor 2 is

$$V_{R_2} = \tau_2 \cdot v_1 \cdot t_{cr} = (1.2248) \cdot (1,600 \text{ liter/min}) \cdot (1.035 \text{ min}) = 2,028 \text{ liter}. \qquad (p)$$

Note that, as expected, we obtained the same value for V_{R2} by both methods. However, the calculation based on selecting Stream 1 as the reference stream is lengthier.

Example 6-5 The elementary, gas-phase reaction A + B → C is carried out in an isothermal-isobaric plug-flow reactor operated at 2 atm and 170°C. At this temperature, k = 90 liter/(mole min), and the vapor pressure of the product, C, is 0.3 atm. The reactor is fed with two gas streams: the first one consists of 80% A, 10% B, 10% inert (I), and is at 2.5 atm and 150°C; the second consists of 80% B, 20% I, and is at 3 atm and 180°C. The first stream is fed at a rate of 100 mole/min and the second at a rate of

120 mole/min. Determine:

a. the conversion of A when C begins to condense,
b. the reactor volume where C starts to condense, and
c. the reactor volume needed for 85% conversion of A.
d. Plot the performance curve.

Solution In the first portion of the reactor, all the species are gaseous, the reaction is

$$A(g) + B(g) \rightarrow C(g),$$

and the stoichiometric coefficients are

$$s_A = -1; \quad s_B = -1; \quad s_C = 1; \quad \text{and} \quad \Delta = -1.$$

First, we have to select a reference stream. In this case, we select a fictitious stream at 2 atm and 170°C that is formed by combining the two feed streams. Hence,

$$F_{A0} = 0.8\,F_1 = (0.8)\,(100) = 80 \text{ mole/min}$$

$$F_{B0} = 0.1\,F_1 + 0.8\,F_2 = (0.1)\,(100) + (0.8)\,(120) = 106 \text{ mole/min}$$

$$F_{I0} = 0.1\,F_1 + 0.2\,F_2 = (0.1)\,(100) + (0.2)\,(120) = 34 \text{ mole/min}$$

$$(F_{tot})_0 = 220 \text{ mole/min}.$$

The composition of the reference stream is $y_{A0} = 0.364$, $y_{B0} = 0.482$, and $y_{I0} = 0.154$. Assuming ideal gas behavior, the total concentration of the reference stream is

$$C_0 = \frac{P_0}{R \cdot T_0} = \frac{(2 \text{ atm})}{(0.08205 \text{ lit - atm/mole}^\circ K) \cdot (443^\circ K)} = 5.50 \cdot 10^{-2} \text{ mole/lit}.$$

The volumetric flow rate of the reference stream is

$$v_0 = \frac{(F_{tot})_0}{C_0} = \frac{220 \text{ mole/min}}{0.055 \text{ mole/liter}} = 4{,}000 \text{ liter/min}.$$

a. Species C starts to condense when its partial pressure in the reactor is 0.3 atm. But, at any point in the reactor,

$$P_C = y_C \cdot P = \frac{F_C}{F_{tot}} P.$$

Using the stoichiometric relations (1.4-6) and (1.4-8),

$$P_C = \frac{s_C \cdot Z}{1 + \Delta \cdot Z} P = \frac{Z}{1 - Z}(2 \text{ atm}) = 0.3 \text{ atm}. \tag{a}$$

Solving (a), we obtain $Z = 0.130$, and using (6.2-6), the conversion is

$$f_A = -\frac{s_A}{y_{A0}} Z = -\frac{(-1)}{0.364} 0.130 = 0.358. \tag{b}$$

b. To determine the reactor volume where $f_A = 0.358$, we use the dimensionless design equation for a plug-flow reactor, (6.2-2),

$$\frac{dZ}{d\tau} = r \cdot \left(\frac{t_{cr}}{C_0}\right).$$
(c)

Since the reaction is elementary, $r = k \cdot C_A \cdot C_B$, and, using (6.2-9),

$$r = k \cdot C_A \cdot C_B = k \cdot C_0^2 \frac{(y_{A0} - Z) \cdot (y_{B0} - Z)}{(1 + \Delta \cdot Z)^2}.$$
(d)

Using (2.5-3), for a second-order reaction, the characteristic reaction time is

$$t_{cr} = \frac{1}{k \cdot C_0} = \frac{1}{(90 \text{ liter/mole min}) \cdot (5.50 \cdot 10^{-2} \text{ mole/liter})} = 0.202 \text{ min}.$$
(e)

Substituting (d) and (e) into (c) and the latter into (c), the dimensionless design equation reduces to

$$\frac{dZ}{d\tau} = \frac{(0.364 - Z) \cdot (0.482 - Z)}{(1 - Z)^2}.$$
(f)

To determine the dimensionless space time for $Z = 0.130$, we use (6.2-3)

$$\tau = \int_0^{0.130} \frac{(1 - Z)^2}{(0.364 - Z) \cdot (0.482 - Z)} \, dZ = 0.930.$$
(g)

Using (6.1-3) and (e), the volume of the reactor for $Z = 0.130$ is

$$V_R = \tau \cdot v_0 \cdot t_{cr} = (0.930) \cdot (4,000 \text{ lit/min}) \cdot (0.202 \text{ min}) = 753.4 \text{ liter}.$$

c. After a dimensionless extent of 0.130 is reached, the following reaction takes place in the reactor:

$$A(g) + B(g) \rightarrow C(liq).$$
(h)

For this reaction, $\Delta_{gas} = -2$. Hence, for $Z > 0.130$, a portion of C is formed in the gas-phase and a portion in the liquid-phase (with negligible volume). The total molar flow rate of the **gas-phase** is now (for $Z > 0.130$)

$$(F_{tot})_{gas} = (F_{tot})_0 \cdot [1 + (-1) \cdot 0.130 + (-2) \cdot (Z - 0.130)] = (F_{tot})_0 \cdot (1.13 - 2 \cdot Z).$$

Using (6.2-9), the concentrations of the two reactants are now (for $Z > 0.130$)

$$C_A = C_0 \cdot \frac{0.364 - Z}{1.13 - 2 \cdot Z} \quad \text{and} \quad C_B = C_0 \cdot \frac{0.482 - Z}{1.13 - 2 \cdot Z}.$$
(i)

Substituting (i) and (d) into (c), the dimensionless design equation is now

$$\frac{dZ}{d\tau} = \frac{(0.364 - Z) \cdot (0.482 - Z)}{(1.13 - 2 \cdot Z)^2}.$$ (j)

For $f_A = 0.85$,

$$Z = -\frac{y_{A0}}{s_A} f_A = -\frac{0.364}{(-1)} 0.85 = 0.3094.$$

To determine the dimensionless space time of the reactor section where product C is formed as a liquid, we integrate (j) between 0.130 and 0.3094,

$$\tau = \int_{0.130}^{0.3094} \frac{(1.13 - 2 \cdot Z)^2}{(0.364 - Z) \cdot (0.482 - Z)} \, dZ = 2.544.$$ (k)

Using (6.1-3) and (e), the volume of the reactor when condensed C is formed is

$$V_R = \tau \cdot v_0 \cdot t_{cr} = (2.544) \cdot (4,000 \text{ lit/min}) \cdot (0.202 \text{ min}) = 2,055.5 \text{ liter}.$$

The total volume of the reactor is $V_R = 753.4 + 2,055.5 = 2,809$ liter.
d. The reactor operation is described by two equations: (f) for $0 \le Z \le 0.130$ and (k) for $Z > 0.13$. The figure below shows the operating curve of the reactor. Note that in the second portion of the reactor, the expansion factor is smaller than in the first stage; consequently, the total reactor volume is smaller than that calculated by ignoring the condensation of species C. If we integrate (g) between 0 and 0.3094, we would obtain $\tau = 4.477$, which corresponds to a total reactor volume of 3,857 liter.

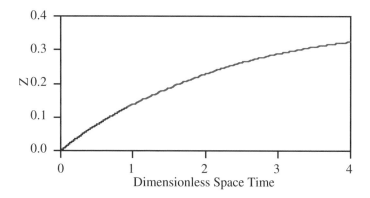

Example 6-6 Design a packed-bed reactor for the gas-phase, catalytic, cracking reaction

$$A \rightarrow B + C.$$

The reactor should produce 20 metric ton per day of B at 75% conversion of A. A process stream at 387°C and 2 atm, consisting of 90% A and 10% inert (I), is available in the plant. Based on the kinetic data below,

a. determine the feed rate of the process stream to the reactor,
b. derive the dimensionless design equation and plot the reactor operating curve,
c. determine the volume of the reactor, and
d. calculate the mass of the catalyst in the bed.

Data: Molecular mass of species B: 28 g/mole;
Bulk density of the catalyst bed: 1.30 kg/liter;
The reaction rate expression:

$$r_w = \frac{k_w \cdot C_A}{1 + K \cdot C_A} \quad \left(\frac{\text{mole extent}}{\text{g - catalyst second}} \right),$$

where $k_w = 0.192$ cm^3/g-cat sec, and $K = 60$ lit/mole.

Solution This example shows the use of the design equation when the rate expression is not a power function of the concentrations. The chemical reaction is $A \rightarrow B + C$, and the stoichiometric coefficients are:

$$s_A = -1; \quad s_B = 1; \quad s_C = 1; \quad s_I = 0; \quad \text{and} \quad \Delta = 1.$$

We select the feed stream as the reference stream; hence, $y_{A0} = 0.9$, $y_{I0} = 0.1$, and $y_{B0} = y_{C0} = 0$. The reference concentration is

$$C_0 = \frac{P_0}{R \cdot T_0} = \frac{(2 \text{ atm})}{(0.08205 \text{ lit - atm/mole}°K) \cdot (660°K)} = 3.69 \cdot 10^{-2} \text{ mole/lit}.$$

a. To determine the feed flow rate, we use stoichiometric relation (1.4-6) to express the production rate of species B in terms of $(F_{tot})_0$,

$$F_{B_{out}} = (F_{tot})_0 (y_{B_0} + s_B \cdot Z_{out}). \tag{a}$$

The extent relates to the conversion by (6.2-6),

$$Z_{out} = -\frac{y_{A0}}{s_A} f_{A_{out}} = -\frac{0.9}{(-1)} 0.75 = 0.675,$$

and the molar flow rate of B at the reactor outlet is

$$F_{B_{out}} = \frac{\dot{m}_B}{MW_B} = \frac{20 \cdot 10^3 \text{ kg/day}}{28 \text{ kg/kmole}} = 8.267 \text{ mole/sec}.$$

Hence, using (a), the feed molar flow rate is

$$(F_{tot})_0 = \frac{F_{B_{out}}}{s_B \cdot Z_{out}} = \frac{(8.267 \text{ mole/sec})}{(1) \cdot 0.675} = 12.25 \text{ mole/sec},$$

and the feed volumetric flow rate is

$$v_0 = \frac{(F_{tot})_0}{C_0} = \frac{12.25 \text{ mole/sec}}{3.69 \cdot 10^{-2} \text{ mole/lit}} = 332.0 \text{ lit/sec}.$$ (b)

b. The rate expression, defined on a catalyst mass basis, is

$$r_w = \frac{k_w \cdot C_A}{1 + K \cdot C_A}.$$ (c)

Using (2.2-4), the rate on a volumetric-based reaction rate is

$$r = \rho_{bed} \cdot \frac{k_w \cdot C_A}{1 + K \cdot C_A} = \frac{k \cdot C_A}{1 + K \cdot C_A},$$ (d)

where $k = \rho_{bed} \cdot k_w = 0.250 \text{ sec}^{-1}$ is the volume-based reaction rate constant. For a gas-phase reaction, using (6.2-9), the rate expression is

$$r = \frac{k \cdot C_0 \dfrac{0.9 - Z}{1 + \Delta \cdot Z}}{1 + K \cdot C_0 \dfrac{0.9 - Z}{1 + \Delta \cdot Z}}.$$ (e)

Substituting in (6.2-2), the differential, dimensionless, design equation is

$$\frac{dZ}{d\tau} = \frac{\dfrac{0.9 - Z}{1 + \Delta \cdot Z}}{1 + K \cdot C_0 \dfrac{0.9 - Z}{1 + \Delta \cdot Z}},$$ (f)

where the characteristic reaction time, defined by (2.5-3), is

$$t_{cr} = \frac{1}{k} = 4 \text{ sec}.$$ (g)

Substituting the values of $\Delta = 1$ and $K \cdot C_0 = 2.21$, (d) becomes

$$\frac{dZ}{d\tau} = \frac{0.9 - Z}{2.99 - 1.21 \cdot Z}.$$ (h)

We solve (h) numerically subject to the initial condition that at $\tau = 0$, $Z = 0$. The dimensionless reaction operating curve is shown in the figure below.

c. To determine the reactor volume, we use either the dimensionless operating curve or the integral form of the dimensionless design equation, (6.2-4),

$$\tau = \int_0^{0.675} \frac{2.99 - 1.21 \cdot Z}{0.9 - Z} \, dZ = 3.572.$$ (i)

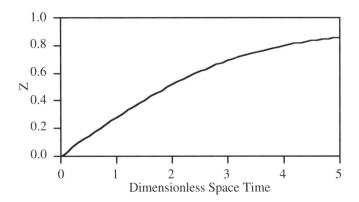

Using (6.1-3) and (g), the volume of the reactor is

$$V_R = \tau \cdot v_0 \cdot t_{cr} = \frac{\tau \cdot v_0}{k'} = \frac{(3.572) \cdot (332.0 \text{ lit/sec})}{(0.250 \text{ sec}^{-1})} = 4,743 \text{ liter}. \tag{j}$$

d. The mass of the catalyst bed is

$$M_{bed} = \rho_{bed} \cdot V_R = (1.30 \text{ kg/lit}) \cdot (4,743 \text{ liter}) = 6,166 \text{ kg}.$$

We conclude the discussion on the applications of the design equation for plug-flow reactors with two comments. First, note that the design equation provides us with the reactor volume needed to obtain a given conversion level. It does not indicate whether we should use a long reactor with a small diameter or a short reactor with a large diameter (provided of course that the plug-flow assumption is valid). The reactor diameter is determined by other considerations such as the pressure drop (pumping cost) and the heat-transfer area needed to provide (or remove) heat to the reactor. The effect of pressure drop on the performance of gaseous plug-flow reactors is discussed in Section 6.5. The effect of heat-transfer on the performance of plug-flow reactors is discussed in Section 6.4. Second, in the examples above, we designed the reactor for a given specified extent (or conversion level). However, the issue of what reaction extent should be specified is not discussed here. The cost of the reactants, value of the products, cost of the equipment, and the operating expenses (including separation costs) dictate the optimal value of the extent. These issues are covered in Chapter 9.

6.2.2 Determination of Reaction Rate

In the preceding section, we described how to apply the design equation when the reaction rate expression is given. Now, we will discuss how to determine the rate expression from data obtained during plug-flow operations. The method is based on measuring the exit composition of the reactor at different space times, and then, by differentiating the data, we obtain the reaction rate at the exit conditions. Different space times are obtained by either withdrawing samples at different points along the reactor (different reactor volumes) or by varying the feed flow rate. The approach is similar to the differential method applied to batch reactors.

To derive a relation between the reaction rate and the extent, we rearrange the design equation (6.2-2) and use (6.1-3) to obtain

$$r = C_0 \frac{dZ}{dt_{sp}}, \qquad (6.2\text{-}16)$$

where $t_{sp} = V_R/v_0$ is the space time. For the reactor outlet,

$$r_{out} = C_0 \frac{dZ_{out}}{dt_{sp}}. \qquad (6.2\text{-}17)$$

Hence, by plotting $C_0 \cdot Z_{out}$ versus t_{sp} and taking the derivatives, we can determine r_{out} as shown schematically in Figure 6-5.

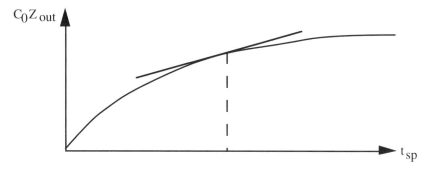

Figure 6-5: Determination of r from Data Obtained on a Plug-Flow Reactor

The main difficulty in using (6.2-17) is that the extent is not a measurable quantity. Therefore, we have to derive a relationship between Z_{out} and an appropriate measured quantity. We do so by using the design equation and relevant stoichiometric relations. In most applications, we measure the concentration of a species at the reactor outlet and calculate the extent by either (6.2-8) for liquid-phase reactions or (6.2-9) for gas-phase reactions. We can then determine the orders of the individual species for power rate expressions.

Example 6-7 A stream of gaseous A at 3 atm and 30°C (C_{A0} = 120 mmole/liter) is fed into a 500 liter plug-flow reactor where it decomposes according to the reaction A $\rightarrow$ 2 B + C. Based on the data below, determine:
a. the rate of reaction at C_A = 60 mmole/lit,
b. the order of the reaction, and

c. the rate constant at 30°C.

Data: v_0 (lit/min)   5   10.0 12.5 16.7 25   50   125   250 ∞

C_{Aout} (mmole/lit)   7.1 13.5 16.2 20. 26.8 40.6 62.6 79.5 120

Solution The reaction is A → 2 B + C, and the stoichiometric coefficients are:

$$s_A = -1; \quad s_B = 2; \quad s_C = 1; \quad \text{and} \quad \Delta = 2.$$

We select the inlet stream as the reference stream; hence, $y_{A0} = 1$, $y_{B0} = y_{C0} = 0$, and $C_0 = C_{A0} = 120$ mmole/liter. For gas-phase reactions, we calculate Z_{out} using (6.2-9),

$$Z_{out} = \frac{C_0 \cdot y_{A0} - C_{Aout}}{\Delta \cdot C_{Aout} - s_A \cdot C_0}. \tag{a}$$

The reactor space time is

$$t_{sp} = \frac{V_R}{v_0}. \tag{b}$$

For each run, we calculate t_{sp} using (b) and Z_{out}; then we plot $C_0 \cdot Z_{out}$ versus t_{sp}.

v_0 (lit/min)	5	10	12.5	16.7	25	50	125	250	∞
C_{Aout} (mmole/lit)	7.1	13.5	16.2	20.2	26.8	40.6	62.6	79.5	120
Z_{out}	0.841	0.724	0.681	0.622	0.537	0.395	0.234	0.145	0
t_{sp} (min)	100	50	40	30	20	10	4	2	0
$C_0 \cdot Z_{out}$ (mmole/lit)	111	86.9	81.7	74.6	64.4	47.4	28.1	17.0	0

Using (a), for $C_{Aout} = 60$ mmole/lit, $Z_{out} = 0.25$ and $C_0 \cdot Z_{out} = 30$ mmole/lit. We determine the slope at this point by fitting the first three data points to a second-order polynomial and then calculating the slope at $t_{sp} = 4.58$ min ($C_0 \cdot Z_{out} = 30$). The slope is $r = 2.91$ mmole/liter min.

b. Assuming the rate expression is in the form $r = k \cdot C_A{}^\alpha$, using (6.2-17), we differentiate the data to obtain the local rate. Then, noting that

$$\ln\left(C_0 \frac{dZ_{out}}{dt_{sp}} \right) = \ln(k) + \alpha \cdot \ln(C_{Aout}), \tag{c}$$

and by plotting $\ln(C_0 \cdot dZ_{out}/dt_{sp})$ versus $\ln(C_{Aout})$, we should get a straight line whose slope is a. We calculate the needed variables in the table below. Note that since the data points are not equally spaced, we calculate the derivative at the midpoints of each two adjacent points.

t_{sp} (min)	$C_0 \cdot Z_{out}$ (mmole/lit)	C_{Aout} (mmole/liter)	$\Delta(C_0 \cdot Z_{out})/\Delta t_{sp}$ (mmole/lit min)	$(C_{Aout})_{ave}$ (mmole/liter)
0	0.0	120.0		
2	17.3	79.5	8.65	99.7
4	28.1	62.6	5.40	71.0
10	47.4	40.6	3.27	51.6
20	64.4	26.8	1.02	23.5
40	81.7	16.2	0.71	16.5
50	86.9	13.5	0.52	14.9
100	111.0	7.1	0.48	10.3

The plot is shown below, and the slope is

$$\text{slope} = \frac{\ln\ 10 - \ln\ 0.35}{\ln\ 100 - \ln\ 10} = 1.48. \tag{d}$$

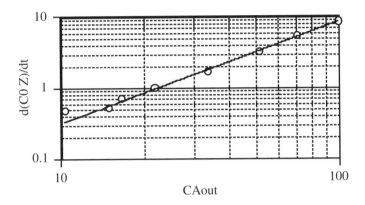

c. Now that the rate expression is known ($\alpha = 1.5$), we can use (6.2-3) to verify the rate expression and determine the value of k. It is similar to the integral method described in Section 5.4.1. For $\alpha = 1.5$, the dimensionless design equation, (6.2-2), is

$$\frac{dZ}{d\tau} = \left(\frac{1-Z}{1+2\cdot Z}\right)^{1.5}, \tag{e}$$

and the characteristic reaction time is

$$t_{cr} = \frac{1}{k \cdot C_{A_0}^{0.5}}. \tag{f}$$

Separating the variables and integrating,

$$\tau = \int_0^{Z_{out}} \frac{(1+2\cdot Z)^{1.5}}{(1-Z)^{1.5}} dZ = G(Z_{out}). \tag{g}$$

The right hand side of (g) is a function of Z_{out}, $G(Z_{out})$. Hence, by plotting $G(Z_{out})$ versus the space time, t_{sp}, we obtain a line whose slope is $1/t_{cr}$. The table below provides the calculated data, and the plot is shown in the figure below.

t_{sp} (min)	100	50	40	30	20	10	4	2	0
Z_{out}	0.841	0.724	0.681	0.622	0.537	0.395	0.234	0.144	0
$G(Z_{out})$	10.00	5.00	4.00	3.00	2.00	1.00	0.40	0.20	0

The slope is

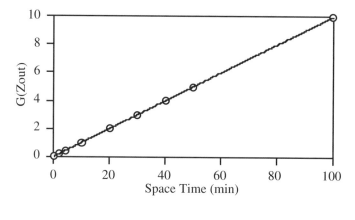

$$\text{slope} = \frac{1}{t_{cr}} = k \cdot C_{A_0}^{0.5} = 0.1 \text{ min}^{-1}, \tag{h}$$

and the rate constant at 30°C is $k = 9.13 \cdot 10^{-3}$ (lit/mmole)$^{0.5}$ min^{-1}.

6.3 ISOTHERMAL OPERATIONS WITH MULTIPLE REACTIONS

When more than one chemical reaction takes place in a plug-flow reactor, we have to address several issues before we start the design. First, we have to determine how many independent reactions there are (and how many design equations are needed) and should select a set of independent reactions. Next, we have to identify all the reactions that actually take place in the reactor (including dependent reactions) and express their rates. As discussed in Chapter 3, we have to write design equation (6.1-1) for each of the independent chemical reactions. To solve the design equations (obtain relationships between Z_m's and τ), we have to express the rates of the individual chemical reactions, r_m's and r_k's, in terms of the Z_m's and τ. The procedure for designing plug-flow reactors with multiple chemical reactions goes as follows:

a. Identify all the chemical reactions that take place in the reactor, and define the stoichiometric coefficients of each species in each reaction.
b. Determine the number of independent chemical reactions.
c. Select a set of independent reactions among the reactions whose rate expressions are given.
d. For each dependent reaction, determine its α_{km} multipliers with each of the independent reactions using (1.6-9).
e. Select a reference stream (determine T_0, C_0, v_0) and its species compositions, y_{j0}'s.
f. Specify the inlet conditions (T_{in}, Z_{min}'s).
g. Write a design equation for each independent chemical reaction.
h. Select a reaction, and define the characteristic reaction time, t_{cr}.

i. Express the reaction rates in terms of the dimensionless extents of the independent reactions, Z_m's, and the dimensionless temperature, θ.
j. Solve the design equations for Z_m's as functions of the dimensionless space time, τ, and obtain the dimensionless reaction operating curves.
k. Calculate the dimensionless species operating curves using (6.1-7).
l. Determine the reactor volume based on the most desirable value of τ obtained from the dimensionless operating curves, using (6.1-3).

Below, we describe the design formulation of isothermal plug-flow reactors with multiple reactions for various types of chemical reactions (reversible, series, parallel, etc.). In most cases, we solve the design equations numerically by applying a numerical technique such as the Runge-Kutta method or using commercial mathematical software such as HiQ, Mathcad, Maple, Mathematica, etc. In some simple cases, we can obtain analytical solutions. Note that, for isothermal operations, $d\theta = 0$, and we do not have to solve the energy balance equation simultaneously with the design equations.

Example 6-8 The reversible gas-phase chemical reaction A $\leftrightarrow$ 2 B takes place in a plug-flow reactor operated at 2 atm and 120°C. The forward reaction is first-order, and the backward reaction is second-order. We want to process a 1,000 mole/min stream of pure A at 120°C and 2 atm and achieve a level of 90% of the equilibrium conversion. At the inlet temperature (120°C), $k_1 = 0.1$ min^{-1} and $k_2 = 0.322$ lit mole^{-1} min^{-1}.
a. Derive the design equation, and plot the dimensionless reaction operating curve for a plug-flow reactor.
b. What is the equilibrium composition at 120°C in a plug-flow reactor for $k_2 \cdot C_0/k_1 = 0.5$?
c. What is the required reactor volume to obtain 90% of the equilibrium conversion?
d. If a stream of Species B is fed into the reactor at a volumetric flow rate of 100 lit/min, derive the dimensionless design equation, and plot the operating curve for $k_2 \cdot C_0/k_1 = 0.5$.
e. For a feed of B, what are the mole fractions of A and B at equilibrium for $k_2 \cdot C_0/k_1 = 0.5$?

Solution We treat a reversible reaction as two separate reactions:

$$\text{Reaction 1:} \quad A \rightarrow 2\,B \qquad r_1 = k_1\,C_A$$

$$\text{Reaction 2:} \quad 2\,B \rightarrow A \qquad r_2 = k_2\,C_B.$$

The stoichiometric coefficients of the chemical reactions are

$$s_{A1} = -1; \; s_{B1} = 2; \; \Delta_1 = 1;$$

$$s_{A2} = 1; \; s_{B2} = -2; \; \Delta_2 = -1.$$

We select the forward reaction (Reaction 1) as the independent reaction and the reverse reaction as the dependent reaction. Hence, the index of the independent reaction is m = 1, the index of the dependent reaction is k = 2, and, since Reaction 2 is the reverse of Reaction 1, $\alpha_{21} = -1$. Since there is only one independent reaction, we can describe the operation by a single design equation. Using (6.1-1), the design equation is

$$\frac{dZ}{d\tau} = (r_1 - r_2)\frac{t_{cr}}{C_0}.$$ (a)

We select the inlet stream as the reference stream; hence y_{A0}, y_{B0}, and C_0 are specified, and $Z_{1in} = 0$. For gas-phase reactions, using (6.1-12), the concentrations of the two species in terms of the extent of the independent reaction are:

$$C_A = C_0 \frac{y_{A0} + s_{A_1} \cdot Z_1}{1 + \Delta_1 \cdot Z_1}$$ (b)

$$C_B = C_0 \frac{y_{B0} + s_{B_1} \cdot Z_1}{1 + \Delta_1 \cdot Z_1}.$$ (c)

The rates of the two reactions are, respectively,

$$r_1 = k_1 \cdot C_0 \frac{y_{A0} + s_{A_1} \cdot Z_1}{1 + \Delta_1 \cdot Z_1}$$ (d)

$$r_2 = k_2 \cdot C_0^2 \left(\frac{y_{B0} + s_{B_1} \cdot Z_1}{1 + \Delta_1 \cdot Z_1} \right)^2.$$ (e)

We select Reaction 1 as the leading reaction and define the characteristic reaction time

$$t_{cr} = \frac{1}{k_1} = 10 \text{ min}.$$ (f)

Substituting (d) and (e) into (a), the design equation is

$$\frac{dZ_1}{d\tau} = \frac{y_{A0} + s_{A_1} \cdot Z_1}{1 + \Delta_1 \cdot Z_1} - \left(\frac{k_2 \cdot C_0}{k_1} \right) \cdot \left(\frac{y_{B0} + s_{B_1} \cdot Z_1}{1 + \Delta_1 \cdot Z_1} \right)^2.$$ (g)

a. When only reactant A is fed into the reactor, $y_{A0} = 1$, $y_{B0} = 0$, and (g) reduces to

$$\frac{dZ_1}{d\tau} = \frac{1 - Z_1}{1 + Z_1} - \left(\frac{k_2 \cdot C_0}{k_1} \right) \cdot \left(\frac{2 \cdot Z_1}{1 + Z_1} \right)^2.$$ (h)

We solve (h) numerically subject to the initial condition that at $\tau = 0$, $Z_1 = 0$ for different values of $k_2 \cdot C_0/k_1$. The figure below shows the operating curves for $k_2 \cdot C_0/k_1 = 0$; 0.5, 1.0, and 2.0. Note that the curve for $k_2 \cdot C_0/k_1 = 0$ represents the solution of an irreversible reaction.

b. At equilibrium, $dZ_1/d\tau = 0$, and we obtain,

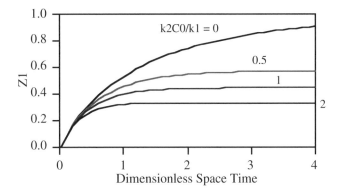

$$Z_{1eq} = \left(1 + 4\frac{k_2 \cdot C_0}{k_1}\right)^{-0.5}. \tag{i}$$

For $k_2 C_0/k_1 = 0.5$, $Z_{1eq} = 0.577$, and, using (6.2-6), the equilibrium conversion is $f_{Aeq} = 0.577$. The species compositions at equilibrium are

$$y_{Aeq} = \frac{y_{A0} - Z_{1eq}}{1 + \Delta_1 \cdot Z_{1eq}} = \frac{1 - 0.577}{1 + 0.577} = 0.268 \tag{j}$$

$$y_{Beq} = \frac{y_{B0} + 2 \cdot Z_{1eq}}{1 + \Delta_1 \cdot Z_{1eq}} = \frac{2 \cdot 0.577}{1 + 0.577} = 0.732. \tag{k}$$

c. The extent for 90% of the equilibrium conversion is $(0.577) \cdot (0.9) = 0.519$. From the operating curve for $k_2 \cdot C_0/k_1 = 0.5$, an extent of 0.519 is reached at $\tau = 1.506$. Using (6.1-3) and (f), the required reactor volume is

$$V_R = \tau \cdot v_0 \cdot t_{cr} = (1.506) \cdot (100 \text{ lit/min}) \cdot (10 \text{ min}) = 1,506 \text{ liter}.$$

d. When species B is fed into the reactor, $y_{A0} = 0$, $y_{B0} = 1$, and (g) reduces to

$$\frac{dZ_1}{d\tau} = \frac{-Z_1}{1 + Z_1} - \left(\frac{k_2 \cdot C_0}{k_1}\right)\left(\frac{1 + 2 \cdot Z_1}{1 + Z_1}\right)^2. \tag{l}$$

We solve (l) numerically for $k_2 \cdot C_0/k_1 = 0.5$ subject to the initial condition that at $\tau = 0$, $Z_1 = 0$. The figure below shows the reaction operating curve for this operation. Note that Z_1 is negative because the selected independent reaction proceeds backward.

e. At equilibrium, $dZ_1/d\tau = 0$, and, from (l), we obtain $Z_{1eq} = -0.211$. The species compositions at equilibrium are

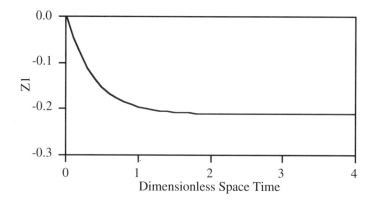

$$y_{A_{eq}} = \frac{y_{A_0} - Z_{1_{eq}}}{1 + \Delta_1 \cdot Z_{1_{eq}}} = \frac{0 - (-0.211)}{1 + (-0.211)} = 0.267 \qquad \text{(m)}$$

$$y_{B_{eq}} = \frac{y_{B_0} + 2 \cdot Z_{1_{eq}}}{1 + \Delta_1 \cdot Z_{1_{eq}}} = \frac{1 + 2 \cdot (-0.211)}{1 + (-0.211)} = 0.733. \qquad \text{(n)}$$

We see that, as expected, the equilibrium composition is the same regardless of what species is fed to the reactor.

Example 6-9 A valuable product B is produced in a plug-flow reactor by the gas-phase, first-order chemical reaction

$$A \rightarrow 2\,B.$$

Under the operating conditions, B also decomposes according to the first-order reaction

$$B \rightarrow C + D.$$

A gaseous stream of reactant A (C_{A0} = 0.04 mole/lit) is fed into a 200 liter tubular reactor at a rate of 100 liter/min. At the reactor temperature, k_1 = 2 min^{-1} and k_2 = 1 min^{-1}.

a. Derive the design equations, and plot the reaction and species operating curves.
b. Determine the conversion of A and the production rate of B and of C for the given feed rate.
c. Determine the optimal reactor volume to maximize B production.
d. Determine the conversion of A, the production rate of B and of C, and the yield of B in the optimal reactor.

Solution This is an example of a series (consecutive) chemical reaction. The chemical reactions taking place in the reactor are:

$$\text{Reaction 1:} \quad A \rightarrow 2\,B,$$

$$\text{Reaction 2:} \quad B \rightarrow C + D,$$

and their stoichiometric coefficients are:

$$s_{A1} = -1; \quad s_{B1} = 2; \quad s_{C1} = 0; \quad s_{D1} = 0; \quad \Delta_1 = 1;$$

$$s_{A2} = 0; \quad s_{B2} = -1; \quad s_{C2} = 1; \quad s_{D2} = 1; \quad \Delta_2 = 1.$$

Since each reaction has a species that does not participate in the other, the two reactions are independent, and there is no dependent reaction. We write the dimensionless design equation (6.1-1) for each of the independent reactions (m = 1, 2),

$$\frac{dZ_1}{d\tau} = r_1 \cdot \left(\frac{t_{cr}}{C_0} \right) \tag{a}$$

$$\frac{dZ_2}{d\tau} = r_2 \cdot \left(\frac{t_{cr}}{C_0} \right). \tag{b}$$

We select the inlet steam as the reference stream; hence, $Z_{1in} = Z_{2in} = 0$. Since only reactant A is fed into the reactor, $C_0 = C_{A0} = 4$ mole/lit, $y_{A0} = 1$, and $y_{B0} = y_{C0} = y_{D0} = 0$. Also,

$$(F_{tot})_0 = v_0 \cdot C_0 = (100 \text{ lit/min}) \cdot (0.04 \text{ mole/lit}) = 4 \text{ mole/min}.$$

Using stoichiometric relation (1.5-5), we express the local molar flow rate of each species in terms of Z_1 and Z_2,

$$F_A = (F_{tot})_0 \cdot (y_{A_0} + s_{A_1} \cdot Z_1 + s_{A_2} \cdot Z_2) = (F_{tot})_0 \cdot (1 - Z_1) \tag{c}$$

$$F_B = (F_{tot})_0 \cdot (y_{B_0} + s_{B_1} \cdot Z_1 + s_{B_2} \cdot Z_2) = (F_{tot})_0 \cdot (2 \cdot Z_1 - Z_2) \tag{d}$$

$$F_C = (F_{tot})_0 \cdot (y_{C_0} + s_{C_1} \cdot Z_1 + s_{C_2} \cdot Z_2) = (F_{tot})_0 \cdot Z_2 \tag{e}$$

$$F_D = (F_{tot})_0 \cdot (y_{D_0} + s_{D_1} \cdot Z_1 + s_{D_2} \cdot Z_2) = (F_{tot})_0 \cdot Z_2. \tag{f}$$

a. For gas-phase reactions, we use (6.1-12) and the rates are

$$r_1 = k_1 \cdot C_0 \frac{1 - Z_1}{1 + Z_1 + Z_2} \tag{g}$$

$$r_2 = k_2 \cdot C_0 \frac{2 \cdot Z_1 - Z_2}{1 + Z_1 + Z_2}. \tag{h}$$

We define the characteristic reaction time on the basis of Reaction 1; hence,

$$t_{cr} = \frac{1}{k_1} = 0.5 \text{ min}. \tag{i}$$

Substituting (g), (h), and (i) into (a) and (b), the design equations reduce to

$$\frac{dZ_1}{d\tau} = \frac{1 - Z_1}{1 + Z_1 + Z_2} \tag{j}$$

$$\frac{dZ_2}{d\tau} = \left(\frac{k_2}{k_1}\right)\frac{2 \cdot Z_1 - Z_2}{1 + Z_1 + Z_2}.$$ (k)

We solve (j) and (k) numerically subject to the initial condition that at $\tau = 0$, $Z_1 = Z_2 = 0$. The solution is shown in the figure below. Once we have Z_1 and Z_2 as a function of dimensionless space time, we use (c) through (f) to calculate the dimensionless molar flow rates of the individual species as a function of the dimensionless space time. The figure below shows these curves.

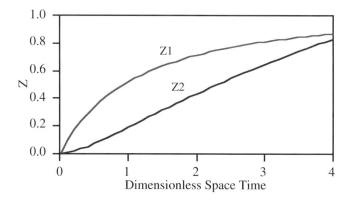

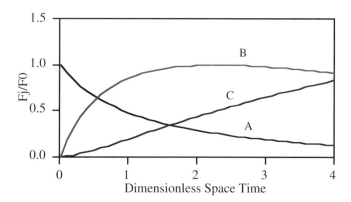

b. For the given feed flow rate, using (6.1-3) and (i), the dimensionless space time is

$$\tau = \frac{V_R}{v_0}k_1 = \frac{200 \text{ lit}}{100 \text{ lit/min}}(2 \text{ min}^{-1}) = 4.$$

At $\tau = 4$, the solutions of (j) and (k) are $Z_1 = 0.874$ and $Z_2 = 0.832$, respectively. Using (c) through (f), the species molar flow rates at the reactor outlet are

$$F_A = (4 \text{ mole/min}) \cdot (1 - 0.874) = 0.504 \text{ mole/min}$$
$$F_B = (4 \text{ mole/min}) \cdot (2 \cdot 0.874 - 0.832) = 3.66 \text{ mole/min}$$
$$F_C = F_D = (4 \text{ mole/min}) \cdot (0.832) = 3.33 \text{ mole/min}.$$

The conversion of A is

$$f_A = \frac{4 - 0.504}{4} = 0.874.$$

c. To determine the reactor volume that gives the highest production rate of B, we use the species operating curves (or tabulated calculated data) and find that maximum F_B is reached at $\tau = 2.30$. Using (6.1-3) and (i), the optimal reactor volume is

$$V_R = \frac{\tau \cdot v_0}{k_1} = \frac{(2.30) \cdot (100 \text{ lit/min})}{(2 \text{ min}^{-1})} = 115 \text{ liter}.$$

d. From the reaction operating curves, at $\tau = 2.30$, the solutions of (j) and (k) are $Z_1 = 0.751$ and $Z_2 = 0.501$. Using (c) through (f), the species molar flow rates at the reactor outlet are

$$F_A = (4 \text{ mole/min}) \cdot (1 - 0.751) = 0.996 \text{ mole/min}$$
$$F_B = (4 \text{ mole/min}) \cdot (2 \cdot 0.751 - 0.501) = 4.00 \text{ mole/min}$$
$$F_C = F_D = (4 \text{ mole/min}) \cdot (0.501) = 2.00 \text{ mole/min}.$$

The conversion of A is

$$f_A = \frac{4 - 0.996}{4} = 0.751.$$

The table below provides a comparison between the given and the optimal reactors:

	Given Reactor	Optimal Reactor
Reactor Volume (liter)	200	115
Volumetric Feed Rate (lit/min)	100	100
Conversion of A	0.950	0.751
Production Rate of B (mole/min)	3.66	3.99
Production Rate of C (mole/min)	3.33	2.00

We see that the volume of the optimal reactor is slightly more than half that of the given reactor. While the feed rate is the same, the conversion of A is slightly lower, the production rate of B is about 10% higher, and the production of C is reduced by 40%.

Example 6-10 Diels-Alder reactions are organic reactions between a hydrocarbon with two double bonds and another unsaturated hydrocarbon to form a cyclic molecule. However, in many cases, the species with the two double bonds reacts with itself to form an undesired product. A gaseous mixture containing 50% butadiene (B) and 50% acrolein (A) is fed at a rate of 1 mole/min into a plug-flow reactor, where the following reactions take place

$$A + B \rightarrow P \qquad r_1 = k_1 \cdot C_A \cdot C_B$$

$$2\,B \rightarrow W \qquad\qquad r_2 = k_2 \cdot C_B^2.$$

The reactor operates at 330°C and 2 atm. At these conditions, $k_1 = 5.86$ liter mole^{-1} min^{-1} and $k_2 = 0.72$ liter mole^{-1} min^{-1}. Assume ideal gas behavior.

a. Derive the design equations, and plot the performance curves for the independent reactions.

b. Plot the performance curves for the individual species.

c. Determine the operating time needed to achieve 80% conversion of B.

Solution The chemical reactions taking place in the reactor are

$$\text{Reaction 1:} \qquad A + B \rightarrow P,$$

$$\text{Reaction 2:} \qquad 2\,B \rightarrow W,$$

and the stoichiometric coefficients are:

$$s_{A1} = -1; \ s_{B1} = -1; \ s_{P1} = 1; \ s_{W1} = 0; \ \Delta_1 = -1;$$

$$s_{A2} = 0; \ s_{B2} = -2; \ s_{P2} = 0; \ s_{W2} = 1; \ \Delta_2 = -1.$$

Since each chemical reaction has a species that does not appear in the other, the two reactions are independent, and there is no dependent reaction in this case. We select the inlet stream as the reference stream; hence, $Z_{1in} = Z_{2in} = 0$,

$$C_0 = \frac{P_0}{R \cdot T_0} = \frac{(2 \text{ atm})}{(82.05 \cdot 10^{-3} \text{ lit - atm/mole}^\circ\text{K}) \cdot (603^\circ\text{K})} = 4.04 \cdot 10^{-2}\,\text{mole/lit} \qquad \text{(a)}$$

$$v_0 = \frac{(F_{tot})_0}{C_0} = \frac{(1 \text{ mole/min})}{(4.04 \cdot 10^{-2}\,\text{mole/lit})} = 24.75 \text{ lit/min}.$$

The feed composition is $y_{A0} = 0.5$ and $y_{B0} = 0.5$.

a. We write design equation (6.1-1) for each independent reaction,

$$\frac{dZ_1}{d\tau} = r_1 \cdot \left(\frac{t_{cr}}{C_0}\right) \qquad\qquad \text{(b)}$$

$$\frac{dZ_2}{d\tau} = r_2 \cdot \left(\frac{t_{cr}}{C_0}\right). \qquad\qquad \text{(c)}$$

Using the stoichiometric relations for flow reactors, (6.1-7) and (1.5-6), the molar flow rates of the different species at any point in the reactor, expressed in terms of the dimensionless extents of the independent reactions, are

$$F_A = (F_{tot})_0 \cdot (y_{A_0} - Z_1) \qquad\qquad \text{(d)}$$

$$F_B = (F_{tot})_0 \cdot (y_{B_0} - Z_1 - 2 \cdot Z_2) \qquad\qquad \text{(e)}$$

$$F_P = (F_{tot})_0 \cdot (y_{P_0} + Z_1) \qquad\qquad \text{(f)}$$

$$F_W = (F_{tot})_0 \cdot (y_{W_0} + Z_2). \qquad\qquad \text{(g)}$$

For gas-phase reactions, we use (6.1-12) to express the concentrations, and the reaction rates are

$$r_1 = k_1 \cdot C_0^2 \frac{(y_{A_0} - Z_1) \cdot (y_{B_0} - Z_1 - 2 \cdot Z_2)}{(1 - Z_1 - Z_2)^2} \qquad \text{(h)}$$

$$r_2 = k_2 \cdot C_0^2 \left(\frac{y_{B_0} - Z_1 - 2 \cdot Z_2}{1 - Z_1 - Z_2} \right)^2 . \qquad \text{(i)}$$

We select Reaction 1 as the leading reaction; hence, the characteristic reaction time is

$$t_{cr} = \frac{1}{k_1 \cdot C_0} = 4.22 \text{ min}. \qquad \text{(j)}$$

Substituting (h), (i), and (j) into (b) and (c), the design equations reduce to

$$\frac{dZ_1}{d\tau} = \frac{(y_{A_0} - Z_1) \cdot (y_{B_0} - Z_1 - 2 \cdot Z_2)}{(1 - Z_1 - Z_2)^2} \qquad \text{(k)}$$

$$\frac{dZ_2}{d\tau} = \left(\frac{k_2}{k_1} \right) \cdot \left(\frac{y_{B_0} - Z_1 - 2 \cdot Z_2}{1 - Z_1 - Z_2} \right)^2 . \qquad \text{(l)}$$

From the given data, $k_2/k_1 = 0.123$, and we solve (k) and (l) numerically subject to the initial conditions that at $\tau = 0$, $Z_1 = Z_2 = 0$. The first figure below shows the reaction operating curves.

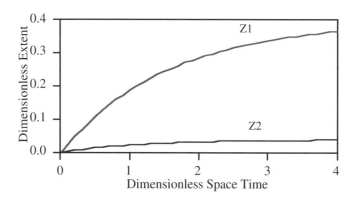

b. We substitute Z_1 and Z_2 in (c) through (f) to calculate the dimensionless molar content of the species as a function of dimensionless space time, τ. The curves of the individual species contents in the reactor as a function of τ are shown in the second figure below.

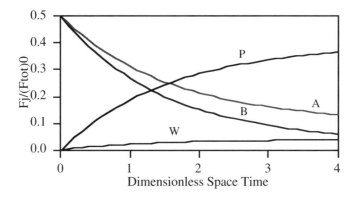

c. Using the conversion definition (1.5-7b), for 80% conversion of B, $F_B/F_{B0} = 0.20$ and $F_B/(F_{tot})_0 = 0.10$. From the species operating curve for B, $F_B/(F_{tot})_0 = 0.10$ is achieved at $\tau = 2.80$. Using (6.1-3) and (j), the required reactor volume is

$$V_R = \tau \cdot v_0 \cdot t_{cr} = (2.80) \cdot (24.75 \text{ lit/min}) \cdot (4.22 \text{ min}) = 292.7 \text{ liter}. \qquad \text{(m)}$$

Example 6-11 The following simplified reversible reactions (mechanism) were proposed for gas-phase cracking of hydrocarbons:

$$A \leftrightarrow 2\,B \qquad\qquad r_1 = k_1 \cdot C_A \qquad r_2 = k_2 \cdot C_B^2$$

$$A + B \leftrightarrow C \qquad\qquad r_3 = k_3 \cdot C_A \cdot C_B \qquad r_4 = k_4 \cdot C_C$$

$$A + C \leftrightarrow D \qquad\qquad r_5 = k_5 \cdot C_A \cdot C_C \qquad r_6 = k_6 \cdot C_D$$

Species B is the desired product. We want to design an isothermal, isobaric tubular reactor (plug-flow reactor) to be operated at 489°C and 5 atm. A stream of pure A at a rate of 1 mole/sec is available in the plant.

a. Derive the design equations, and plot the reaction operating curves.

b. Plot the species operating curves.

c. Determine the maximum production rate of B and the required reactor volume to achieve it.

d. Determine the mole fractions of the species at equilibrium ($\tau = \infty$).

e. Repeat parts a, b, and d for a feed stream that consists only of species D.

Data: At 489°C, $k_1 = 2$ min^{-1}; $k_2 = 20$ lit mole^{-1} min^{-1}; $k_3 = 50$ lit mole^{-1} min^{-1}; $k_4 = 0.8$ min^{-1}; $k_5 = 125$ lit mole^{-1} min^{-1}; and $k_6 = 2$ min^{-1}.

Solution This is an example of series-parallel reversible reactions. The reactor design formulation of these chemical reactions was discussed in Example 3-1. Here, we complete the design for an isothermal plug-flow reactor and obtain the dimensionless reaction and species operating curves. Recall that these are three reversible reactions, and, following the heuristic rule of Section 3.4, we select Reactions 1, 3, and 5 as a set of independent reactions. Hence, the indices of the independent reactions are m = 1, 3, and 5, the indices of the dependent reactions are k = 2. 4, and 6, and we express the design equations in terms of Z_1, Z_3 and Z_5. The stoichiometric coefficients of the three independent reactions are:

$$s_{A_1} = -1; \quad s_{B_1} = 2; \quad s_{C_1} = 0; \quad s_{D_1} = 0; \quad \Delta_1 = -1;$$
$$s_{A_3} = -1; \quad s_{B_3} = -1; \quad s_{C_3} = 1; \quad s_{D_3} = 0; \quad \Delta_3 = -1;$$
$$s_{A_5} = -1; \quad s_{B_5} = 0; \quad s_{C_5} = -1; \quad s_{D_5} = 1; \quad \Delta_5 = -1.$$

The multipliers of the independent reactions to obtain the dependent reactions are:

$$\alpha_{21} = -1; \; \alpha_{23} = 0; \; \alpha_{25} = 0;$$

$$\alpha_{41} = 0; \; \alpha_{43} = -1; \; \alpha_{45} = 0;$$

$$\alpha_{61} = 0; \; \alpha_{63} = 0; \; \alpha_{65} = -1.$$

We select the inlet stream as the reference stream; hence, $Z_{1in} = Z_{2in} = Z_{3in} = 0$,

$$C_0 = \frac{P_0}{R \cdot T_0} = \frac{(5 \text{ atm})}{(82.05 \cdot 10^{-3} \text{ lit - atm/mole}°\text{K}) \cdot (762°\text{K})} = 8.00 \cdot 10^{-2} \text{ mole/lit}, \qquad \text{(a)}$$

and

$$v_0 = \frac{(F_{tot})_0}{C_0} = \frac{(1 \text{ mole/min})}{(4.04 \cdot 10^{-2} \text{ mole/lit})} = 24.75 \text{ lit/min}.$$

We write design equation (6.1-1) for each of the independent reactions,

$$\frac{dZ_1}{d\tau} = (r_1 - r_2) \cdot \left(\frac{t_{cr}}{C_0} \right) \qquad \text{(b)}$$

$$\frac{dZ_3}{d\tau} = (r_3 - r_4) \cdot \left(\frac{t_{cr}}{C_0} \right) \qquad \text{(c)}$$

$$\frac{dZ_5}{d\tau} = (r_5 - r_6) \cdot \left(\frac{t_{cr}}{C_0} \right). \qquad \text{(d)}$$

Using stoichiometric relation (1.5-5), we express the molar flow rate of each species at any point in the reactor in terms of Z_1, Z_3, and Z_5:

$$F_A = (F_{tot})_0 \cdot (y_{A_0} - Z_1 - Z_3 - Z_5) \qquad \text{(e)}$$
$$F_B = (F_{tot})_0 \cdot (y_{B_0} + 2 \cdot Z_1 - Z_3) \qquad \text{(f)}$$
$$F_C = (F_{tot})_0 \cdot (y_{C_0} + Z_3 - Z_5) \qquad \text{(g)}$$
$$F_D = (F_{tot})_0 \cdot (y_{D_0} + Z_5). \qquad \text{(h)}$$

For gas-phase reactions, we use (6.1-12) to express the species concentrations, and the rates of the six chemical reactions are:

$$r_1 = k_1 \cdot C_0 \frac{y_{A_0} - Z_1 - Z_3 - Z_5}{1 + Z_1 - Z_3 - Z_5} \tag{i}$$

$$r_2 = k_2 \cdot C_0^2 \left(\frac{y_{B_0} + 2 \cdot Z_1 - Z_3}{1 + Z_1 - Z_3 - Z_5} \right)^2 \tag{j}$$

$$r_3 = k_3 \cdot C_0^2 \frac{(y_{A_0} - Z_1 - Z_3 - Z_5) \cdot (y_{B_0} + 2 \cdot Z_1 - Z_3)}{(1 + Z_1 - Z_3 - Z_5)^2} \tag{k}$$

$$r_4 = k_4 \cdot C_0 \frac{y_{C_0} + Z_3 - Z_5}{1 + Z_1 - Z_3 - Z_5} \tag{l}$$

$$r_5 = k_5 \cdot C_0^2 \frac{(y_{A_0} - Z_1 - Z_3 - Z_5) \cdot (y_{C_0} + Z_3 - Z_5)}{(1 + Z_1 - Z_3 - Z_5)^2} \tag{m}$$

$$r_6 = k_6 \cdot C_0 \frac{y_{D_0} + Z_5}{1 + Z_1 - Z_3 - Z_5}. \tag{n}$$

We select Reaction 1 and define the characteristic reaction time by

$$t_{cr} = \frac{1}{k_1} = 0.5 \text{ min}. \tag{o}$$

Substituting (i) through (n) and (o) into (b), (c) and (d), the design equations reduce to

$$\frac{dZ_1}{d\tau} = \frac{y_{A_0} - Z_1 - Z_3 - Z_5}{1 + Z_1 - Z_3 - Z_5} - \left(\frac{k_2 \cdot C_0}{k_1} \right) \cdot \left(\frac{y_{B_0} + 2 \cdot Z_1 - Z_3}{1 + Z_1 - Z_3 - Z_5} \right)^2 \tag{p}$$

$$\frac{dZ_3}{d\tau} = \left(\frac{k_3 \cdot C_0}{k_1} \right) \cdot \frac{(y_{A_0} - Z_1 - Z_3 - Z_5) \cdot (y_{B_0} + 2 \cdot Z_1 - Z_3)}{(1 + Z_1 - Z_3 - Z_5)^2} -$$
$$- \left(\frac{k_4}{k_1} \right) \cdot \frac{y_{C_0} + Z_3 - Z_5}{1 + Z_1 - Z_3 - Z_5} \tag{q}$$

$$\frac{dZ_5}{d\tau} = \left(\frac{k_5 \cdot C_0}{k_1} \right) \cdot \frac{(y_{A_0} - Z_1 - Z_3 - Z_5) \cdot (y_{C_0} + Z_3 - Z_5)}{(1 + Z_1 - Z_3 - Z_5)^2} -$$
$$- \left(\frac{k_6}{k_1} \right) \cdot \frac{y_{D_0} + Z_5}{1 + Z_1 - Z_3 - Z_5}. \tag{r}$$

For the given kinetic data,

$$\frac{k_2 \cdot C_0}{k_1} = 0.8; \quad \frac{k_3 \cdot C_0}{k_1} = 2; \quad \frac{k_4}{k_1} = 0.4; \quad \frac{k_5 \cdot C_0}{k_1} = 5; \quad \frac{k_6}{k_1} = 1. \quad \text{(s)}$$

a. When a stream of species A is fed into the reactor, $y_{A0} = 1$, $y_{B0} = y_{C0} = y_{D0} = 0$. We substitute these values into (p), (q), and (r), and solve them numerically for Z_1, Z_3, and Z_5, subject to the initial conditions that at $\tau = 0$, $Z_1 = Z_3 = Z_5 = 0$. The figure below shows the solution – the reaction operating curves for the three independent chemical reactions.

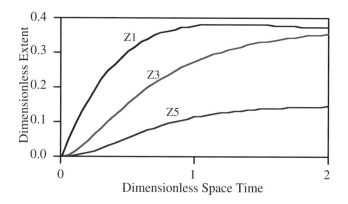

b. Using the solutions of Z_1, Z_3, and Z_5, we apply (e) through (h) to calculate the dimensionless molar flow rates of the individual species as functions of τ. The figure below shows the species operating curves.

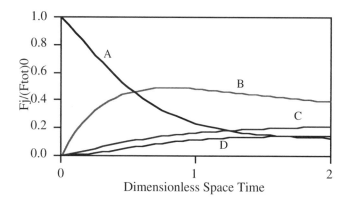

c. From the species operating curve, for this operation, the maximum value of $F_{Bout}/(F_{tot})_0$ is 0.480, and it is reached at $\tau = 0.75$. The production rate of species B is $F_{Bout} =$

0.48 $(F_{tot})_0$ = 28.8 mole/min. Using (6.1-3) and (o), the reactor volume required to maximize B production is

$$V_R = \tau \cdot v_0 \cdot t_{cr} = (0.75) \cdot (12.5 \text{ lit/sec}) \cdot (30 \text{ sec}) = 281.2 \text{ liter}.$$

d. To determine the equilibrium composition, we first calculate the extents of the independent reactions at equilibrium by equating (p), (q), and (r) to zero. We obtain a set of nonlinear algebraic equations whose solutions are

$$Z_{1_{eq}} = 0.3588; \quad Z_{3_{eq}} = 0.3786; \quad \text{and} \quad Z_{5_{eq}} = 0.1518.$$

Substituting these values into (e) through (h), we obtain

$$\frac{F_{A_{eq}}}{(F_{tot})_0} = 0.1116; \quad \frac{F_{B_{eq}}}{(F_{tot})_0} = 0.3388; \quad \frac{F_{C_{eq}}}{(F_{tot})_0} = 0.2268; \quad \text{and} \quad \frac{F_{D_{eq}}}{(F_{tot})_0} = 0.1518.$$

The corresponding mole fractions are

$$y_{A_{eq}} = 0.1339; \quad y_{B_{eq}} = 0.4091; \quad y_{C_{eq}} = 0.2738; \quad \text{and} \quad y_{D_{eq}} = 0.1832.$$

e. The reactor design calculations when the feed stream consists of species D proceeds in the same way as in Parts a, b, and d. The only difference is that now $y_{A0} = y_{B0} = y_{C0} = 0$ and $y_{D0} = 1$. Substituting these values into (p), (q), and (r), we solve them numerically. The figure below shows the reaction operating curves for this case. Note that in

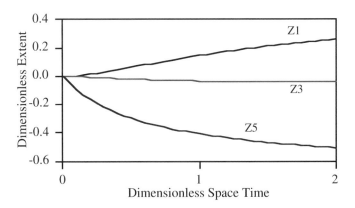

this case, the extents of independent reactions 3 and 5 are negative since they proceed in the reverse direction. For these solutions of Z_1, Z_3, and Z_5, we use (e) through (h) to calculate the dimensionless molar flow rates of the individual species as a function of τ. The curves are shown in the graph below. To determine the equilibrium composition, we equate (p), (q), and (r) to zero and obtain a set of nonlinear algebraic equations whose solutions are

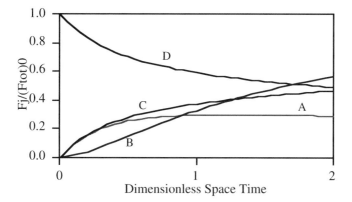

$$Z_{1_{eq}} = 0.3969; \quad Z_{3_{eq}} = -0.0536; \quad \text{and} \quad Z_{5_{eq}} = -0.6206.$$

Substituting these values into (e) through (h), we obtain

$$\frac{F_{A_{eq}}}{(F_{tot})_0} = 0.2772; \quad \frac{F_{B_{eq}}}{(F_{tot})_0} = 0.7275; \quad \frac{F_{C_{eq}}}{(F_{tot})_0} = 0.5670; \quad \text{and} \quad \frac{F_{D_{eq}}}{(F_{tot})_0} = 0.3795.$$

The corresponding mole fractions at equilibrium are

$$y_{A_{eq}} = 0.1339; \quad y_{B_{eq}} = 0.4091; \quad y_{C_{eq}} = 0.2738; \quad \text{and} \quad y_{D_{eq}} = 0.1832.$$

Note that, as expected, the equilibrium composition is independent of the feed composition.

Example 6-12 Ammonia oxidation is carried out in an isothermal plug-flow reactor. The following gaseous chemical reactions take place in the reactor:

$$4\,NH_3 + 5\,O_2 \rightarrow 4\,NO + 6\,H_2O \qquad r_1 = k_1 \cdot C_{NH3} \cdot C_{O2}^2$$

$$4\,NH_3 + 3\,O_2 \rightarrow 2\,N_2 + 6\,H_2O \qquad r_2 = k_2 \cdot C_{NH3} \cdot C_{O2}$$

$$2\,NO + O_2 \rightarrow 2\,NO_2 \qquad r_3 = k_3 \cdot C_{O2} \cdot C_{NO}^2$$

$$4\,NH_3 + 6\,NO \rightarrow 5\,N_2 + 6\,H_2O \qquad r_4 = k_4 \cdot C_{NH3}^{2/3} \cdot C_{NO}$$

The desired product is NO. A stream consisting of 50% NH_3 and 50% O_2 (mole %) at 609°K and 2 atm is fed into the reactor at a rate of 240 liter/min. Based on the rate data below,

a. derive the design equations and plot the reaction and species operating curves and
b. determine the volume of the plug-flow reactor for optimal production of NO; what is the flow rate of each species at the reactor exit?

Data: At 609°K, $k_1 = 20$ (liter/mole)^2 min^{-1}; $k_2 = 0.04$ (liter/mole) min^{-1};

$$k_3 = 40 \text{ (liter/mole)}^2 \text{ min}^{-1}; \quad k_4 = 0.0274 \text{ (liter/mole)}^{2/3} \text{ min}^{-1}$$

Solution The reactor design formulation of these chemical reactions was discussed in Example 3-3. Here, we complete the design for an isothermal plug-flow reactor and obtain the dimensionless reaction and species operating curves. Recall that there are three independent reactions and one dependent reaction, and, following the heuristic rule, we select a set of three reactions from the given reactions. We select Reactions 1, 2, and 3 as a set of independent reactions; hence, $m = 1, 2, 3$, $k = 4$, and we express the design equations in terms of Z_1, Z_2 and Z_3. The stoichiometric coefficients of the independent reactions are:

$$(s_{NH3})_1 = -4; \; (s_{O2})_1 = -5; \; (s_{NO})_1 = 4; \; (s_{H2O})_1 = 6; \; (s_{N2})_1 = 0; \; (s_{NO2})_1 = 0; \; \Delta_1 = 1,$$

$$(s_{NH3})_2 = -4; \; (s_{O2})_2 = -3; \; (s_{NO})_2 = 0; \; (s_{H2O})_2 = 6; \; (s_{N2})_2 = 2; \; (s_{NO2})_2 = 0; \; \Delta_2 = 1,$$

$$(s_{NH3})_3 = 0; \; (s_{O2})_3 = -1; \; (s_{NO})_3 = -2; \; (s_{H2O})_3 = 0; \; (s_{N2})_3 = 0; \; (s_{NO2})_3 = 2; \; \Delta_3 = -1.$$

Recall from Example 3-2 that the multipliers α_{km}'s of the dependent reaction (Reaction 4) and the three independent reactions are: $\alpha_{43} = 0$, $\alpha_{42} = 2.5$, and $\alpha_{41} = -1.5$. We select the feed stream as the reference stream; hence, $Z_{10} = Z_{20} = 0 \; Z_{30} = 0$, and the reference concentration is

$$C_0 = \frac{P_0}{R \cdot T_0} = \frac{(2 \text{ atm})}{(0.08206 \text{ lit - atm/mole}^\circ K) \cdot (609^\circ K)} = 0.04 \text{ mole/lit}. \qquad (a)$$

The molar flow rate of the reference stream is

$$(F_{tot})_0 = v_0 \cdot C_0 = (240 \text{ lit/mole}) \cdot (0.04 \text{ mole/liter}) = 9.6 \text{ mole/min}.$$

For the selected reference stream, $y_{NH30} = 0.5$, $y_{O20} = 0.5$, $y_{NO0} = y_{H2O0} = y_{N20} = y_{NO20} = 0$. Using stoichiometric relation (1.5-5), the molar flow rate of each species at any point in the reactor expressed in terms of Z_1, Z_2, and Z_3 is

$$F_{NH3} = (F_{tot})_0 \cdot (y_{NH30} - 4 \cdot Z_1 - 4 \cdot Z_2) \qquad (b)$$

$$F_{O2} = (F_{tot})_0 \cdot (y_{O20} - 5 \cdot Z_1 - 3 \cdot Z_2 - Z_3) \qquad (c)$$

$$F_{NO} = (F_{tot})_0 \cdot (y_{NO0} + 4 \cdot Z_1 - 2 \cdot Z_3) \qquad (d)$$

$$F_{H2O} = (F_{tot})_0 \cdot (y_{H2O0} + 6 \cdot Z_1 + 6 \cdot Z_2) \qquad (e)$$

$$F_{N2} = (F_{tot})_0 \cdot (y_{N20} + 2 \cdot Z_2) \qquad (f)$$

$$F_{NO2} = (F_{tot})_0 \cdot (y_{NO20} + 2 \cdot Z_3). \qquad (g)$$

a. To design a plug-flow reactor, we write design equation (6.1-1) for each of the independent chemical reactions,

$$\frac{dZ_1}{d\tau} = (r_1 - 1.5 \cdot r_4) \cdot \left(\frac{t_{cr}}{C_0} \right) \qquad (h)$$

$$\frac{dZ_2}{d\tau} = (r_2 + 2.5 \cdot r_4) \cdot \left(\frac{t_{cr}}{C_0}\right) \tag{i}$$

$$\frac{dZ_3}{d\tau} = r_3 \cdot \left(\frac{t_{cr}}{C_0}\right). \tag{j}$$

We select Reaction 1, the leading reaction, as the reference reaction and define characteristic reaction time by

$$t_{cr} = \frac{C_0}{k_1 \cdot C_0{}^3} = \frac{1}{k_1 \cdot C_0{}^2} = \frac{1}{(20) \cdot (0.04)^2} = 31.25 \text{ min}. \tag{k}$$

For gas-phase reactions, we express the species concentrations by (6.1-12), and the rates of the four chemical reactions, expressed in terms of dimensionless extents, are:

$$r_1 = k_1 \cdot C_0{}^3 \left(\frac{y_{NH_{30}} - 4 \cdot Z_1 - 4 \cdot Z_2}{1 + Z_1 + Z_2 - Z_3}\right) \cdot \left(\frac{y_{O_{20}} - 5 \cdot Z_1 - 3 \cdot Z_2 - Z_3}{1 + Z_1 + Z_2 - Z_3}\right)^2 \tag{l}$$

$$r_2 = k_2 \cdot C_0{}^2 \left(\frac{y_{NH_{30}} - 4 \cdot Z_1 - 4 \cdot Z_2}{1 + Z_1 + Z_2 - Z_3}\right) \cdot \left(\frac{y_{O_{20}} - 5 \cdot Z_1 - 3 \cdot Z_2 - Z_3}{1 + Z_1 + Z_2 - Z_3}\right) \tag{m}$$

$$r_3 = k_3 \cdot C_0{}^3 \left(\frac{y_{O_{20}} - 5 \cdot Z_1 - 3 \cdot Z_2 - Z_3}{1 + Z_1 + Z_2 - Z_3}\right) \cdot \left(\frac{y_{NO_0} + 4 \cdot Z_1 - 2 \cdot Z_3}{1 + Z_1 + Z_2 - Z_3}\right)^2 \tag{n}$$

$$r_4 = k_4 \cdot C_0{}^{5/3} \left(\frac{y_{NH_{30}} - 4 \cdot Z_1 - 4 \cdot Z_2}{1 + Z_1 + Z_2 - Z_3}\right)^{2/3} \cdot \left(\frac{y_{NO_0} + 4 \cdot Z_1 - 2 \cdot Z_3}{1 + Z_1 + Z_2 - Z_3}\right). \tag{p}$$

Substituting (l) through (p) and (k) into (h), (i), and (j), the design equations become

$$\frac{dZ_1}{d\tau} = \frac{(y_{NH_{30}} - 4 \cdot Z_1 - 4 \cdot Z_2) \cdot (y_{O_{20}} - 5 \cdot Z_1 - 3 \cdot Z_2 - Z_3)^2}{(1 + Z_1 + Z_2 - Z_3)^3} -$$

$$-1.5 \left(\frac{k_4}{k_1 \cdot C_0{}^{4/3}}\right) \frac{(y_{NH_{30}} - 4 \cdot Z_1 - 4 \cdot Z_2)^{2/3} \cdot (y_{NO_0} + 4 \cdot Z_1 - 2 \cdot Z_3)}{(1 + Z_1 + Z_2 - Z_3)^{5/3}} \tag{q}$$

$$\frac{dZ_2}{d\tau} = \left(\frac{k_2}{k_1 \cdot C_0}\right) \frac{(y_{NH_{30}} - 4 \cdot Z_1 - 4 \cdot Z_2) \cdot (y_{O_{20}} - 5 \cdot Z_1 - 3 \cdot Z_2 - Z_3)}{(1 + Z_1 + Z_2 - Z_3)^2} +$$

$$+2.5 \left(\frac{k_4}{k_1 \cdot C_0{}^{4/3}}\right) \frac{(y_{NH_{30}} - 4 \cdot Z_1 - 4 \cdot Z_2)^{2/3} \cdot (y_{NO_0} + 4 \cdot Z_1 - 2 \cdot Z_3)}{(1 + Z_1 + Z_2 - Z_3)^{5/3}} \tag{r}$$

$$\frac{dZ_3}{d\tau} = \left(\frac{k_3}{k_1}\right) \frac{(y_{O_{20}} - 5 \cdot Z_1 - 3 \cdot Z_2 - Z_3) \cdot (y_{NO_0} + 4 \cdot Z_1 - 2 \cdot Z_3)^2}{(1 + Z_1 + Z_2 - Z_3)^3}. \qquad (s)$$

For the given data, the parameters of the dimensionless design equations are:

$$\left(\frac{k_2}{k_1 \cdot C_0}\right) = 0.05; \qquad \left(\frac{k_3}{k_1}\right) = 2; \qquad \left(\frac{k_4}{k_1 \cdot C_0^{4/3}}\right) = 0.1.$$

We substitute these values into (q), (r), and (s), and solve them numerically for Z_1, Z_2, and Z_3, subject to the initial conditions that at $\tau = 0$, $Z_1 = Z_2 = Z_3 = 0$, using a mathematical software. The figure below shows the solutions of the design equations--the reaction operating curves. Based on these solutions of Z_1, Z_2, and Z_3, we use (b) through

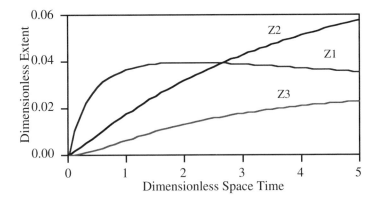

(g) to calculate the dimensionless species operating curves (dimensionless contents as functions of τ). The figure below shows the species operating curves.

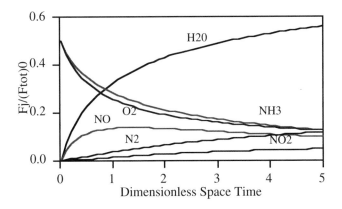

b. From the species operating curve of NO (or the table of calculated data), maximum production is achieved at $\tau = 1.36$, and, using the definition of the dimensionless space time, (6.1-3), and (k), the required reactor volume is

$$V_R = \tau \cdot v_0 \cdot t_{cr} = (1.36) \cdot (240 \text{ lit/min}) \cdot (31.25 \text{ min}) = 10,200 \text{ liter}.$$

At $\tau = 1.36$, $Z_1 = 0.0387$, $Z_2 = 0.0232$, and $Z_3 = 0.0089$. Substituting these values into (b) through (g), the species molar flow rates in the outlet are: $F_{NH3out} = 2.42$, $F_{O2out} = 2.19$, $F_{NOout} = 1.32$, $F_{H2Oout} = 3.57$, $F_{N2out} = 0.445$, and $F_{NO2out} = 0.171$ mole/min.

6.4 NON-ISOTHERMAL OPERATIONS

The design formulation of non-isothermal plug-flow reactors with multiple reactions follows the same procedure outlined in the previous section — we write design equation (6.1-1) for each independent reaction. Since the temperature varies along the reactor, the design equations have an additional variable, and we should solve the design equations simultaneously with the energy balance equation, (6.1-13).

The energy balance equation contains another variable — the temperature of the heating (or cooling) fluid, θ_F, which may also vary along the reactor. (Note that θ_F is constant only when either evaporation or condensation takes place in the shell.) We distinguish between two configurations: co-current flow, shown in Figure 6-6a, and counter-current flow, shown in Figure 6-6b. To derive an equation for θ_F, we write the

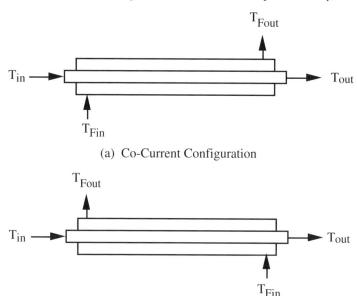

(a) Co-Current Configuration

(b) Counter-Current Configuration

Figure 6-6: Configuration of Heating/Cooling in Plug-Flow Reactors

energy balance equation over the heating fluid in the shell element. For **co-current** flow,

$$d\dot{Q} = U \cdot (T_F - T) \cdot dS = -\dot{m}_F \cdot \bar{c}_{pF} \cdot dT_F, \tag{6.4-1}$$

where $\dot{m}_F$ and $\bar{c}_{pF}$ are, respectively, the mass flow rate and the heat capacity of the heating (or cooling) fluid. Using (4.2-51), $dS = (S/V) \cdot dV_R$, and (6.4-1) reduces to

$$\frac{dT_F}{dV_R} = -\left(\frac{S}{V}\right) \cdot \frac{U \cdot (T_F - T)}{\dot{m}_F \cdot \bar{c}_{pF}}. \tag{6.4-2}$$

To convert (6.4-3) to dimensionless form, we use the definition of the dimensionless space time, (6.1-3), and divide both sides by the reference thermal energy, $(F_{tot})_0 \cdot \hat{c}_{p0} \cdot T_0$,

$$\frac{d\theta_F}{d\tau} = -HTN \cdot \left(\frac{(F_{tot})_0 \cdot \hat{c}_{p0}}{\dot{m}_F \cdot \bar{c}_{pF}}\right) \cdot (\theta_F - \theta), \tag{6.4-3a}$$

where HTN is the heat-transfer number, defined by (6.1-4). The first term in parentheses on the right is the dimensionless heat-transfer number, and the second term in the parentheses is the ratio of the heat capacity of the reference stream to that of the heating/cooling fluid.

For **counter-current** flow, the energy balance over the heating fluid is

$$dQ = U \cdot (T_F - T) \cdot dS = \dot{m}_m \cdot \hat{c}_{pF} \cdot dT_F.$$

Following a similar procedure to the one we use for co-current flow, we obtain

$$\frac{d\theta_F}{d\tau} = HTN \cdot \left(\frac{(F_{tot})_0 \cdot \hat{c}_{p0}}{\dot{m}_m \cdot \bar{c}_{pF}}\right) \cdot (\theta_F - \theta). \tag{6.4-3b}$$

Hence, when designing a plug-flow reactor with a heating/cooling fluid whose temperature varies, we have to solve either (6.4-3a) or (6.4-3b) simultaneously with the design equations and the energy balance equation of the reactor.

The procedure for setting up the dimensionless energy balance equation goes as follows:

a. Define the reference state, and identify T_0, C_0, $(F_{tot})_0$, and y_{j0}'s.
b. Determine the specific molar heat capacity of the reference state, $\hat{c}_{p0}$.
c. Determine the dimensionless activation energies, γ_i's, of **all** chemical reactions.
d. Determine the dimensionless heat of reactions, DHR_m's, of the **independent** reactions.
e. Determine the correction factor of the heat capacity, $CF(Z_m, \theta)$.
f. Specify the dimensionless heat-transfer number, HTN.

g. Determine (or specify) the inlet dimensionless temperature of the heating/cooling fluid, θ_F.
h. Determine (or specify) the inlet dimensionless temperature, $\theta(0)$.
i. Solve the energy balance equation simultaneously with the design equations to obtain Z_m's and θ as functions of the dimensionless operating time, τ.

The design formulation of non-isothermal plug-flow reactors consists of $(n_{ind} + 2)$ nonlinear first-order differential equations of the form,

$$\frac{dZ_1}{d\tau} = G_1(\tau, Z_1, Z_2, ..., Z_{n_{ind}}, \theta, \theta_F),$$

$$\frac{dZ_2}{d\tau} = G_2(\tau, Z_1, Z_2, ..., Z_{n_{ind}}, \theta, \theta_F),$$

$$\vdots \qquad\qquad (6.4\text{-}4)$$

$$\frac{dZ_{n_{ind}}}{d\tau} = G_{n_{ind}}(\tau, Z_1, Z_2, ..., Z_{n_{ind}}, \theta, \theta_F),$$

$$\frac{d\theta}{d\tau} = G_{n_{ind}+1}(\tau, Z_1, Z_2, ..., Z_{n_{ind}}, \theta, \theta_F),$$

$$\frac{d\theta_F}{d\tau} = G_{n_{ind}+1}(\tau, Z_1, Z_2, ..., Z_{n_{ind}}, \theta, \theta_F),$$

where τ is the dimensionless space time defined by (6.1-3). The solutions of these equations provide Z_m's, θ, and θ_F as functions of τ. Note that when heating (or cooling) fluid either evaporates of condenses in the shell, θ_F is constant, and $d\theta_F/d\tau = 0$. Also note that usually the inlet temperature of the heating/cooling fluid, T_{Fin}, is known whereas the outlet temperature, T_{Fout}, should be calculated. Hence, the co-current configuration can be readily solved, but the counter-current configuration is solved iteratively by trial-and-error. In the latter, we guess the value of T_{Fout} as an initial value and solve the formulation equations. We then check if the value of T_{Fin} at the end of the reactor is in agreement with the given value of T_{Fin}. If not, we guess another value of T_{Fout} and repeat the calculation.

The examples below illustrate the design of non-isothermal plug-flow reactors.

Example 6-13 The first-order, gas-phase reaction $A \rightarrow B + C$ is carried out in a plug-flow reactor. The reactor is made of a 1 ft tube, and it operates at 2 atm. A stream consisting of 80% A and 20% I (by mole) is fed into the reactor at 800°R at the rate of 10 ft³/min. Based on the data below, derive the reaction operating curve, and determine the reactor length and the outlet temperature for 60% conversion at the following operating conditions:

a. The reactor is operated isothermally at 800°R.
b. Calculate the overall heating (or cooling) load in (a).
c. The reactor is operated adiabatically.
d. The reactor is placed in a furnace at 820°R, and the heat-transfer coefficient is 1 Btu/

hr ft^2 °R.

Data: At 800°R, $k(T_0) = 15.0$ hr^{-1}; $\Delta H_R(T_0) = 8,000$ Btu/lbmole A; $E_a/R = 4,000$°R;

$\hat{c}_{PA} = 20$ Btu/lbmole°R; $\hat{c}_{PB} = 15$ Btu/lbmole°R;

$\hat{c}_{PC} = 12$ Btu/lbmole °R; $\hat{c}_{PI} = 10$ Btu/lbmole °R;

Solution The reaction is A $\rightarrow$ B + C, and its stoichiometric coefficients are $s_A = -1$, $s_B = 1$, $s_C = 1$, $s_I = 0$, and $\Delta = 1$. The rate expression is $r = k\, C_A$. Using (6.1-1), for a plug-flow reactor with a single reaction, the dimensionless design equation is

$$\frac{dZ}{d\tau} = r \cdot \left(\frac{t_{cr}}{C_0} \right). \tag{a}$$

We select the inlet stream as the reference stream, $Z_{in} = 0$,

$$C_0 = \frac{P}{R \cdot T_0} = \frac{2}{(0.7302) \cdot (800)} = 3.42 \cdot 10^{-3} \text{ lbmole/ft}^3$$

$$(F_{tot})_0 = v_0 \cdot C_0 = (10 \text{ ft}^3/\text{min}) \cdot (3.42 \cdot 10^{-3} \text{ lbmole/ft}^3) \cdot (60 \text{ min/hr}) =$$
$$= 2.05 \text{ lbmole/hr}. \tag{b}$$

For the given feed composition, $y_{A0} = 0.80$, $y_{I0} = 0.20$, and $y_{B0} = y_{C0} = 0$. Using the stoichiometric relations,

$$F_A = (F_{tot})_0 \cdot (y_{A0} - Z) = (F_{tot})_0 \cdot (0.8 - Z) \tag{c}$$

$$F_B = (F_{tot})_0 \cdot (y_{B0} + Z) = (F_{tot})_0 \cdot Z \tag{d}$$

$$F_C = (F_{tot})_0 \cdot (y_{C0} + Z) = (F_{tot})_0 \cdot Z. \tag{e}$$

For gas-phase reactions, using (6.1-12), the reaction rate is

$$r = k(T_0) \cdot e^{\gamma \frac{\theta-1}{\theta}} \cdot C_0 \frac{0.8 - Z}{(1+Z) \cdot \theta}. \tag{f}$$

We define a characteristic reaction time,

$$t_{cr} = \frac{1}{k(T_0)} = 0.0667 \text{ hr} = 4 \text{ min}. \tag{g}$$

Substituting (f) and (g) into (a), the design equation reduces to

$$\frac{dZ}{d\tau} = \frac{0.8 - Z}{(1+Z) \cdot \theta} \cdot e^{\gamma \frac{\theta-1}{\theta}}. \tag{h}$$

a. For isothermal operation, $\theta = 1$, and (h) reduces to

$$\frac{dZ}{d\tau} = \frac{0.8 - Z}{1 + Z}.$$

(i)

The graph below shows the reaction operating curve for isothermal operation. Using

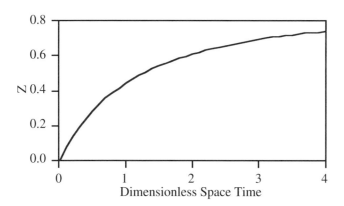

(1.5-11), for 60% conversion, the extent is

$$Z = -\frac{y_{A0}}{s_A} f_A = -\frac{0.8}{(-1)} 0.6 = 0.48,$$

and, from the graph, $Z = 0.48$ is reached at $\tau = 1.17$. Using (g) and (6.1-3), the required reactor volume is

$$V_R = v_0 \cdot \tau \cdot t_{cr} = (10 \text{ ft}^3/\text{min}) \cdot (1.17) \cdot (4 \text{ min}) = 46.8 \text{ ft}^3,$$

(j)

and the length of the reactor is 59.6 ft.

b. For isothermal operation, $d\theta/d\tau = 0$, and the energy balance equation reduces to

$$\frac{d}{d\tau}\left(\frac{\dot{Q}}{T_0 \cdot (F_{tot})_0 \cdot \hat{c}_{p0}}\right) = \left(\frac{\Delta H_R(T_0)}{T_0 \cdot \hat{c}_{p0}}\right)\frac{dZ}{d\tau},$$

which, upon integration, becomes

$$\dot{Q} = (F_{tot})_0 \cdot \Delta H_R(T_0) \cdot Z_{out} =$$
$$= (2.06 \text{ lb - mole/hr}) \cdot (8,000 \text{ Btu/lb - mole}) \cdot (0.48) = 7.91 \cdot 10^3 \text{ Btu/min} \cdot$$

c. For non-isothermal operations we have to solve (h) together with the energy balance equation. For adiabatic operation, $(S/V) = 0$; hence, $HTN = 0$, and (6.1-13) becomes

$$\frac{d\theta}{d\tau} = \frac{-1}{CF(Z_m, \theta)} \cdot DHR \cdot \frac{dZ}{d\tau}.$$

(k)

For this case, using (4.2-59) and (6.1-15),

$$\hat{c}_{P0} = \sum_j^{\text{all}} y_{j0} \cdot \hat{c}_{Pj}(1) = y_{A0} \cdot \hat{c}_{PA}(1) + y_{I0} \cdot \hat{c}_{PI}(1) = 18 \text{ Btu/lbmole}^\circ\text{R}, \qquad (1)$$

$$\text{DHR} = \left(\frac{\Delta H_R(T_0)}{T_0 \cdot \hat{c}_{P0}} \right) = 0.556,$$

and g = 5. Using (4.2-61), the correction factor of the heat capacity is

$$\text{CF}(Z_m, \theta) = 1 + \frac{1}{\hat{c}_{P0}} \left[\hat{c}_{PA}(\theta) \cdot (-Z) + \hat{c}_{PB}(\theta) \cdot Z + \hat{c}_{PC}(\theta) \cdot Z \right] = \frac{18 + 7 \cdot Z}{18}. \qquad (m)$$

Substituting (m) and (h) into (k), the energy balance equation reduces to

$$\frac{d\theta}{d\tau} = -\left(\frac{18}{18 + 7 \cdot Z} \right) \cdot (0.556) \underbrace{\frac{0.8 - Z}{(1+Z) \cdot \theta} \cdot e^{\gamma \frac{\theta-1}{\theta}}}_{dz/d\tau} \qquad (n)$$

We solve equations (h) and (n) simultaneously, subject to the initial conditions that at τ = 0, Z = 0 and θ = 1. The graph below shows the reaction operating curve and the θ curve as a function of τ. From the calculated data, an extent of Z = 0.48 is reached at τ

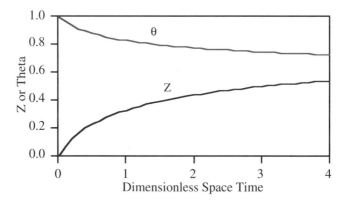

= 2.70, and, at that time, θ = 0.755. Using (g) and (6.1-3), the required reactor volume for this case is

$$V_R = v_0 \cdot \tau \cdot t_{cr} = (10 \text{ ft}^3/\text{min}) \cdot (2.70) \cdot (4 \text{ min}) = 108 \text{ ft}^3,$$

and the length of the reactor is 149.1 ft. Using the definition of the dimensionless temperature, the exit temperature is T = $T_0 \cdot \theta$ = (800)·(0.755) = 587.2°R.

d. For operations with heat-transfer, design equation (h) applies, but the energy balance equation should include the heat-transfer term. Using (6.1-13), the energy balance equation becomes

$$\frac{d\theta}{d\tau} = \frac{1}{CF(Z_m, \theta)} \cdot \left[HTN \cdot (\theta_F - \theta) - DHR \cdot \frac{dZ}{d\tau} \right]. \tag{o}$$

For this case, the heat-transfer number

$$HTN = \frac{U \cdot t_{cr}}{C_0 \cdot \hat{c}_{p0}} \cdot \left(\frac{S}{V} \right) = 1.11,$$

$\theta_F = 820/800 = 1.025$, and (o) reduces to

$$\frac{d\theta}{d\tau} = \frac{18}{18 + 7 \cdot Z} \cdot \left[(1.11) \cdot (1.025 - \theta) - (0.556) \cdot \frac{dZ}{d\tau} \right]. \tag{p}$$

We solve equations (h) and (p) simultaneously, using the initial conditions that at $\tau = 0$, $Z = 0$ and $\theta = 1$. The figure below shows the solution. From the calculated data, an

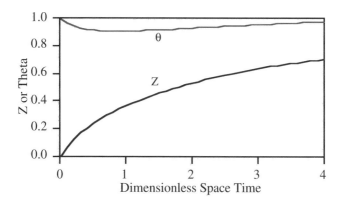

extent of 0.48 is reached at $\tau = 1.64$, and the dimensionless temperature at that space time is $\theta = 0.922$. Using (g) and (6.1-3), the required reactor volume for this case is

$$V_R = v_0 \cdot \tau \cdot t_{cr} = (10 \text{ ft}^3/\text{min}) \cdot (1.64) \cdot (4 \text{ min}) = 65.6 \text{ ft}^3,$$

and the length of the reactor is 90.6 ft. Using the definition of the dimensionless temperature, the exit temperature is

$$T = T_0 \cdot \theta = (800) \cdot (0.922) = 737.6°R.$$

Example 6-14 The gas-phase elementary reactions

$$A + B \rightarrow C$$

$$C + B \rightarrow D$$

are carried out in a tubular reactor. A gas mixture of 40% A, 40% B, and 20% I (inert) is fed into a 10 cm ID reactor at a rate of 0.4 m^3/min. The inlet temperature is 150°C, and the reactor operates at a constant pressure of 2 atm. Based on the data below, derive the design equations, and plot the operating curves for the operations indicated. In each case, calculate the reactor volume needed to obtain 80% conversion of B, the molar flow rate of product C, and the reactor outlet temperature.
a. Isothermal operation.
b. Determine the heating/cooling load in (a).
c. Adiabatic operation.
d. Operation with heat-transfer. The reactor is cooled by condensing steam in the shell side at $T_F = 130°C$, and the heat-transfer coefficient is 1 kcal/m^2 hr °K.
Data: At 150°C $k_1 = 0.2$ liter mole^{-1}sec^{-1}; $k_2 = 0.1$ liter mole^{-1}sec^{-1};
   At 150°C $\Delta H_{R1} = -12,000$ cal/mole extent; $\Delta H_{R2} = -9,000$ cal/mole extent;
   $E_{a1} = 4,950$ cal/mole; $E_{a2} = 7,682$ cal/mole;

$\hat{c}_{PA} = 16$ cal/mole°K; $\hat{c}_{PB} = 8$ cal/mole°K; $\hat{c}_{PC} = 20$ cal/mole°K;

$\hat{c}_{PD} = 26$ cal/mole°K; $\hat{c}_{PI} = 10$ cal/mole°K.

Solution The chemical reactions are:

$$\text{Reaction 1:} \quad A + B \rightarrow C$$

$$\text{Reaction 2:} \quad C + B \rightarrow D,$$

and the stoichiometric coefficients are:

$$s_{A1} = -1; \quad s_{B1} = -1; \quad s_{C1} = 1; \quad s_{D1} = 0; \quad \Delta_1 = -1;$$

$$s_{A2} = 0; \quad s_{B1} = -1; \quad s_{C1} = -1; \quad s_{D1} = 1; \quad \Delta_2 = -1.$$

Since each reaction has a species that does not appear in the other reaction, the two reactions are independent, and there is no dependent reaction. We write the dimensionless design equation (6.1-13) for the two independent reactions,

$$\frac{dZ_1}{d\tau} = r_1 \cdot \left(\frac{t_{cr}}{C_0} \right) \tag{a}$$

$$\frac{dZ_2}{d\tau} = r_2 \cdot \left(\frac{t_{cr}}{C_0} \right). \tag{b}$$

We select the inlet stream as the reference stream; hence, $Z_{in} = 0$,

$$C_0 = \frac{P}{R \cdot T_0} = 5.76 \cdot 10^{-2} \text{ mole/liter}$$

$(F_{tot})_0 = v_0 \cdot C_0 = (400 \text{ lit/min}) (5.76 \ 10^{-2} \text{ mole/liter}) = 23.05 \text{ mole/min}.$

For the given feed composition, $y_{A0} = 0.40$, $y_{B0} = 0.40$, $y_{I0} = 0.20$, and $y_{C0} = y_{D0} = 0$. Using the stoichiometric relation (6.1-7),

$$F_A = (F_{tot})_0 \cdot (y_{A0} - Z_1) \tag{c}$$

$$F_B = (F_{tot})_0 \cdot (y_{B0} - Z_1 - Z_2) \tag{d}$$

$$F_C = (F_{tot})_0 \cdot (y_{C0} + Z_1 - Z_2) \tag{e}$$

$$F_D = (F_{tot})_0 \cdot (y_{D0} + Z_2). \tag{f}$$

For gas-phase reactions, the species concentrations are expressed by (6.1-12), and the rates of the two reactions are

$$r_1 = k_1 \cdot C_A \cdot C_B = k_1(T_0) \cdot e^{\gamma_1 \frac{\theta-1}{\theta}} \cdot C_0^2 \frac{(y_{A0} - Z_1) \cdot (y_{B0} - Z_1 - Z_2)}{[(1 - Z_1 - Z_2) \cdot \theta]^2} \tag{g}$$

$$r_2 = k_2 \cdot C_C \cdot C_B = k_2(T_0) \cdot e^{\gamma_2 \frac{\theta-1}{\theta}} \cdot C_0^2 \frac{(Z_1 - Z_2) \cdot (y_{B0} - Z_1 - Z_2)}{[(1 - Z_1 - Z_2) \cdot \theta]^2}. \tag{h}$$

Defining a characteristic reaction time,

$$t_{cr} = \frac{1}{k_1(T_0) \cdot C_0} = 86.6 \text{ sec} = 1.45 \text{ min}. \tag{i}$$

Substituting (g) and (h) into (a) and (b), the two design equations become

$$\frac{dZ_1}{d\tau} = \frac{(y_{A0} - Z_1) \cdot (y_{B0} - Z_1 - Z_2)}{[(1 - Z_1 - Z_2) \cdot \theta]^2} \cdot e^{\gamma_1 \frac{\theta-1}{\theta}} \tag{j}$$

$$\frac{dZ_2}{d\tau} = \left(\frac{k_2(T_0)}{k_1(T_0)} \right) \cdot \frac{(Z_1 - Z_2) \cdot (y_{B0} - Z_1 - Z_2)}{[(1 - Z_1 - Z_2) \cdot \theta]^2} \cdot e^{\gamma_2 \frac{\theta-1}{\theta}}. \tag{k}$$

Using (4.2-59), the specific molar heat capacity of the reference stream is

$$\hat{c}_{P0} = \sum_j^{all} y_{j0} \cdot \hat{c}_{Pj}(1) = y_{A0} \cdot \hat{c}_{PA}(1) + y_{B0} \cdot \hat{c}_{PB}(1) + y_{I0} \cdot \hat{c}_{PI}(1) =$$

$$= (0.4) \cdot (16) + (0.4) \cdot (8) + (0.2) \cdot (10) = 11.6 \text{ cal/mole}°K.$$

Using (4.2-61), the correction factor of the heat capacity is

$$CF(Z_m, \theta) = 1 + \frac{1}{\hat{c}_{P0}} \sum_m^{n_{ind}} \hat{c}_{Pj}(\theta) \sum_m^{n_{ind}} (s_j)_m \cdot Z_m =$$

$$= 1 + \frac{1}{\hat{c}_{P0}} \left[\hat{c}_{PA} \cdot (-Z_1) + \hat{c}_{PB} \cdot (-Z_1 - Z_2) + \hat{c}_{PC} \cdot (Z_1 - Z_2) + \hat{c}_{PD} \cdot Z_2 \right] =$$

$$= 1 + \frac{1}{11.6} \left[(16) \cdot (-Z_1) + (8) \cdot (-Z_1 - Z_2) + (20) \cdot (Z_1 - Z_2) + (26) \cdot Z_2 \right],$$

and
$$CF(Z_m, \theta) = \frac{11.6 - 4 \cdot Z_1 - 2 \cdot Z_2}{11.6}. \tag{1}$$

Substituting (l) into (6.1-13), the energy balance equation reduces to

$$\frac{d\theta}{d\tau} = \left(\frac{11.6}{11.6 - 4 \cdot Z_1 - 2 \cdot Z_2} \right) \left[HTN \cdot (\theta_F - \theta) - DHR_1 \cdot \frac{dZ_1}{d\tau} - DHR_2 \cdot \frac{dZ_2}{d\tau} \right]. \tag{m}$$

We have to solve (j), (k), and (m) simultaneously subject to the initial condition that at $\tau = 0$, $Z_1 = Z_2 = 0$ and $\theta = 1$. The numerical values of the various parameters are:

$$\gamma_1 = \frac{E_{a1}}{R \cdot T_0} = 5.89; \quad DHR_1 = \left(\frac{\Delta H_{R_1}(T_0)}{T_0 \cdot \hat{c}_{P0}} \right) = -2.45;$$

$$\gamma_2 = \frac{E_{a2}}{R \cdot T_0} = 9.14; \quad DHR_2 = \left(\frac{\Delta H_{R_2}(T_0)}{T_0 \cdot \hat{c}_{P0}} \right) = -1.834,$$

and, for operation with heat-transfer,

$$HTN = \frac{U \cdot t_{cr}}{C_0 \cdot \hat{c}_{P0}} \cdot \left(\frac{S}{V} \right) = 1.71.$$

a. For isothermal operation, $\theta = 1$, and the two design equations, (j) and (k), reduce to

$$\frac{dZ_1}{d\tau} = \frac{(0.4 - Z_1) \cdot (0.4 - Z_1 - Z_2)}{(1 - Z_1 - Z_2)^2} \tag{n}$$

$$\frac{dZ_2}{d\tau} = (0.5) \frac{(Z_1 - Z_2) \cdot (0.4 - Z_1 - Z_2)}{(1 - Z_1 - Z_2)^2}. \tag{o}$$

We solve (n) and (o) numerically subject to the initial conditions that at $\tau = 0$, $Z_1 = Z_2 = 0$. The figures below show, respectively, the dimensional operating curves of the two reactions and the species operating curves. Using (1.5-7a), for $f_B = 0.80$, $F_B/F_{B0} = 0.2$ and $F_B/(F_{tot})_0 = 0.08$. From the species operating curve, $F_B/(F_{tot})_0 = 0.08$ is reached at $\tau = 3.44$ and, at that time, $Z_1 = 0.256$ and $Z_2 = 0.064$. Hence, using (i) and (6.1-3), the reactor volume is

$$V_R = v_0 \cdot t_{cr} \cdot \tau = (400 \text{ lit/min}) \cdot (1.45 \text{ min}) \cdot (3.44) = 1,995 \text{ liter}.$$

Using (c) through (f), the species molar flow rates at the reactor outlet are:

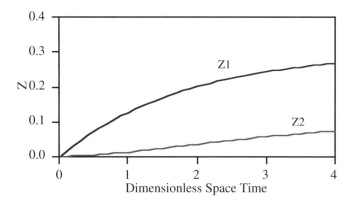

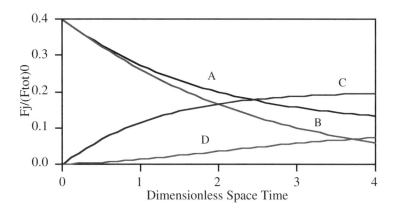

$$F_A = (23.05) \cdot (0.4 - 0.256) = 3.32 \text{ mole/min}$$

$$F_B = (23.05) \cdot (0.4 - 0.256 - 0.0640) = 1.84 \text{ mole/min}$$

$$F_C = (23.05) \cdot (0.256 - 0.0640) = 4.43 \text{ mole/min}$$

$$F_D = (23.05) \cdot (0.0640) = 1.48 \text{ mole/min}.$$

b. For isothermal operation, the energy balance equation reduces to

$$\frac{d}{d\tau} \left(\frac{\dot{Q}}{T_0 \cdot (F_{tot})_0 \cdot \hat{c}_{p0}} \right) = \left(\frac{\Delta H_{R_1}(T_0)}{T_0 \cdot \hat{c}_{p0}} \right) \frac{dZ_1}{d\tau} + \left(\frac{\Delta H_{R_2}(T_0)}{T_0 \cdot \hat{c}_{p0}} \right) \frac{dZ_2}{d\tau},$$

which, upon integration, becomes

$$\dot{Q} = (F_{tot})_0 \cdot \left[\Delta H_{R_1}(T_0) \cdot Z_1 + \Delta H_{R_2}(T_0) \cdot Z_2 \right] =$$

$$= (23.05 \text{ mole/min}) \cdot \left[(-12,000) \cdot (0.256) + (-9,000) \cdot (0.0640)\right] = -84.09 \text{ kcal/min}.$$

The negative sign indicates that heat is removed from the reactor.

c. For adiabatic operation, $(S/V) = 0$; hence, $HTN = 0$, and the energy balance equation, (m), reduces to

$$\frac{d\theta}{d\tau} = \left(\frac{11.6}{11.6 - 4 \cdot Z_1 - 2 \cdot Z_2}\right) \cdot \left[(2.45) \cdot \frac{dZ_1}{d\tau} + (1.834) \cdot \frac{dZ_2}{d\tau}\right]. \qquad (p)$$

We solve (j), (k), and (p) numerically subject to the initial condition that at $\tau = 0$, $Z_1 = Z_2 = 0$, and $\theta = 1$. The figures below show, respectively, the reaction operating curves, the temperature profile, and the species operating curves. We use the calculated data to determine τ for $F_B/(F_{tot})_0 = 0.08$ and find that it is $\tau = 1.064$ and, at that time, $Z_1 = 0.220$

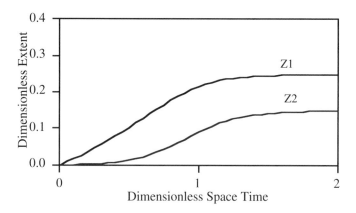

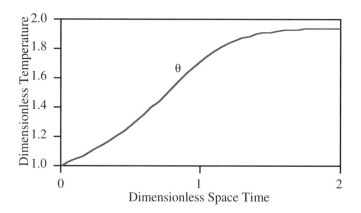

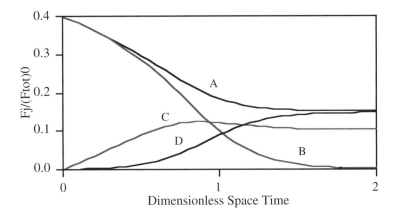

$Z_2 = 0.102$ and $\theta = 1.758$. Hence, using (i) and (6.1-3), the needed reactor volume is

$$V_R = v_0 \cdot t_{cr} \cdot \tau = (400 \text{ lit/min}) \cdot (1.45 \text{ min}) \cdot (1.064) = 617.1 \text{ liter}.$$

Using (c) through (f), the species molar flow rates at the reactor outlet are

$$F_A = (23.05) \cdot (0.4 - 0.220) = 4.16 \text{ mole/min}$$

$$F_B = (23.05) \cdot (0.4 - 0.220 - 0.100) = 1.84 \text{ mole/min}$$

$$F_C = (23.05) \cdot (0.220 - 0.100) = 2.73 \text{ mole/min}$$

$$F_D = (23.05) \cdot (0.100) = 2.32 \text{ mole/min}.$$

The exit temperature is $T = T_0\,\theta = (423) \cdot (1.758) = 743.6°K$.

d. For operations with heat-transfer, we substitute the numerical parameters, and (m) reduces to

$$\frac{d\theta}{d\tau} = \left(\frac{11.6}{11.6 - 4 \cdot Z_1 - 2 \cdot Z_2}\right) \cdot \left[(1.71) \cdot (0.953 - \theta) + (2.45) \cdot \frac{dZ_1}{d\tau} + (1.83) \cdot \frac{dZ_2}{d\tau}\right]. \quad (q)$$

We solve (j), (k), and (q) numerically, subject to the initial condition that at $\tau = 0$, $Z_1 = Z_2 = 0$, and $\theta = 1$. The figures below show, respectively, the reaction operating curves, the temperature profile, and the species operating curves. We use the calculated data to determine τ for which $F_B/(F_{tot})_0 = 0.08$ and find that it is $\tau = 1.924$, and, at that time, $Z_1 = 0.241$ and $Z_2 = 0.079$. Hence, using (i) and (6.1-3), the needed reactor volume is

$$V_R = v_0 \cdot t_{cr} \cdot \tau = (400 \text{ lit/min}) \cdot (1.45 \text{ min}) \cdot (1.924) = 1,115.9 \text{ liter}.$$

Using (c) through (f), the species molar flow rates at the reactor outlet are

$$F_A = (23.05) \cdot (0.4 - 0.2383) = 3.67 \text{ mole/min}$$

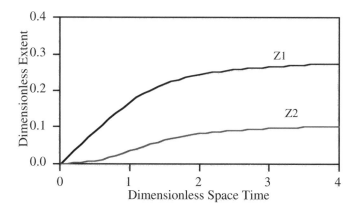

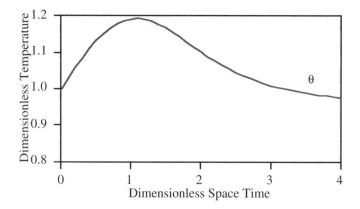

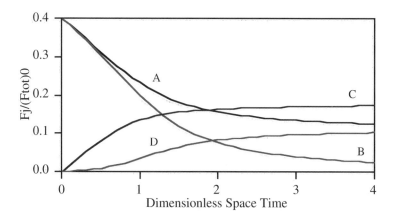

$$F_B = (23.05)\cdot(0.4 - 0.2383 - 0.0817) = 1.84 \text{ mole/min}$$

$$F_C = (23.05)\cdot(0.2383 - 0.0817) = 3.73 \text{ mole/min}$$

$$F_D = (23.05)\cdot(0.0817) = 1.824 \text{ mole/min}.$$

Example 6-15 The gas-phase cracking of a hydrocarbon is described by the following simplified reversible chemical reactions:

$A \leftrightarrow 2\,B$	$r_1 = k_1 \cdot C_A$	$r_2 = k_2 \cdot C_B^2$
$A + B \leftrightarrow C$	$r_3 = k_3 \cdot C_A \cdot C_B$	$r_4 = k_4 \cdot C_C$
$A + C \leftrightarrow D$	$r_5 = k_5 \cdot C_A \cdot C_C$	$r_6 = k_6 \cdot C_D$

Species B is the desired product. We want to design an isobaric tubular reactor (plug-flow reactor) to process a stream (at 641°C and 3 atm) of species A that is fed to the reactor at a rate of 1 mol/sec. Based on the data below, derive the design equations, plot the reaction and species operating curves, and plot the temperature profile for each of the operations indicated. In each case, calculate the reactor volume needed to obtain maximum production of product B, and determine the species molar flow rates at the reactor outlet.

a. Isothermal operation.
b. Determine the heating/cooling load in (a).
c. Adiabatic operation.
d. Operation with heat-transfer. The 0.1 m ID reactor is heated by a fluid in the shell side at $T_F = 957°K$, and the heat-transfer coefficient is 90 cal/m^2 hr °K.

Data: At 641°C, $k_1 = 2 \text{ min}^{-1}$, $k_2 = 10 \text{ lit mole}^{-1} \text{ min}^{-1}$, $k_3 = 150 \text{ lit mole}^{-1} \text{ min}^{-1}$, $k_4 = 0.08 \text{ min}^{-1}$, $k_5 = 200 \text{ lit mole}^{-1} \text{ min}^{-1}$, $k_6 = 0.8 \text{ min}^{-1}$.
$\Delta H_{R1} = 80 \text{ kcal/mole}$; $\Delta H_{R3} = -60 \text{ kcal/mole}$; $\Delta H_{R5} = -40 \text{ kcal/mole}$;
$E_{a1} = 110 \text{ kcal/mole}$; $E_{a2} = 30 \text{ kcal/mole}$; $E_{a3} = 20 \text{ kcal/mole}$;
$E_{a4} = 80 \text{ kcal/mole}$; $E_{a5} = 30 \text{ kcal/mole}$; $E_{a6} = 70 \text{ kcal/mole}$;
$\hat{c}_{PA} = 30 \text{ cal/mole°K}$; $\hat{c}_{PB} = 20 \text{ cal/mole°K}$; $\hat{c}_{PC} = 40 \text{ cal/mole°K}$;
$\hat{c}_{PD} = 60 \text{ cal/mole°K}$.

Solution The reaction scheme in this case is the same as that of Example 6-11. We select the three forward reactions as the set of independent reactions. Hence, the indices of the independent reactions are $m_1 = 1$, $m_2 = 3$, $m_5 = 5$, and those of the dependent reactions are $k_1 = 2$, $k_2 = 4$, $k_3 = 6$. The values of the α_{km}'s factors are: $\alpha_{21} = -1$, $\alpha_{43} = -1$, $\alpha_{65} = -1$, and all the other are equal to zero. The stoichiometric coefficients of the independent reactions are:

$$(s_A)_1 = -1; \ (s_B)_1 = 2; \ (s_C)_1 = 0; \ (s_D)_1 = 0; \ \Delta_1 = 1;$$

$$(s_A)_3 = -1; \ (s_B)_3 = -1; \ (s_C)_3 = 1; \ (s_D)_3 = 0; \ \Delta_3 = -1;$$

$$(s_A)_5 = -1; \ (s_B)_5 = 0; \ (s_C)_5 = -1; \ (s_D)_5 = 1; \ \Delta_5 = -1.$$

We write design equation (6.1-1) for each of the independent reactions,

$$\frac{dZ_1}{d\tau} = (r_1 - r_2) \cdot \left(\frac{t_{cr}}{C_0} \right) \tag{a}$$

$$\frac{dZ_3}{d\tau} = (r_3 - r_4) \cdot \left(\frac{t_{cr}}{C_0} \right) \tag{b}$$

$$\frac{dZ_5}{d\tau} = (r_5 - r_6) \cdot \left(\frac{t_{cr}}{C_0} \right). \tag{c}$$

The energy balance equation (6.1-13) for this case is

$$\frac{d\theta}{d\tau} = \frac{1}{CF(Z_m, \theta)} \cdot \left[HTN \cdot (\theta_F - \theta) - DHR_1 \cdot \frac{dZ_1}{d\tau} - DHR_3 \cdot \frac{dZ_3}{d\tau} - DHR_5 \cdot \frac{dZ_5}{d\tau} \right]. \tag{d}$$

We select the inlet stream as the reference stream; hence, $T_0 = 914°K$, $y_{A0} = 1$, $y_{B0} = y_{C0} = y_{D0} = 0$, and $Z_{1in} = Z_{2in} = Z_{3in} = 0$. Also,

$$C_0 = \frac{P_0}{R \cdot T_0} = \frac{(3 \text{ atm})}{(82.05 \cdot 10^{-3} \text{ lit atm/mole°K}) \cdot (914°K)} = 4.00 \cdot 10^{-2} \text{ mole/lit} \tag{e}$$

$$v_0 = \frac{(F_{tot})_0}{C_0} = \frac{(60 \text{ mole/min})}{(4.00 \cdot 10^{-2} \text{ mole/lit})} = 1,500 \text{ lit/min}$$

$$\hat{c}_{p0} = \hat{c}_{pA}(T_0) = 30 \text{ cal/mole°K}.$$

The numerical values of the various parameters are:

$$\gamma_1 = \frac{E_{a1}}{R \cdot T_0} = 60.59; \quad \gamma_2 = \frac{E_{a2}}{R \cdot T_0} = 16.52; \quad \gamma_3 = \frac{E_{a3}}{R \cdot T_0} = 11.01;$$

$$\gamma_4 = \frac{E_{a4}}{R \cdot T_0} = 44.05; \quad \gamma_5 = \frac{E_{a5}}{R \cdot T_0} = 16.52; \quad \gamma_6 = \frac{E_{a6}}{R \cdot T_0} = 33.04;$$

$$DHR_1 = \left(\frac{\Delta H_{R1}(T_0)}{T_0 \cdot \hat{c}_{p0}} \right) = 2.92; \quad DHR_3 = \left(\frac{\Delta H_{R3}(T_0)}{T_0 \cdot \hat{c}_{p0}} \right) = -2.19;$$

$$DHR_5 = \left(\frac{\Delta H_{R5}(T_0)}{T_0 \cdot \hat{c}_{p0}} \right) = -1.46.$$

Using stoichiometric relation (1-5.5), we express the molar flow rate of each species at any point in the reactor in terms of Z_1, Z_3, and Z_5,

$$F_A = (F_{tot})_0 \cdot (y_{A0} - Z_1 - Z_3 - Z_5) \tag{f}$$

$$F_B = (F_{tot})_0 \cdot (y_{B0} + 2 \cdot Z_1 - Z_3) \tag{g}$$

$$F_C = (F_{tot})_0 \cdot (y_{C0} + Z_3 - Z_5) \tag{h}$$

$$F_D = (F_{tot})_0 \cdot (y_{D0} + Z_5). \tag{i}$$

For gas-phase reactions, we use (6.1-12) to express the species concentrations, and the rates of the six chemical reactions are:

$$r_1 = k_1(T) \cdot C_0 \frac{y_{A0} - Z_1 - Z_3 - Z_5}{(1 + Z_1 - Z_3 - Z_5) \cdot \theta} \cdot e^{\gamma_1 \frac{\theta - 1}{\theta}} \tag{j}$$

$$r_2 = k_2(T) \cdot C_0^2 \left[\frac{y_{B0} + 2 \cdot Z_1 - Z_3}{(1 + Z_1 - Z_3 - Z_5) \cdot \theta} \right]^2 \cdot e^{\gamma_2 \frac{\theta - 1}{\theta}} \tag{k}$$

$$r_3 = k_3(T) \cdot C_0^2 \frac{(y_{A0} - Z_1 - Z_3 - Z_5) \cdot (y_{B0} + 2 \cdot Z_1 - Z_3)}{\left[(1 + Z_1 - Z_3 - Z_5) \cdot \theta \right]^2} \cdot e^{\gamma_3 \frac{\theta - 1}{\theta}} \tag{l}$$

$$r_4 = k_4(T) \cdot C_0 \frac{y_{C0} + Z_3 - Z_5}{(1 + Z_1 - Z_3 - Z_5) \cdot \theta} \cdot e^{\gamma_4 \frac{\theta - 1}{\theta}} \tag{m}$$

$$r_5 = k_5(T) \cdot C_0^2 \frac{(y_{A0} - Z_1 - Z_3 - Z_5) \cdot (y_{C0} + Z_3 - Z_5)}{\left[(1 + Z_1 - Z_3 - Z_5) \cdot \theta \right]^2} \cdot e^{\gamma_5 \frac{\theta - 1}{\theta}} \tag{n}$$

$$r_6 = k_6(T) \cdot C_0 \frac{y_{D0} + Z_5}{(1 + Z_1 - Z_3 - Z_5) \cdot \theta} \cdot e^{\gamma_6 \frac{\theta - 1}{\theta}}. \tag{o}$$

We select Reaction 1 and define the characteristic reaction time by

$$t_{cr} = \frac{1}{k_1(T_0)} = 0.5 \text{ min}. \tag{p}$$

Substituting (i) through (n) and (o) into (b), (c), and (d), the design equations reduce to

$$\frac{dZ_1}{d\tau} = \frac{y_{A0} - Z_1 - Z_3 - Z_5}{(1 + Z_1 - Z_3 - Z_5) \cdot \theta} \cdot e^{\gamma_1 \frac{\theta - 1}{\theta}} -$$

$$- \left(\frac{k_2(T_0) \cdot C_0}{k_1(T_0)} \right) \cdot \left[\frac{(y_{B0} + 2 \cdot Z_1 - Z_3)}{(1 + Z_1 - Z_3 - Z_5) \cdot \theta} \right]^2 \cdot e^{\gamma_2 \frac{\theta - 1}{\theta}} \tag{q}$$

$$\frac{dZ_3}{d\tau} = \left(\frac{k_3(T_0) \cdot C_0}{k_1(T_0)} \right) \cdot \frac{(y_{A0} - Z_1 - Z_3 - Z_5) \cdot (y_{B0} + 2 \cdot Z_1 - Z_3)}{\left[(1 + Z_1 - Z_3 - Z_5) \cdot \theta \right]^2} \cdot e^{\gamma_3 \frac{\theta - 1}{\theta}} -$$

$$-\left(\frac{k_4(T_0)}{k_1(T_0)}\right) \cdot \frac{y_{C_0} + Z_3 - Z_5}{(1 + Z_1 - Z_3 - Z_5) \cdot \theta} \cdot e^{\gamma_4 \frac{\theta-1}{\theta}} \tag{r}$$

$$\frac{dZ_5}{d\tau} = \left(\frac{k_5(T_0) \cdot C_0}{k_1(T_0)}\right) \cdot \frac{(y_{A_0} - Z_1 - Z_3 - Z_5) \cdot (y_{C_0} + Z_3 - Z_5)}{\left[(1 + Z_1 - Z_3 - Z_5) \cdot \theta\right]^2} \cdot e^{\gamma_5 \frac{\theta-1}{\theta}} -$$

$$-\left(\frac{k_6(T_0)}{k_1(T_0)}\right) \cdot \frac{y_{D_0} + Z_5}{(1 + Z_1 - Z_3 - Z_5) \cdot \theta} \cdot e^{\gamma_6 \frac{\theta-1}{\theta}}. \tag{s}$$

For the given kinetic data,

$$\frac{k_2(T_0) \cdot C_0}{k_1(T_0)} = 0.2; \quad \frac{k_3(T_0) \cdot C_0}{k_1(T_0)} = 3; \quad \frac{k_4(T_0)}{k_1(T_0)} = 0.03;$$

$$\frac{k_5(T_0) \cdot C_0}{k_1(T_0)} = 4; \quad \frac{k_6(T_0)}{k_1(T_0)} = 0.4.$$

a. For isothermal operation, $\theta = 1$, the design equations, (q), (r), and (s), reduce to

$$\frac{dZ_1}{d\tau} = \frac{1 - Z_1 - Z_3 - Z_5}{1 + Z_1 - Z_3 - Z_5} - \left(\frac{k_2(T_0) \cdot C_0}{k_1(T_0)}\right) \cdot \left(\frac{2 \cdot Z_1 - Z_3}{1 + Z_1 - Z_3 - Z_5}\right)^2 \tag{t}$$

$$\frac{dZ_3}{d\tau} = \left(\frac{k_3(T_0) \cdot C_0}{k_1(T_0)}\right) \cdot \frac{(1 - Z_1 - Z_3 - Z_5) \cdot (2 \cdot Z_1 - Z_3)}{(1 + Z_1 - Z_3 - Z_5)^2} -$$

$$-\left(\frac{k_4(T_0)}{k_1(T_0)}\right) \cdot \frac{Z_3 - Z_5}{1 + Z_1 - Z_3 - Z_5} \tag{u}$$

$$\frac{dZ_5}{d\tau} = \left(\frac{k_5(T_0) \cdot C_0}{k_1(T_0)}\right) \cdot \frac{(1 - Z_1 - Z_3 - Z_5) \cdot (Z_3 - Z_5)}{(1 + Z_1 - Z_3 - Z_5)^2} -$$

$$-\left(\frac{k_6(T_0)}{k_1(T_0)}\right) \cdot \frac{Z_5}{1 + Z_1 - Z_3 - Z_5}. \tag{v}$$

We solve these equations numerically for Z_1, Z_3, and Z_5 subject to the initial conditions that at $\tau = 0$, $Z_1 = Z_3 = Z_5 = 0$. The figures below show, respectively, the reaction operating curves and the species operating curves. From the species operating curve, for this operation, the maximum value of $F_{B_{out}}/(F_{tot})_0$ is 0.449, and it is reached at $\tau =$

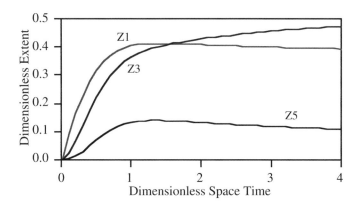

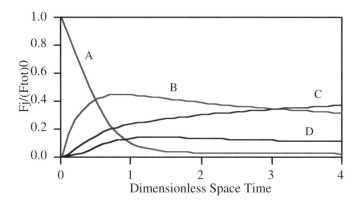

0.865. The production rate of species B is $F_{Bout} = 0.449 \cdot (F_{tot})_0 = 26.94$ mole/min. Using (p) and (6.1-3), the reactor volume is

$$V_R = v_0 \cdot t_{cr} \cdot \tau = (1{,}500 \text{ lit/min}) \cdot (0.5 \text{ min}) \cdot (0.865) = 642.8 \text{ liter}.$$

b. For isothermal operation, $d\theta/dt = 0$, and the energy balance equation reduces to

$$\frac{d}{d\tau}\left(\frac{\dot{Q}}{T_0 \cdot (F_{tot})_0 \cdot \hat{c}_{p0}}\right) = \left(\frac{\Delta H_{R_1}(T_0)}{T_0 \cdot \hat{c}_{p0}}\right)\frac{dZ_1}{d\tau} + \left(\frac{\Delta H_{R_3}(T_0)}{T_0 \cdot \hat{c}_{p0}}\right)\frac{dZ_3}{d\tau} + \left(\frac{\Delta H_{R_5}(T_0)}{T_0 \cdot \hat{c}_{p0}}\right)\frac{dZ_5}{d\tau},$$

which, upon integration, becomes

$$\dot{Q} = (F_{tot})_0 \cdot \left[\Delta H_{R_1}(T_0) \cdot Z_1 + \Delta H_{R_3}(T_0) \cdot Z_3 + \Delta H_{R_5}(T_0) \cdot Z_5\right].$$

At $\tau = 0.865$, $Z_1 = 0.395$, $Z_3 = 0.341$, and $Z_5 = .123$, and the heating/cooling load is

$$\dot{Q} = (60) \cdot [(18) \cdot (0.3935) + (-18) \cdot (0.1215)] = 1.94 \text{ kcal/min} \cdot$$

Since Reaction 1 is endothermic and the other two are exothermic, a relatively small amount of heat should be added to the reactor.

c. For adiabatic operations, $(S/V) = 0$; hence HTN = 0, and the energy balance equation, (m), reduces to

$$\frac{d\theta}{d\tau} = \frac{1}{CF(Z_m, \theta)} \cdot \left[(0.655) \cdot \frac{dZ_1}{d\tau} + (-0.655) \cdot \frac{dZ_3}{d\tau} + (-0.291) \cdot \frac{dZ_5}{d\tau} \right]. \qquad (w)$$

Using, (4.2-34), the correction factor of the specific heat of the reacting fluid is

$$CF(Z_m, \theta) =$$

$$= 1 + \frac{1}{30} \left[(30) \cdot (-Z_1 - Z_3 - Z_5) + (20) \cdot (2 \cdot Z_1 - Z_3) + (40) \cdot (Z_1 - Z_3) + (60) \cdot Z_5 \right]$$

$$CF(Z_m, \theta) = \frac{30 - 2 \cdot Z_1 - Z_3 - 2 \cdot Z_5}{30}. \qquad (x)$$

We substitute (q), (r), and (s) into (w) and solve it simultaneously with (q), (r), and (s) numerically, subject to the initial condition that at $\tau = 0$, $Z_1 = Z_3 = Z_5 = 0$ and $\theta = 1$. The figures show, respectively, the reaction operating curves, the dimensionless temperature profile, and the species operating curves. From the species operating curve, the

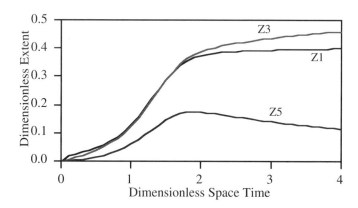

maximum value of $F_{Bout}/(F_{tot})_0$ is 0.365 and it is reached at $\tau = 2.2$. The production rate of species B is $F_{Bout} = 0.365 \cdot (F_{tot})_0 = 21.9$ mole/hr. Using (6.1-3) and (o), the required reactor volume for optimal production of B is

$$V_R = \tau \cdot v_0 \cdot t_{cr} = (2.20) \cdot (1,500 \text{ lit/min}) \cdot (0.5 \text{ min}) = 1,650 \text{ liter}.$$

d. For the operation with heat-transfer,

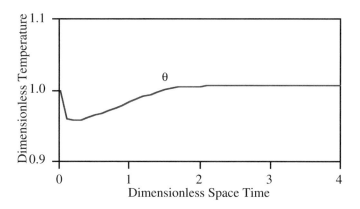

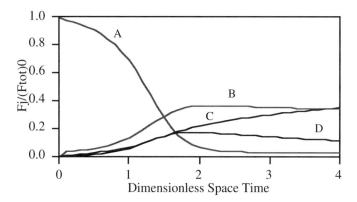

$$\text{HTN} = \frac{U \cdot t_{cr}}{C_0 \cdot \hat{c}_{P0}} \cdot \left(\frac{S}{V}\right) = 0.025 \quad \text{and} \quad \theta_F = \frac{T_F}{T_0} = 1.047.$$

We substitute the numerical parameters into (m), and the energy balance equation reduces to

$$\frac{d\theta}{d\tau} = \frac{30}{30 - 2 \cdot Z_1 - Z_3 - 2 \cdot Z_5}(0.025) \cdot (1.047 - \theta) -$$

$$-\frac{30}{30 - 2 \cdot Z_1 - Z_3 - 2 \cdot Z_5}\left[(0.656)\frac{dZ_1}{d\tau} + (-0.656)\frac{dZ_3}{d\tau} + (-0.292)\frac{dZ_5}{d\tau}\right]. \quad \text{(y)}$$

We substitute (q), (r), and (s) into (y) and solve it simultaneously with (q), (r), and (s), subject to the initial condition that at $\tau = 0$, $Z_1 = Z_3 = Z_5 = 0$ and $\theta = 1$. The figures below show, respectively, the reaction operating curves, the dimensionless temperature profile, and the species operating curves. From the species operating curve, the maximum value of $F_{Bout}/(F_{tot})_0$ is 0.663, and it is reached at $\tau = 0.70$. The production rate of species B is $F_{Bout} = 0.5 \cdot (F_{tot})_0 = 39.8$ mole/hr. Using (6.1-3) and (p), the required

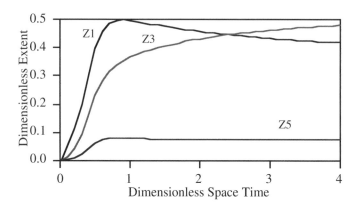

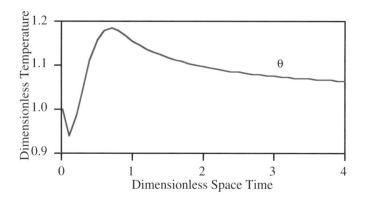

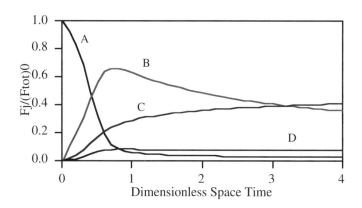

reactor volume for optimal production of B is

$$V_R = \tau \cdot v_0 \cdot t_{cr} = (0.7) \cdot (1,500 \text{ lit/min}) \cdot (0.5 \text{ min}) = 525 \text{ liter}.$$

6.5 EFFECT OF PRESSURE DROP

In the preceding sections, we discussed the operation of gas-phase plug-flow re-actors under the assumption that the pressure does not vary along the reactor – that P/P_0 = 1 in (6.1-12). However, in many applications, the pressure significantly changes and, therefore, affects the reaction rates. In this section, we incorporate the variation in pressure into the design equations. For convenience, we divide the discussion into two parts: tubular tube with uniform diameter and packed-bed reactors.

We consider first a cylindrical reactor of uniform diameter D. To derive an ex-pression for the pressure drop, we write the momentum balance equation for a reactor element of length dL and cross-sectional area A,

$$-A \cdot dP - F_f = \dot{m} \cdot du, \tag{6.5-1}$$

where -(A·dP) is the pressure force, and F_f is the friction force. Expressing the friction force in terms of a friction factor, f, and the kinetic energy, for an empty cylindrical tube of diameter D,

$$F_f = f \, (\pi \cdot D \cdot dL) \, \frac{1}{2} \rho \cdot u^2,$$

and (6.5-1) becomes

$$-\frac{dP}{dL} = 4 \, f \, \frac{\dot{m}}{D \cdot A} \, \frac{1}{2} \cdot u + \frac{\dot{m}}{A} \frac{du}{dL}. \tag{6.5-2}$$

The first term on the right indicates the pressure drop due to friction, and the second indicates the pressure drop due to change in velocity (kinetic energy). In many appli-cations, the second term in (6.5-2) is small in comparison to the first, and the momentum balance equation reduces to

$$-\frac{dP}{dL} \approx 2 \cdot f \cdot \frac{\dot{m}}{D \cdot A^2} \cdot v, \tag{6.5-3}$$

where v is the local volumetric flow rate. Using (6.1-11) and noting from (6.1-3) that dL = $(v_0 \cdot t_{cr}/A)$ dτ, (6.5-3) becomes

$$-\frac{d\left(\dfrac{P}{P_0}\right)}{d\tau} \approx \left(2 \, f \, \frac{\dot{m} \cdot v_0^2 \cdot t_{cr}}{D \cdot A^2 \cdot P_0}\right) \cdot \left(1 + \sum_{m}^{n_{ind}} \Delta_m \cdot Z_m\right) \cdot \theta \cdot \left(\frac{P_0}{P}\right). \tag{6.5-4}$$

Equation (6.5-4) provides an **approximate** relation for the changes in pressure along a plug-flow reactor, expressed in terms of dimensionless extents and temperature. It is applicable when the velocity does not exceed 80 to 90% of the sound velocity. For these situations, we solve (6.5-4) simultaneously with the design equations (6.1-1) and energy balance equation (6.1-13), subject to specified initial conditions. Note that the first parenthesis on the right is a dimensionless friction number for the reference stream.

When the gas velocity approaches the sound velocity, the kinetic energy and viscous work terms in the energy balance equation are not negligible (as assumed in Chapter 4). For these cases, we write the general energy balance equation for a differential plug-flow reactor with length dL (see Eq. 4.2-44),

$$d\dot{Q} - d\dot{W}_{vis} = \dot{m} \cdot d\left(h + \frac{1}{2} \cdot u^2 + g \cdot z\right). \tag{6.5-5}$$

Assuming negligible change in the potential energy, differentiating the right hand side, and rearranging, we obtain

$$\frac{d\dot{Q}}{\dot{m}} - \frac{d\dot{W}_{vis}}{\dot{m}} = dh + u \cdot du. \tag{6.5-6}$$

For flow in cylindrical conduits, the viscous work per unit mass of fluid in reactor length dL is expressed in terms of a friction factor and a specific kinetic energy by

$$\frac{d\dot{W}_{vis}}{\dot{m}} = 4 \, f \, \frac{dL}{D} \frac{1}{2} \cdot u^2, \tag{6.5-7}$$

and the energy balance equation becomes

$$d\dot{Q} - 4 \, f \, \dot{m} \, \frac{dL}{D} \frac{1}{2} \cdot u^2 = \dot{m} \cdot dh + \dot{m} \cdot u \cdot du. \tag{6.5-8}$$

Multiplying (6.5-8) by ρ and then subtracting (6.5-2), we obtain

$$-\frac{dP}{dL} = v \cdot \left(\frac{d\dot{Q}}{dL} - \dot{m} \cdot \frac{dh}{dL}\right). \tag{6.5-9}$$

For cylindrical reactors, where the heat is transferred through the wall,

$$d\dot{Q} = U \cdot (\pi \cdot D \cdot dL) \cdot (T_F - T), \tag{6.5-10}$$

and, using (4.2-47),

$$\dot{m} \cdot dh = d\dot{H} = \sum_{m}^{n_{ind}} \Delta H_{Rm}(T_0) \, d\dot{X}_m + (\Sigma F_j \cdot \hat{c}_{p_j}) \cdot dT. \tag{6.5-11}$$

Substituting these relationships into (6.5-8) and using (4.2-54) and (6.1-11), we obtain

$$-\frac{dP}{dL} =$$

$$= \frac{\left(\dfrac{P}{P_0}\right) \cdot \left(U \cdot \pi \cdot D \cdot (T_F - T) - \displaystyle\sum_{m}^{n_{ind}} \Delta H_{R\,m}(T_0) \dfrac{d\dot{X}_m}{dL} - (F_{tot})_0 \cdot \hat{c}_{p0} \cdot CF \dfrac{dT}{dL}\right)}{v_0 \cdot \left(1 + \displaystyle\sum_{m}^{n_{ind}} \Delta_m \cdot Z_m\right) \cdot \theta} . \quad (6.5\text{-}12)$$

From (6.1-3), $dL = (v_0 \cdot t_{cr}/A)\, d\tau$, using (2.3-4), (6.1-2), (6.1-4), (4.2-22), and (4.2-23), and, noting that for an ideal gas, $R \cdot C_0 = P_0/T_0$, (6.5-12) reduces to

$$-\frac{d\left(\dfrac{P}{P_0}\right)}{d\tau} = \frac{\left(\dfrac{\hat{c}_{p0}}{R}\right) \cdot \left(\dfrac{P}{P_0}\right) \cdot \left(HTN \cdot (\theta_F - \theta) - \displaystyle\sum_{m}^{n_{ind}} DHR_m \dfrac{dZ_m}{d\tau} - CF \dfrac{d\theta}{d\tau}\right)}{\left(1 + \displaystyle\sum_{m}^{n_{ind}} \Delta_m \cdot Z_m\right) \cdot \theta} . \quad (6.5\text{-}13)$$

Equation (6.5-13) provides the changes in pressure along a plug-flow reactor, expressed in terms of dimensionless pressure. It should be solved simultaneously with the design equations for the independent reactions, $dZ_m/d\tau$, and the energy balance equation, $d\theta/d\tau$. Note that the expression inside the parenthesis on the numerator on right hand side of (6.5-13) is the energy balance equation derived under the assumption that the kinetic energy term and the friction work term are negligible (see Section 4.2). Indeed, under these assumptions, $d(P/P_0)/d\tau = 0$. To design a plug-flow reactor, we have to express the energy balance equation with terms for kinetic energy and the friction work.

To derive an expression for the temperature variation in a tubular plug-flow reactor, we rearrange the general energy balance equation (6.5-8),

$$\frac{d\dot{Q}}{dL} - 2\, f\, \dot{m}\, \frac{1}{D}\, u^2 = \dot{m} \cdot \frac{dh}{dL} + \dot{m}\, u \cdot \frac{du}{dL}. \quad (6.5\text{-}14)$$

For a reactor with uniform cross-sectional area A, $u = v/A$, and, using (6.1-11),

$$u = \frac{v_0}{A}\left(1 + \sum_{m}^{n_{ind}} \Delta_m \cdot Z_m\right) \cdot \theta \cdot \left(\frac{P_0}{P}\right). \quad (6.5\text{-}15)$$

Differentiating (6.5-15),

$$\frac{du}{dL} = \frac{v_0}{A}\left(1 + \sum_{m}^{n_{ind}} \Delta_m \cdot Z_m\right) \cdot \left(\frac{P_0}{P}\right)\frac{d\theta}{dL} + \frac{v_0}{A}\,\theta \cdot \left(\frac{P_0}{P}\right)\sum_{m}^{n_{ind}} \Delta_m \cdot \frac{dZ_m}{dL} -$$

$$-\frac{v_0}{A}\left(1+\sum_m^{n_{ind}}\Delta_m\cdot Z_m\right)\cdot\left(\frac{P_0}{P^2}\right)\cdot\theta\cdot\frac{dP}{dL}.\qquad(6.5\text{-}16)$$

We substitute (6.5-13) and (6.5-15) into (6.5-14) and note from (6.1-3) that $dL = (v_0{\cdot}t_{cr}/A)\,d\tau$. Then, using (2.3-4), (6.1-2), (6.1-4), (4.2-22), (4.2-23), and noting that, for an ideal gas, $R{\cdot}C_0 = P_0/T_0$, the four terms in (6.5-14) reduce to:

$$\frac{d\dot{Q}}{dL}=\left(\frac{v_0\cdot t_{cr}}{A}\right)\frac{d\dot{Q}}{d\tau}=\left(\frac{A}{v_0\cdot t_{cr}}\right)\cdot HTN\cdot F_0\cdot\hat{c}_{p0}\cdot T_0\cdot(\theta_F-\theta),$$

$$2\,f\,\dot{m}\,\frac{1}{D}\,u^2=2\,f\,\dot{m}\,\frac{1}{D}\,\frac{v_0^2}{A^2}\left(1+\sum_m^{n_{ind}}\Delta_m\cdot Z_m\right)^2\cdot\theta^2\cdot\left(\frac{P_0}{P}\right)^2,$$

$$\dot{m}\,\frac{dh}{dL}=\left(\frac{A}{v_0\cdot t_{cr}}\right)\frac{d\dot{H}}{d\tau}=\left(\frac{A}{v_0\cdot t_{cr}}\right)\cdot F_0\cdot\hat{c}_{p0}\cdot T_0\cdot\left(\sum_m^{n_{ind}}DHR_m\frac{dZ_m}{d\tau}+CF\frac{d\theta}{d\tau}\right),$$

$$\dot{m}\cdot u\,\frac{du}{dL}=\left(\frac{A}{v_0\cdot t_{cr}}\right)\cdot\dot{m}\cdot u\,\frac{du}{d\tau}=\left(\frac{A}{v_0\cdot t_{cr}}\right)\cdot\dot{m}\cdot\frac{v_0}{A}\left(1+\sum_m^{n_{ind}}\Delta_m\cdot Z_m\right)\cdot\theta\cdot\left(\frac{P_0}{P}\right)\frac{du}{d\tau}=$$

$$=\left(\frac{\dot{m}\cdot v_0}{A\cdot t_{cr}}\right)\cdot\left(1+\sum_m^{n_{ind}}\Delta_m\cdot Z_m\right)^2\cdot\theta\cdot\left(\frac{P_0}{P}\right)^2\frac{d\theta}{d\tau}+$$

$$+\left(\frac{\dot{m}\cdot v_0}{A\cdot t_{cr}}\right)\cdot\left(1+\sum_m^{n_{ind}}\Delta_m\cdot Z_m\right)^2\cdot\theta^2\cdot\left(\frac{P_0}{P}\right)^2\sum_m^{n_{ind}}\Delta_m\frac{dZ_m}{d\tau}-$$

$$-\left(\frac{\dot{m}\cdot v_0}{A\cdot t_{cr}}\right)\cdot\left(1+\sum_m^{n_{ind}}\Delta_m\cdot Z_m\right)^2\cdot\theta^2\cdot\left(\frac{P_0}{P}\right)^3\frac{d(P/P_0)}{d\tau}.$$

Substituting these terms into (6.5-14) and using (6.5-13), we obtain

$$\frac{d\theta}{d\tau}=\frac{(TERM)}{(Denom.)},\qquad(6.5\text{-}17)$$

where

$$(\text{TERM}) = \text{HTN} \cdot \left[1 - \frac{\dot{m} \cdot v_0^2}{A^2 \cdot F_0 \cdot \hat{c}_{p0} \cdot T_0} \left(1 + \sum_m^{n_{ind}} \Delta_m \cdot Z_m \right)^2 \cdot \theta \cdot \left(\frac{P_0}{P} \right)^2 \right] (\theta_F - \theta) -$$

$$-2\, f\, \frac{\dot{m} \cdot v_0^3 \cdot t_{cr}}{D \cdot A^3 \cdot F_0 \cdot \hat{c}_{p0} \cdot T_0} \left(1 + \sum_m^{n_{ind}} \Delta_m \cdot Z_m \right)^2 \cdot \theta^2 \cdot \left(\frac{P_0}{P} \right)^2 -$$

$$- \sum_m^{n_{ind}} \text{DHR}_m \cdot \left[1 - \frac{\dot{m} \cdot v_0^2}{A^2 \cdot F_0 \cdot R \cdot T_0} \left(1 + \sum_m^{n_{ind}} \Delta_m \cdot Z_m \right) \cdot \theta \cdot \left(\frac{P_0}{P} \right)^2 \right] \frac{dZ_m}{d\tau} -$$

$$- \left(\frac{\dot{m} \cdot v_0^2}{A^2 \cdot F_0 \cdot \hat{c}_{p0} \cdot T_0} \right) \left(1 + \sum_m^{n_{ind}} \Delta_m \cdot Z_m \right) \cdot \theta^2 \cdot \left(\frac{P_0}{P} \right)^2 \sum_m^{n_{ind}} \Delta_m \cdot \frac{dZ_m}{d\tau}.$$

and

$$\text{Denom.} = \text{CF}(Z_m, \theta) \cdot \left[1 - \frac{\dot{m} \cdot v_0^2}{A^2 \cdot F_0 \cdot R \cdot T_0} \left(1 + \sum_m^{n_{ind}} \Delta_m \cdot Z_m \right) \cdot \theta \cdot \left(\frac{P_0}{P} \right)^2 \right] +$$

$$+ \frac{\dot{m} \cdot v_0^2}{A^2 \cdot F_0 \cdot \hat{c}_{p0} \cdot T_0} \left(1 + \sum_m^{n_{ind}} \Delta_m \cdot Z_m \right)^2 \cdot \theta \cdot \left(\frac{P_0}{P} \right)^2.$$

Eq. (6.5-17) is the dimensionless, differential energy balance equation for cylindrical tubular flow reactors, relating the temperature, θ, to the extents of the independent reactions, Z_m's, and P/P_0 as functions of space time τ. To design a plug-flow reactor, we have to solve design equations (6.1-1), the energy balance equation, (6.5-17), and the momentum balance, (6.5-13), simultaneously subject to specified initial conditions.

Note that (6.5-17) was derived from the energy balance equation (first law of thermodynamics) that does not impose a limit on the value of θ. However, the second law of thermodynamics imposes a restriction on the conversion of thermal energy to kinetic energy. For compressible flows in tubes with uniform cross-sectional area, the velocity cannot exceed the sound velocity; hence,

$$u \le \left(\frac{1}{MW} \frac{\hat{c}_p \cdot R \cdot T}{\hat{c}_p - R} \right)^{0.5}. \tag{6.5-18}$$

Using (6.5-15), the second law implies that

$$\left(\frac{P}{P_0} \right) \ge \left[\left(\frac{v_0}{A} \right)^2 \left(\frac{MW_0}{R \cdot T_0} \right) \left(\frac{\text{CF}(Z_m, \theta) \cdot \hat{c}_{p0}}{\text{CF}(Z_m, \theta) \cdot \hat{c}_{p0} - R} \right) \left(1 + \sum_m^{n_{ind}} \Delta_m \cdot Z_m \right)^3 \cdot \theta \right]^{0.5}. \tag{6.5-19}$$

For flow through a packed-bed reactor, the pressure drop is expressed in terms of the Ergun's equation,

$$-\frac{dP}{dL} = 150 \frac{(1-\varepsilon)^2}{\varepsilon^3} \frac{\mu \cdot u}{(\phi \cdot d_p)^2} + 1.75 \frac{(1-\varepsilon)}{\varepsilon^3} \frac{\rho \cdot u^2}{\phi \cdot d_p} , \qquad (6.5\text{-}20)$$

where d_p is the particle diameter, f is the shape factor, and ε is the bed porosity. Noting that $\rho \cdot u = G$, the mass velocity is constant, and, from (6.1-3), $dL = (v_0 \cdot t_{cr}/A) \, d\tau$, (6.5-20) reduces to

$$-\frac{d\left(\dfrac{P}{P_0}\right)}{d\tau} =$$

$$= \frac{v_0^2 \cdot t_{cr}}{A \cdot P_0} \left(150 \frac{(1-\varepsilon)^2}{\varepsilon^3} \frac{\mu}{(\phi \cdot d_p)^2} + 1.75 \frac{(1-\varepsilon) \cdot G}{\varepsilon^3 \cdot \phi \cdot d_p} \right) \left(1 + \sum_m^{n_{ind}} \Delta_m \cdot Z_m \right) \cdot \theta \cdot \left(\frac{P_0}{P} \right).$$

$$(6.5\text{-}21)$$

Equation (6.5-21) provides a relation for the change in pressure along a packed-bed reactor. In most applications, the gas velocity in packed-bed reactors is much lower than the sound velocity. Hence, to formulate the reactor design, we should solve design equations (6.1-1), energy balance equation (6.1-13), and momentum balance equation (6.5-21) simultaneously, subject to specified initial conditions. The solutions provide Z_m, θ, and P/P_0 as functions of dimensionless space time, τ.

6.6 SUMMARY

In this chapter, we analyzed the operation of plug-flow reactors. We covered the following topics:

a. The underlying assumptions of the plug-flow reactor model and when they are satisfied in practice.

b. The dimensionless, species-based design equations, the energy balance equation, and the auxiliary relations for species concentrations.

c. The dimensionless reaction operating curves and species operating curves.

d. Design and operation of isothermal reactors with single reactions.

e. Methods to determine parameters of the rate expression experimentally.

f. Design and operation of isothermal reactors with multiple reactions.

g. Design and operation of non-isothermal reactors with multiple reactions.

h. Design and operation of gas-phase plug-flow reactors with multiple reactions where the pressure drop along the reactor is not negligible.

PROBLEMS

6-1$_2$ At 650°C, phosphine vapor decomposes according to the first-order, chemical reaction

$$4 \text{ PH}_3 \rightarrow \text{P}_4 \text{ (g)} + 6 \text{ H}_2.$$

Phosphine is fed into a plug-flow reactor at a rate of 10 mole/hr in a stream which consists of 2/3 phosphine and 1/3 inert. The reactor is operated at 650°C and 11.4 atm. Based on the data below,
a. derive the dimensionless design equation,
b. plot the dimensionless reactor performance curve and the operating curves, and
c. determine the needed reactor volumes for 75% and 90% conversion.
Data: at 650°C, k = 10 hr^{-1}.

6-2$_2$ An aqueous feed containing A and B ($C_{A0} = 0.1$ mole/lit, $C_{B0} = 0.2$ mole/lit) is fed at a rate of 400 liter/min into a plug-flow reactor, where the reaction A + B $\rightarrow$ R takes place. The rate expression of the reaction is $(-r_A) = 200 \cdot C_A \cdot C_B$ mole/lit min.
a. Derive the dimensionless design equation.
b. Plot the dimensionless operating curves for each species.
c. Determine the needed reactor volumes for 90%, 99%, and 99.9% conversion.

6-3$_2$ The homogeneous, gas-phase chemical reaction A $\rightarrow$ 3 R follows second-order kinetics. For a feed rate of 4 m^3/hr of pure A at 5 atm and 350°C, an experimental reactor consisting of a 2.5 cm ID pipe 2 m long gives 60% conversion of feed. A commercial plant is designed to treat 320 m^3/hr of feed consisting of 50% A, 50% inert at 35 atm, and 350°C to obtain 80% conversion.
a. How many 3-m lengths of 2.5 cm ID pipe are required?
b. Should they be placed in parallel or in series?
Assume plug-flow in pipe, negligible pressure drop, and ideal gas behavior.

6-4$_2$ Formic acid are fed at a rate of 8 mole/hr into a 10 liter plug-flow reactor operated at 150°C and 1 atm. The acid decomposes according to the first-order, chemical reaction

$$\text{HCOOH} \rightarrow \text{H}_2\text{O} + \text{CO}.$$

Based on the data below,
a. derive the dimensionless design equation,
b. plot the dimensionless reactor performance curve and the species operating curves, and
c. determine the conversion of the formic acid.
Data: at 150°C, k = 2.46 min^{-1}.

6-5$_2$ The gas leaving an ammonia oxidation plant consists of 10% nitric oxide (NO), 1% nitrogen oxide (NO$_2$), 8% oxygen, and the rest is inert. This gas is allowed to oxidize in a tubular reactor according to the reaction 2 NO + O$_2$ $\rightarrow$ 2 NO$_2$ until the

NO_2:NO ratio reaches 8:1. (In the process, the product gas is then absorbed in water in a separate vessel to produce nitric acid according to the reaction $3\ NO_2 + H_2O \rightarrow 2\ HNO_3 + NO$.) Calculate the size of the tubular reactor (assume plug-flow) operating at 20°C and 1 atm needed for the oxidation for a gas feed rate of 10,000 m^3/hr (measured at 0°C and 1 atm). According to Bodenstein in <u>Phys. Chem.</u> 100, 87 (1922) the rate expression of the reaction is $(r_{NO2}) = 14{,}000\ C_{NO2}\ C_{O2}$ (mole/liter sec) (Adapted from Denbigh and Turner, p. 57).

6-6$_4$ Methanol is produced by the gas-phase reaction $CO + 2\ H_2 \rightarrow CH_3OH$. However, at the reactor operating conditions, the undesirable reaction $CO + 3H_2 \rightarrow CH_4 + H_2O$ is also taking place. Both reactions are second-order (each is first-order with respect to each reactant) and $k_2/k_1 = 1.2$. A synthesis gas stream is fed into a plug-flow reactor operated at 450°C and 5 atm. Plot the reaction and species operating curves for isothermal operation with a feed consisting of 50% CO and 50% H_2 (mole basis).

6-7$_4$ Below is a simplified kinetic model of the cracking of propane to produce ethylene.

$$C_3H_8 \rightarrow C_2H_4 + CH_4 \qquad\qquad r_1 = k_1 \cdot C_{C3H8}$$

$$C_3H_8 \leftrightarrow C_3H_6 + H_2 \qquad\qquad r_2 = k_2 \cdot (C_{C3H8} - (1/K_2) \cdot C_{C3H6} \cdot C_{H2})$$

$$C_3H_8 + C_2H_4 \rightarrow C_2H_6 + C_3H_6 \qquad r_3 = k_3 \cdot C_{C3H8}\ C_{C2H4}$$

$$2\ C_3H_6 \rightarrow 3\ C_2H_4 \qquad\qquad r_4 = k_4 \cdot C_{C3H6}$$

$$C_3H_6 \leftrightarrow C_2H_2 + CH_4 \qquad\qquad r_5 = k_5 \cdot (C_{C3H8} - (1/K_5) \cdot C_{C2H2} \cdot C_{CH4})$$

$$C_2H_4 + C_2H_2 \rightarrow C_4H_6 \qquad\qquad r_6 = k_6\ C_{C2H4}\ C_{C2H2}.$$

At 800°C, the values of the rate constants are: $k_1 = 2.341$ sec^{-1}, $k_2 = 2.12$ sec^{-1}, $K_2 = 1000$, $k_3 = 23.63$ m^3/kmole sec, $k_4 = 0.816$ sec^{-1}, $k_5 = 0.305$ sec^{-1}, $K_5 = 2000$, and $k_6 = 4.06\ 10^3$ m^3/kmole sec. You are asked to design a PFR for cracking of propane to be operated at 2 atm. Plot the performance curves of the independent reactions and the curve for the generation of ethylene and propylene for an isothermal reactor operating at 800°C.

6-8$_4$ The liquid-phase chemical reactions

$$A + B \leftrightarrow C$$

$$C + B \rightarrow D$$

take place in a plug-flow reactor, operating isothermally at 90°C. A 200 liter/min stream is to be processed in the reactor. Based on the data below, derive and plot the reaction and species operating curves.

a. Determine the reactor volume needed for 70% conversion of B.

b. Determine the production rates of species A, B, C, and D when $f_B = 0.7$.

c. If we want to maximize the production of C, what should be the reactor volume?

d. What is the maximum yield of C?

Data: The rate expressions of the chemical reactions are:

$r_1 = k_1 \cdot C_A \cdot C_B$; $r_2 = k_2 \cdot C_C$; $r_3 = k_3 \cdot C_C \cdot C_B$;

At 90°C, $k_1 = 3$ lit mole^{-1} min.$^{-1}$; $k_2 = 0.5$ min.$^{-1}$; $k_3 = 1$ lit mole^{-1} min.$^{-1}$

Feed position: $C_{A0} = 2$ mole/lit; $C_{B0} = 2$ mole/lit; $C_{C0} = C_{D0} = 0$.

Feed rate: 120 lit/min.

6-9$_4$ The first-order gas-phase reaction $A \rightarrow B + C$ takes place in a plug-flow reactor. A stream consisting of 90% A and 10% I (% mole) is fed into a 200 liter reactor at a rate of 50 lit/sec. The feed is at 731°K and 3 atm. Based on the data below, derive the reaction operating curves and the temperature curve for each of the operations below. Determine:

a. the conversion of A when the reactor is operated isothermally,

b. the heating/cooling load in (a),

c. the conversion of A when the reactor is operated adiabatically, and

d. the reactor temperature in (c).

Data: At 731°K $k = 0.2$ sec^{-1}; $E_a = 12,000$ cal/mole;

At 731°K $DH_R = -10,000$ cal/mole extent;

$\hat{c}_{PA} = 25$ cal/mole°K; $\hat{c}_{PB} = 15$ cal/mole°K; $\hat{c}_{PC} = 18$ cal/mole°K;

$\hat{c}_{PI} = 9$ cal/mole°K

6-10$_4$ The elementary liquid-phase reactions

$$A + B \rightarrow C$$

$$C + B \rightarrow D$$

take place in a plug-flow reactor with a diameter of 10 cm. A solution ($C_{A0} = 2$ mole/lit, $C_{B0} = 2$ mole/lit) at 80°C is fed into the reactor at a rate of 200 lit/min. Based on the data below, calculate:

a. The length for maximum production of C if the reactor is operated isothermally.

b. The heating/cooling load in (a).

c. The length for maximum production of C if the reactor is operated adiabatically.

d. The reactor temperature profile along the reactor in (c).

e. The length for maximum production of C if the shell of the reactor is maintained at 80°C, and the heat-transfer coefficient is 3 cal/cm^{2}min°C.

f. The reactor temperature profile along the reactor in (e).

Data: At 80°C $k_1 = 0.1$ liter mole^{-1}min^{-1}; $E_{a1} = 12,000$ cal/mole;

At 80°C $k_2 = 0.2$ liter mole^{-1}min^{-1}; $E_{a2} = 16,000$ cal/mole;

At 80°C $\Delta H_{R1} = -15,000$ cal/mole extent; $\Delta H_{R2} = -10,000$ cal/mole extent;

Density = 900 g/liter: Heat Capacity = 0.8 cal/g°C.

6-11$_4$ The irreversible gas-phase reactions (both are first-order)

$$A \rightarrow 2V$$

$C_{P_0} = -65$

$D_{HR_1} = 0.126$

$D_{HR_2} = 0.1894$

$\delta_1 = 5.51$

$\delta_2 = 8.26$

$$V \rightarrow 2\,W$$

are carried out in a tubular reactor (plug-flow). A stream of species A is fed into a 10 cm ID reactor at a rate of 20 liter/min. The feed is at 3 atm and 731°K ($C_{A0} = 0.05$ mole/liter). Based on the data below, calculate:

a. The length of an isothermal reactor needed for 40% conversion of A.
b. The production rate of V in (a).
c. The heating/cooling load in (a).
d. The length of an adiabatic reactor needed for 40% conversion of A.
e. The production rate of V in (d).
f. The length of a reactor whose wall temperature is 750°K and U = 20 cal/cm²min°K needed for 40% conversion of A.

Data: At 731°K $k_1 = 2$ min⁻¹; $k_2 = 0.5$ min⁻¹;
$E_{a1} = 8,000$ cal/mole; $E_{a2} = 12,000$ cal/mole;
At 731°K $\Delta H_{R1} = 3,000$ cal/mole of V; $\Delta H_{R2} = 4,500$ cal/mole of W;
$\hat{c}_{PA} = 65$ cal/mole°K; $\hat{c}_{PV} = 40$ cal/mole°K; $\hat{c}_{PW} = 25$ cal/mole°K.

6-12₄ Cracking of naphtha cut to produce olefins is a common process in the petrochemical industry. The cracking reactions are represented by the simplified elementary gas-phase reactions:

$$C_{10}H_{22} \rightarrow C_4H_{10} + C_6H_{12}$$

$$C_4H_{10} \rightarrow C_3H_6 + CH_4$$

$$C_6H_{12} \rightarrow C_2H_4 + C_4H_8$$

$$C_4H_8 \rightarrow 2\,C_2H_4$$

The reactions take place in a 90 m long 10 cm ID tubular reactor (plug-flow) placed in a furnace chamber. The reactor wall is maintained at 900°C. A gas mixture at T = 700°C and P = 5 atm., consisting of 90% naphtha ($C_{10}H_{22}$) and 10% I, is fed into the reactor at a rate of 100 lit/sec. Based on the data below:

a. Plot the reaction and species operating curves and the temperature profile.
b. What should the reactor length be to optimize the production rate of ethylene?
c. Calculate the production rate of ethylene and propylene.
d. Calculate the heating/cooling load of the reactor.

Data: $r_1 = k_1 C_{C10H22}$; at 700°C $k_1 = 2.0$ sec⁻¹; $E_{a1} = 25,000$ cal/mole
$r_2 = k_2 C_{C4H10}$; at 700°C $k_2 = 0.5$ sec⁻¹; $E_{a2} = 35,000$ cal/mole
$r_3 = k_3 C_{C6H12}$; at 700°C $k_3 = 0.01$ sec⁻¹; $E_{a3} = 40,000$ cal/mole
$r_4 = k_4 C_{C4H8}$; at 700°C $k_4 = 0.001$ sec⁻¹; $E_{a4} = 45,000$ cal/mole
At 700°C, $\Delta H_{R1}(T_0) = 20,000$ cal/mole extent;
$\Delta H_{R2}(T_0) = 35,000$ cal/mole extent;
$\Delta H_{R3}(T_0) = 45,000$ cal/mole extent;
$\Delta H_{R4}(T_0) = 55,000$ cal/mole extent;

$\hat{c}_{pC10H22} = 280$ cal/mole°C; $\hat{c}_{pC6H12} = 180$ cal/mole°C; $\hat{c}_{pCH4} = 20$ cal/mole°C;

$\hat{c}_{pC4H10} = 150$ cal/mole°C; $\hat{c}_{pC4H8} = 140$ cal/mole°C; $\hat{c}_{pI} = 10$ cal/mole°C; $\hat{c}_{pC3H6} = 120$ cal/mole°C; $\hat{c}_{pC2H4} = 30$ cal/mole°C.

The average heat-transfer coefficient in the reactor is $U = 2$ cal/cm^{2}sec°K.

6-13$_4$ The first-order gas-phase reaction A $\rightarrow$ B + C takes place in a plug-flow reactor. A stream consisting of species 90% A and 10% I (% mole) is fed into a 200 liter reactor at a rate of 20 lit/sec. The feed is at 731°K and 3 atm. Based on the data below, calculate:
a. The conversion of A when the reactor is operated isothermally.
b. The heating/cooling load in (a).
c. The conversion of A when the reactor is operated adiabatically.
d. The reactor temperature in (c).
Data: At 731°K $k = 0.2$ sec^{-1}; $E_a = 12,000$ cal/mole;
At 731°K $\Delta H_R = -10,000$ cal/mole extent;
$\hat{c}_{PA} = 25$ cal/mole°K; $\hat{c}_{PB} = 15$ cal/mole°K; $\hat{c}_{PC} = 18$ cal/mole°K;
$\hat{c}_{PI} = 9$ cal/mole°K

6-14$_2$ The gas-phase reaction A $\rightarrow$ B + C is carried out in a cascade of two plug-flow reactors connected in series. Pure A is fed at a rate of 100 lbmole/hr into the first reactor, whose volume is 100 ft^3. The molar flow rate of A at the exit of the first reactor is 60 lbmole/hr, and that at the exit of the second reactor is 20 lbmole/hr. The temperature throughout the system is 150°C, and the pressure is 2 atm. The reaction is second-order. Determine the volume of Reactor 2 by:
a. taking the inlet stream into the system as the reference stream and
b taking the inlet stream into Reactor 2 as the reference stream.

6-15$_2$ Pure gaseous A at about 3 atm and 30°C (120 mmole/lit) is fed into a 1 liter plug-flow reactor at various flow rates. There A decomposes according to the reaction A $\rightarrow$ 2 R, and the exit concentration of A is measured for each flow rate. From the data below, find:
a. a rate expression of the reaction. Assume that reactant A alone affects the rate.
b. What volumetric inflow rate v_0 and outflow rate v_F of a $C_{A0} = 320$ mmole/lit pure A feed can be fed into a 560 liter plug-flow reactor to achieve a 50% converted of A?

Data:

v_0, lit/min.	0.06	0.48	1.5	8.1
C_{Aout}, mmole/lit	30	60	80	105

(Adapted from Levenspiel's Omnibook)

6-16$_4$ The elementary liquid-phase reactions

$$A + B \rightarrow C$$

$$C + B \rightarrow D$$

$$D + B \rightarrow E$$

take place in a 0.1 m ID tubular reactor. A solution ($C_{A0} = 2$ mole/lit, $C_{B0} = 2$ mole/lit) at 80°C is fed into the reactor at a rate of 100 lit/min. Based on the data below, derive the reaction operating curves and the temperature curve for each of the operations below. For each case, determine the reactor length needed for maximum production of C and the production rate of C and D.

a. Isothermal operation at 80°C.
b. The heating/cooling load on the reactor in (a).
c. Adiabatic operation. What is the outlet temperature?
d. Heat is removed with U = 20 kcal/m² hr °K, and $T_F = 80°C$

Data: At 80°C $k_1 = 0.1$ lit mol⁻¹ min⁻¹; $E_{a1} = 6,000$ cal/mole;
$k_2 = 0.2$ lit mol⁻¹ min⁻¹; $E_{a2} = 8,000$ cal/mole;
$k_3 = 0.3$ lit mol⁻¹ min⁻¹; $E_{a2} = 10,000$ cal/mole;
$\Delta H_{R1} = -15,000$ cal/mole extent; $\Delta H_{R2} = -10,000$ cal/mole extent;
$\Delta H_{R3} = -8,000$ cal/mole extent;

Density of the solution = 1,000 g/liter;   Heat Capacity of the solution = 1 cal/g°C.

CHAPTER SEVEN

CONTINUOUS STIRRED TANK REACTOR (CSTR)

The continuous stirred tank reactor (CSTR) is a mathematical model that describes an important class of continuous reactors – continuous, steady, well-agitated tank reactors. The CSTR model is based on two assumptions:
• steady state operation, and
• the same conditions exist everywhere inside the reactor (due to good mixing).
These conditions are readily achieved in small scale agitated reactors. However, in large industrial reactors, it is not easy to achieve good mixing, and special care should be given to the design of the tank and agitator. Furthermore, even when the same conditions exist in most sections of the reactor, the conditions near the reactor inlet and near the reactor wall are different from those in the remainder of the reactor. Since these zones usually represent a small portion of the reactor, the CSTR model provides a reasonable estimate of the well-agitated large reactors. Figure 7-1 shows schematically a liquid-phase CSTR.

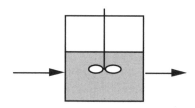

Figure 7-1: Schematic Description of a Liquid-Phase CSTR

7.1 DESIGN EQUATIONS AND AUXILIARY RELATIONS

The reaction-based, dimensionless design equation of a CSTR, written for the m-th independent reaction, derived in Chapter 3, is

$$Z_{m_{out}} - Z_{m_{in}} = \left(r_{m_{out}} + \sum_{k}^{n_{dep}} \alpha_{km} \cdot r_{k_{out}} \right) \cdot \tau \cdot \left(\frac{t_{cr}}{C_0} \right), \qquad (7.1\text{-}1)$$

where Z_m is the dimensionless extent of the m-th independent reaction defined by

$$Z_m = \frac{\dot{X}_m}{(F_{tot})_0}, \qquad (7.1\text{-}2)$$

τ is the dimensionless space time of the reactor defined by

$$\tau = \frac{V_R}{v_0 \cdot t_{cr}}, \qquad (7.1\text{-}3)$$

where t_{cr} is a conveniently-selected characteristic reaction time (see Section 2.5), and C_0 is a conveniently-selected reference concentration defined by

$$C_0 = \frac{(F_{tot})_0}{v_0}, \qquad (7.1\text{-}4)$$

where $(F_{tot})_0$ and v_0 are, respectively, the total molar flow rate and the volumetric flow rate of the reference stream.

As discussed in Chapter 3, to describe the operation of a CSTR with multiple reactions, we have to write (7.1-1) for each of the independent chemical reactions. The solution of the design equations (the relationships between $Z_{m_{out}}$'s and τ) are the reaction operating curves of the reactor that completely describe the reactor operation. To solve the design equations, we have to express the rates of the individual chemical reactions that take place in the reactor (r_m's and r_k's) in terms of Z_m's and τ. Below, we discuss the auxiliary relations used to express the design equations explicitly in terms of Z_m's and τ.

The volume-based rate expression of the i-th reaction (see Section 2-3) is

$$r_i = k_i(T_0) \cdot e^{\gamma_i \frac{\theta - 1}{\theta}} \cdot h_i(C_j's), \qquad (7.1\text{-}5)$$

where $k_i(T_0)$ is the reaction rate constant at reference temperature T_0, γ_i is the dimensionless activation energy ($\gamma_i = Ea_i/R \cdot T_0$), and $h_i(C_j's)$ is a function of the species concentrations, given by the rate expression. To express the rates of the chemical reactions in terms of Z_m's and τ, we have to relate the species concentrations to Z_m's . For a CSTR, the concentration of species j is the same everywhere inside the reactor and is equal to the outlet concentration,

$$C_{j_{out}} = \frac{F_{j_{out}}}{v_{out}}, \tag{7.1-6}$$

where $F_{j_{out}}$ and v_{out} are, respectively, the molar flow rate of species j and the volumetric flow rate at the reactor outlet. Using stoichiometric relation (1.5-5),

$$F_{j_{out}} = (F_{tot})_0 \left(y_{j_0} + \sum_{m}^{n_{ind}} (s_j)_m \cdot Z_{m_{out}} \right), \tag{7.1-7}$$

and the outlet concentration of species j is

$$C_{j_{out}} = \frac{(F_{tot})_0}{v_{out}} \left(y_{j_0} + \sum_{m}^{n_{ind}} (s_j)_m \cdot Z_{m_{out}} \right). \tag{7.1-8}$$

For **liquid-phase** reactions, the density of the reacting fluid is assumed to be constant; hence, $v_{out} = v_0$, and (7.1-8) reduces to

$$C_{j_{out}} = C_0 \left(y_{j_0} + \sum_{m}^{n_{ind}} (s_j)_m \cdot Z_{m_{out}} \right). \tag{7.1-9}$$

Eq. (7.1-9) provides the species concentrations in terms of the extents of the independent reactions for liquid-phase reactions. For **gas-phase** reactions, the volumetric flow rate depends on the total molar flow rate and the temperature and pressure in the reactor. Assuming ideal gas behavior, the volumetric flow rate is

$$v_{out} = v_0 \cdot \left(\frac{(F_{tot})_{out}}{(F_{tot})_0} \right) \cdot \left(\frac{T_{out}}{T_0} \right) \cdot \left(\frac{P_0}{P_{out}} \right). \tag{7.1-10}$$

Using stoichiometric relation (1.5-6) to express the total molar flow rate in terms of the extents of the independent reactions, (7.1-10) becomes

$$v_{out} = v_0 \cdot \left(1 + \sum_{m}^{n_{ind}} \Delta_m \cdot Z_{m_{out}} \right) \cdot \theta_{out} \cdot \left(\frac{P_0}{P_{out}} \right), \tag{7.1-11}$$

where θ_{out} is the dimensionless temperature at the reactor outlet, T_{out}/T_0. Substituting (7.1-11) into (7.1-8), we obtain

$$C_{j_{out}} = C_0 \frac{y_{j_0} + \sum_{m}^{n_{ind}} (s_j)_m \cdot Z_{m_{out}}}{\left(1 + \sum_{m}^{n_{ind}} \Delta_m \cdot Z_{m_{out}} \right) \cdot \theta_{out}} \left(\frac{P_{out}}{P_0} \right). \tag{7.1-12}$$

Eq. (7.1-12) provides the species concentrations in terms of the extents of the independent reactions for gas-phase reactions in CSTRs. Table 7.1 provides a summary of the design equations and auxiliary relations used in the design formulation of CSTRs.

Note that (7.1-5) and (7.1-12) have another dependent variable, θ_{out}, the reactor dimensionless temperature. Since the reactor temperature may be different than the inlet temperature, to design the reactor we have express it by applying the energy balance equation. For CSTRs with negligible viscous and shaft work, the dimensionless energy balance equation, derived in Chapter 4, is

$$HTN \cdot \tau \cdot (\theta_F - \theta_{out}) =$$

$$= \sum_m^{n_{ind}} DHR_m \cdot (Z_{m_{out}} - Z_{m_{in}}) + \int_1^{\theta_{out}} CF(Z,\theta)_{out} \, d\theta - \int_1^{\theta_{in}} CF(Z,\theta)_{in} \, d\theta, \quad (7.1\text{-}13)$$

where HTN is the dimensionless heat-transfer number defined by (4.2-22),

$$HTN = \frac{U \cdot t_{cr}}{C_0 \cdot \hat{c}_{p0}} \left(\frac{S}{V} \right), \quad (7.1\text{-}14)$$

DHR_m is the dimensionless heat of reaction of the m-th independent chemical reaction, defined by (4.2-23),

$$DHR_m = \frac{\Delta H_{Rm}(T_0)}{T_0 \cdot \hat{c}_{p0}}, \quad (7.1\text{-}15)$$

and $CF(Z_m, \theta)$ is the correction factor of the heat capacity, defined by (4.2-54). The quantity $\hat{c}_{p0}$ is the specific molar heat capacity of the reference stream, defined for gas-phase reactions by (4.2-59) and for liquid-phase reactions by (4.2-60). Note that the term on the left hand side of (7.1-13) indicates the dimensionless heat transfer rate to the reactor,

$$\frac{\dot{Q}}{(F_{tot})_0 \cdot T_0 \cdot \hat{c}_{p0}} = \frac{U \cdot t_{cr}}{C_0 \cdot \hat{c}_{p0}} \left(\frac{S}{V} \right) \cdot \tau \cdot (\theta_F - \theta_{out}). \quad (7.1\text{-}16)$$

To design non-isothermal CSTRs, we have to solve the design equations simultaneously with (7.1-13) and determine $Z_{m_{out}}$ and θ_{out} for given τ. Table 7.2 provides a summary of the energy balance equation and auxiliary relations used in the design formulation of CSTRs.

Below, we discuss how to apply the design equations and the energy balance equations to determine various quantities concerning the operations of CSTRs. In Section 7.2 we examine isothermal operations with **single** reactions to illustrate how the rate expressions are incorporated into the design equation and how rate expressions are determined. In Section 7.3, we expand the analysis to isothermal operations with **multiple** reactions. In Section 7.4, we consider non-isothermal operations with multiple reactions.

Table 7.1: Design Equations and Related Quantities

Design Equation	**For the m-th Independent Reaction** $$Z_{m_{out}} - Z_{m_{in}} = \left(r_{m_{out}} + \sum_{k}^{n_{dep}} \alpha_{km} \cdot r_{k_{out}} \right) \cdot \tau \cdot \left(\frac{t_{cr}}{C_0} \right) \quad (A)$$
Definitions	**Dimensionless Space Time** $$\tau \equiv \frac{V_R}{v_0 \cdot t_{cr}} \quad (B)$$ **Dimensionless Extent of the m-th Independent Reaction** $$Z_m = \frac{\dot{X}_m}{(F_{tot})_0} \quad (C)$$ **Reference Concentration** $$C_0 = \frac{(F_{tot})_0}{v_0} \quad (D)$$ **Characteristic Reaction Time** $$t_{cr} = \frac{\text{Characteristic Concentration}}{\text{Characteristic Reaction Rate}} = \frac{C_0}{r_0} \quad (E)$$
Species Concentrations	**For Liquid-Phase Reactions** $$C_{j_{out}} = C_0 \cdot \left(y_{j0} + \sum_{m}^{n_{ind}} (s_j)_m \cdot Z_{m_{out}} \right) \quad (F)$$ **For Gas-Phase Reactions** $$C_{j_{out}} = C_0 \frac{y_{j0} + \sum_{m}^{n_{ind}} (s_j)_m \cdot Z_{m_{out}}}{\left(1 + \sum_{m}^{n_{ind}} \Delta_m \cdot Z_{m_{out}} \right) \cdot \theta} \cdot \left(\frac{P_{out}}{P_0} \right) \quad (G)$$

Table 7.2: Energy Balance Equation and Related Quantities

Energy Balance Equation

$$HTN \cdot \tau \cdot (\theta_F - \theta_{out}) =$$

$$= \sum_{m}^{n_{ind}} DHR_m \cdot (Z_{m_{out}} - Z_{m_{in}}) + \int_1^{\theta_{out}} CF(Z,\theta)_{out} \, d\theta - \int_1^{\theta_{in}} CF(Z,\theta)_{in} \, d\theta \quad (A)$$

Definitions and Auxiliary Relations

Dimensionless Temperature

$$\theta \equiv \frac{T}{T_0} \quad (B)$$

Specific Molar Heat Capacity of Reference Stream

$$\hat{c}_{P0} = \sum_{j}^{all} y_{j0} \cdot \hat{c}_{P_j}(T_0) \quad \text{or} \quad \hat{c}_{P0} = \frac{\dot{m}}{(F_{tot})_0} \bar{c}_P \quad (C)$$

Dimensionless Heat of Reaction of the m-th Independent Reaction

$$DHR_m = \frac{\Delta H_{R_m}(T_0)}{T_0 \cdot \hat{c}_{P0}} \quad (D)$$

Dimensionless Heat Transfer Number

$$HTN = \frac{U \cdot t_{cr}}{C_0 \cdot \hat{c}_{P0}} \left(\frac{S}{V}\right) \quad (E)$$

Dimensionless Heat Rate

$$\frac{\dot{Q}}{(F_{tot})_0 \cdot T_0 \cdot \hat{c}_{P0}} \quad (F)$$

Correction Factor of Heat Capacity (Gas-Phase Reactions)

$$CF(Z_m, \theta) = \frac{\sum_{j}^{all} y_{j0} \cdot \hat{c}_{P_j}(\theta)}{\sum_{j}^{all} y_{j0} \cdot \hat{c}_{P_j}(1)} + \frac{\sum_{j}^{all} \hat{c}_{P_j}(\theta) \cdot \sum_{m}^{n_{ind}} (s_j)_m \cdot Z_m}{\sum_{j}^{all} y_{j0} \cdot \hat{c}_{P_j}(1)} \quad (G)$$

7.2 ISOTHERMAL OPERATIONS WITH SINGLE REACTIONS

We start the analysis of CSTRs by considering isothermal operations with **single** chemical reactions. Isothermal CSTRs are defined as those where $\theta_{out} = \theta_{in}$. Hence, we do not have to determine the reactor temperature, θ_{out}, and only have to solve the design equations. The energy balance equation provides the heating (or cooling) load necessary to maintain isothermal conditions. Furthermore, for isothermal operations, the individual reaction rates depend only on the species concentrations, and, when the reactor temperature is taken as the reference temperature, $T = T_0$, and (7.1-5) reduces to

$$r_i = k_i(T_0) \cdot h_i(C_j' s). \tag{7.2-1}$$

When a **single** chemical reaction takes place in a CSTR, there is only one independent reaction and no dependent reactions, and the operation is described by a single design equation. Hence, (7.1-1) reduces to

$$Z_{out} - Z_{in} = r_{out} \cdot \tau \cdot \left(\frac{t_{cr}}{C_0}\right), \tag{7.2-2}$$

where Z_{out} and Z_{in} are the dimensionless extents of the reaction at the reactor outlet and inlet, respectively, and r_{out} is its rate. We can rearrange (7.2-2),

$$\tau = \left(\frac{C_0}{t_{cr}}\right) \frac{Z_{out} - Z_{in}}{r_{out}}. \tag{7.2-3}$$

Note that the values of Z_{in} and Z_{out} depend on the selection of the reference stream. Also note that if we use the definition of the characteristic reaction time, (2.5-1), (7.2-3) reduces to

$$\tau = \left(\frac{r_0}{r_{out}}\right) \cdot (Z_{out} - Z_{in}). \tag{7.2-4}$$

The solution of the design equation, Z_{out} versus τ, is the dimensionless reaction operating curve of the reactor. It describes the progress of the chemical reaction inside the CSTR. Furthermore, once Z_{out} is known, we can apply stoichiometric relation (7.1-7) to obtain the molar flow rates of the individual species at the reactor outlet. Note that for given inlet conditions, design equations (7.2-2) and (7.2-3) have three variables: the dimensionless space time, τ, the reaction extent at the reactor outlet, Z_{out}, and the reaction rate, r_{out}. The design equation is applied to determine any one of these variables when the other two are provided. A typical design problem is to determine the reactor volume needed to obtain a specified extent (or conversion) for a given feed rate and reaction rate. The second application is to determine the extent (or conversion) at the reactor outlet for a given reactor volume and reaction rate. The third application is to determine the reaction rate when the extent (or conversion) at the reactor outlet is provided for a given feed rate.

If one prefers to express the design equation in terms of the reactor volume, V_R, rather than the dimensionless space time τ, by substituting the definition of the dimensionless space time, (7.1-3), into (7.2-2), we obtain

$$V_R = (F_{tot})_0 \frac{Z_{out} - Z_{in}}{r_{out}}.$$ (7.2-5)

For CSTRs with **single** chemical reactions, the common practice has been to express the design equation in terms of the conversion of the limiting reactant, f_A. We can easily do so since, for single reactions, the conversion is proportional to the extent. Using stoichiometric relation (1.5-11),

$$Z = -\frac{y_{A0}}{s_A} f_A,$$

and design equation (7.2-5) becomes

$$V_R = F_{A_0} \frac{f_{A_{out}} - f_{A_{in}}}{(-r_A)_{out}},$$ (7.2-6)

where $(-r_A)_{out} = -s_A \cdot r_{out}$, is the depletion rate of reactant A.

To solve design equation (7.2-3), we have to express the reaction rate, r_{out}, in terms of the dimensionless extent, Z_{out}. To do so, we have to express the species concentrations in terms of Z_{out}. For single **liquid-phase** reactions, (7.1-9) reduces to

$$C_{j_{out}} = C_0(y_{j0} + s_j \cdot Z_{out}).$$ (7.2-7)

For single **gas-phase** reactions, (7.1-12) reduces to

$$C_{j_{out}} = C_0 \frac{y_{j0} + s_j \cdot Z_{out}}{1 + \Delta \cdot Z_{out}}.$$ (7.2-8)

Below, we analyze the operation of isothermal CSTRs with **single** reactions for different types of chemical reactions. For convenience, we divide the analysis into two sections: reactor design and determination of the rate expression. In the former, we determine the size of the reactor or the production rate of a given reactor for a known reaction rate, and, in the latter, we determine the rate expression from reactor operating data.

7.2.1 Design of a Single CSTR

In general, the reaction rate is provided either in the form of experimental data or as an algebraic expression. First, we discuss the design of a CSTR when the reaction rate is provided in the form of experimental data without a rate expression. Examining the structure of design equation (7.2-4), we see that if we know how r varies with Z, we can then plot r_0/r versus Z, and the area of the rectangle between Z_{in} and Z_{out} is equal to the dimensionless space time, τ (see Figure 7-2). By determining the area of the rectangle for different values of Z_{out}, we can obtain numerically a relationship between Z_{out} and τ and plot the dimensionless operating curve. From the operating curve, we can determine the necessary τ for any specified extent. The required reactor volume, V_R, is readily determined from the dimensionless space time by $V_R = v_0 \cdot t_{cr} \cdot \tau$.

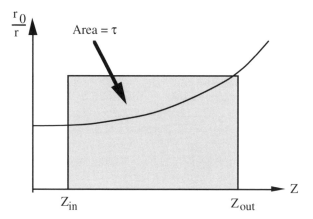

Figure 7-2: Determination of Dimensionless Space Time from Rate Experimental Data

Recall from Figure 6.2 that the dimensionless space time, τ, for a plug-flow reactor is equal to the area under the curve. Hence, for common rate expressions, the volume of a CSTR needed to achieve a certain extent is larger than that of a plug-flow reactor. This is illustrated by comparing the operating curves of CSTR and plug-flow reactors, shown in Figure 7-3. To understand the reason for the different performance

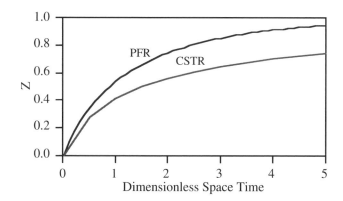

Figure 7-3: Comparison of Operating Curves of Plug-Flow Reactor and CSTR

of a plug-flow reactor and a CSTR, we show schematically in Figure 7-4 the concentration profile of the reactant A in each reactor. In a plug-flow reactor, the concentration decreases gradually along the reactor, whereas in a CSTR, the outlet concentration is maintained everywhere in the reactor. For common kinetics, the reaction rate is faster at higher concentrations. Hence, a plug-flow reactor with the same volume as a CSTR (same space time) provides a higher conversion (or extent) than the CSTR. In fact, from a mixing perspective, a plug-flow reactor represents the best reactor configuration, whereas a CSTR represents the worse configuration. The zone between the two curves in Figure 7-3 is where operating curves of actual reactors (neither plug-like nor are they well-mixed) would be located.

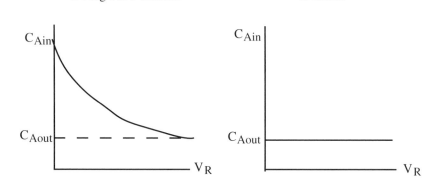

Figure 7-4: Concentration Profile in PFR and CSTR

Next, we describe how to obtain the operating curve of a CSTR when the reaction rate is given in the form of experimental data. In many applications, the experimental rate data of a chemical reaction are provided in terms of the formation (or depletion) rate of species j, (r_j), at different concentrations of species j, C_j. To obtain a relation between r_{out} and Z_{out}, we have to relate r to (r_j) and C_j to Z. Using relation (2.2-5),

$$r_{out} = \frac{(r_j)_{out}}{s_j}. \tag{7.2-9}$$

As for the relationship between C_{jout} to Z_{out}, we distinguish between liquid-phase and gas-phase reactions. For **liquid-phase** reactions, we use (7.1-9) and obtain

$$Z_{out} = \frac{C_{jout} - y_{j0} \cdot C_0}{s_j \cdot C_0}. \tag{7.2-10}$$

For **gas-phase** reactions, we use (7.1-12) and obtain

$$Z_{out} = \frac{C_{jout} - y_{j0} \cdot C_0}{s_j \cdot C_0 - \Delta \cdot C_{jout}}. \tag{7.2-11}$$

The two examples below illustrate how these relations are used to design CSTRs with liquid-phase and gas-phase reactions when the reaction rate is given in the form of experimental data.

Example 7-1 The liquid-phase reaction A $\rightarrow$ B + C is to be carried out in an isothermal CSTR. A solution with a concentration of $C_{A0} = 7$ mole/liter is fed into the reactor at a rate of 50 liter/min. The experimental rate data, obtained in a batch reactor, are provided below.

a. Plot the dimensionless operating curve of a plug-flow reactor, and
b. determine the reactor volume needed to obtain an outlet stream with $C_{Aout} = 1$ mole/liter.
c. If the stream is fed into an existing 200 liter reactor, determine C_{Aout}.

Data:

C_A (mole/lit)	0.0	1.0	2.0	3.0	4.0	5.0	6.0	7.0	8.0
$(-r_A)$ (mole/lit. min)	0.0	0.118	0.222	0.316	0.400	0.476	0.546	0.609	0.667

Solution The chemical reaction is A $\rightarrow$ B + C, and the stoichiometric coefficients of the reaction are: $s_A = -1$ and $s_B = 1$. We select the reference stream as a stream with $C_0 = 8$ mole/liter and a flow rate of $v_0 = 50$ liter/min. For this basis, $(F_{tot})_0 = 400$ mole/min., $y_{A0} = (F_A)_0/(F_{tot})_0 = 0.875$, $y_{B0} = 0$, and $y_{C0} = 0$. We use (2.2-5) to determine the reaction rate,

$$r = -\frac{(-r_A)}{s_A} = (-r_A),$$

(a)

and, using (2.5-1), the characteristic reaction time is

$$t_{cr} = \frac{C_0}{r_0} = \frac{8.0}{0.667} = 12 \text{ min}.$$

(b)

For liquid-phase reactions, we use (7.2-10) to relate the dimensionless extent Z to each corresponding value of the C_A,

$$Z = \frac{C_A - y_{A0} \cdot C_0}{s_A \cdot C_0} = \frac{C_A - 7.0}{(-1) \cdot 8.0} = \frac{7.0 - C_A}{8.0}.$$

(c)

The calculated values of r and Z are shown in the table below (note that these values are calculated the same way in Example 6-1):

C_A (mole/lit)	0.0	1.0	2.0	3.0	4.0	5.0	6.0	7.0	8.0
Z (dimensionless)	0.875	0.750	0.625	0.500	0.375	0.250	0.125	0.00	—
r (mole/lit. min.)	0.0	0.118	0.222	0.316	0.400	0.476	0.546	0.609	0.667
(r_0/r) (dimensionless)	∞	5.65	3.00	2.11	1.667	1.40	1.221	1.095	1.00

The first figure below shows the plot of (r_0/r) versus Z for this reaction.

a. Now that the curve is known, we can determine the needed dimensionless operating time, τ, for any value of Z by calculating the area of the rectangle between the origin and any value of Z. The area for the i-th point is

$$\tau_i = \left(\frac{r_0}{r}\right)_i (Z_i - 0).$$

The calculated values of Z and τ are shown in the table below:

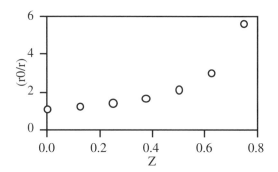

Z (dimensionless) 0.0   0.125   0.250 0.375   0.500 0.625 0.750 0.875
τ (dimensionless) 0.0   0.153   0.350 0.625   1.055 1.875 4.24   ∞

The figure below shows the plot of $Z(\tau)$ versus τ — the dimensionless reaction operating curve.

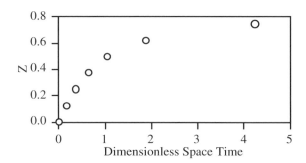

b. For $C_A = 1$ mole/lit, $Z = 0.75$, and, using the dimensionless operating curve, $\tau = 4.24$. Hence, using the definition of the dimensionless space time, (7.1-3), the volume of the reactor is

$$V_R = v_0 \cdot t_{cr} \cdot \tau = (50 \text{ liter/min}) \cdot (12 \text{ min}) \cdot (4.24) = 2{,}550 \text{ liter.} \qquad \text{(f)}$$

Note that the volume of a CSTR is much larger than that of a plug-flow reactor (Example 6-1).

c. Using (7.1-3), for a 200 liter reactor, the dimensionless space time is

$$\tau = \frac{V_R}{v_0 \cdot t_{cr}} = \frac{200}{50 \cdot 12} = 0.333. \qquad \text{(g)}$$

From the reaction operating curve (or interpolation of the values in the table above), for

$\tau = 0.333$, $Z = 0.235$. Using (7.2-10), the outlet concentration is

$$C_A = C_0(y_{j0} - Z_{out}) = 8.0 \cdot (0.875 - 0.235) = 5.12 \text{ mole/liter}.$$

Example 7-2 The gas-phase reaction $A + B \rightarrow C$ is investigated in an isobaric batch reactor. The reactor is fed with a stream consisting of 40% A and 60% B (% mole) at a rate of 20 mole/min. The inlet pressure is 3 atm, and the temperature is 400°K. The following data was obtained in an isothermal batch reactor at 400°K:

$C_A \cdot 10^2$ (mole/lit)   3.656 3.199 2.742 2.285 1.828 1.371 0.914 0.457
$(-r_A) \cdot 10^2$ (mole/lit. min) 3.146 2.700 2.270 1.865 1.483 1.123 0.787 0.463

We now want to design an isothermal CSTR to process a stream at 3 atm and 400°K with the same composition. Based on the experimental rate data,

a. plot the dimensionless operating curve of a plug-flow reactor, and
b. determine the reactor volume needed to obtain 80% conversion.
c. If the stream is fed into an existing 1,500 liter CSTR, determine the conversion.

Solution The chemical reaction is $A + B \rightarrow C$, and the stoichiometric coefficients are: $s_A = -1$, $s_B = -1$, $s_C = 1$, and $\Delta = -1$. We select the inlet stream as a reference stream; hence, $y_{A0} = 0.4$, $y_{B0} = 0.6$, $y_{C0} = 0$, the reference concentration is

$$C_0 = \frac{P_0}{R \cdot T_0} = \frac{(3 \text{ atm})}{(0.08206 \text{ lit atm/mole}°K) \cdot (400°K)} = 9.14 \cdot 10^{-2} \text{ mole/liter}, \quad (a)$$

and the volumetric flow rate of the reference stream is

$$v_0 = \frac{(F_{tot})_0}{C_0} = \frac{(20 \text{ mole/min})}{(9.14 \cdot 10^{-2} \text{ mole/liter})} = 219 \text{ liter/min}. \quad (b)$$

For gas-phase reactions, we use (7.2-11) to relate Z to C_A,

$$Z = \frac{C_A - y_{A0} \cdot C_0}{s_A \cdot C_0 - \Delta_{gas} \cdot C_A} = \frac{0.4 \cdot C_0 - C_A}{C_0 - C_A}. \quad (c)$$

To obtain the reaction rate at each point, we use (7.2-9),

$$r = -\frac{(-r_A)}{s_A} = (-r_A). \quad (d)$$

At the reference stream, $r_0 = 3.146$, and the characteristic reaction time is

$$t_{cr} = \frac{C_0}{r_0} = \frac{0.0914}{0.0315} = 2.90 \text{ min}. \quad (e)$$

The calculated values of r and Z for the given data are shown in the table below:

$C_A \cdot 10^2$ (mole/lit)	3.656	3.199	2.742	2.285	1.828	1.371	0.914	0.457
Z (dimensionless)	0.0	0.077	0.143	0.200	0.250	0.294	0.333	0.368
(r_0/r) (dimensionless)	1.00	1.165	1.386	1.687	2.121	2.801	4.00	6.795

The figure below shows the plot of (r_0/r) versus Z for this operation.

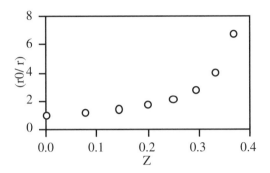

a. Now that the curve is constructed, we can construct the dimensionless operating curve by determining the area of a rectangle for any value of Z. The area for the i-th point is

$$\tau_i = \left(\frac{r_0}{r}\right)_i (Z_i - 0).$$

The calculated values of Z and τ are shown in the table below:

Z	0.0	0.0769	0.143	0.200	0.250	0.294	0.333	0.368
τ (dimensionless)	0	0.090	0.198	0.337	0.530	0.823	1.333	2.500

The figure below shows the plot of Z versus τ, the dimensionless reaction operating curve.

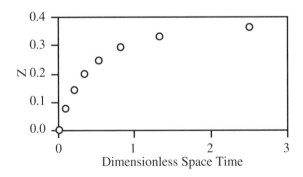

b. We use the operating curve to design the reactor. Applying (7.2-5), for 80% conversion, the extent is

$$Z = -\frac{y_{A0}}{s_A}f_A = -\frac{0.4}{(-1)}0.8 = 0.32.$$ \hfill (f)

From the curve (or interpolation of tabulated values), $Z = 0.32$ is reached at $\tau = 1.30$. Applying the definition of the dimensionless time, (7.1-3),

$$V_R = v_0 \cdot t_{cr} \cdot \tau = (219 \text{ liter/min}) \cdot (2.90 \text{ min}) \cdot (1.30) = 826 \text{ liter}.$$

c. For a reactor with a volume of 1,500 liter, the dimensionless operating time is

$$\tau = \frac{V_R}{v_0 \cdot t_{cr}} = \frac{(1,500 \text{ liter})}{(219 \text{ liter/min}) \cdot (2.90 \text{ min})} = 2.36.$$

From the operating curve, at $\tau = 2.36$, $Z = 0.355$, and the outlet conversion is

$$f_A = -\frac{s_A}{y_{A0}} Z = -\frac{(-1)}{0.4} 0.355 = 0.888. \tag{g}$$

We continue the analysis of isothermal CSTRs with **single** chemical reactions and consider applications of the design equation when the reaction rate is provided in the form of algebraic expressions. For these cases, we use either (7.2-7) or (7.2-8) to express the species concentrations in terms of dimensionless extent Z, substitute these relations into the rate expression, and substitute the latter into the design equation, (7.2-2). We then select a reference stream (usually the inlet stream) and determine Z_{in}. Next, we solve the design equation for different values of dimensionless space time to construct the reaction operating curve (a plot of Z versus τ).

Consider, for example, the first-order gas-phase chemical reaction A $\rightarrow$ Products and a stream with a given composition to be fed to the reactor. For this case, $s_A = -1$, Δ is determined by the reaction's stoichiometry, $r = k \cdot C_A$, and y_{A0}, y_{B0}, and y_{C0} are specified. Using (7.2-8), the reaction rate, expressed in terms of extent, is

$$r = k \cdot C_0 \frac{y_{A0} - Z}{1 + \Delta \cdot Z}. \tag{7.2-12}$$

Substituting (7.2-12) into (7.2-2), the design equation becomes

$$Z_{out} - Z_{in} = k \cdot t_{cr} \cdot \tau \frac{y_{A0} - Z_{out}}{1 + \Delta \cdot Z_{out}}. \tag{7.2-13}$$

Using (2.5-3), for first-order reactions, the characteristic reaction time is $t_{cr} = 1/k$, and, when only species A is fed into the reactor ($y_{A0} = 1$) and the feed stream is the reference stream ($Z_{in} = 0$), the design equation reduces to

$$Z_{out} = \tau \frac{1 - Z_{out}}{1 + \Delta \cdot Z_{out}}. \tag{7.2-14}$$

The solution of this design equation for different values of Δ is shown in Figure 7-5. The figure illustrates the effect of the change in the volumetric flow rate of gas-phase reactions. Note that for larger Δ, a larger reactor volume (V_R is proportional to τ) is required to achieve a given level of extent.

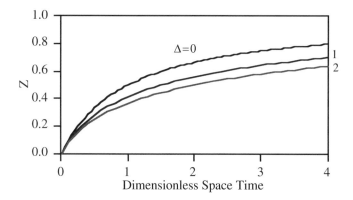

Figure 7-5: Operating Curves of a CSTR with a Single, Gas-Phase, First-Order Chemical Reaction of the Form A → Products

---

Example 7-3 A gas-phase, decomposition reaction, A → B + C, is carried out in an isothermal CSTR. The reaction is first-order, and its rate constant at the operating temperature is k = 10 sec^{-1} = 600 min^{-1}. A gaseous stream of reactant A ($C_{A0} = 4 \cdot 10^{-3}$ lbmole/ft^3) is fed into the reactor at a rate of 100 lbmole/min. Determine:

a. the volume of the reactor needed to achieve 80% conversion and
b. the volume of the reactor needed to achieve 80% conversion if, by mistake, we use $\Delta = 0$.
c. What would the actual conversion be if we use a CSTR with the volume calculated in (b)?
d. What should be the feed flow rate if we want to maintain 80% on CSTR with the volume calculated in (b)?

Solution The chemical reaction is A → B + C, and the stoichiometric coefficients are:

$$s_A = -1; \quad s_B = 1; \quad s_C = 1; \quad \text{and} \quad \Delta = 1.$$

We select the inlet stream as the reference stream ($Z_{in} = 0$) and, since only species A is fed into the reactor, $y_{A0} = 1$, $y_{B0} = y_{C0} = 0$, $C_0 = C_{A0} = 4 \cdot 10^{-3}$ lbmole/ft3, and $F_{A0} = (F_{tot})_0$ = 100 lbmole/min. The volumetric feed rate is

$$v_0 = \frac{(F_{tot})_0}{C_0} = \frac{(100 \text{ lbmole/min})}{(4 \cdot 10^{-3} \text{ lbmole/ft}^3)} = 25 \cdot 10^3 \text{ ft}^3/\text{min},$$

and, using (7.2-5),

$$Z_{out} = -\frac{y_{A0}}{s_A} f_{A_{out}} = 0.80.$$

For a CSTR with a single chemical reaction, the design equation (7.2-2) is

$$Z_{out} - Z_{in} = r \cdot \tau \cdot \left(\frac{t_{cr}}{C_0}\right). \tag{a}$$

For gas-phase reactions, we express C_A in terms of Z by (7.2-8), and the reaction rate is

$$r = k \cdot C_0 \frac{1 - Z}{1 + \Delta \cdot Z}. \tag{b}$$

For a first-order reaction, the characteristic reaction time is

$$t_{cr} = \frac{1}{k} = 10 \sec. \tag{c}$$

Substituting (b) and (c) into (a), the dimensionless design equation is

$$Z_{out} = \tau \cdot \frac{1 - Z_{out}}{1 + Z_{out}}. \tag{d}$$

For $Z_{out} = 0.8$, the solution is $\tau = 7.20$. Using the definition of the dimensionless space time, (7.1-3), the reactor volume needed is

$$V_R = \tau \cdot v_0 \cdot t_{cr} = \frac{\tau \cdot v_0}{k} = \frac{(7.20) \cdot (25 \cdot 10^3 \text{ ft}^3/\text{min})}{(600 \text{ min}^{-1})} = 300 \text{ ft}^3. \tag{e}$$

b. If we do not account for the change in the volumetric flow rate ($\Delta = 0$), the dimensionless design equation is

$$Z_{out} = \tau \cdot (1 - Z_{out}), \tag{f}$$

and, for $Z_{out} = 0.8$, the solution is $\tau = 4.0$. Hence, the calculated reactor volume is

$$V_R = \tau \cdot v_0 \cdot t_{cr} = \frac{\tau \cdot v_0}{k} = \frac{(4.0) \cdot (25 \cdot 10^3 \text{ ft}^3/\text{min})}{(600 \text{ min}^{-1})} = 166.7 \text{ ft}^3. \tag{g}$$

Note that, by using a wrong species concentration expression, we specify a reactor volume that is only 55% of the required volume.

c. We now calculate the actual outlet conversion on a CSTR with volume of 166.7 ft³. For $V_R = 166.7$ ft³ and the given feed rate, the dimensionless space time is 4.0. To determine the outlet conversion, we should solve (d) for $\tau = 4.0$, and the solution is $Z_{out} = f_{Aout} = 0.701$. Hence, if we use the wrongly-specified reactor volume, we will obtain only 70.1% conversion instead of the desired 80%.

d. To attain a conversion of 0.80 on the 166.7 ft³ reactor, the feed flow rate should be reduced. To determine the feed rate, we solve the dimensionless design equation (e) and then use the wrongly-specified reactor volume. We saw above that for $Z_{out} = 0.80$, $\tau = 7.20$; hence,

$$v_0 = \frac{V_R}{\tau \cdot t_{cr}} = \frac{V_R \cdot k}{\tau} = \frac{(166.7 \text{ ft}^3) \cdot (600 \text{ min}^{-1})}{7.20} = 13.89 \cdot 10^3 \text{ ft}^3/\text{min}, \quad \text{(h)}$$

and

$$(F_{tot})_0 = v_0 C_0 = (13.89 \cdot 10^3 \text{ ft}^3/\text{min})(4 \cdot 10^{-3} \text{ lbmole/ft3}, = 55.7 \text{ lbmole/min}.$$

Hence, if we use the wrong expression for the species concentration but want to maintain the specified conversion, we can process only 55.6 lbmole/min instead of 100 lbmole/min.

The design of CSTRs and the values of calculated quantities (reactor volume, species production rates, etc.) do not depend on the specific reference stream selected. Example 7-4 below illustrates how to use a fictitious stream as a reference stream and the behavior of a CSTR when one reactant is condensable. Example 7-5 illustrates the design of a CSTR with a chemical reaction whose rate expression is not a power function of the concentrations.

Example 7-4 The elementary, gas-phase reaction A + B → C is carried out in an isothermal-isobaric CSTR operated at 2 atm and 170°C. At this temperature, k = 90 liter/(mole min), and the vapor pressure of the product, C, is 0.3 atm. The reactor is fed with two gas streams: the first one consists of 80% A, 10% B, 10% inert (I), and is at 2.5 atm and 150°C; the second consists of 80% B, 20% I, and is at 3 atm and 180°C. The first stream is fed at a rate of 100 mole/min and the second at a rate of 120 mole/min. Determine:

a. the conversion of reactant A when product C begins to condense.
b. What is the reactor volume if C just starts to condense and
c. the reactor volume needed for 85% conversion of A?
d. Plot the performance curve.

Solution At low conversion of reactant A, all the species are gaseous, the reaction is

$$A(g) + B(g) \rightarrow C(g), \quad \text{(a)}$$

and the stoichiometric coefficients are

$$s_A = -1; \quad s_B = -1; \quad s_C = 1; \quad \text{and} \quad \Delta = \Delta_{gas} = -1.$$

First, we have to select a reference stream. In this case, we select a fictitious stream at 2 atm and 170°C that is formed by combining the two feed streams. Hence,

$$F_{A0} = 0.8 F_1 = (0.8)(100) = 80 \text{ mole/min}$$

$$F_{B0} = 0.1 F_1 + 0.8 F_2 = (0.1)(100) + (0.8)(120) = 106 \text{ mole/min}$$

$$F_{I0} = 0.1 F_1 + 0.2 F_2 = (0.1)(100) + (0.2)(120) = 34 \text{ mole/min}$$

$$(F_{tot})_0 = 220 \text{ mole/min}.$$

The composition of the reference stream is: $y_{A0} = 0.364$, $y_{B0} = 0.482$, $y_{I0} = 0.154$, and

$Z_{in} = 0$. Assuming ideal gas behavior, the total concentration of the reference stream is

$$C_0 = \frac{P_0}{R \cdot T_0} = \frac{(2 \text{ atm})}{(443°K) \cdot (0.08205 \text{ lit atm/mole}°K)} = 5.50 \cdot 10^{-2} \text{ mole/liter}.$$

The volumetric flow rate of the reference stream is

$$v_0 = \frac{(F_{tot})_0}{C_0} = \frac{(220 \text{ mole/min})}{(0.055 \text{ mole/liter})} = 4,000 \text{ liter/min}.$$

a. Species C starts to condense when its partial pressure in the reactor is 0.3 atm,

$$P_C = y_C \cdot P = \left(\frac{F_C}{F_{tot}}\right) \cdot P.$$

Using stoichiometric relations (1.5-5) and (1.5-6),

$$P_C = \left(\frac{s_C \cdot Z_{out}}{1 + \Delta_{gas} \cdot Z_{out}}\right) \cdot P = \frac{Z_{out}}{1 - Z_{out}} \cdot (2 \text{ atm}) = 0.3 \text{ atm}. \tag{b}$$

Solving (b), we obtain $Z_{out} = 0.130$, and, using (7.2-5), the conversion is

$$f_{A_{out}} = -\frac{s_A}{y_{A0}} Z_{out} = -\frac{(-1)}{0.364} 0.130 = 0.358.$$

b. To determine the reactor volume where $f_{A out} = 0.358$ (or $Z_{out} = 0.130$), we use the design equation for a CSTR, (7.2-2), with $Z_{in} = 0$,

$$Z_{out} = r_{out} \cdot \tau \cdot \left(\frac{t_{cr}}{C_0}\right). \tag{c}$$

Since the reaction is elementary, the species orders are equal to the absolute values of the stoichiometric coefficients. Hence, $r = k \cdot C_A \cdot C_B$, and, using (7.2-8),

$$r_{out} = k \cdot C_{A_{out}} \cdot C_{B_{out}} = k \cdot C_0^2 \frac{(y_{A0} - Z_{out}) \cdot (y_{B0} - Z_{out})}{(1 + \Delta_{gas} \cdot Z_{out})^2}. \tag{d}$$

Using (2.5-3), for second-order reactions, the characteristic reaction time is

$$t_{cr} = \frac{1}{k \cdot C_0} = \frac{1}{(90 \text{ liter/mole min}) \cdot (5.50 \cdot 10^{-2} \text{ mole/liter})} = 0.202 \text{ min}. \tag{e}$$

Substituting (d) and (e) into (c) and the latter into (c), the design equation reduces to

$$\tau = \frac{Z_{out} \cdot (1 - Z_{out})^2}{(0.364 - Z_{out}) \cdot (0.482 - Z_{out})} = 1.195. \tag{f}$$

Using (7.1-3) and (e), the volume of the reactor for $Z_{out} = 0.130$ is

$$V_R = \tau \cdot v_0 \cdot t_{cr} = (1.195) \cdot (4000 \text{ lit/min}) \cdot (0.202 \text{ min}) = 965.2 \text{ liter}.$$

c. If the reactor is larger than 965.2 liter, the dimensionless extent is larger than 0.130, and a portion of product C is formed by reaction (a) and a portion by the following reaction:

$$A(g) + B(g) \rightarrow C(liq). \tag{g}$$

For this reaction, we assume that the volume of liquid C formed is negligible, and Δ_{gas} = -2. Using (1.5-6), the total **gas-phase** molar flow rate at the reactor outlet is now

$$(F_{tot})_{gas} = (F_{tot})_0 \cdot [1 + (-1) \cdot 0.130 + (-2) \cdot (Z - 0.130)] = 1.13 - 2 \cdot Z, \tag{h}$$

and, modifying (7.1-12), the concentration of species j in the gas phase is now

$$C_j = C_0 \cdot \frac{y_{j0} + s_j \cdot Z}{1.13 - 2 \cdot Z}. \tag{i}$$

Substituting (i) into the rate expression (d) and the latter into (c), for $Z > 0.130$ the design equation is

$$Z_{out} = \tau \cdot \frac{(y_{A0} - Z_{out}) \cdot (y_{B0} - Z_{out})}{(1.13 - 2 \cdot Z)^2}. \tag{j}$$

Using (7.2-5), for $f_A = 0.85$,

$$Z_{out} = -\frac{y_{A0}}{s_A} f_{A out} = -\frac{0.364}{(-1)} \cdot 0.85 = 0.3094. \tag{k}$$

Substituting $Z_{out} = 0.3094$ into (j), we obtain

$$\tau = \frac{Z_{out} \cdot (1.13 - 2 \cdot Z)^2}{(0.364 - Z_{out}) \cdot (0.482 - Z_{out})} = 8.58. \tag{l}$$

Using (7.1-3) and (e), the volume of the reactor when condensed C is formed,

$$V_R = \tau \cdot v_0 \cdot t_{cr} = (8.58) \cdot (4000 \text{ lit/min}) \cdot (0.202 \text{ min}) = 6,932 \text{ liter}.$$

d. The reactor operation is described by two design equations: (f) for $0 < Z_{out} < 0.130$ and (j) for $Z > 0.130$. The figure below shows the operating curve of the reactor. Note that for $\tau < 1.195$, Δ_{gas} = -1, and the operating curve is described by the first section. For $\tau > 1.195$, Δ_{gas} = -2, and the operating curve is described by the second section.

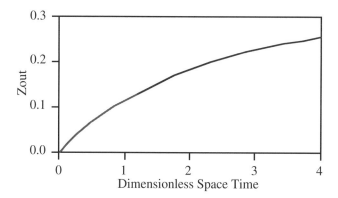

Example 7-5 A biological waste, A, is decomposed by an enzymatic reaction A $\to$ B + C in aqueous solution. The rate expression of the reaction is

$$r = \frac{k \cdot C_A}{K_m + C_A}.$$

An aqueous solution with a concentration of 2 mole A/liter is fed into a CSTR at a rate of 200 liter/min. For the enzyme type and concentration used, k = 0.1 mole/lit min and K_m = 4 mole/lit.

a. Derive and plot the dimensionless operating curve of the reactor.

b. What should be the reactor volume to achieve 80% conversion?

Solution The chemical reaction is A $\to$ B + C, and the stoichiometric coefficients are: s_A = -1, s_B = 1, s_C = 1, Δ = 1, and, since this is a liquid phase reaction, Δ_{gas} = 0. We select the inlet stream as the reference stream; hence, Z_{in} = 0. Since reactant A is the only species fed into the reactor, $C_0 = C_{A0}$, y_{A0} = 1, y_{B0} = 1, and y_{C0} = 1. The dimensionless design equation of a CSTR, (7.2-2), is

$$Z_{out} = \tau \cdot r_{out} \cdot \left(\frac{t_{cr}}{C_0} \right). \tag{a}$$

In this case,

$$r = \frac{k \cdot C_A}{K_m + C_A}. \tag{b}$$

For liquid-phase reactions, using (7.2-7), (b) becomes

$$r = k \frac{1 - Z}{\dfrac{K_m}{C_0} + (1 - Z)}. \tag{c}$$

Applying the procedure described in Chapter 2, the reference reaction rate is r_0 = k, and the characteristic reaction time is

$$t_{cr} = \frac{C_0}{k} = \frac{(2 \text{ mole/liter})}{(0.1 \text{ mole/lit min})} = 20 \text{ min} . \tag{d}$$

The dimensionless space time is

$$\tau = \frac{V_R}{v_0 \cdot t_{cr}} = \frac{k \cdot V_R}{v_0 \cdot C_0} . \tag{e}$$

Substituting (c) and (d) into (a), for $K_m/C_0 = 2$, the design equation becomes

$$\tau = \frac{Z_{out} \cdot (3 - Z_{out})}{1 - Z_{out}} . \tag{f}$$

The figure below shows the dimensionless reaction operating curve.

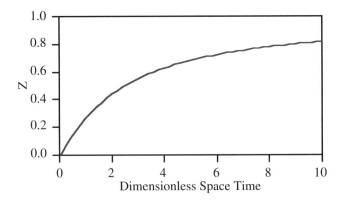

b. Using stoichiometric relation (1.5-5), the dimensionless extent for $f_{Aout} = 0.8$ is

$$Z_{out} = -\frac{y_A(0)}{s_A} f_{A \, out} = -\frac{1}{(-1)}(0.80) = 0.80 .$$

From (f), the dimensionless operating time needed for $Z_{out} = 0.80$ is $\tau = 8.8$. We calculate the required reactor volume from (e),

$$V_R = \frac{\tau \cdot v_0 \cdot C_0}{k} = \frac{(8.8) \cdot (200 \text{ lit/min}) \cdot (2 \text{ mole/lit})}{(0.1 \text{ mole/lit min})} = 35,200 \text{ liter} .$$

7.2.2 Determination of the Reaction Rate Expression

It is easy to determine the rate expression from data obtained on CSTRs since we can readily calculate the reaction rate, r_{out}, at the outlet conditions from operating data. Selecting the inlet stream as the reference stream, $Z_{in} = 0$, we can write (7.2-2) as

$$r_{out} = \frac{1}{\tau} \cdot \left(\frac{C_0}{t_{cr}} \right) \cdot Z_{out}. \qquad (7.2\text{-}15)$$

If we substitute the definition of the dimensionless space time, (7.1-3), we obtain

$$r_{out} = \frac{(F_{tot})_0}{V_R} \cdot Z_{out}. \qquad (7.2\text{-}16)$$

Hence, for known $(F_{tot})_0$ and V_R and for experimentally-determined Z_{out}, we calculate r_{out}. Usually, the species concentrations at the reactor outlet are measured experimentally, and, for n-th order reactions,

$$r = k \cdot C_A{}^n.$$

Substituting in (7.2-16), we obtain

$$k \cdot C_{A_{out}}{}^n = \frac{(F_{tot})_0}{V_R} \cdot Z_{out}.$$

To determine the reaction order, we take the logarithm of both sides and obtain

$$\ln(k) + n \cdot \ln(C_{A_{out}}) = \ln\left(\frac{C_0 \cdot v_0}{V_R} \cdot Z_{out} \right). \qquad (7.2\text{-}17)$$

Modifying the term on the right hand side, we can write (7.2-17) as

$$\ln(k) + n \cdot \ln(C_{A_{out}}) = \ln\left(\frac{C_0}{V_R} \right) + \ln\left[v_0 \cdot Z_{out} \right]. \qquad (7.2\text{-}18)$$

Thus, by operating a CSTR isothermally and measuring $C_{A_{out}}$ for different values of v_0, we can plot $\ln(v_0 \cdot Z_{out})$ versus $\ln(C_{A_{out}})$ and determine the reaction order from the slope of the line. This is illustrated in Example 7-6 below. Once the order is known, we can determine the value of the reaction rate constant from the design equation. We repeat this procedure at different reactor temperatures to determine the activation energy.

Example 7-6 The gas-phase dimerization of reactant A, $2\,A \rightarrow R$, is investigated on a CSTR. A stream of species A at 3 atm and 30°C (120 mmole/lit) is fed into a 1 liter CSTR at different flow rates, and the exit concentration of A is measured. From the data below, determine the order of the reaction and the value of the rate constant, k.

v_0 (lit/min)	0.035	0.18	0.45	1.7
$C_{A_{out}}$ (mmole/lit)	30	60	80	105

Solution The chemical reaction is $2\,A \rightarrow R$, and the stoichiometric coefficients are:

$$s_A = -2; \quad s_R = 1; \quad \text{and} \quad \Delta = -1.$$

We select the inlet stream as the reference stream; hence, $Z_{in} = 0$, and $y_{A0} = 1$ and $y_{B0} =$

0. The rate expression is

$$r = k \cdot C_A{}^\alpha, \tag{a}$$

and we want to determine α and k. For gas-phase reactions, we use (7.2-8),

$$C_{Aout} = C_0 \cdot \frac{1 - 2 \cdot Z_{out}}{1 + \Delta \cdot Z_{out}}. \tag{b}$$

Substituting $\Delta = -1$ into (b), the outlet extent relates to C_{Aout} by

$$Z_{out} = \frac{C_0 - C_{Aout}}{2 \cdot C_0 - C_{Aout}}. \tag{c}$$

We use (c) to calculate Z_{out} for each run, and then we calculate ($v_0 \cdot Z_{out}$) for each run. The table below shows the calculated values.

v_0 (lit/min)	0.035	0.180	0.450	0.700
C_{Aout} (mmole/lit)	30	60	80	105
Z_{out}	0.429	0.333	0.250	0.111
$v_0 \cdot Z_{out}$ (lit/min)	0.0150	0.060	0.1125	0.189

Using (7.2-18),

$$\alpha \cdot \ln C_{Aout} = \ln\left(\frac{C_{A0}}{2 \cdot k \cdot V_R}\right) + \ln(v_0 \cdot Z_{out}). \tag{d}$$

Thus, by plotting $\ln(v_0 \cdot Z_{out})$ versus $\ln(C_{Aout})$, we should get a straight line whose slope is α. The plot is shown in the figure below. From the figure, the slope is

$$\alpha = \frac{\ln(0.17) - \ln(0.0155)}{\ln(100) - \ln(30)} = 2.06 \approx 2.$$

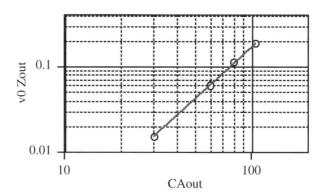

Now that we know that $\alpha = 2$, we use the dimensionless design equation (7.2-3) to calculate the dimensionless space time for each run. In this case, the characteristic reaction time is

$$t_{cr} = \frac{1}{k \cdot C_0},$$ (e)

and (7.2-3) reduces to

$$\tau = Z_{out} \cdot \left(\frac{1 - Z_{out}}{1 - 2 \cdot Z_{out}} \right)^2.$$ (f)

Using (7.2-3) and (e), we calculate the rate constant for each run by

$$k = \frac{v_0}{2 \cdot V_R \cdot C_0} \cdot Z_{out} \cdot \left(\frac{1 - Z_{out}}{1 - 2 \cdot Z_{out}} \right)^2.$$ (g)

The average value of k is 0.002 liter/mmole min.

7.2.3 Cascade of CSTRs Connected in Series

Consider several CSTRs (not necessarily of the same volume) connected such that the effluent from one reactor is fed into the next reactor, as shown schematically in Figure 7-6. We select the inlet stream to the cascade as the reference stream and write the dimensionless design equation, (7.2-3), for each reactor

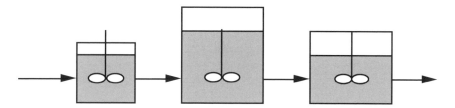

Figure 7-6: Cascade of CSTRs

$$\tau = \left(\frac{C_0}{t_{cr}} \right) \cdot \frac{Z_{out} - Z_{in}}{r_{out}}.$$ (7.2-3)

Thus, if we know how the reaction rate, r, relates to the extent, Z, we can present the design equation graphically by plotting (r_0/r) versus Z, as shown schematically in Figure 7-7. Here, the dimensionless space time of each reactor is represented by a rectangle. Note that the total volume of the cascade is smaller than the volume of a single CSTR used for the same conversion. Furthermore, note that when a cascade consisting of numerous small CSTRs is used to obtain a given conversion, the rectangular areas approach the area under the curve in Figure 7-7. Hence, the total volume of the cascade converges to the volume of a plug-flow reactor. We will prove this behavior algebraically below for first-order, liquid-phase reactions.

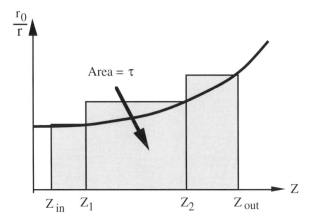

Figure 7-7: Graphical Determination of Reactor Volumes in a Cascade of CSTRs

Consider a liquid-phase, first-order reaction of the form $A \rightarrow P + R$ in an isothermal cascade of CSTRs where only reactant A is fed to the system. Taking the inlet stream to the cascade as the reference stream and since only reactant A is fed, $y_{A0} = 1$. For liquid-phase, first-order reactions, the dimensionless design equation, (7.2-3), for the n-th CSTR is

$$\tau_n = \frac{Z_{n_{out}} - Z_{n_{in}}}{1 - Z_{n_{out}}}, \tag{7.2-19}$$

where τ_n is the dimensionless space time of the n-th reactor in the cascade, $\tau_n = V_{Rn}/(v_0 \cdot t_{cr})$. For the first reactor in the cascade,

$$\tau_1 = \frac{Z_1}{1 - Z_1},$$

or
$$Z_1 = \frac{\tau_1}{1 - \tau_1}. \tag{7.2-20}$$

For the second reactor in the cascade,

$$\tau_2 = \frac{Z_2 - Z_1}{1 - Z_2},$$

or
$$Z_2 = \frac{\tau_1 + \tau_2 \cdot (1 + \tau_1)}{(1 - \tau_1) \cdot (1 - \tau_2)}. \tag{7.2-21}$$

Similarly, for the n-th reactor in the cascade,

$$Z_n = \frac{\tau_1 + \tau_2 \cdot (1 + \tau_1) + \ldots + \tau_n \cdot (1 + \tau_1) \cdots (1 + \tau_{n-1})}{(1 - \tau_1) \cdot (1 - \tau_2) \cdots (1 - \tau_n)}. \qquad (7.2\text{-}22)$$

For the special case that each reactor has the same volume, $\tau_1 = \tau_2 = \ldots = \tau_n = \tau$, (7.2-22) reduces to

$$Z_n = 1 + \frac{1}{(1 + \tau)^n}. \qquad (7.2\text{-}23)$$

If we keep the **total** volume of the cascade constant and increase the number of reactors, the volume of each becomes smaller, $\tau = \tau_{tot}/n$. It can be shown that for a very large number of units,

$$1 - Z_n = \lim_{n \to \infty} \left(1 + \frac{\tau_{tot}}{n} \right)^{-n} = e^{-\tau_{tot}}. \qquad (7.2\text{-}24)$$

The right hand side of (7.2-24) is the expression for 1 - Z at the outlet of a plug-flow reactor with a dimensionless space time τ_{tot}.

In practice, the cascade usually consists of either equal-size reactors or a set of specified numbers of CSTRs in which the volume of each is selected such that total volume of the cascade, for a given outlet conversion, is the smallest. A cascade of equal-size CSTRs is represented in Figure 7-7 by rectangles whose areas are the same. A cascade with the total smallest volume is represented in Figure 7-7 by a given number of rectangles whose combined area is the smallest. Below, we consider the performance of a cascade of CSTRs when the reaction rate is given.

Example 7-7 A gas-phase decomposition reaction, $A \to 2\,B$, is carried out in a cascade of CSTRs. The reaction is first-order, and a stream of reactant A is fed into the system at a rate of 100 lbmole/min. The feed concentration is $C_{A0} = 4 \cdot 10^{-3}$ lbmole/ft^3, and, at the reactor temperature, $k = 10$ sec$^{-1} = 600$ min^{-1}. Determine:
a. the volume of each reactor in a series of two equal-size CSTRs needed for 80% conversion,
b. the volume of each reactor in a series of two optimized CSTRs needed for 80% conversion,
c. the volume of each reactor in a series of three equal-size CSTRs needed for 80% conversion, and
d. the volume of each reactor in a series of three optimized CSTRs needed for 80% conversion.
e. Compare the results in (a) and (b) to the values obtained in Examples 6-3 and 7-3.
f. Plot the performance curves of the reactor for (a) and (c) and compare them to the corresponding curves for PFR (Example 6-1) and a single CSTR (Example 7-3).
Data: $k = 10$ sec$^{-1} = 600$ min.$^{-1}$, $C_{A0} = 4 \cdot 10^{-3}$ lbmole/ft^3.
Solution The chemical reaction is $A \to 2\,B$, and its stoichiometric coefficients are: $s_A = -1$, $s_B = 2$, and $\Delta = 1$. Taking the inlet stream to the cascade as the reference stream, since only reactant A is fed, $C_0 = C_{A0}$, $y_{A0} = 1$, and $y_{B0} = 0$. For gas-phase reactions, using (7.2-8),

$$r_{out} = k \cdot C_0 \frac{1 - Z_{out}}{1 + \Delta \cdot Z_{out}}, \tag{a}$$

and the characteristic reaction time is

$$t_{cr} = \frac{1}{k}. \tag{b}$$

Substituting (a) and (b) into (7.2-3), the design equation of each reactor is

$$\tau = \frac{(Z_{out} - Z_{in}) \cdot (1 + Z_{out})}{1 - Z_{out}}. \tag{c}$$

a. In a cascade of two CSTRs, for the first reactor,

$$\tau_1 = \frac{Z_1 \cdot (1 + Z_1)}{1 - Z_1}, \tag{d}$$

for the second reactor, $Z_{in} = Z_1$, $Z_{out} = Z_2$, and (c) reduces to

$$\tau_2 = \frac{(Z_2 - Z_1) \cdot (1 + Z_2)}{1 - Z_2}. \tag{e}$$

For a cascade of equal-size reactors, $\tau_1 = \tau_2$, and we combine (d) and (e) to obtain

$$\frac{(Z_1) \cdot (1 + Z_1)}{1 - Z_1} = \frac{(Z_2 - Z_1) \cdot (1 + Z_2)}{1 - Z_2}. \tag{f}$$

Substituting $Z_2 = 0.8$, we solve (f) for Z_1 and obtain $Z_1 = 0.569$. Now that we know Z_1 and Z_2, we can use either (d) or (e) to calculate the dimensionless space time of each reactor, $\tau_1 = \tau_2 = 2.075$. To calculate the volume of each CSTR in the cascade, we use (7.1-3) and (b),

$$V_R = \frac{\tau \cdot v_0}{k} = \frac{(2.075) \cdot (25,000 \text{ ft}^3/\text{min})}{(600 \text{ min}^{-1})} = 86.46 \text{ ft}^3. \tag{g}$$

The total volume of the cascade is $V_{R1} + V_{R2} = 86.46 + 86.46 = 172.92$ ft³.
b. To determine Z_1 for an optimized cascade of two CSTRs, we take the derivative of $\tau_1 + \tau_2$ with respect to Z_1 and equate to zero,

$$\frac{d(\tau_1 + \tau_2)}{dZ_1} = 0. \tag{h}$$

Substituting (d) and (e), (h) reduces to

$$(1 + 2Z_1) \cdot (1 - Z_1) + Z_1 \cdot (1 + Z_1) - \left(\frac{1 + Z_2}{1 - Z_2} \right) \cdot (1 - Z_1)^2 = 0. \tag{i}$$

Substituting $Z_2 = 0.8$, we solve (i) for Z_1 and obtain $Z_1 = 0.528$. Now that we know the values of Z_1 and Z_2, we use (d) and (e) to calculate the dimensionless space time of each reactor. We find that $\tau_1 = 1.919$ and $\tau_2 = 2.225$. Using (g), the volume of the two reactors are: $V_{R1} = 79.96$ and $V_{R2} = 92.71$ ft³, and the total reactor volume in the cascade is $V_{R1} + V_{R2} = 172.67$ ft³.

c. In a cascade of three equal-size CSTRs, we write the dimensionless design equation for each reactor as we did above. For the first reactor,

$$\tau_1 = \frac{Z_1 \cdot (1 + Z_1)}{1 - Z_1}, \tag{j}$$

for the second reactor,

$$\tau_2 = \frac{(Z_2 - Z_1) \cdot (1 + Z_2)}{1 - Z_2}, \tag{k}$$

and for the third reactor,

$$\tau_3 = \frac{(Z_3 - Z_2) \cdot (1 + Z_3)}{1 - Z_3}. \tag{l}$$

For a cascade of equal-size reactors, $\tau_1 = \tau_2$ and $\tau_2 = \tau_3$. We combine (j), (k), and (l) and obtain two equations with two unknowns, Z_1 and Z_2,

$$\frac{Z_1 \cdot (1 + Z_1)}{1 - Z_1} = \frac{(Z_3 - Z_2) \cdot (1 + Z_3)}{1 - Z_3} \tag{m}$$

$$\frac{(Z_2 - Z_1) \cdot (1 + Z_2)}{1 - Z_2} = \frac{(Z_3 - Z_2) \cdot (1 + Z_3)}{1 - Z_3}. \tag{n}$$

Substituting $Z_3 = 0.8$, we solve (m) and (n) for Z_1 and Z_2 (using a mathematical software) and obtain $Z_1 = 0.4445$ and $Z_2 = 0.6716$. Now that we know the values of Z_1 and Z_2, we use either (j), (k), or (l) to calculate the dimensionless space time of each reactor. We obtain $\tau_1 = \tau_2 = \tau_3 = 1.156$. To calculate the volume of each reactor, we use (7.1-3) and (b),

$$V_R = \frac{\tau \cdot v_0}{k} = \frac{(1.156) \cdot (25,000 \ \text{ft}^3/\text{min})}{(600 \ \text{min}^{-1})} = 48.15 \ \text{ft}^3.$$

The total volume of the cascade is $3 \cdot 48.15 = 144.45$ ft³.

d. To determine Z_1 and Z_2 for an optimized cascade of three CSTRs, we take the derivative of $(\tau_1 + \tau_2 + \tau_3)$ with respect to Z_1 and Z_2 and equate them to zero,

$$\frac{\partial(\tau_1 + \tau_2 + \tau_3)}{\partial Z_1} = 0; \quad \text{and} \quad \frac{\partial(\tau_1 + \tau_2 + \tau_3)}{\partial Z_2} = 0. \tag{o}$$

Substituting (j), (k), and (l) and taking the derivatives, the two equations in (o) become

$$(1 + 2 \cdot Z_1) \cdot (1 - Z_1) + Z_1 \cdot (1 + Z_1) - \left(\frac{1 + Z_2}{1 - Z_2}\right) \cdot (1 - Z_1)^2 = 0, \tag{p}$$

$$(1 + 2 \cdot Z_2 - Z_1) \cdot (1 - Z_2) + (Z_2 - Z_1) \cdot (1 + Z_2) - \left(\frac{1 + Z_3}{1 - Z_3}\right) \cdot (1 - Z_2)^2 = 0. \tag{q}$$

We substitute $Z_3 = 0.8$, solve (p) and (q) simultaneously and obtain $Z_1 = 0.4152$ and $Z_2 = 0.6580$. Now that we know the values of Z_1 and Z_2, we use (j), (k), and (l) to calculate the dimensionless space time of each reactor. We find that $\tau_1 = 1.005$, $\tau_2 = 1.177$, and $\tau_3 = 1.278$. Using (g), the volumes of the three reactors are: $V_{R1} = 41.88$, $V_{R2} = 49.04$, and $V_{R3} = 53.25$ ft^3, and the total volume of the cascade is $V_{R1} + V_{R2} + V_{R3} = 144.17$ ft^3.

e. The total volume of the cascade for the different cases is summarized below:

A single CSTR (from Example 7-3)	300.00 ft^3.
A cascade of two equal-size CSTRs	172.92 ft^3.
An optimized cascade of two CSTRs	172.67 ft^3.
A cascade of three equal-size CSTRs	144.45 ft^3.
An optimized cascade of three CSTRs	144.17 ft^3.
A plug-flow reactor (from Example 6-3)	100.08 ft^3.

7.3 ISOTHERMAL OPERATIONS WITH MULTIPLE REACTIONS

When more than one chemical reaction takes place in the reactor, several issues should be addressed before we start the design procedure. First, we have to determine how many independent reactions there are (and how many design equations are needed) and select a set of independent reactions. Next, we have to identify all the reactions that actually take place (including dependent reactions) and express their rates. To determine the reactor compositions and all other state quantities, we have to write design equation (7.1-1) for each of the independent chemical reactions. To solve the design equations (obtain relationships between Z_m's and τ), we have to express the rates of the individual chemical reactions, r_m's and r_k's, in terms of the Z_m's and τ. The procedure for designing CSTRs with multiple reactions goes as follows:

a. Identify all the chemical reactions that take place in the reactor, and define the stoichiometric coefficients of each species in each reaction.
b. Determine the number of **independent** chemical reactions.
c. Select a set of **independent** reactions among the reactions whose rate expressions are given.
d. For each **dependent** chemical reaction, determine the α_{km} multipliers with each of the independent reactions.
e. Select a **reference stream** (determine T_0, C_0, v_0) and the reference species compositions, y_{j0}'s.
f. Specify the inlet conditions (T_{in}, Z_{min}'s).

g. Write design equation (7.1-1) for **each** independent chemical reaction.

h. Select a leading (or desirable) reaction and determine the expression form of the characteristic reaction time, t_{cr}, and its numerical value.

i. Express the reaction rates of all chemical reactions in terms of the dimensionless extents of the independent reactions, Z_m's, and the dimensionless temperature, θ.

j. Solve the design equations for Z_{mout}'s as functions of the dimensionless space time, τ, and obtain the dimensionless reaction operating curves.

k. Calculate the dimensionless species operating curves, using (1.5-5).

l. Determine the reactor volume based on the most desirable value of τ obtained from the dimensionless operating curves.

Below, we describe the design formulation of isothermal CSTRs with multiple reactions for various types of chemical reactions (reversible, series, parallel, etc.). In most cases, we solve the equations numerically by applying a numerical technique such as the Newton-Rhaphson method or using a mathematical software such as HiQ, Mathcad, Maple, etc. In some simple cases, we can obtain analytical solutions. Note that, for isothermal operations, $\theta_{out} = \theta_{in}$, and we do not have to solve the energy balance equation simultaneously with the design equations.

Example 7-8 The reversible, gas-phase chemical reaction A $\leftrightarrow$ 2 B takes place in a CSTR operated at 2 atm and 120°C. The forward reaction is first-order, and the backward reaction is second-order. We want to process a 100 mole/min stream of pure A and achieve a level of 90% of the equilibrium conversion. At the operating conditions (120°C), $k_1 = 0.1$ min^{-1} and $k_2 = 0.322$ lit mole^{-1} min^{-1}.

a. Derive the design equation, and plot the dimensionless reaction operating curve.

b. What is the equilibrium composition at 120°C in a CSTR for $k_2 \cdot C_0/k_1 = 0.5$?

c. What is the required reactor volume needed to reach 80% of the equilibrium conversion?

Solution We treat a reversible reaction as two separate reactions:

Reaction 1	A $\rightarrow$ 2 B	$r_1 = k_1 \cdot C_A$
Reaction 2	2 B $\rightarrow$ A	$r_2 = k_2 \cdot C_B$

The stoichiometric coefficients of the chemical reactions are

$$s_{A1} = -1; \ s_{B1} = 2; \ \Delta_1 = 1, \quad \text{and} \quad s_{A2} = 1; \ s_{B2} = -2, \text{ and } \Delta_2 = -1.$$

We select the forward reaction (Reaction 1) as the independent reaction and the reverse reaction as the dependent reaction. Hence, the index of the independent reaction is m = 1, the index of the dependent reaction is k = 2, and, since Reaction 2 is the reverse of Reaction 1, $\alpha_{21} = -1$. Since there is only one independent reaction, we can describe the operation by a single design equation. We select the inlet stream as the reference stream; hence $Z_{1in} = 0$,

$$C_0 = \frac{P_0}{R \cdot T_0} = \frac{(2 \text{ atm})}{(82.05 \cdot 10^{-3} \text{ lit atm/mole}°K) \cdot (393°K)} = 6.02 \cdot 10^{-2} \text{ mole/lit},$$

$$v_0 = \frac{(F_{tot})_0}{C_0} = \frac{(100 \text{ mole/min})}{(6.202 \cdot 10^{-2} \text{ mole/lit})} = 1{,}612 \text{ lit/min}. \tag{a}$$

a. To design a CSTR, we write (7.1-1) for the independent reaction

$$Z_{1_{out}} - Z_{1_{in}} = (r_{1_{out}} - r_{2_{out}}) \cdot \tau \cdot \left(\frac{t_{cr}}{C_0}\right). \tag{b}$$

For gas-phase reactions, we use (6.1-12) to express the concentrations of the two species in terms of the extent of the independent reaction, and the rates of the two reactions are, respectively,

$$r_1 = k_1 \cdot C_0 \frac{y_{A0} + s_{A1} \cdot Z_1}{1 + \Delta_1 \cdot Z_1} \tag{c}$$

$$r_2 = k_2 \cdot C_0^2 \left(\frac{y_{B0} + s_{B1} \cdot Z_1}{1 + \Delta_1 \cdot Z_1}\right)^2. \tag{d}$$

We select Reaction 1 and define the characteristic reaction time as

$$t_{cr} = \frac{1}{k_1} = 10 \text{ min}. \tag{e}$$

Substituting (c), (d), and (e) into (b), the design equation is

$$Z_{1_{out}} = \left[\frac{y_{A0} - Z_{1_{out}}}{1 + Z_{1_{out}}} - \left(\frac{k_2 \cdot C_0}{k_1}\right) \left(\frac{y_{B0} + 2 \cdot Z_{1_{out}}}{1 + Z_{1_{out}}}\right)^2\right] \cdot \tau. \tag{f}$$

When only reactant A is fed into the reactor, $y_{A0} = 1$ and $y_{B0} = 0$, and (f) becomes

$$Z_{1_{out}} - \left[\frac{1 - Z_{1_{out}}}{1 + Z_{1_{out}}} - \left(\frac{k_2 \cdot C_0}{k_1}\right) \cdot \left(\frac{2 \cdot Z_{1_{out}}}{1 + Z_{1_{out}}}\right)^2\right] \cdot \tau = 0. \tag{g}$$

We solve (g) numerically for different values of τ by a mathematical software. The figure below shows the operating curves for various values of $k_2 \cdot C_0/k_1$. Note that the curve for $k_2 \cdot C_0/k_1 = 0$ represents the solution of the irreversible reaction.

b. At equilibrium, $r_1 = r_2$, and, from (c) and (d), for $y_{A0} = 1$ and $y_{B0} = 0$ we obtain

$$Z_{1_{eq}} = \left(1 + 4\frac{k_2 \cdot C_0}{k_1}\right)^{-0.5}. \tag{h}$$

For $k_2 \cdot C_0/k_1 = 0.5$, $Z_{1_{eq}} = 0.577$, and, using (7.2-6), the equilibrium conversion is f_{Aeq}

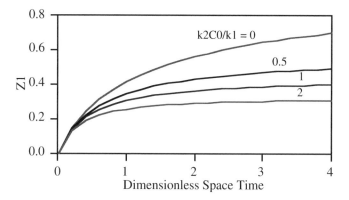

$= 0.577$. The species compositions at equilibrium are

$$y_{A_{eq}} = \frac{y_{A_0} - Z_{1_{eq}}}{1 + \Delta_1 \cdot Z_{1_{eq}}} = \frac{1 - 0.577}{1 + 0.577} = 0.268 \tag{j}$$

$$y_{B_{eq}} = \frac{y_{B_0} + 2 \cdot Z_{1_{eq}}}{1 + \Delta_1 \cdot Z_{1_{eq}}} = \frac{2 \cdot 0.577}{1 + 0.577} = 0.732. \tag{k}$$

c. The extent for 80% of the equilibrium conversion is 0.4616. From the operating curve for $k_2 \cdot C_0 / k_1 = 0.5$, an extent of 0.4616 is reached at $\tau = 2.74$. Using (7.1-3) and (e), the required reactor volume is

$$V_R = \tau \cdot v_0 \cdot t_{cr} = (2.74) \cdot (100 \text{ lit/min}) \cdot (10 \text{ min}) = 2,740 \text{ liter.}$$

Example 7-9 An organic solution containing reactant A ($C_{A0} = 2.0$ mole/lit) is fed into a 1,200 liter CSTR, where A reacts according to the second-order reaction $2 A \rightarrow B$. Species B is the desired product, but, at the reactor operating conditions, it decomposes according to the first-order reaction $B \rightarrow C + 2 D$. The feed rate is 100 lit./min, and, at the operating conditions, $k_1 = 10$ liter mole^{-1} hr^{-1} and $k_2 = 4$ hr^{-1}.
a. Derive the design equations, and plot the reaction and species operating curves.
b. Determine the conversion of A and the production rates of B and C for the given feed flow rate.
c. Determine what should be the feed flow rate to maximize the production of B.
d. Calculate the conversion of A and production rate of B at optimal feed rate.
Solution The reactions taking place in the reactor are:

$$\text{Reaction 1:} \qquad 2 A \rightarrow B$$

$$\text{Reaction 2:} \qquad B \rightarrow C + 2 D,$$

and their stoichiometric coefficients are:

$$s_{A1} = -2; \quad s_{B1} = 1; \quad s_{C1} = 0; \quad s_{D1} = 0; \quad \Delta_1 = -1;$$

$$s_{A2} = 0; \quad s_{B2} = -1; \quad s_{C2} = 1; \quad s_{D2} = 2; \quad \Delta_2 = 2.$$

Since each reaction has at least one species that does not appear in the other, the two reactions are independent, and there are no dependent reactions. We select the inlet steam as the reference stream; hence, $Z_{1in} = Z_{2in} = 0$. Since only reactant A is fed into the reactor, $C_0 = C_{A0} = 2$ mole/lit, $y_{A0} = 1$, and $y_{B0} = y_{C0} = y_{D0} = 0$. Also,

$$(F_{tot})_0 = v_0 \cdot C_0 = (100 \text{ lit/min}) (2 \text{ mole/lit}) = 200 \text{ mole/min}.$$

Using stoichiometric relation (1.5-5), the species molar flow rates are:

$$F_{A_{out}} = (F_{tot})_0 \cdot (1 - 2 \cdot Z_{1_{out}}) \tag{a}$$

$$F_{B_{out}} = (F_{tot})_0 \cdot (Z_{1_{out}} - Z_{2_{out}}) \tag{b}$$

$$F_{C_{out}} = (F_{tot})_0 \cdot Z_{2_{out}} \tag{c}$$

$$F_{D_{out}} = (F_{tot})_0 \cdot 2 \cdot Z_{2_{out}}. \tag{d}$$

a. To design a CSTR, we write design equation (7.1-1) for each independent reaction,

$$Z_{1_{out}} = r_{1_{out}} \cdot \tau \cdot \left(\frac{t_{cr}}{C_0} \right) \tag{e}$$

$$Z_{2_{out}} = r_{2_{out}} \cdot \tau \cdot \left(\frac{t_{cr}}{C_0} \right). \tag{f}$$

For liquid-phase reactions, we use (7.1-9) to express the species concentrations and the reaction rates are

$$r_{1_{out}} = k_1 \cdot C_0^2 \cdot (1 - 2 \cdot Z_{1_{out}})^2 \tag{g}$$

$$r_{2_{out}} = k_2 \cdot C_0 \cdot (Z_{1_{out}} - Z_{2_{out}}). \tag{h}$$

We select Reaction 1 and define the characteristic reaction time by

$$t_{cr} = \frac{1}{k_1 \cdot C_0} = 0.05 \text{ hr} = 3 \text{ min}. \tag{i}$$

Substituting (g), (h), and (i) into (e) and (f), the design equations become

$$Z_{1_{out}} - (1 - 2 \cdot Z_{1_{out}})^2 \cdot \tau = 0 \tag{j}$$

$$Z_{2_{out}} - \left(\frac{k_2}{k_1 \cdot C_0} \right) \cdot (Z_{1_{out}} - Z_{2_{out}}) \cdot \tau = 0. \tag{k}$$

For the given data, $k_2/k_1 \cdot C_0 = 0.2$. In this case, the two design equations are not coupled, and we can solve them sequentially. We solve (j), discarding the root that does not converge to 0 for $\tau \rightarrow 0$, and obtain

$$Z_{1_{out}} = \frac{(1 + 4 \cdot \tau) - \sqrt{1 + 8 \cdot \tau}}{8 \cdot \tau}. \tag{l}$$

Substituting (m) into (k), we obtain

$$Z_{2_{out}} = \left(\frac{0.2 \cdot \tau}{1 + 0.2 \cdot \tau} \right) \cdot \frac{(1 + 4 \cdot \tau) - \sqrt{1 + 8 \cdot \tau}}{8 \cdot \tau}. \tag{m}$$

The two reaction operating curves are shown in the figure below. Using stoichiometric relations (a) through (d), we calculate the species operating curves, shown in the second figure below.

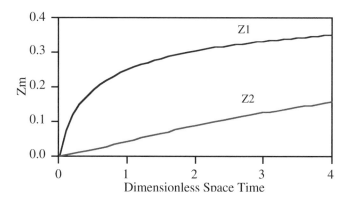

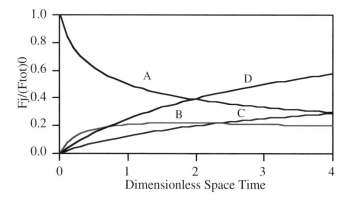

b. Using (7.1-3) and (i), for the given volumetric feed rate, the dimensionless space time is

$$\tau = \frac{V_R}{v_0 \cdot t_{cr}} = \frac{(1200 \text{ lit})}{(100 \text{ lit/min}) \cdot (3 \text{ min})} = 4.$$

Solving (l) and (m) for $\tau = 4$, we obtain, $Z_{1out} = 0.352$ and $Z_{2out} = 0.156$. Now that we know Z_{1out} and Z_{2out}, we calculate the molar flow rates of all the species at the reactor outlet using (a) through (d),

$$F_{Aout} = (F_{tot})_0 (1 - 2 \cdot Z_1) = (200) (1 - 2 \cdot 0.352) = 59.2 \text{ mole/min}$$

$$F_{Bout} = (F_{tot})_0 (Z_1 - Z_2) = (200) (0.352 - 0.156) = 39.2 \text{ mole/min}$$

$$F_{Cout} = (F_{tot})_0 Z_2 = (200) (0.156) = 31.2 \text{ mole/min}$$

$$F_{Dout} = (F_{tot})_0 (2 \cdot Z_2) = (200) (2 \cdot 0.156) = 62.4 \text{ mole/min}.$$

The conversion of reactant A is

$$f_{Aout} = \frac{F_{A0} - F_{Aout}}{F_{A0}} = \frac{200 - 59.2}{200} = 0.704.$$

c. From the species operating curve, maximum F_{Bout} is reached at $\tau = 1.74$. (Note that, in this case, we could have obtained this value analytically by deriving explicit expressions for F_{Bout} in terms of τ and then taking the derivative of F_{Bout} with respect to τ and equating it to zero.) Using (7.1-3) and (m), for $\tau = 1.74$, the volumetric flow rate of the feed is

$$v_0 = \frac{V_R}{\tau \cdot t_{cr}} = \frac{(1200 \text{ lit})}{(1.74) \cdot (3 \text{ min})} = 229.9 \text{ lit/min}.$$

The optimal molar flow rate of the feed is $(F_{tot})_0 = v_0 \cdot C_0 = 459.8 \text{ mole/min}$.
d. To determine the production rate of B for the optimal feed rate, we first solve (l) and (m) for $\tau = 1.74$. The solutions are $Z_{1out} = 0.294$ and $Z_{2out} = 0.0759$. Now that we know Z_{1out} and Z_{2out}, we calculate the molar flow rates of all the species at the reactor outlet using (a) through (d),

$$F_{Aout} = (F_{tot})_0 (1 - 2 \cdot Z_{1out}) = (459.8) (1 - 2 \cdot 0.294) = 189.4 \text{ mole/min}$$

$$F_{Bout} = (F_{tot})_0 (Z_1 - Z_2) = (459.8) (0.294 - 0.0759) = 100.3 \text{ mole/min}$$

$$F_{Cout} = (F_{tot})_0 Z_2 = (459.8) (0.0759) = 34.9 \text{ mole/min}$$

$$F_{Dout} = (F_{tot})_0 (2 \cdot Z_2) = (459.8) (2 \cdot 0.0759) = 69.8 \text{ mole/min}.$$

The conversion of A at optimal feed rate is

$$f_{A_{out}} = \frac{F_{A0} - F_{A_{out}}}{F_{A0}} = \frac{459.8 - 189.4}{459.8} = 0.588.$$

The table below provides a comparison between the given and the optimal operations:

	Given Feed Rate	Optimal Feed Rate
Reactor Volume (liter)	1200	1200
Volumetric Feed Rate (lit/min)	100	229.9
Conversion of A	0.704	0.588
Production Rate of B (mole/min)	39.2	100.3
Production Rate of C (mole/min)	31.2	34.9
Production Rate of D (mole/min)	62.4	69.8

We see that at optimal feed rate, we process more than twice the material through the reactor than in the given feed rate. While the conversion of A is slightly lower, the production rate of B is about 2.5 times higher, while the production rates of C and D are only slightly higher.

Example 7-10 A valuable product B is produced in a CSTR by the gas-phase, first-order chemical reaction

$$A \rightarrow 2\,B.$$

Under the operating conditions, B decomposes according to the first-order reaction

$$B \rightarrow C + D.$$

A gaseous stream of reactant A ($C_{A0} = 0.04$ mole/lit) is fed into a 200 liter CSTR at a rate of 100 liter/min. At the reactor temperature, $k_1 = 2$ min^{-1} and $k_2 = 1$ min^{-1}.
a. Derive the design equations, and plot the reaction and species operating curves.
b. Determine the conversion of A and the production rate of B and C for the given feed rate.
c. Determine the optimal reactor volume to maximize B production.
d. Determine the conversion of A, the production rate of B and C, and the yield of B in the optimal reactor.

Solution This is an example of series (consecutive) chemical reactions. The chemical reactions taking place in the reactor are:

Reaction 1: $A \rightarrow 2\,B$

Reaction 2: $B \rightarrow C + D,$

and their stoichiometric coefficients are:

$$s_{A1} = -1;\ s_{B1} = 2;\ s_{C1} = 0;\ s_{D1} = 0;\ \Delta_1 = 1;$$

$$s_{A2} = 0;\ s_{B2} = -1;\ s_{C2} = 1;\ s_{D2} = 1;\ \Delta_2 = 1.$$

Since each reaction has a species that does not participate in the other, the two reactions are independent, and there is no dependent reaction. We select the inlet steam as the reference stream; hence, $Z_{1in} = Z_{2in} = 0$. Since only reactant A is fed into the reactor, C_0

$= C_{A0} = 0.04$ mole/lit, $y_{A0} = 1$, and $y_{B0} = y_{C0} = y_{D0} = 0$. Also,

$$(F_{tot})_0 = v_0 \cdot C_0 = (100 \text{ lit/min}) (0.04 \text{ mole/lit}) = 4 \text{ mole/min}.$$

Using (1.5-5), the species molar flow rates at the reactor outlet are:

$$F_{A_{out}} = (F_{tot})_0 \cdot (y_{A0} + s_{A1} \cdot Z_{1_{out}} + s_{A2} \cdot Z_{2_{out}}) = (F_{tot})_0 \cdot (1 - Z_{1_{out}}) \qquad (a)$$

$$F_{B_{out}} = (F_{tot})_0 \cdot (y_{B0} + s_{B1} \cdot Z_{1_{out}} + s_{B2} \cdot Z_{2_{out}}) = (F_{tot})_0 \cdot (2 \cdot Z_{1_{out}} - Z_{2_{out}}) \qquad (b)$$

$$F_{C_{out}} = (F_{tot})_0 \cdot (y_{C0} + s_{C1} \cdot Z_{1_{out}} + s_{C2} \cdot Z_{2_{out}}) = (F_{tot})_0 \cdot Z_{2_{out}} \qquad (c)$$

$$F_{D_{out}} = (F_{tot})_0 \cdot (y_{D0} + s_{D1} \cdot Z_{1_{out}} + s_{D2} \cdot Z_{2_{out}}) = (F_{tot})_0 \cdot Z_{2_{out}}. \qquad (d)$$

We write the design equation (7.1-1) for each of the independent reactions,

$$Z_{1_{out}} = r_{1_{out}} \cdot \tau \cdot \left(\frac{t_{cr}}{C_0}\right) \qquad (e)$$

$$Z_{2_{out}} = r_{2_{out}} \cdot \tau \cdot \left(\frac{t_{cr}}{C_0}\right). \qquad (f)$$

a. For gas-phase reactions, we use (7.1-12), and the rates are

$$r_{1_{out}} = k_1 \cdot C_0 \cdot \frac{1 - Z_{1_{out}}}{1 + Z_{1_{out}} + Z_{2_{out}}} \qquad (g)$$

$$r_{2_{out}} = k_2 \cdot C_0 \cdot \frac{2 \cdot Z_{1_{out}} - Z_{2_{out}}}{1 + Z_{1_{out}} + Z_{2_{out}}}. \qquad (h)$$

We define the characteristic reaction time on the basis of Reaction 1; hence,

$$t_{cr} = \frac{1}{k_1} = 0.5 \text{ min}. \qquad (i)$$

Substituting (g), (h), and (i) into (a) and (b), the design equations reduce to

$$Z_{1_{out}} - \left(\frac{1 - Z_{1_{out}}}{1 + Z_{1_{out}} + Z_{2_{out}}}\right) \cdot \tau = 0 \qquad (j)$$

$$Z_{2_{out}} - \left(\frac{k_2}{k_1}\right) \cdot \left(\frac{2 \cdot Z_{1_{out}} - Z_{2_{out}}}{1 + Z_{1_{out}} + Z_{2_{out}}}\right) \cdot \tau = 0. \qquad (k)$$

To obtain the reaction operating curves, we solve (j) and (k) simultaneously for different values of τ, applying numerical methods. The operating curves are shown in the figure below. Once we have Z_1 and Z_2 as a function of τ, we use (a) through (d) to calculate the dimensionless molar flow rates of the individual species as a function of the dimensionless space time. The figures below show these curves.

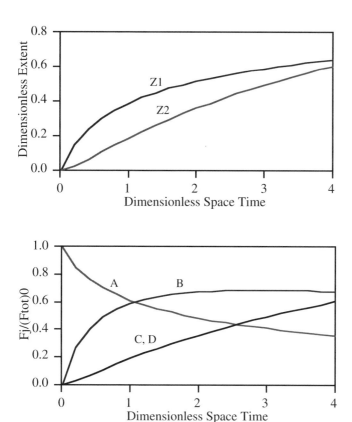

b. For the given feed flow rate, using (7.1-3) and (i), the dimensionless space time is

$$\tau = \frac{V_R}{v_0 \cdot t_{cr}} = \frac{(200 \text{ lit})}{(100 \text{ lit/min}) \cdot (0.5 \text{ min})} = 4.$$

At $\tau = 4$, the solutions of (j) and (k) are $Z_1 = 0.6406$ and $Z_2 = 0.6037$. Using (a) through (d), the species molar flow rates at the reactor outlet are: $F_{Aout} = 1.438$ mole/min, $F_{Bout} = 2.71$ mole/min, and $F_{Cout} = F_{Dout} = 2.42$ mole/min.
The conversion of A is

$$f_{Aout} = \frac{F_{A0} - F_{Aout}}{F_{A0}} = \frac{4 - 1.438}{4} = 0.641.$$

c. To determine the reactor volume that gives the highest production rate of B, we use the species operating curves (or tabulated calculated data) and find that maximum $F_{B_{out}}$ is reached at $\tau = 3.0$. Using (6.1-3) and (i), the optimal reactor volume is

$$V_R = \tau \cdot v_0 \cdot t_{cr} = (3.0) \cdot (100 \text{ lit/min}) \cdot (0.5 \text{ min}) = 150 \text{ liter}.$$

d. From the reaction operating curves, at $\tau = 3.0$, $Z_1 = 0.5901$ and $Z_2 = 0.4939$. Using (a) through (d), the species molar flow rates at the reactor outlet are,

$$F_{A_{out}} = (4 \text{ mole/min}) \cdot (1 - 0.5901) = 1.64 \text{ mole/min}$$

$$F_{B_{out}} = (4 \text{ mole/min}) \cdot (2 \cdot 0.5901 - 0.4939) = 2.75 \text{ mole/min}$$

$$F_{C_{out}} = F_{D_{out}} = (4 \text{ mole/min}) \cdot 0.4939 = 1.98 \text{ mole/min}.$$

The conversion of A is

$$f_{A_{out}} = \frac{F_{A0} - F_{A_{out}}}{F_{A0}} = \frac{4 - 0.5901}{4} = 0.590.$$

The table below provides a comparison between the given and the optimal reactors:

	Given Reactor	Optimal Reactor
Reactor Volume (liter)	200	150
Volumetric Feed Rate (lit/min)	100	100
Conversion of A	0.641	0.590
Production Rate of B (mole/min)	2.71	2.75
Production Rate of C (mole/min)	2.42	1.98

We see that the volume of the optimal reactor is 25% smaller than the given reactor. While processing the same feed rate, the conversion of A is slightly lower, the production rate of B is slightly higher, and the production rate of C is reduced by about 20%.

Example 7-11 A chemical plant generates a stream of species A and a stream of species B that presently are being discarded. The engineering department suggested to use an available 2,000 liter CSTR to produce valuable product, C, by combining these streams. The reactor operates at 184°C and 1.5 atm, and, at these conditions, the following gas-phase reactions take place,

$$A + B \leftrightarrow C$$

$$C + B \rightarrow D.$$

The reactions are elementary; thus $r_1 = k_1 \cdot C_A \cdot C_B$; $r_2 = k_2 \cdot C_C$; and $r_3 = k_1 \cdot C_C \cdot C_B$. The available pumping equipment in the plant can provide a maximum feed rate of 800 lit/min (at 184°C and 1.5 atm). Based on the data below, determine:

a. the proportion of A and B in the feed to maximize the production of C,
b. the rates stream A and stream B are fed to the reactor,
c. the flow rates of all species at the reactor outlet (at optimal feed composition), and

d. the conversions of A and B.

Data: At 184°C, $k_1 = 20$ lit/mole min; $k_2 = 0.16$ min^{-1}; $k_3 = 40$ lit/mole min.

Solution Using the methodology developed in Chapter 1, we represent the given chemical reactions as three individual reactions:

$$\text{Reaction 1:} \qquad A + B \rightarrow C \tag{a}$$

$$\text{Reaction 2:} \qquad C \rightarrow A + B \tag{b}$$

$$\text{Reaction 3:} \qquad C + B \rightarrow D. \tag{c}$$

The stoichiometric coefficients of the different species are:

$$s_{A1} = -1; \ s_{B1} = -1; \ s_{C1} = 1; \ s_{D1} = 0; \ \Delta_1 = -1;$$

$$s_{A2} = 1; \ s_{B2} = 1; \ s_{C2} = -1; \ s_{D2} = 0; \ \Delta_2 = 1;$$

$$s_{A3} = 0; \ s_{B3} = -1; \ s_{C3} = -1; \ s_{D3} = 1; \ \Delta_3 = -1.$$

To determine the number of independent reactions, we construct the stoichiometric matrix as described in Section 1-6,

$$\begin{array}{cccc} A & D & C & D \end{array}$$
$$\begin{bmatrix} -1 & -1 & 1 & 0 \\ 1 & 1 & -1 & 0 \\ 0 & -1 & -1 & 1 \end{bmatrix}.$$

Using elementary row operations, we reduce the matrix to

$$\begin{bmatrix} -1 & -1 & 1 & 0 \\ 0 & -1 & -1 & 1 \\ 0 & 0 & 0 & 0 \end{bmatrix}. \tag{d}$$

Since the reduced matrix has two nonzero rows, the system has two independent chemical reactions, and the nonzero rows in (d) provide a set of independent reactions. In this case, an examination of the nonzero rows indicates that the two independent reactions are Reaction 1 and Reaction 3. Hence, among the three chemical reactions with given rates, two are independent and one is dependent. The indices of the independent reactions are $m = 1$ and $m = 3$; the index of the dependent reaction is $k = 2$. To determine the multipliers, α_{21} and α_{23}, relating the two independent reactions to the dependent reaction, we use (1.6-9),

$$\alpha_{21} \cdot (s_j)_1 + \alpha_{23} \cdot (s_j)_3 = (s_j)_2. \tag{e}$$

We write (e) for species A and obtain $\alpha_{21} = -1$; we write (e) for species B and obtain $\alpha_{23} = 0$; we then verify these values by writing (e) for species C.

We write the design equation, (7.1-1), for each of the two independent reactions,

$$Z_{1_{out}} = (r_{1_{out}} - r_{2_{out}}) \cdot \tau \cdot \left(\frac{t_{cr}}{C_0}\right) \tag{f}$$

$$Z_{3_{out}} = r_{3_{out}} \cdot \tau \cdot \left(\frac{t_{cr}}{C_0}\right). \tag{g}$$

We select the inlet steam as the reference stream and denote $(y_A)_0$ and $(y_B)_0$ to be the mole fractions of the two reactants in the feed stream $(y_{C0} = y_{D0} = 0)$. Using (7.1-7), we express the molar flow rate of each species in terms of the dimensionless extents,

$$F_{A_{out}} = (y_{A0} - Z_{1_{out}}) \cdot (F_{tot})_0 \tag{h}$$

$$F_{B_{out}} = (y_{B0} - Z_{1_{out}} - Z_{3_{out}}) \cdot (F_{tot})_0 \tag{i}$$

$$F_{C_{out}} = (Z_{1_{out}} - Z_{3_{out}}) \cdot (F_{tot})_0 \tag{j}$$

$$F_{D_{out}} = Z_{3_{out}} \cdot (F_{tot})_0. \tag{k}$$

Using (7.1-11), for gas-phase reactions, the rates of the reactions are:

$$r_{1_{out}} = k_1 \cdot C_0^2 \cdot \frac{(y_{A0} - Z_{1_{out}}) \cdot (y_{B0} - Z_{1_{out}} - Z_{3_{out}})}{(1 - Z_{1_{out}} - Z_{3_{out}})^2} \tag{l}$$

$$r_{2_{out}} = k_2 \cdot C_0 \cdot \frac{Z_{1_{out}} - Z_{3_{out}}}{1 - Z_{1_{out}} - Z_{3_{out}}} \tag{m}$$

$$r_{3_{out}} = k_3 \cdot C_0^2 \cdot \frac{(Z_{1_{out}} - Z_{3_{out}}) \cdot (y_{B0} - Z_{1_{out}} - Z_{3_{out}})}{(1 - Z_{1_{out}} - Z_{3_{out}})^2}. \tag{n}$$

We select Reaction 1 as the leading reaction; hence, the characteristic reaction time is

$$t_{cr} = \frac{1}{k_1 \cdot C_0}. \tag{o}$$

Substituting (l), (m), (n), and (o) into (a) and (b), the design equations reduce to

$$Z_{1_{out}} =$$

$$= \left(\frac{(y_{A0} - Z_{1_{out}}) \cdot (y_{B0} - Z_{1_{out}} - Z_{3_{out}})}{(1 - Z_{1_{out}} - Z_{3_{out}})^2} - \left(\frac{k_2}{k_1 \cdot C_0}\right) \frac{Z_{1_{out}} - Z_{3_{out}}}{1 - Z_{1_{out}} - Z_{3_{out}}}\right) \cdot \tau \tag{p}$$

$$Z_{3_{out}} = \left(\frac{k_3}{k_1}\right) \cdot \frac{(Z_{1_{out}} - Z_{3_{out}}) \cdot (y_{B0} - Z_{1_{out}} - Z_{3_{out}})}{(1 - Z_{1_{out}} - Z_{3_{out}})^2} \cdot \tau. \tag{q}$$

Using the given data,

$$C_0 = \frac{P}{R \cdot T} = \frac{1.5 \text{ atm}}{(0.08203 \text{ lit atm/mole}^\circ K) \cdot (457^\circ K)} = 0.04 \text{ mole/lit}$$

$$(F_{tot})_0 = v_0 \cdot C_0 = (800 \text{ lit/min}) \cdot (0.04 \text{ mole/lit}) = 32 \text{ mole/min}$$

$$\tau = \frac{V_R}{v_0} \cdot k_1 \cdot C_0 = \frac{2000 \text{ lit}}{800 \text{ lit/min}} \cdot (20 \text{ lit/min mole}) \cdot (0.04 \text{ mole/lit}) = 2$$

$$\left(\frac{k_2}{k_1 \cdot C_0}\right) = 0.2, \text{ and } \left(\frac{k_3}{k_1}\right) = 2 .$$

Substituting these values and noting that $y_{B0} = 1 - y_{A0}$, the design equations reduce to

$$Z_{1_{out}} -$$

$$-2 \cdot \left(\frac{(y_{A0} - Z_{1_{out}}) \cdot (1 - y_{A0} - Z_{1_{out}} - Z_{3_{out}})}{(1 - Z_{1_{out}} - Z_{3_{out}})^2} - 0.2 \cdot \frac{Z_{1_{out}} - Z_{3_{out}}}{1 - Z_{1_{out}} - Z_{3_{out}}} \right) = 0 \quad (r)$$

$$Z_{3_{out}} - 4 \cdot \frac{(Z_{1_{out}} - Z_{3_{out}}) \cdot (1 - y_{A0} - Z_{1_{out}} - Z_{3_{out}})}{(1 - Z_{1_{out}} - Z_{3_{out}})^2} = 0 . \quad (s)$$

a. To find the optimal feed composition, we solve (r) and (s) for different values of y_{A0} and calculate $F_{Cout}/(F_{tot})_0$ by (j). The table below provides the results of theses calculations (using a mathematical software):

$(y_A)_0$	Z_1	Z_3	$F_{Cout}/(F_{tot})_0$
0.50	0.1956	0.1197	0.0759
0.55	0.1856	0.1039	0.0817
0.60	0.1726	0.0873	0.0853
0.65	0.1571	0.0708	0.0863
0.70	0.1398	0.0550	0.0848
0.75	0.1207	0.0405	0.0802
0.80	0.1000	0.0276	0.0724

The graph below shows the plot of $F_{Cout}/(F_{tot})_0$ versus y_{A0}. From the graph, maximum production rate of C is achieved for $(y_A)_0 = 0.65$.

b. The optimal feed rate of reactant A is

$$F_{A0} = (0.65) \cdot (32 \text{ mole/min}) = 20.8 \text{ mole/min} ,$$

and the optimal feed rate of reactant B is

$$F_{B0} = (0.35) (32 \text{ mole/min}) = 11.2 \text{ mole/min}.$$

c. For the optimal feed composition, the solution of (r) and (s) is $Z_1 = 0.157$ and $Z_3 = 0.0708$. From (h) through (k), the flow rates of the individual species at the reactor

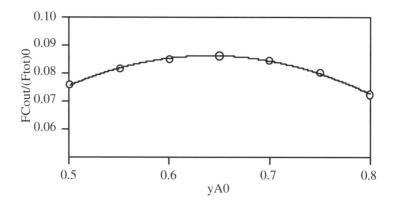

outlet are: $F_{Aout} = 15.8$ mole/min, $F_{Bout} = 3.91$ mole/min, $F_{Cout} = 2.76$ mole/min, and $F_{Dout} = 2.27$ mole/min.

d. The conversion of A is

$$f_A = \frac{20.8 - 15.8}{20.8} = 0.24 \,,$$

and the conversion of B is

$$f_B = \frac{11.2 - 3.91}{11.2} = 0.651 \,.$$

Example 7-12 The following reversible reactions (mechanism) were proposed for gas-phase cracking of hydrocarbons:

$$A \leftrightarrow 2\,B \qquad r_1 = k_1 \cdot C_A \qquad r_2 = k_2 \cdot C_B^2$$

$$A + B \leftrightarrow C \qquad r_3 = k_3 \cdot C_A \cdot C_B \qquad r_4 = k_4 \cdot C_C$$

$$A + C \leftrightarrow D \qquad r_5 = k_5 \cdot C_A \cdot C_C \qquad r_6 = k_6 \cdot C_D$$

B is the desired product. We want to design an isothermal CSTR to be operated at 489°C and 5 atm. A stream of pure A at a rate of 1 mole/sec is available in the plant.

a. Derive the design equations, and plot the operating curves for the reactions.

b. Plot the operating curves for the individual species.

c. Determine the volume of the reactor needed to produce 2.16 kmole/hr of B.

d. Repeat parts a and b for a feed stream that consists only of species D.

Data: At 489°C, $k_1 = 2$ min^{-1}; $k_2 = 5$ liter mole^{-1} min^{-1}; $k_3 = 50$ liter mole^{-1} min^{-1}; $k_4 = 2$ min^{-1}; $k_5 = 100$ liter mole^{-1} min^{-1}; and $k_6 = 4$ min^{-1}.

Solution The reaction scheme in this case is the same as that of Example 6-11. We select the three forward reactions as the set of independent reactions. Hence, the indices of the independent reactions are $m_1 = 1$, $m_2 = 3$, and $m_5 = 5$, and those of the dependent reactions are $k_1 = 2$, $k_2 = 4$, and $k_3 = 6$, and the factors α_{km}'s are: $\alpha_{21} = -1$,

$\alpha_{43} = -1$, and $\alpha_{65} = -1$, and all the others are zero. The stoichiometric coefficients of the independent reactions are:

$$s_{A1} = -1; \quad s_{B1} = 2; \quad s_{C1} = 0; \quad s_{D1} = 0; \quad \Delta_1 = 1;$$

$$s_{A3} = -1; \quad s_{B3} = -1; \quad s_{C3} = 1; \quad s_{D3} = 0; \quad \Delta_3 = -1;$$

$$s_{A5} = -1; \quad s_{A5} = 0; \quad s_{C3} = -1; \quad s_{D5} = 1; \quad \Delta_5 = -1.$$

a. We write design equation (7.1-1) for each of the independent reactions,

$$Z_{1_{out}} - Z_{1_{in}} = (r_{1_{out}} - r_{2_{out}}) \cdot \tau \cdot \left(\frac{t_{cr}}{C_0} \right) \tag{a}$$

$$Z_{3_{out}} - Z_{3_{in}} = (r_{3_{out}} - r_{4_{out}}) \cdot \tau \cdot \left(\frac{t_{cr}}{C_0} \right) \tag{b}$$

$$Z_{5_{out}} - Z_{5_{in}} = (r_{5_{out}} - r_{6_{out}}) \cdot \tau \cdot \left(\frac{t_{cr}}{C_0} \right). \tag{c}$$

We select the inlet stream as the reference stream; hence, $T_0 = 762°K$, $y_{A0} = 1$, $y_{B0} = y_{C0} = y_{D0} = 0$, and $Z_{10} = Z_{20} = Z_{30} = 0$. Also,

$$C_0 = \frac{P_0}{R \cdot T_0} = \frac{(5 \text{ atm})}{(82.05 \cdot 10^{-3} \text{ lit-atm/mole°K}) \cdot (762°K)} = 8.00 \cdot 10^{-2} \text{ mole/lit} \tag{d}$$

$$v_0 = \frac{(F_{tot})_0}{C_0} = \frac{(60 \text{ mole/min})}{(8.00 \cdot 10^{-2} \text{ mole/lit})} = 750 \text{ lit/min}.$$

Using stoichiometric relation (1.5-5), the species molar flow rates are:

$$F_{A_{out}} = (F_{tot})_0 \cdot (y_{A0} - Z_{1_{out}} - Z_{3_{out}} - Z_{5_{out}}) \tag{e}$$

$$F_{B_{out}} = (F_{tot})_0 \cdot (y_{B0} + 2 \cdot Z_{1_{out}} - Z_{3_{out}}) \tag{f}$$

$$F_{C_{out}} = (F_{tot})_0 \cdot (y_{C0} + Z_{3_{out}} - Z_{5_{out}}) \tag{g}$$

$$F_{D_{out}} = (F_{tot})_0 \cdot (y_{D0} + Z_{5_{out}}). \tag{h}$$

For gas-phase reactions, we use (7.1-12), and the rates of the chemical reactions are:

$$r_1 = k_1 \cdot C_0 \cdot \frac{y_{A0} - Z_{1_{out}} - Z_{3_{out}} - Z_{5_{out}}}{1 + Z_{1_{out}} - Z_{3_{out}} - Z_{5_{out}}} \tag{i}$$

$$r_2 = k_2 \cdot C_0^2 \cdot \left(\frac{y_{B0} + 2 \cdot Z_{1_{out}} - Z_{3_{out}}}{1 + Z_{1_{out}} - Z_{3_{out}} - Z_{5_{out}}} \right)^2 \tag{j}$$

$$r_3 = k_3 \cdot C_0^2 \cdot \frac{(y_{A0} - Z_{1out} - Z_{3out} - Z_{5out}) \cdot (y_{B0} + 2 \cdot Z_{1out} - Z_{3out})}{(1 + Z_{1out} - Z_{3out} - Z_{5out})^2} \tag{k}$$

$$r_4 = k_4 \cdot C_0 \cdot \frac{y_{C0} + Z_{3out} - Z_{5out}}{1 + Z_{1out} - Z_{3out} - Z_{5out}} \tag{l}$$

$$r_5 = k_5 \cdot C_0^2 \cdot \frac{(y_{A0} - Z_{1out} - Z_{3out} - Z_{5out}) \cdot (y_{C0} + Z_{3out} - Z_{5out})}{(1 + Z_{1out} - Z_{3out} - Z_{5out})^2} \tag{m}$$

$$r_6 = k_6 \cdot C_0 \cdot \frac{y_{D0} + Z_{5out}}{1 + Z_{1out} - Z_{3out} - Z_{5out}}. \tag{n}$$

We select Reaction 1 as the leading reaction; hence the characteristic reaction time is

$$t_{cr} = \frac{1}{k_1} = 0.5 \text{ min}. \tag{o}$$

Substituting (i) through (n) and using (o), the design equations reduce to

$$Z_{1out} - \tau \cdot \left(\frac{y_{A0} - Z_{1out} - Z_{3out} - Z_{5out}}{1 + Z_{1out} - Z_{3out} - Z_{5out}} \right) +$$

$$+ \tau \cdot \left(\frac{k_2 \cdot C_0}{k_1} \right) \left(\frac{y_{B0} + 2 \cdot Z_{1out} - Z_{3out}}{1 + Z_{1out} - Z_{3out} - Z_{5out}} \right)^2 = 0 \tag{p}$$

$$Z_{3out} - \tau \cdot \left(\frac{k_3 \cdot C_0}{k_1} \right) \frac{(y_{A0} - Z_{1out} - Z_{3out} - Z_{5out}) \cdot (y_{B0} + 2 \cdot Z_{1out} - Z_{3out})}{(1 + Z_{1out} - Z_{3out} - Z_{5out})^2} +$$

$$+ \tau \cdot \left(\frac{k_4}{k_1} \right) \cdot \left(\frac{y_{C0} + Z_{3out} - Z_{5out}}{1 + Z_{1out} - Z_{3out} - Z_{5out}} \right) = 0 \tag{q}$$

$$Z_{5out} - \tau \cdot \left(\frac{k_5 \cdot C_0}{k_1} \right) \frac{(y_{A0} - Z_{1out} - Z_{3out} - Z_{5out}) \cdot (y_{C0} + Z_{3out} - Z_{5out})}{(1 + Z_{1out} - Z_{3out} - Z_{5out})^2} +$$

$$+ \tau \cdot \left(\frac{k_6}{k_1} \right) \cdot \left(\frac{y_{D0} + Z_{5out}}{1 + Z_{1out} - Z_{3out} - Z_{5out}} \right) = 0. \tag{r}$$

For the given data, $y_{A0} = 1$ and $y_{B0} = y_{C0} = y_{D0} = 0$, and the parameters are:

$$\frac{k_2 \cdot C_0}{k_1} = 0.2, \quad \frac{k_3 \cdot C_0}{k_1} = 2, \quad \frac{k_4}{k_1} = 1, \quad \frac{k_5 \cdot C_0}{k_1} = 4, \quad \frac{k_6}{k_1} = 2. \tag{s}$$

We substitute these values into (p), (q), and (r) and solve them numerically for Z_1, Z_3, and Z_5 for different values of τ. The first figure below shows the reaction operating curves.

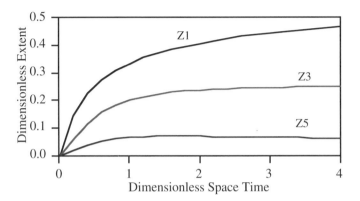

b. For the solutions of Z_1, Z_3, and Z_5, we use (e) through (h) to calculate the dimensionless species operating curves, shown in the second figure below.

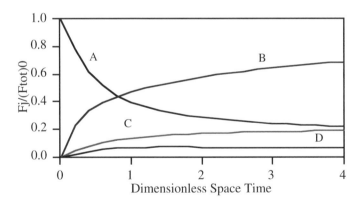

c. For $F_{Bout} = 2.16$ kmole/hr,

$$\frac{F_{Bout}}{(F_{tot})_0} = \left(\frac{2.16 \text{ kmole/hr}}{1 \text{ mole/sec}}\right) \cdot \left(\frac{1 \text{ hr}}{3600 \text{ sec}}\right) \cdot \left(\frac{1000 \text{ mole}}{1 \text{ kmole}}\right) = 0.60.$$

From the species operating curve, $F_{Bout}/(F_{tot})_0 = 0.85$ is reached at $\tau = 2.31$. Using (7.1-3) and (o), the needed reactor volume is

$$V_R = \frac{\tau \cdot v_0}{k_1} = \frac{(2.31) \cdot (12.5 \text{ lit/sec})}{(2 \text{ min}^{-1})} \cdot \left(\frac{60 \text{ sec}}{1 \text{ min}} \right) = 866.2 \text{ liter}.$$

d. When the feed stream consists of species D, the reactor calculations proceed in the same way as in parts (a), (b), and (d). The only difference is that, now, $(y_A)_0 = (y_B)_0 = (y_C)_0 = 0$ and $(y_D)_0 = 1$. Substituting these values into (p), (q), and (r), we solve them numerically using the parameters in (s). The figures below show the reaction and species operating curves, respectively. Note that in this case, the extents of independent reactions 3 and 5 are negative, since their reverse reactions actually proceed.

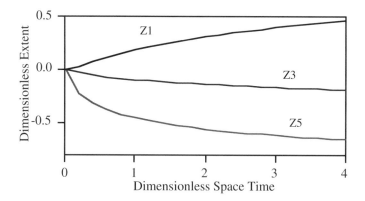

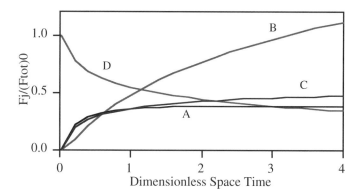

7.4 NON-ISOTHERMAL OPERATIONS

The design formulation of non-isothermal CSTRs with multiple reactions follows the same procedure outlined in the previous section — we write design equation (7.1-1) for each independent reaction. However, since the reactor temperature, T_{out}, is not known, we should solve the design equations simultaneously with the energy balance equation, (7.1-13).

By definition, the temperature of a CSTR is the same everywhere; therefore, all CSTRs are, in principle, "isothermal." Usually, the inlet temperature is specified, and, as a part of the design, we have to determine the reactor temperature. When the reactor temperature is the same as the inlet temperature, $T_{out} = T_{in}$, we refer to a CSTR as an "isothermal reactor." But when $T \neq T_{in}$, we do not know, a priori, what is the reactor temperature, and we have to determine it. Also, note that the energy balance equation contains another variable — the temperature of the heating (or cooling) fluid, θ_F, (it is constant only when the fluid either evaporates or condenses). Because of the complex geometry of the heat-transfer surface (shell or a coil) in a CSTR, an average of the fluid's inlet and outlet temperatures is usually used.

The procedure for setting up the dimensionless energy balance equation goes as follows:

a. Define the **reference stream** and identify T_0, C_0, $(F_{tot})_0$, and y_{j0}'s.
b. Determine the specific molar heat capacity of the reference stream, $\hat{c}_{p0}$.
c. Determine the dimensionless activation energies, γ_i's, of **all** chemical reactions.
d. Determine the dimensionless heat of reactions, DHR_m's, of the **independent** reactions.
e. Determine the correction factor of the heat capacity, $CF(Z_m, \theta)$.
f. Specify the dimensionless heat-transfer number, HTN.
g. Determine (or specify) the dimensionless temperature of the heating/cooling fluid, θ_F.
h. Determine (or specify) the inlet dimensionless temperature, $\theta(0)$.
i. Solve the energy balance equation simultaneously with the design equations to obtain Z_m's and θ_{out} as functions of the dimensionless space time, τ.

Hence, the design formulation of non-isothermal CSTRs consists of $(n_{ind} + 1)$ simultaneous, nonlinear algebraic equations,

$$G_1(\tau, Z_{1_{out}}, Z_{2_{out}}, ..., Z_{n_{out}}, \theta_{out}) = 0$$
$$G_2(\tau, Z_{1_{out}}, Z_{2_{out}}, ..., Z_{n_{out}}, \theta_{out}) = 0$$

$$\vdots \qquad\qquad (7.4\text{-}1)$$

$$G_{n_{ind}}(\tau, Z_{1_{out}}, Z_{2_{out}}, ..., Z_{n_{out}}, \theta_{out}) = 0$$
$$G_{n_{ind}+1}(\tau, Z_{1_{out}}, Z_{2_{out}}, ..., Z_{n_{out}}, \theta_{out}) = 0.$$

We have n_{ind} design equations, and the $(n_{ind}+1)$-th equation is the energy balance equation (4.2-68). We have to solve them for different values of dimensionless space time, t, defined by (7.1-3). Below, we illustrate how to design non-isothermal CSTRs.

Example 7-13 The first-order reaction A → 2 B takes place in an aqueous solution. A solution with a concentration of $C_{A0} = 0.8$ mole/liter is fed at a rate of 200 liter/min into a cascade of two equal-size 100 liter CSTRs connected in series. Based on the data below, for the indicated operations, determine the conversion of reactant A and the outlet temperature of each reactor:

a. Derive the reaction operating curve of each reactor for isothermal operation.
b. Determine the heating/cooling load of each reactor in (a).
c. Determine the extent and the temperature of each reactor for adiabatic operation.
d. Derive the reaction operating curve of each reactor for adiabatic operation.

Data: $T_0 = 47°C = 320°K$; $\Delta H_R(T_0) = -20$ kcal/mole B;
At 47°C, $k = 0.4$ min^{-1}; $E_a = 9,000$ cal/mole;
$\rho = 1.0$ kg/liter; $\bar{c}_p = 1.0$ kcal/kg°K.

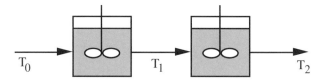

Solution The reaction is A → 2 B, and the stoichiometric coefficients are: $s_A = -1$ and $s_B = 2$. The rate expression is $r = k \cdot C_A$, and, for this reaction, the heat of reaction is

$$\Delta H_R (320°K) = \left(\frac{-20 \text{ kcal}}{\text{mole B}} \right) \left(\frac{2 \text{ mole B}}{\text{mole extent}} \right) = -40 \ \frac{\text{kcal}}{\text{mole extent}} .$$

We select the inlet stream to the first reactor as the reference stream; hence, $C_0 = C_{A0}$, $y_{A0} = 1$, $y_{B0} = 0$ and $Z_{1\text{in}} = 0$. Also,

$$(F_{tot})_0 = v_0 \cdot C_0 = (200 \text{ liter/min}) (0.8 \text{ mole/liter}) = 160 \text{ mole/min}. \tag{a}$$

For liquid-phase reactions, the specific molar heat capacity of the reference stream, defined by (4.2-60), is

$$\hat{c}_{p0} = \frac{\dot{m}}{(F_{tot})_0} \cdot \bar{c}_p = \frac{200 \text{ kg/min}}{160 \text{ mole/min}} \left(1 \ \frac{\text{kcal}}{\text{kg °K}} \right) = 1,250 \ \frac{\text{cal}}{\text{mole °K}} . \tag{b}$$

The dimensionless heat of reaction is

$$\text{DHR} = \left(\frac{\Delta H_R (T_0)}{T_0 \cdot \hat{c}_{p0}} \right) = \frac{-40,000}{(320) \cdot (1250)} = -0.10, \tag{c}$$

and the dimensionless activation energy is

$$\gamma = \frac{E_a}{R \cdot T_0} = \frac{9,000}{1.987 \cdot 320} = 14.154. \tag{d}$$

Since only one reaction takes place, we need only one design equation to describe the reactor operation, and, using (7.1-1), the design equation for each reactor is

$$Z_{out} - Z_{in} = r_{out} \cdot \tau \cdot \left(\frac{t_{cr}}{C_0} \right). \tag{e}$$

For a liquid-phase reaction, using (7.1-9), the reaction rate is

$$r_{out} = k(T_0) \cdot C_0 \cdot (1 - Z_{out}) \cdot e^{\gamma \cdot \frac{\theta - 1}{\theta}}. \tag{f}$$

We define the characteristic reaction time as

$$t_{cr} = \frac{1}{k(T_0)} = \frac{1}{0.4} = 2.5 \text{ min}. \tag{g}$$

Substituting (f) and (g) into (e), the design equation for the first reactor is

$$Z_{out_1} - (1 - Z_{out_1}) \cdot \tau_1 \cdot e^{\gamma \cdot \frac{\theta_1 - 1}{\theta_1}} = 0, \tag{h}$$

and the design equation for the second reactor is

$$Z_{out_2} - Z_{out_1} - (1 - Z_{out_2}) \cdot \tau_2 \cdot e^{\gamma \cdot \frac{\theta_2 - 1}{\theta_2}} = 0. \tag{i}$$

For the given reactors, the dimensionless space time of each reactor is

$$\tau_1 = \tau_2 = \frac{V_R}{v_0 \cdot t_{cr}} = \frac{(100 \text{ lit})}{(200 \text{ lit/min}) \cdot (25 \text{ min})} = 0.2. $$

a. For isothermal operation, $\theta_1 = \theta_2 = 1$, and we solve (h) and then (i) to obtain

$$Z_{out_1} = 0.1667 \text{ and } Z_{out_2} = 0.3056.$$

b. Using (7.1-13), for isothermal operation, the heating load in the first reactor is

$$\frac{\dot{Q}_1}{(F_{tot})_0 \cdot T_0 \cdot \hat{c}_{p0}} = \left(\frac{\Delta H_R(T_0)}{T_0 \cdot \hat{c}_{p0}} \right) \cdot (Z_{out_1} - 0), \tag{j}$$

and, in the second reactor, it is

$$\frac{\dot{Q}_2}{(F_{tot})_0 \cdot T_0 \cdot \hat{c}_{p0}} = \left(\frac{\Delta H_R (T_0)}{T_0 \cdot \hat{c}_{p0}} \right) \cdot (Z_{out_2} - Z_{out_1}). \tag{k}$$

In this case, the heating loads of the reactors are

$$\dot{Q}_1 = (F_{tot})_0 \cdot \Delta H_R (T_0) \cdot Z_{out_1} = (160) \cdot (-40) \cdot (0.167) = -1{,}066.9 \text{ kcal/min}.$$

$$\dot{Q}_2 = (160) \cdot (-40) \cdot (0.306 - 0.167) = -889 \text{ kcal/min}.$$

The negative sign indicates that heat is being removed from the reactor.

c. For liquid-phase reactions, assuming constant heat capacity, from (4.2-62), $CF(Z_m, \theta)$ = 1. For adiabatic operations, $(S/V) = 0$, and, for the first reactor, (7.1-13) becomes

$$DHR \cdot (Z_{out_1} - 0) + (1) \cdot (\theta_1 - 1) = 0. \tag{l}$$

We solve (h) and (l) simultaneously for $\tau = 0.2$ and obtain

$$Z_{out_1} = 0.2114 \text{ and } \theta_1 = 1.0211.$$

For the second reactor, (7.1-13) becomes

$$DHR \cdot (Z_{out_2} - Z_{out_1}) + (1) \cdot (\theta_2 - \theta_1) = 0. \tag{m}$$

We use the values of $(Z_{out})_1$ and θ_1 to solve (i) and (m) simultaneously for $\tau = 0.2$ and obtain $(Z_{out})_2 = 0.4165$ and $\theta_2 = 1.0415$. Hence, for adiabatic operation, the temperature of the first reactor is $T_1 = (1.0211)(320°K) = 326.7°K$, and the temperature of the second reactor is $T_2 = (1.0415)(320°K) = 333.3°K$.

d. To obtain the reaction operating curve of the first reactor, we solve (h) and (l) simultaneously for different values of τ and obtain $(Z_{out})_1$ and θ_1. For each value of τ, we use the calculated values of $(Z_{out})_1$ and θ_1 to solve (i) and (m) for the same value of τ to obtain the values of $(Z_{out})_2$ and θ_2. The figures show, respectively, the reaction operat-

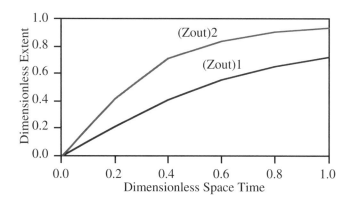

ing curve of each reactor and the temperature curves. Note that, for a given value of τ, we can solve the equations by trial-and error as follows: for the first reactor, we guess $(Z_{out})_1$, next calculate θ_1 by (1), and then check if (h) is satisfied for these two values. If not, we repeat the procedure. Once we have the values of $(Z_{out})_1$ and θ_1, we follow the procedure for (i) and (m) to determine the values of $(Z_{out})_2$ and θ_2.

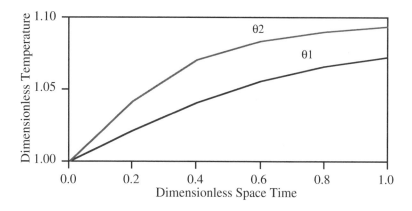

Example 7-14 The elementary gas-phase reactions

$$A + B \rightarrow C$$

$$C + B \rightarrow D$$

are carried out in a CSTR. A gas mixture of 40% A, 40% B, and 20% I (inert) is fed into a reactor at a rate of 0.4 m³/min. The inlet temperature is 150°C, and the reactor operates at a pressure of 2 atm. Based on the data below, derive the design equations, and plot the operating curves for the operations indicated. Calculate the reactor volume needed to obtain 60% conversion of B, the molar flow rate of product C, and the temperature at the reactor outlet.
a. Isothermal operation.
b. Determine the heating/cooling load in (a).
c. Adiabatic operation.
d. Operation with heat-transfer. The reactor is cooled by condensing steam in the shell side at $T_F = 130°C$, $(S/V) = 2$ m⁻¹, and the heat-transfer coefficient is 6.9 kcal/ m² hr °K.
Data: At 150°C $k_1 = 0.2$ liter mole⁻¹sec⁻¹; $k_2 = 0.1$ liter mole⁻¹sec⁻¹;
At 150°C $\Delta H_{R1} = -12,000$ cal/mole extent; $\Delta H_{R2} = -9,000$ cal/mole extent;
$E_{a1} = 4,950$ cal/mole; $E_{a2} = 7,682$ cal/mole;
$\hat{c}_{PA} = 16$ cal/mole°K; $\hat{c}_{PB} = 8$ cal/mole°K; $\hat{c}_{PC} = 20$ cal/mole°K;
$\hat{c}_{PD} = 26$ cal/mole°K; $\hat{c}_{PI} = 10$ cal/mole°K.
Solution The reactions are:

$$\text{Reaction 1:} \qquad A + B \rightarrow C$$

$$\text{Reaction 2:} \qquad C + B \rightarrow D,$$

and the stoichiometric coefficients are:

$$s_{A1} = -1; \quad s_{B1} = -1; \quad s_{C1} = 1; \quad s_{D1} = 0; \quad \Delta_1 = -1;$$

$$s_{A2} = 0; \quad s_{B1} = -1; \quad s_{C1} = -1; \quad s_{D1} = 1; \quad \Delta_2 = -1.$$

Since each reaction has a species that does not appear in the other reaction, the two reactions are independent, and there is no dependent reaction. We select the inlet stream as the reference stream; hence, $Z_{1in} = Z_{2in} = 0$, and $y_{A0} = 0.40$, $y_{B0} = 0.40$, $y_{I0} = 0.20$, and $y_{C0} = y_{D0} = 0$. Also,

$$C_0 = \frac{P}{R \cdot T_0} = 5.76 \ 10^{-2} \ \text{mole/liter}$$

$$(F_{tot})_0 = v_0 \cdot C_0 = (400 \ \text{lit/min}) \cdot (5.76 \ 10^{-2} \ \text{mole/liter}) = 23.05 \ \text{mole/min} \,.$$

The dimensionless activation energies of the two reactions are:

$$\gamma_1 = \frac{E_{a1}}{R \cdot T_0} = 5.89 \qquad \gamma_2 = \frac{E_{a2}}{R \cdot T_0} = 9.14 \,.$$

To determine the specific heat capacity of the reference stream, $\hat{c}_{p0}$, we use (4.2-58),

$$\hat{c}_{p0} = \sum_{j}^{all} y_{j0} \cdot \hat{c}_{pj}(1) = y_{A0} \cdot \hat{c}_{PA}(1) + y_{B0} \cdot \hat{c}_{PB}(1) + y_{I0} \cdot \hat{c}_{PI}(1)$$

$$\hat{c}_{p0} = (0.4) \cdot (16) + (0.4) \cdot (8) + (0.2) \cdot (10) = 11.6 \ \text{cal/mole}^{\circ}K \,.$$

The dimensionless heat of reactions are:

$$DHR_1 = \left(\frac{\Delta H_{R1}(T_0)}{T_0 \cdot \hat{c}_{p0}} \right) = -2.45 \quad DHR_2 = \left(\frac{\Delta H_{R2}(T_0)}{T_0 \cdot \hat{c}_{p0}} \right) = -1.834 \,.$$

Using (4.2-61), the correction factor of the heat capacity is

$$CF(Z_m, \theta) = 1 + \frac{1}{\hat{c}_{p0}} \sum_{j}^{all} \hat{c}_{pj}(\theta) \sum_{m}^{n_{ind}} (s_j)_m \cdot Z_m =$$

$$= 1 + \frac{1}{\hat{c}_{p0}} \left[\hat{c}_{PA} \cdot (-Z_1) + \hat{c}_{PB} \cdot (-Z_1 - Z_2) + \hat{c}_{PC} \cdot (Z_1 - Z_2) + \hat{c}_{PD} \cdot Z_2 \right] =$$

$$= 1 + \frac{1}{11.6} \left[(16) \cdot (-Z_1) + (8) \cdot (-Z_1 - Z_2) + (20) \cdot (Z_1 - Z_2) + (26) \cdot Z_2 \right]$$

$$CF(Z_m, \theta) = \frac{11.6 - 4 \cdot Z_1 - 2 \cdot Z_2}{11.6}.$$

We write the dimensionless design equation (7.1-13) for the two independent reactions

$$Z_{1_{out}} = r_{1_{out}} \cdot \tau \cdot \left(\frac{t_{cr}}{C_0} \right) \tag{a}$$

$$Z_{2_{out}} = r_{2_{out}} \cdot \tau \cdot \left(\frac{t_{cr}}{C_0} \right). \tag{b}$$

Using the stoichiometric relation (1.5-5),

$$F_{A_{out}} = (F_{tot})_0 \cdot \left(y_{A_0} - Z_{1_{out}} \right) \tag{c}$$

$$F_{B_{out}} = (F_{tot})_0 \cdot \left(y_{B_0} - Z_{1_{out}} - Z_{2_{out}} \right) \tag{d}$$

$$F_{C_{out}} = (F_{tot})_0 \cdot \left(y_{C_0} + Z_{1_{out}} - Z_{2_{out}} \right) \tag{e}$$

$$F_{D_{out}} = (F_{tot})_0 \cdot \left(y_{D_0} + Z_{2_{out}} \right). \tag{f}$$

For gas-phase reactions, using (7.1-12), the rates of the two reactions are

$$r_{1_{out}} = k_1(T_0) \cdot C_0^2 \cdot \frac{(y_{A_0} - Z_{1_{out}}) \cdot (y_{B_0} - Z_{1_{out}} - Z_{2_{out}})}{\left[(1 - Z_{1_{out}} - Z_{2_{out}}) \cdot \theta \right]^2} \cdot e^{\gamma_1 \cdot \frac{\theta - 1}{\theta}} \tag{g}$$

$$r_{2_{out}} = k_2(T_0) \cdot C_0^2 \cdot \frac{(Z_{1_{out}} - Z_{2_{out}}) \cdot (y_{B_0} - Z_{1_{out}} - Z_{2_{out}})}{\left[(1 - Z_{1_{out}} - Z_{2_{out}}) \cdot \theta \right]^2} \cdot e^{\gamma_2 \cdot \frac{\theta - 1}{\theta}}. \tag{h}$$

Defining a characteristic reaction time,

$$t_{cr} = \frac{1}{k_1(T_0) \cdot C_0} = 86.8 \ sec = 1.45 \ min. \tag{i}$$

Substituting (g) and (h) into (a) and (b), the two design equations become

$$Z_{1_{out}} - \frac{(y_{A_0} - Z_{1_{out}}) \cdot (y_{B_0} - Z_{1_{out}} - Z_{2_{out}})}{\left[(1 - Z_{1_{out}} - Z_{2_{out}}) \cdot \theta \right]^2} \cdot \tau \cdot e^{\gamma_1 \cdot \frac{\theta - 1}{\theta}} = 0 \tag{j}$$

$$Z_{2_{out}} - \left(\frac{k_2(T_0)}{k_1(T_0)} \right) \cdot \frac{(Z_{1_{out}} - Z_{2_{out}}) \cdot (y_{B_0} - Z_{1_{out}} - Z_{2_{out}})}{\left[(1 - Z_{1_{out}} - Z_{2_{out}}) \cdot \theta \right]^2} \cdot \tau \cdot e^{\gamma_2 \cdot \frac{\theta - 1}{\theta}} = 0. \tag{k}$$

Substituting $CF(Z_m, \theta)$ into (7.1-13), the energy balance equation reduces to

$$HTN \cdot \tau \cdot (\theta_F - \theta) - DHR_1 \cdot Z_1 - DHR_2 \cdot Z_2 - \frac{11.6 - 4 \cdot Z_1 - 2 \cdot Z_2}{11.6}(\theta - 1) = 0. \quad (1)$$

We have to solve (j), (k), and (l) simultaneously for different values of τ. We do so numerically, using a mathematical software.

a. For isothermal operation ($\theta = 1$), the two design equations, (j) and (k), reduce to

$$Z_{1_{out}} - \frac{(0.4 - Z_{1_{out}}) \cdot (0.4 - Z_{1_{out}} - Z_{2_{out}})}{(1 - Z_{1_{out}} - Z_{2_{out}})^2} \cdot \tau = 0 \quad (m)$$

$$Z_{2_{out}} - (0.5) \cdot \frac{(Z_{1_{out}} - Z_{2_{out}}) \cdot (0.4 - Z_{1_{out}} - Z_{2_{out}})}{(1 - Z_{1_{out}} - Z_{2_{out}})^2} \cdot \tau = 0. \quad (n)$$

We solve (m) and (n) numerically for different values of τ. The figures below show,

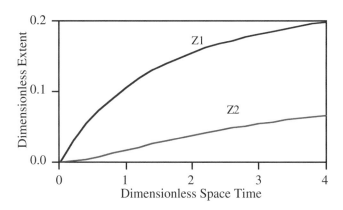

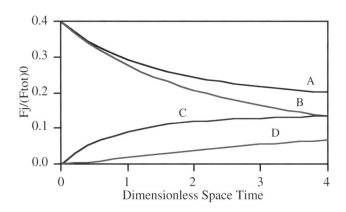

respectively, the operating curves of the two reactions and the species operating curves. For 60% conversion of B, $F_B/F_{B0} = 0.4$ or $F_B/(F_{tot})_0 = 0.16$. We use the calculated data to determine τ for which $F_B/(F_{tot})_0 = 0.16$. We find that $\tau = 3.1$, and, at that time, $Z_1 = 0.1847$ and $Z_2 = 0.0554$. Hence, using (i) and (7.1-3), the reactor volume is

$$V_R = v_0 \cdot t_{cr} \cdot \tau = (400 \text{ lit/min}) \cdot (1.45 \text{ min}) \cdot (3.1) = 1,798 \text{ liter.}$$

Using (c) through (f), the species molar flow rates at the reactor outlet are: $F_A = 4.96$ mole/min, $F_B = 3.64$ mole/min, $F_C = 2.98$ mole/min, and $F_D = 1.28$ mole/min.
b. For isothermal operation $\theta = 1$, the energy balance equation, (7.1-3), becomes

$$\left(\frac{\dot{Q}}{T_0 \cdot (F_{tot})_0 \cdot \hat{c}_{p0}} \right) = \left(\frac{\Delta H_{R_1}(T_0)}{T_0 \cdot \hat{c}_{p0}} \right) \cdot Z_{1_{out}} + \left(\frac{\Delta H_{R_2}(T_0)}{T_0 \cdot \hat{c}_{p0}} \right) \cdot Z_{2_{out}}, \qquad (o)$$

which reduces to

$$\dot{Q} = (F_{tot})_0 \cdot \left[\Delta H_{R_1}(T_0) \cdot Z_{1_{out}} + \Delta H_{R_2}(T_0) \cdot Z_{2_{out}} \right] =$$

$$= (23.05) \cdot \left[(-12,000) \cdot (0.1847) + (-9,000) \cdot (0.0554) \right] = -16.6 \text{ kcal/min.}$$

The negative sign indicates that heat is removed from the reactor.
c. For adiabatic operation, $(S/V) = 0$; hence, HTN = 0, and the energy balance equation, (l), reduces to

$$(-2.45) \cdot Z_{1_{out}} + (-1.834) \cdot Z_{2_{out}} + \frac{11.6 - 4 \cdot Z_{1_{out}} - 2 \cdot Z_{2_{out}}}{11.6} \cdot (\theta - 1) = 0. \qquad (p)$$

We solve (j), (k), and (p) numerically for different values of τ. The figures below show, respectively, the operating curves of the reactions, the temperature, and the species. We use the calculated data to determine τ for which $F_B/(F_{tot})_0 = 0.16$. We find that it is $t = 0.61$, and, at that time, $Z_1 = 0.1553$, $Z_2 = 0.0835$, and $\theta = 1.68$. Hence, using (i) and (7.1-3), the reactor volume is

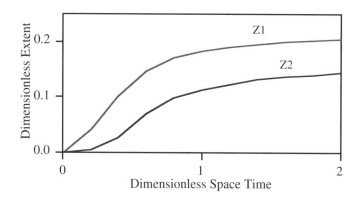

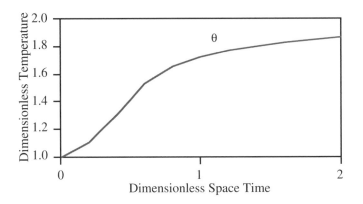

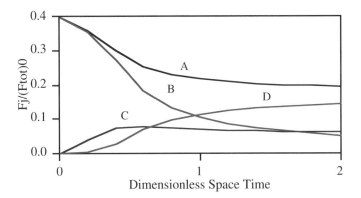

$$V_R = v_0 \cdot t_{cr} \cdot t = (400 \text{ lit/min}) \cdot (1.45 \text{ min}) \cdot (0.61) = 354 \text{ liter},$$

and the reactor temperature is $T = (1.68)(423) = 710.6°K$. Using (c) through (f), the species molar flow rates at the reactor outlet are: $F_A = 5.64$ mole/min, $F_B = 3.72$ mole/min, $F_C = 3.08$ mole/min, and $F_D = 1.92$ mole/min.

d. For operation with heat-transfer, the dimensionless heat-transfer number is

$$\text{HTN} = \frac{U \cdot t_{cr}}{C_0 \cdot \hat{c}_{p0}} \cdot \left(\frac{S}{V}\right) = 1.$$

Substituting into (l), the energy balance equation reduces to

$$(1) \cdot \tau \cdot (0.9527 - \theta) + (2.45) \cdot Z_{1_{out}} + (1.834) \cdot Z_{2_{out}} -$$

$$- \frac{11.6 - 4 \cdot Z_{1_{out}} - 2 \cdot Z_{2_{out}}}{11.6} \cdot (\theta - 1) = 0. \tag{q}$$

We solve (j), (k), and (q) numerically for different values of τ. The figures below show, respectively, the operating curves of the reaction, the temperature, and the species. We

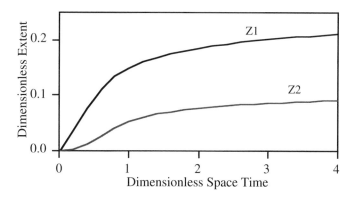

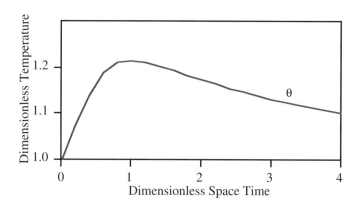

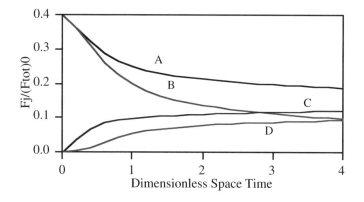

use the calculated data to determine τ for which $F_B/(F_{tot})_0 = 0.16$. We find that it is $\tau = 1.30$, and, at that time, $Z_1 = 0.1705$, $Z_2 = 0.0707$, and $\theta = 1.24$. Hence, the needed reactor volume is

$$V_R = v_0 \cdot t_{cr} \cdot \tau = (400 \text{ lit/min}) \cdot (1.45 \text{ min}) \cdot (1.3) = 754 \text{ liter.}$$

Using (c) through (f), the species molar flow rates at the reactor outlet are: $F_A = 5.29$ mole/min, $F_B = 3.66$ mole/min, $F_C = 2.30$ mole/min, and $F_D = 1.63$ mole/min. The reactor temperature is

$$T = (423°K)(1.24) = 525°K.$$

7.5 SUMMARY

In this chapter we discussed the operation of continuous stirred-tank reactors. We covered the following topics:

a. The underlying assumptions of the CSTR model and when they are satisfied in practice.
b. The dimensionless, species-based design equations, the energy balance equation, and the auxiliary relations for species concentrations.
c. The dimensionless reaction operating curves and species operating curves.
d. Design and operation of isothermal CSTRs with single reactions.
e. Operation and optimization of a cascade of CSTRs.
f. Experimental methods to determine parameters of the rate expression.
g. Design and operation of isothermal CSTRs with multiple reactions.
h. Design and operation of non-isothermal CSTRs with multiple reactions.

PROBLEMS

7.1₁ Gaseous reactant A decomposes according to the reaction $2A \rightarrow R$. The rate expression is $(-r_A) = 0.05 \, C_A^2$ mole/lit sec. A stream consisting of 50% A - 50% inert (by mole) is fed at a rate of 180 lit/min. into a 1 m³ CSTR. Plot the dimensionless reaction and species operating curves. If $C_{A0} = 300$ mmole/lit, what is the conversion of A? (Adapted from Levenspiel's Omnibook)

7.2₂ A gas stream at 2 atm and 677°K contains reactant A, and an inert at a proportion of 1/6 A and 5/6 inert is fed at a rate of 2 lit/min. A second gas stream of pure B at 1.95 atm and 330°K is fed at a rate of 0.5 lit/min. into a CSTR. The volume of the reactor is 0.75 lit, and it is kept at 440°K and 1.3 atm, where A and B react according to the reaction $A + 2B \rightarrow 4R$. The rate expression is $(-r_A) = k \, C_A \, C_B$. Plot the dimensionless reaction and species operating curves. The partial pressure of A in the reactor and in the exit stream is 0.029 atm. Determine
a. the fractional conversion of A and of B in the reactor and

b. the reaction rate constant (m³/mole min) of the reaction.
(Adapted from Levenspiel's Omnibook)

7.3₁ An aqueous stream of A and B (C_{A0} = 100 mmole/lit and C_{B0} = 200 mmole/lit) is fed at a rate of 400 lit/min. into a CSTR, where the following reaction takes place: A + B → R. The rate expression of the reaction is $(-r_A) = 200\, C_A\, C_B$ mole/lit min. Find the volume needed for 90% conversion of A.
(Adapted from Levenspiel's Omnibook)

7.4₂ In the presence of an enzyme of fixed concentration, reactant A in aqueous solution decomposes into product R at the following rates, (C_A alone affects the rate)

C_A, mole / lit	1	2	3	4	5	6	8	10
$(-r_A)$, mole / lit min.	1	2	3	4	4.7	4.9	5	5

We plan to carry out this reaction in a large-scale CSTR at the same fixed enzyme concentration. Plot the dimensionless reaction and species operating curves for C_{A0} = 10 mole/lit. Find the volumetric rate of the stream (C_{A0} = 10 mole/lit) we can feed to 250 lit CSTR if we want to obtain 80% conversion of A.
(Adapted from Levenspiel's Omnibook)

7.5₂ A combined aqueous feed stream of A and B is fed to a 1 liter CSTR, and the following data are obtained

Feed composition (mole/lit)	Flow rate (lit/min)	Output (mmole/lit)
$C_{A0} = C_{B0}$ = 100	v_0 = 1	C_{AF} = 50
$C_{A0} = C_{B0}$ = 200	v_0 = 9	C_{AF} = 150
C_{A0} = 200, C_{B0} = 100	v_0 = 3	C_{AF} = 150

If the stoichiometry of the reaction is A + B → R + S, find the rate expression.
(Adapted from Levenspiel's Omnibook)

7.6₂ Pure gaseous A at about 3 atm and 30°C (120 mmole/lit) is fed into a 1 liter CSTR at various flow rates. There A decomposes according to the reaction A → 3 R, and the exit concentration of A is measured for each flow rate. From the data below, find:
a. a rate expression of the reaction. Assume that reactant A alone affects the rate.
b. What volumetric inflow rate v_0 and outflow rate v_F of a C_{A0} = 320 mmole/lit pure A feed can be fed into a 560 liter CSTR to achieve a 50% converted of A?

Data:	v_0, lit/min.	0.06	0.48	1.5	8.1
	C_{Aout}, mmole/lit	30	60	80	105

(Adapted from Levenspiel's Omnibook)

7.7₂ A stream of a high molecular weight hydrocarbon A is fed continuously to a heated high temperature CSTR where it thermally cracks (homogeneous gas reaction) into lower molecular weight materials, collectively called R, by a stoichiometry approximated by A → 5 R. By changing the feed rate, different extents of cracking are obtained as follows:

F_{A0}, mmole/hr	300	1000	3000	5000

C_{Aout}, mmole/liter   16   30   50   60

The internal void volume of the reactor is $V_R = 0.1$ liter, and, at the temperature of the reactor, the feed concentration is $C_{A0} = 1000$ mmole/liter. Find a rate expression to fit the cracking reaction. (Adapted from Levenspiel's Omnibook)

7.8$_2$ A liquid stream is fed at different temperatures into a CSTR, where A decomposes according to the first-order reaction A $\rightarrow$ R + P. The feed flow rate is adjusted to keep the composition of the exit stream constant. Assuming constant feed density, find the activation energy of the reaction from the following results:

T, °C	19	27	31	37
v_0, arbitrary units	1	2	3	5

(Adapted from Levenspiel's Omnibook)

7.9$_2$ The gas-phase reaction A $\rightarrow$ B + C is carried out in a cascade of two CSTRs connected in series. Pure A is fed at a rate of 100 lbmole/hr into the first reactor, whose volume is 100 ft^3. The molar flow rate of A at the exit of the first reactor is 60 lbmole/hr, and that at the exit of the second reactor is 20 lbmole/hr. The temperature throughout the system is 150°C, and the pressure is 2 atm. The reaction is second-order. Determine the volume of Reactor 2 by:

a. taking the inlet stream into the system as the reference stream and

b   taking the inlet stream into Reactor 2 as the reference stream.

7.10$_2$ Water containing a short-lived radioactive species flows continuously through a well-mixed holdup tank. This gives time for the radioactive material to decay into harmless waste. As it now operates, the activity of the exit stream is 1/7 of the feed stream. This is not bad, but we'd like to lower it still more. One of our office secretaries suggests that we insert a baffle down the middle of the tank so that the holdup tank acts as two CSTRs in series. Do you think this would help?   If not, tell why; if so, calculate the expected activity of the exit stream compared to the entering stream. (Adapted from Levenspiel's Omnibook)

7.11$_2$ We wish to treat 10 lit/min. of a waste liquid stream containing A ($C_{A0} = 1$ mole/lit.)   A decomposes according to the reaction A $\rightarrow$ R + P, and its rate expression is

$$(r_A) = \frac{C_A}{0.2 + C_A} \text{ mole/lit min.}$$

We want to achieve 99% conversion of A. Suggest a good arrangement for doing so using two CSTRs, and find the size of the two units needed. Sketch the final design chosen. (Adapted from Levenspiel's Omnibook)

7.12$_2$ Originally, we had planned to lower the activity of a gas stream containing radioactive Xe-138 (half life - 14 min.) by passing it through two holdup tanks in series; both are well-mixed and the mean residence time of the gas is 2 weeks in each tank. It has been suggested that we replace the two tanks with a long tube (assume plug-flow). What must the size of this tube be compared to the two original stirred tanks, and what should be the mean residence time of gas in this tube for the same extent of radioactive decay? (Adapted from Levenspiel's Omnibook)

7-13$_4$ The first-order gas-phase reaction $A \rightarrow B + C$ takes place in a CSTR. A stream consisting of species 90% A and 10% I (% mole) is fed into a 200 liter reactor at a rate of 20 lit/sec. The feed is at 731°K and 3 atm. Based on the data below, calculate:
a. the conversion of A when the reactor is operated isothermally,
b. the heating/cooling load in (a),
c. the conversion of A when the reactor is operated adiabatically, and
d. the reactor temperature in (c).
Data: At 731°K $k = 0.2$ sec^{-1}; $E_a = 12,000$ cal/mole;
    At 731°K $DH_R = -10,000$ cal/mole extent;
    $\hat{c}_{PA} = 25$ cal/mole°K; $\hat{c}_{PB} = 15$ cal/mole°K; $\hat{c}_{PC} = 18$ cal/mole°K;
    $\hat{c}_{PI} = 9$ cal/mole°K.

7-14$_4$ The elementary liquid-phase reactions

$$A + B \rightarrow C$$

$$C + B \rightarrow D$$

take place in two 100 liter CSTRs connected in series. A solution ($C_{A0} = 2$ mole/lit and $C_{B0} = 2$ mole/lit) at 80°C is fed into the first reactor at a rate of 200 lit/min. Based on the data below, calculate:
a. the conversion of A and the production of C at the exit of the second reactor when both reactors are operated isothermally,
b. the heating/cooling load on each reactor in (a),
c. the conversion of A and the production of C at the exit of the second reactor when both reactors are operated adiabatically, and
d. the reactor temperature of each reactor in (c).
Data: At 80°C $k_1 = 0.1$ liter mole^{-1}min^{-1}; $E_{a1} = 12,000$ cal/mole;
    At 80°C $k_2 = 0.2$ liter mole^{-1}min^{-1}; $E_{a1} = 16,000$ cal/mole;
    At 80°C $\Delta H_{R1} = -15,000$ cal/mole extent; $\Delta H_{R2} = -10,000$ cal/mole extent;
    Density of the solution = 900 g/liter. Heat Capacity = 0.8 cal/g°C.

7-15$_4$ Methane is chlorinated in a gas phase at 400°C to produce mono-, di-, tri-, and tetrachloro methane. The desired products are CH_2Cl_2 and CCl_4. The following reactions take place:

$$CH_4 + Cl_2 \rightarrow CH_3Cl + HCl$$

$$CH_3Cl + Cl_2 \rightarrow CH_2Cl_2 + HCl$$

$$CH_2Cl_2 + Cl_2 \rightarrow CHCl_3 + HCl$$

$$CHCl_3 + Cl_2 \rightarrow CCl_4 + HCl.$$

The reactions are elementary and the rate constants are
$$k_1 = 30 \text{ lit/(mole min)}; \quad k_2/k_1 = 3; \quad k_3/k_1 = 1.5; \quad k_4/k_1 = 0.375.$$
A gas stream containing CH_4 and Cl_2 in the proportion of 1:1.2 is fed at 1.2 atm into a CSTR operated at 400°C. The feed rate is 1 mole/min.
a. Plot the dimensionless reaction and species operating curves.
b. What is the reactor volume needed to maximize the production of CH_2Cl_2?
c. What are the conversions of CH_4 and Cl_2?
d. What are the production rates of CH_3Cl, CH_2Cl_2, $CHCl_3$, and CCl_4 in (c)?

7-16$_4$ Methanol is produced by the gas-phase reaction $CO + 2H_2 \rightarrow CH_3OH$. However, at the reactor operating conditions, the undesirable reaction $CO + 3H_2 \rightarrow CH_4 + H_2O$ is also taking place. Assuming both reactions are second-order (each is first-order with respect to each reactant) and $k_2/k_1 = 1.2$. A synthesis gas stream is fed into a CSTR operated at 450°C and 5 atm. Plot the operating curves for:
a. a feed consisting of 1/3 CO and 2/3 H_2 (mole basis), and
b. a feed consisting of 50% CO and 50% H_2 (mole basis).

7-17$_4$ Below is a simplified kinetic model of the cracking of propane to produce ethylene.

$C_3H_8 \rightarrow C_2H_4 + CH_4$	$r_1 = k_1 \, C_{C3H8}$
$C_3H_8 \leftrightarrow C_3H_6 + H_2$	$r_2 = k_2 \, (C_{C3H8} - (1/K_2) \, C_{C3H6} \, C_{H2})$
$C_3H_8 + C_2H_4 \rightarrow C_2H_6 + C_3H_6$	$r_3 = k_3 \, C_{C3H8} \, C_{C2H4}$
$2 \, C_3H_6 \rightarrow 3 \, C_2H_4$	$r_4 = k_4 \, C_{C3H6}$
$C_3H_6 \leftrightarrow C_2H_2 + CH_4$	$r_5 = k_5 \, (C_{C3H8} - (1/K_5) \, C_{C2H2} \, C_{CH4})$
$C_2H_4 + C_2H_2 \rightarrow C_4H_6$	$r_6 = k_6 \, C_{C2H4} \, C_{C2H2}$

At 800°C, the values of the rate constants are $k_1 = 2.341$ sec^{-1}, $k_2 = 2.12$ sec^{-1}, $K_2 = 1000$, $k_3 = 23.63$ m^3/kmole sec, $k_4 = 0.816$ sec^{-1}, $k_5 = 0.305$ sec^{-1}, $K_5 = 2000$, and $k_6 = 4.06 \; 10^3$ m^3/kmole sec. You are asked to design a CSTR for the cracking of propane to be operated at 2 atm. Plot the performance curves of the independent reactions and the curve for the generation of ethylene at 800°C.

CHAPTER EIGHT

OTHER REACTOR CONFIGURATIONS

In this chapter, we expand the analysis of chemical reactors with multiple reactions to additional reactor configurations. In Section 8.1, we analyze semi-batch reactors; in Section 8.2, we examine the operation of one-stage distillation reactors. Section 8.3 covers the operation of plug-flow reactors with continuous injection along their length, and Section 8.4 covers the operation of recycle reactors. In each case, we first derive the reaction-based design equations, expressed in terms of the extents of independent reactions. Next, we convert the equations to dimensionless forms, and then we derive auxiliary equations to express the species concentrations in terms of the extents of the independent reactions.

8.1 SEMI-BATCH REACTORS

A semi-batch reactor is a batch reactor into which one or more reactants are added continuously, and no material is withdrawn during the operation. For convenience, we divide semi-batch reactors into two categories: reactors with liquid-phase reactions where the volume of the reacting fluid changes, shown schematically in the Figure 8-1a, and reactors with gas-phase reactions where the volume does not change, shown schematically in the Figure 8-1b. Semi-batch reactors are used when, in order to maintain high yield of the desirable product, it is desirable to maintain low concentration of the injected reactant.

To derive the design equation of a semi-batch reactor, we write a species balance equation for any species, say species j, that is **not** fed continuously into the reactor. Since the species is not fed or withdrawn, its molar balance equation is

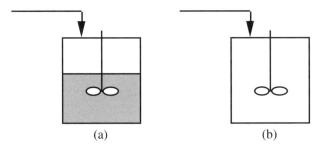

Figure 8-1: Semi-Batch Reactors: (a) Liquid-Phase, (b) Gas-Phase

$$\frac{dN_j}{dt} = (r_j) \cdot V_R(t), \tag{8.1-1}$$

where $V_R(t)$ is the volume of the reactor at time t. We follow the same procedure used in Chapter 3 to derive the reaction-based design equation of an ideal batch reactor and obtain

$$\frac{dX_m}{dt} = \left(r_m + \sum_{k}^{n_{dep}} \alpha_{km} \cdot r_k \right) \cdot V_R(t), \tag{8.1-2}$$

where $X_m(t)$ is the extent of the m-th independent reaction in operating time t. Eq. (8.1-2) is the differential reaction-based design equation for a semi-batch reactor, written for the m-th independent chemical reaction. Note that (8.1-2) is identical to the design equation of an ideal batch reactor. The only differences between batch and semi-batch operations are the way the reactor volume, $V_R(t)$, varies during the operation and the relations of the species concentrations. To express the latter, we have to consider the way in which the species are injected into the reactor.

We consider first semi-batch reactors with **liquid-phase** reactions. To derive a relation for the change in the volume of the reactor, we write an overall material balance over the reactor,

$$\rho_0(t) \cdot v_0(t) = \frac{d(\rho \cdot V_R)}{dt}, \tag{8.1-3}$$

where $v_0(t)$ is the volumetric feed rate of the injected stream, and $\rho_0(t)$ and ρ are the densities of the inlet stream and the reactor, respectively. For liquid-phase reactions, the density of the feed is constant and is the same as that of liquid in the reactor; hence, $\rho(t) = \rho_0(t) = \rho_0$, and (8.1-3) reduces to

$$\frac{dV_R}{dt} = v_0(t). \tag{8.1-4}$$

Since the injection rate, $v_0(t)$, is a function of time, we have to solve (8.1-4) simultaneously with the design equation (8.1-2). Alternatively, we can separate the variables and integrate (8.1-4)

$$V_R(t) = V_R(0) + \int_0^t v_0(x) \cdot dx, \qquad (8.1\text{-}5)$$

and then substitute (8.1-5) into (8.1-2). For simplicity, we consider here the case where the injection rate is uniform. For such operations, $v_0(t) = v_0$, and (8.1-5) reduces to

$$V_R(t) = V_R(0) + v_0 \cdot t, \qquad (8.1\text{-}6)$$

where $V_R(0)$ is the initial volume of the reactor. Substituting (8.1-6) into (8.1-2), the design equation of semi-batch reactors with a uniform injection rate is

$$\frac{dX_m}{dt} = \left(r_m + \sum_k^{n_{dep}} \alpha_{km} \cdot r_k \right) \cdot \left(V_R(0) + v_0 \cdot t \right). \qquad (8.1\text{-}7)$$

To reduce reaction-based design equation (8.1-7) to a dimensionless form, we have to define a dimensionless extent of the m-th independent. For semi-batch reactors, we should select a reference state that applies to all operations, including those with initially empty reactors. Also, we usually compare the operation of a semi-batch reactor to that of a batch reactor; therefore, we would like to formulate their design on the same basis. To do so, we select the total amount of reactants as a reference state and define the dimensionless extent on the basis of a fictitious reference state by

$$Z_m(t) = \frac{\text{extent of the m - th independent reaction in operating time t}}{\text{total number of moles added to the reactor during the operation}}. \qquad (8.1\text{-}8)$$

The total number of moles added to a semi-batch reactor during the entire operation is

$$(N_{tot})_0 = N_{tot}(0) + \int_0^{t_{op}} v_0(x) \cdot (C_0)_{inj} \cdot dx, \qquad (8.1\text{-}9)$$

where $N_{tot}(0)$ is the total number of moles initially charged into the reactor, $(C_0)_{inj}$ is the total concentration of the injected stream, and t_{op} is the total operating time. For semi-batch reactors with a uniform injection rate,

$$(N_{tot})_0 = N_{tot}(0) + v_0 \cdot (C_0)_{inj} \cdot t_{op} = N_{tot}(0) + (F_{tot})_{inj} \cdot t_{op}, \qquad (8.1\text{-}10)$$

and the dimensionless extent is

$$Z_m(t) = \frac{X_m(t)}{(N_{tot})_0} = \frac{X_m(t)}{N_{tot}(0) + (F_{tot})_{inj} \cdot t_{op}}. \qquad (8.1\text{-}11)$$

We also define the reference total molar concentration, C_0, on the basis of the reactor volume when all the reactants are injected. Hence, for semi-batch reactors with a uniform injection rate,

$$C_0 = \frac{N_{tot}(0) + (F_{tot})_{inj} \cdot t_{op}}{V_R(0) + v_0 \cdot t_{op}}. \tag{8.1-12}$$

As we did with batch reactors, we define the dimensionless operating time by

$$\tau \equiv \frac{t}{t_{cr}}, \tag{8.1-13}$$

where t_{cr} is the characteristic reaction time, defined by (2.5-1). Differentiating (8.1-11) and (8.1-13),

$$dX_m = \left(N_{tot}(0) + (F_{tot})_{inj} \cdot t_{op}\right) \cdot dZ_m \tag{8.1-14}$$

$$dt = t_{cr} \cdot d\tau. \tag{8.1-15}$$

Substituting these into (8.1-7) and using (8.1-12), the reaction-based design equation reduces to

$$\frac{dZ_m}{d\tau} = \left(r_m + \sum_{k}^{n_{dep}} \alpha_{km} \cdot r_k\right) \cdot \left(\frac{\dfrac{V_R(0)}{v_0 \cdot t_{op}} + \dfrac{\tau}{\tau_{op}}}{\dfrac{V_R(0)}{v_0 \cdot t_{op}} + 1}\right) \cdot \left(\frac{t_{cr}}{C_0}\right), \tag{8.1-16}$$

where t_{op} is the dimensionless operating time,

$$\tau_{op} \equiv \frac{t_{op}}{t_{cr}}. \tag{}$$

Note that in (8.1-16), by definition, $0 < \tau < \tau_{op}$. Eq. (8.1-16) is the dimensionless, reaction-based design equation of liquid-phase semi-batch reactors with uniform injection rate, written for the m-th independent reaction. To describe the reactor operation, we have to write a design equation for each independent chemical reaction.

For constant volume, semi-batch reactors with **gas-phase** reactions, $V_R(t) = V_R(0)$, and the reaction-based design equation, (8.1-2), reduces to

$$\frac{dX_m}{dt} = \left(r_m + \sum_{k}^{n_{dep}} \alpha_{km} \cdot r_k\right) \cdot V_R(0). \tag{8.1-17}$$

For uniform injection rate, using (8.1-12), the reference total molar concentration is

$$C_0 = \frac{N_{tot}(0) + (F_{tot})_{inj} \cdot t_{op}}{V_R(0)}. \tag{8.1-18}$$

Substituting (8.1-14) and (8.1-15) into (8.1-17) and using (8.1-18), we obtain

$$\frac{dZ_m}{d\tau} = \left(r_m + \sum_{k}^{n_{dep}} \alpha_{km} \cdot r_k \right) \cdot \left(\frac{t_{cr}}{C_0} \right).$$

(8.1-19)

Eq. (8.1-19) is the dimensionless, reaction-based design equation of gas-phase, semi-batch reactors with constant volume and uniform injection rate, written for the m-th independent reaction.

To solve the design equations, we should express the rates of all the chemical reactions (r_m's and r_k's) in terms of the extents of the independent reactions. To obtain the concentrations of the individual species, we use stoichiometric relations (1.4-2) and (8.1-9) to write the molar content of species j in the reactor at operating time t (for uniform injection rate),

$$N_j(t) = N_j(0) + v_0 \cdot (C_j)_{inj} \cdot t + \sum_{m}^{n_{ind}} (s_j)_m \cdot X_m(t),$$

(8.1-20)

where $N_j(0)$ is the number of moles of species j charged into the reactor initially, and

$$(y_j)_{inj} = \frac{(F_j)_{inj}}{(F_{tot})_{inj}} = \frac{(C_j)_{inj}}{(C_0)_{inj}}$$

is the mole fraction of the species j in the injection stream. For a species that is not charged initially into the reactor, $N_j(0) = 0$, and for a species that is not fed continuously, $(y_j)_{inj} = 0$. Using (8.1-11) and (8.1-13), (8.1-20) reduces to

$$N_j(\tau) =$$

$$= \left(N_{tot}(0) + (F_{tot})_{inj} \cdot t_{op} \right) \left(\frac{N_j(0) + (F_{tot})_{inj} \cdot t_{op} \cdot (y_j)_{inj} \cdot \left(\dfrac{\tau}{\tau_{op}} \right)}{N_{tot}(0) + (F_{tot})_{inj} \cdot t_{op}} + \sum_{m}^{n_{ind}} (s_j)_m \cdot Z_m(\tau) \right).$$

(8.1-21)

For **liquid-phase** semi-batch reactors with uniform injection rate, using (8.1-21), (8.1-6), and (8.1-12), the concentration of species j at dimensionless operating time τ is

$$C_j(\tau) =$$

$$= C_0 \left(\frac{\dfrac{V_R(0)}{v_0 \cdot t_{op}} + 1}{\dfrac{V_R(0)}{v_0 \cdot t_{op}} + \dfrac{\tau}{\tau_{op}}} \right) \left(\frac{\dfrac{N_j(0)}{(F_{tot})_{inj} \cdot t_{op}} + (y_j)_{inj} \cdot \left(\dfrac{\tau}{\tau_{op}} \right)}{\dfrac{N_{tot}(0)}{(F_{tot})_{inj} \cdot t_{op}} + 1} + \sum_{m}^{n_{ind}} (s_j)_m \cdot Z_m(\tau) \right). \quad (8.1\text{-}22)$$

For constant-volume, **gas-phase** semi-batch reactors with uniform injection rate, using (8.1-21) and (8.1-18), the concentration of species j at dimensionless operating time τ is

$$C_j(\tau) = C_0 \left(\frac{\dfrac{N_j(0)}{(F_{tot})_{inj} \cdot t_{op}} + (y_j)_{inj} \cdot \left(\dfrac{\tau}{\tau_{op}} \right)}{\dfrac{N_{tot}(0)}{(F_{tot})_{inj} \cdot t_{op}} + 1} + \sum_{m}^{n_{ind}} (s_j)_m \cdot Z_m(\tau) \right). \qquad (8.1\text{-}23)$$

For non-isothermal operations, we have to incorporate the energy balance equation to express variation in the reactor temperature. For semi-batch reactors, assuming no expansion work, we modify energy balance equation (4.2-8) and obtain

$$\Delta H(t) = Q(t) + \int_0^t \dot{m}_{in} \cdot h_{in} dt - W_{sh}(t), \qquad (8.1\text{-}24)$$

where $\dot{m}_{in} h_{in}$ is the rate enthalpy is added to the reactor by the injection stream. Differentiating (8.1-17), we obtain

$$\frac{dH}{dt} = \dot{Q} + \dot{m}_{in} \cdot h_{in} - \dot{W}_{sh}, \qquad (8.1\text{-}25)$$

where $\dot{W}_{sh}$ is the rate of mechanical (shaft) work done by the reactor on the surrounding, and $\dot{Q}$ is the rate heat is added to the reactor, expressed by (see Chapter 4)

$$\dot{Q} = U \cdot \left(\frac{S}{V} \right) \cdot V_R(t) \cdot (T_F - T). \qquad (8.1\text{-}26)$$

Assuming no phase change,

$$\frac{dH}{dt} = \sum_{m}^{n_{ind}} \Delta H_{R_m}(T_0) \cdot \frac{dX_m}{dt} + (\Sigma N_j \cdot \hat{c}_{p_j})_t \frac{dT}{dt}, \qquad (8.1\text{-}27)$$

where $(\Sigma N_j \cdot \hat{c}_{p_j})_t$ indicates the heat capacity of the reacting fluid at time t. Combining (8.1-25), (8.1-26), and (8.1-27), for reactors with negligible shaft work,

$$\frac{dT}{dt} = \frac{1}{(\Sigma N_j \cdot \hat{c}_{p_j})_t} \left[U \cdot \left(\frac{S}{V} \right) \cdot V_R(t) \cdot (T_F - T) + \dot{m}_{in} \cdot h_{in} - \sum_{m}^{n_{ind}} \Delta H_{R_m}(T_0) \cdot \frac{dX_m}{dt} \right].$$
$$(8.1\text{-}28)$$

Eq. (8.1-28) describes the change in the reactor temperature as a function of time. We relate the heat capacity to that of the reference state by a "correction factor,"

$$(\Sigma N_j \cdot \hat{c}_{p_j})_t = CF(Z, \theta)_t \cdot (N_{tot})_0 \cdot \hat{c}_{p0}. \qquad (8.1\text{-}29)$$

To reduce (8.1-28) to dimensionless form, we select a reference state and define dimensionless temperature by

$$\theta = \frac{T}{T_0}. \qquad (8.1\text{-}30)$$

In many cases, we select the reference state at the conditions of the injection stream; then, $h_{in} = 0$.

For **liquid-phase** semi-batch reactors, the enthalpy of the injected stream is

$$\dot{m}_{in} \cdot h_{in} = v_0 \cdot \rho \cdot \bar{c}_p \cdot (T_{inj} - T_0), \qquad (8.1\text{-}31)$$

where $\bar{c}_p$ is the mass-based heat capacity of the liquid. For uniform injection rate, the reference thermal energy is

$$(N_{tot})_0 \cdot \hat{c}_{p0} \cdot T_0 = \left(N_{tot}(0) + (F_{tot})_{inj} \cdot t_{op}\right) \cdot \hat{c}_{p0} \cdot T_0,$$

where $\hat{c}_{p0}$ is the specific molar heat capacity of the reference state,

$$\hat{c}_{p0} = \frac{M_{tot} \cdot \bar{c}_p}{(N_{tot})_0} = \frac{M(t_{op}) \cdot \bar{c}_p}{N_{tot}(0) + (F_{tot})_{inj} \cdot t_{op}}, \qquad (8.1\text{-}32)$$

where M_{tot} is the total mass of the reacting fluid that is being processed, and $\bar{c}_p$ is the mass-based heat capacity. Multiplying both sides of (8.1-28) by the reference thermal energy and using (8.1-13), (8.1-11), and (8.1-6), we obtain

$$\frac{d\theta}{d\tau} = \frac{1}{CF(Z_m, \theta)} HTN \left(\frac{\dfrac{V_R(0)}{v_0 \cdot t_{op}} + \dfrac{\tau}{\tau_{op}}}{\dfrac{V_R(0)}{v_0 \cdot t_{op}} + 1} \right) (\theta_F - \theta) +$$

$$+ \frac{1}{CF(Z_m, \theta)} \left[\frac{1}{\tau_{op}} \left(\frac{\dfrac{\rho \cdot \bar{c}_p}{(C_0)_{inj} \cdot \hat{c}_{p0}}}{\dfrac{N_{tot}(0)}{(F_{tot})_{inj} \cdot t_{op}} + 1} \right) (\theta_{inj} - 1) - \sum_m^{n_{ind}} DHR_m \cdot \frac{dZ_m}{d\tau} \right], \qquad (8.1\text{-}33)$$

where HTN is the heat-transfer number, defined by (4.2-23)

$$HTN = \frac{U \cdot t_{cr}}{C_0 \cdot \hat{c}_{p0}} \left(\frac{S}{V} \right).$$

DHR_m is the dimensionless heat of reaction of the m-th independent reaction, defined by (4.2-24),

$$DHR_m = \frac{\Delta H_{R_m}(T_0)}{T_0 \cdot \hat{c}_{p0}},$$

and $CF(Z_m, \theta)$ is the correction factor of the heat capacity. For liquid-phase reactions, assuming constant density and constant specific heat capacity,

$$CF(Z_m, \theta) = \frac{M(\tau) \cdot \overline{c}_p}{M_{tot} \cdot \overline{c}_p} = \frac{\dfrac{V_R(0)}{v_0 \cdot t_{op}} + \dfrac{\tau}{\tau_{op}}}{\dfrac{V_R(0)}{v_0 \cdot t_{op}} + 1}. \tag{8.1-34}$$

For **gas-phase** reactions with uniform injection rate,

$$\dot{m}_{in} \cdot h_{in} = (F_{tot})_{inj} \cdot \hat{c}_{p_{inj}} \cdot (T_{inj} - T_0), \tag{8.1-35}$$

where $\hat{c}_{p_{inj}}$ is the specific molar heat capacity of the injection stream. Multiplying both sides of (8.1-28) by the reference thermal energy and using (8.1-13), (8.1-11), and (8.1-6), for **constant volume** semi-batch reactors,

$$\frac{d\theta}{d\tau} = \frac{1}{CF(Z_m, \theta)} HTN \cdot (\theta_F - \theta) +$$

$$+ \frac{1}{CF(Z_m, \theta)} \frac{1}{\tau_{op}} \left[\left(\frac{\dfrac{\hat{c}_{p_{inj}}}{\hat{c}_{p0}}}{\dfrac{N_{tot}(0)}{(F_{tot})_{inj} \cdot t_{op}} + 1} \right) (\theta_{inj} - 1) - \sum_m^{n_{ind}} DHR_m \cdot \frac{dZ_m}{d\tau} \right], \tag{8.1-36}$$

where the correction factor of the heat capacity is now

$$CF(Z_m, \theta) = \frac{1}{\hat{c}_{p0}} \sum_j^{all} \left(\frac{\dfrac{N_j(0)}{(F_{tot})_{inj} \cdot t_{op}} + (y_j)_{inj} \cdot \left(\dfrac{\tau}{\tau_{op}} \right)}{\dfrac{N_{tot}(0)}{(F_{tot})_{inj} \cdot t_{op}} + 1} + \sum_m^{n_{ind}} (s_j)_m \cdot Z_m(\tau) \right) \cdot \hat{c}_{p_j}(\theta), \tag{8.1-37}$$

and the specific molar heat capacity of the reference state, $\hat{c}_{p0}$, is

$$\hat{c}_{p0} = \frac{\sum\limits_{j}^{all}\left(\dfrac{N_j(0)}{(F_{tot})_{inj} \cdot t_{op}} + (y_j)_{inj}\right) \cdot \hat{c}_{pj}(1)}{\dfrac{N_{tot}(0)}{(F_{tot})_{inj} \cdot t_{op}} + 1},$$

(8.1-38)

where $\hat{c}_{pj}(1)$ is the molar heat capacity of species j at the reference temperature. Note that since the reference state is the total charge to the reactor, for both gas-phase and liquid-phase reactions, the correction factor is a function of time.

We illustrate the application of these relations in the examples below.

Example 8-1 A valuable reagent V is produced by the first-order chemical reaction A $\rightarrow$ V. However, at the operating conditions, reactant A also reacts according to the second-order chemical reaction 2 A $\rightarrow$ W. To improve the yield of the desirable product, it was suggested to carry out the operation in a semi-batch reactor. An aqueous solution of A with a concentration of 4 mole/liter at 60°C is to be processed in a 200 liter reactor.

a. Derive the design equations, and plot the reaction and species operating curves for isothermal semi-batch operation.
b. Derive the design equations, and plot the reaction and species operating curves for adiabatic semi-batch operation.
c. Compare the semi-batch operations in (a) and (b) to the corresponding batch operations.
d. Determine the operating time needed to achieve 50% conversion of A in each of the cases above.

Data: $k_1 = 0.2$ min^{-1}, $k_2 = 0.4$ liter mole^{-1} min^{-1}.
   $\Delta H_{R1} = -8{,}000$ cal mole^{-1}; $\Delta H_{R2} = -12{,}000$ cal mole^{-1};
   $E_{a1} = 9{,}000$ cal mole^{-1} °K^{-1}; $E_{a2} = 18{,}000$ cal mole^{-1} °K^{-1}.
   Density and heat capacity are the same as water (1,000 g/lit and 1 cal/g °K)

Solution The following chemical reactions take place in the reactor:

   Reaction 1:   A $\rightarrow$ V

   Reaction 2:   2 A $\rightarrow$ W,

and the stoichiometric coefficients are:

$$s_{A1} = -1; \quad s_{V1} = 1; \quad s_{W1} = 0; \quad \Delta_1 = 0;$$

$$s_{A2} = -2; \quad s_{V2} = 0; \quad s_{W2} = 1; \quad \Delta_2 = -1.$$

Since each reaction has a species that does not appear in the other, the two reactions are independent, and there is no dependent reaction. We select the total amount introduced into the reactor during the operation as the reference state. Since the reactor is empty initially, $V_R(0) = 0$, $N_{tot}(0) = N_A(0) = N_V(0) = 0$, and $N_W(0) = 0$. For constant injection rate, $v_0\, t_{op} = 200$ liter, and, since only reactant A is injected, $(C_0)_{inj} = C_{A0} = 4$ mole/liter,

$(F_{tot})_{inj} \, t_{op} = 800$ mole, $(y_A)_{inj} = 1$, $(y_V)_{inj} = 0$, and $(y_W)_{inj} = 0$. Using (8.1-12), the reference concentration is

$$C_0 = \frac{N_{tot}(0) + (F_{tot})_{inj} \cdot t_{op}}{V_R(0) + v_0 \cdot t_{op}} = 4 \text{ mole/liter}.$$

We write the design equations for liquid-phase semi-batch reactors with uniform injection rate, (8.1-16), for each independent reaction,

$$\frac{dZ_1}{d\tau} = r_1 \cdot \left(\frac{\tau}{\tau_{op}}\right) \cdot \left(\frac{t_{cr}}{C_0}\right) \tag{a}$$

$$\frac{dZ_2}{d\tau} = r_2 \cdot \left(\frac{\tau}{\tau_{op}}\right) \cdot \left(\frac{t_{cr}}{C_0}\right). \tag{b}$$

Using (8.1-22), the concentration of reactant A in the reactor is

$$C_A(\tau) = C_0 \left(\frac{\tau_{op}}{\tau}\right) \cdot \left(\frac{\tau}{\tau_{op}} + (-1) \cdot Z_1(\tau) + (-2) \cdot Z_2(\tau)\right). \tag{c}$$

The rates of the two reactions are:

$$r_1 = k_1(T_0) \cdot C_0 \left(\frac{\tau_{op}}{\tau}\right) \cdot \left(\frac{\tau}{\tau_{op}} - Z_1 - 2 \cdot Z_2\right) \cdot e^{\gamma_1 \frac{\theta-1}{\theta}} \tag{d}$$

$$r_2 = k_2(T_0) \cdot C_0{}^2 \left(\frac{\tau_{op}}{\tau}\right)^2 \left(\frac{\tau}{\tau_{op}} - Z_1 - 2 \cdot Z_2\right)^2 \cdot e^{\gamma_2 \frac{\theta-1}{\theta}}. \tag{e}$$

We select Reaction 1 as the leading reaction; hence, the characteristic reaction time is

$$t_{cr} = \frac{1}{k_1 \cdot (T_0)} = 5 \text{ min}. \tag{f}$$

Substituting (d), (e), and (f) into (a) and (b), the design equations reduce to

$$\frac{dZ_1}{d\tau} = \left(\frac{\tau}{\tau_{op}} - Z_1 - 2 \cdot Z_2\right) \cdot e^{\gamma_1 \frac{\theta-1}{\theta}} \tag{g}$$

$$\frac{dZ_2}{d\tau} = \left(\frac{k_2(T_0) \cdot C_0}{k_1(T_0)}\right) \cdot \left(\frac{\tau_{op}}{\tau}\right) \cdot \left(\frac{\tau}{\tau_{op}} - Z_1 - 2 \cdot Z_2\right)^2 \cdot e^{\gamma_2 \frac{\theta-1}{\theta}}. \tag{h}$$

a. For isothermal operation, $\theta = 1$, and the design equations are

$$\frac{dZ_1}{d\tau} = \frac{\tau}{\tau_{op}} - Z_1 - 2 \cdot Z_2 \tag{i}$$

$$\frac{dZ_2}{d\tau} = \left(\frac{k_2(T_0) \cdot C_0}{k_1(T_0)} \right) \cdot \left(\frac{\tau_{op}}{\tau} \right) \cdot \left(\frac{\tau}{\tau_{op}} - Z_1 - 2 \cdot Z_2 \right)^2. \tag{j}$$

We solve (j) and (k) subject to the initial conditions that $Z_1(0) = Z_2(0) = 0$ for different values of operating times, τ_{op}. Then, we determine the final extents at each τ_{op}. The first plot shows how the dimensionless reaction extents vary during the operation for $\tau_{op} = 4$. Note that this is not the operating curve but rather shows the extents' variation during a given operation. The second figure shows the reaction operating curves (Z_m versus τ_{op}) for semi-batch operations. Using (8.1-20), we determine the dimensionless species operating curves, $N_j(\tau_{op})/(N_{tot})_0$ versus τ_{op}. These curves are shown in the

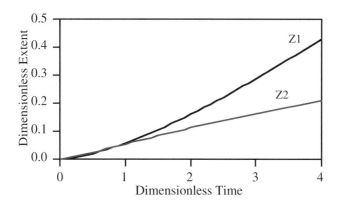

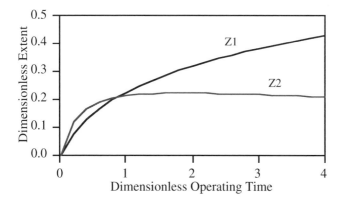

figure below. The species operating curves of the two products are compared to those of an ideal batch reactor (see design equations below). The first figure shows the curves of the desired product, V, and the second shows the curves of the undesirable

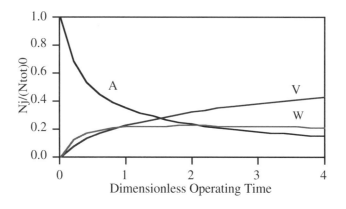

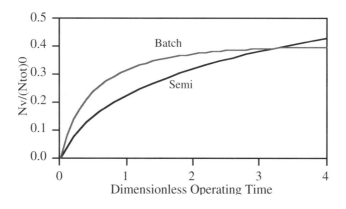

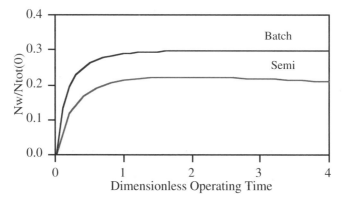

product, W. An examination of the two figures indicates that the production of product V is about the same for both semi-batch and batch operations, but the production of W is much lower in semi-batch operation.

b. For adiabatic semi-batch operation, we have to solve design equations (h) and (i) simultaneously with the energy balance equation. Using (8.1-32), the specific molar heat capacity of the reference state is

$$\hat{c}_{p0} = \frac{(200 \text{ lit}) \cdot (1,000 \text{ g/lit}) \cdot (1 \text{ cal/mole}^\circ K)}{(800 \text{ mole})} = 250 \text{ cal/mole}^\circ K.$$

The reference temperature is $T_0 = 333^\circ K$, and the dimensionless heat of reactions are

$$DHR_1 = \frac{\Delta H_{R_1}(T_0)}{T_0 \cdot \hat{c}_{p0}} = -0.0961 \qquad DHR_2 = \frac{\Delta H_{R_2}(T_0)}{T_0 \cdot \hat{c}_{p0}} = -0.1441.$$

The dimensionless activation energies of the two reactions are

$$\gamma_1 = \frac{E_{a1}}{R \cdot T_0} = 13.6 \qquad \gamma_2 = \frac{E_{a2}}{R \cdot T_0} = 27.2.$$

For adiabatic operation, $(S/V) = 0$, and, therefore, $HTN = 0$. Using (8.1-34), the heat capacity correction factor is

$$CF(Z_m, \theta) = \frac{\tau}{\tau_{op}}. \tag{r}$$

Since $T_{inj} = T_0$, energy balance equation (8.1-33) reduces to

$$\frac{d\theta}{d\tau} = \left(\frac{\tau_{op}}{\tau}\right) \cdot \left[(0.0961) \cdot \frac{dZ_1}{d\tau} + (0.1441) \cdot \frac{dZ_2}{d\tau}\right]. \tag{s}$$

We solve (s) simultaneously with (g) and (h) subject to the initial conditions that $Z_1(0) = Z_2(0) = 0$, $\theta(0) = 1$ for different values of operating times, τ_{op}, and then determine the final extents and θ at each τ_{op}. The plot below shows the reaction operating curves (Z_m versus τ_{op}) for semi-batch operation. Using (8.1-20), we determine the species operating curves, $N_j/(N_{tot})_0$ versus τ_{op}. They are compared to the product operating curves of a batch reactor in figures below. Note that since the activation energy of the second reaction (W production) is higher than that of the first, and both reactions are exothermic, isothermal operation is preferable.

c. For batch operation, we write the design equations for constant-volume batch reactors for each independent reaction,

$$\frac{dZ_1}{d\tau} = r_1 \cdot \left(\frac{t_{cr}}{C_0}\right) \tag{t}$$

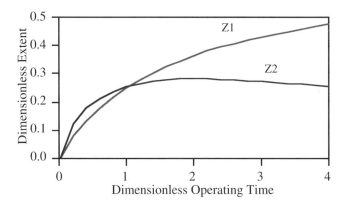

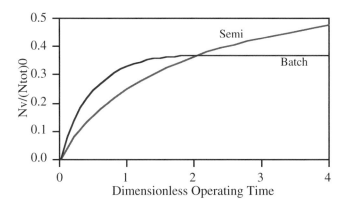

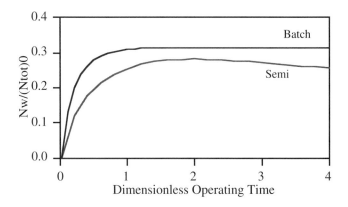

$$\frac{dZ_2}{d\tau} = r_2 \cdot \left(\frac{t_{cr}}{C_0} \right). \tag{u}$$

For a constant-volume batch reactor, the species concentrations are expressed by (5.1-6), and the reaction rates are:

$$r_1 = k_1(T_0) \cdot C_0 (1 - Z_1 - 2 \cdot Z_2) \cdot e^{\gamma_1 \frac{\theta - 1}{\theta}} \tag{v}$$

$$r_2 = k_2(T_0) \cdot C_0^2 (1 - Z_1 - 2 \cdot Z_2)^2 \cdot e^{\gamma_2 \frac{\theta - 1}{\theta}}. \tag{w}$$

Substituting (d), (e), and (a) into (b) and (c), the design equations become

$$\frac{dZ_1}{d\tau} = (1 - Z_1 - 2 \cdot Z_2) \cdot e^{\gamma_1 \frac{\theta - 1}{\theta}} \tag{x}$$

$$\frac{dZ_2}{d\tau} = \left(\frac{k_2(T_0) \cdot C_0}{k_1(T_0)} \right) \cdot (1 - Z_1 - 2 \cdot Z_2)^2 \cdot e^{\gamma_2 \frac{\theta - 1}{\theta}}. \tag{y}$$

The energy balance equation is

$$\frac{d\theta}{d\tau} = (0.0961) \cdot \frac{dZ_1}{d\tau} + (0.1441) \cdot \frac{dZ_2}{d\tau}. \tag{z}$$

We solve (x), (y), and (z) subject to the initial conditions that $Z_1(0) = Z_2(0) = 0$ for different values of operating times, τ_{op}, and then we determine the species compositions at each τ_{op}. The operating curves of the desired product, V, and the undesirable product, W, are compared to those of semi-batch reactors in the figures above.

Example 8-2 A valuable product V is produced in a semi-batch reactor. The following simultaneous, liquid-phase chemical reactions take place in the reactor:

$$A + B \rightarrow V \qquad\qquad r_1 = k_1 \cdot C_A \cdot C_B;$$

$$2\,A \rightarrow W \qquad\qquad r_2 = k_2 \cdot C_A^2.$$

The reactor is initially charged with a 200 liter solution of reactant B with a concentration of $C_B(0) = 3$ mole/liter, and a 200 liter solution of reactant A ($C_{A0} = 3$ mole/liter) is continuously fed into the reactor at a constant rate. Both the injected stream temperature and the initial reactor temperature are 60°C.

a. Derive the design equations, and plot the reaction and species operating curves for isothermal semi-batch operation.
b. Compare the operating curves in (a) to those of isothermal batch operation (both solutions are charged initially into the reactor).
c. Derive the design equations, and plot the reaction and species operating curves for

adiabatic semi-batch operation.

d. Compare the operating curves in (c) to those of adiabatic batch operation.

Data: At 60°C, $k_1 = 0.02$ min^{-1} mole^{-1} min^{-1}; $k_2 = 0.02$ liter mole^{-1} min^{-1};

$\Delta H_{R1} = -9,000$ cal mole^{-1}; $\Delta H_{R2} = -13,000$ cal mole^{-1};

$E_{a1} = 12,000$ cal mole^{-1}°K^{-1}; $E_{a2} = 20,000$ cal mole^{-1}°K^{-1}.

Assume that heat capacity of the solution is independent of temperature and composition and is 0.9 cal g^{-1} °K^{-1}. The density of the solution is 0.85 kg/liter.

Solution The chemical reactions taking place in the reactor are

$$\text{Reaction 1:} \qquad\qquad A + B \rightarrow V$$

$$\text{Reaction 2:} \qquad\qquad 2\,A \rightarrow W,$$

and the stoichiometric coefficients are:

$$s_{A1} = -1;\ s_{B1} = -1;\ s_{V1} = 1;\ s_{W1} = 0;\ \Delta_1 = -1;$$

$$s_{A2} = -2;\ s_{B2} = 0;\ s_{V2} = 0;\ s_{W2} = 1;\ \Delta_2 = -1.$$

Since each reaction has a species that does not appear in the other, the two reactions are independent, and there is no dependent reaction. We select the total amount introduced into the reactor during the operation as the reference state. Hence, $V_R(0) = 200$ liter, and, for constant injection rate, $v_0 \cdot t_{op} = 200$ liter. Since only reactant B is charged initially, $N_{tot}(0) = N_B(0) = 600$ mole, $N_A(0) = 0$, $N_V(0) = 0$, and $N_W(0) = 0$. Since only reactant A is injected, $(C_0)_{inj} = C_{A0} = 3$ mole/liter, $(y_A)_{inj} = 1$, $(y_B)_{inj} = 0$, $(y_V)_{inj} = 0$, $(y_W)_{inj} = 0$, and $v_0 \cdot (C_0)_{inj} \cdot t_{op} = 600$ mole. Hence, $(N_{tot})_0 = 1200$ mole. Using (8.1-12), the reference concentration is

$$C_0 = \frac{N_{tot}(0) + v_0 \cdot (C_0)_{inj} \cdot t_{op}}{V_R(0) + v_0 \cdot t_{op}} = \frac{1,200\ \text{mole}}{400\ \text{liter}} = 3\ \text{mole/liter}.$$

a. We write the design equations for liquid-phase, semi-batch reactors with uniform injection rate, (8.1-16), for each independent reaction,

$$\frac{dZ_1}{d\tau} = r_1 \cdot \left(\frac{\tau_{op} + \tau}{2 \cdot \tau_{op}} \right) \cdot \left(\frac{t_{cr}}{C_0} \right) \tag{a}$$

$$\frac{dZ_2}{d\tau} = r_2 \cdot \left(\frac{\tau_{op} + \tau}{2 \cdot \tau_{op}} \right) \cdot \left(\frac{t_{cr}}{C_0} \right). \tag{b}$$

Using (8.1-22), the concentration of the reactants in the reactor are

$$C_A(\tau) = C_0 \left(\frac{2 \cdot \tau_{op}}{\tau_{op} + \tau} \right) \cdot \left[(0.5) \cdot \frac{\tau}{\tau_{op}} + (-1) \cdot Z_1(\tau) + (-2) \cdot Z_2(\tau) \right] \tag{c}$$

$$C_B(\tau) = C_0\left(\frac{2 \cdot \tau_{op}}{\tau_{op} + \tau}\right) \cdot \left[(0.5) + (-1) \cdot Z_1(\tau)\right]. \tag{d}$$

The rate of the two reactions are

$$r_1 = k_1(T_0) \cdot C_0^2\left(\frac{2 \cdot \tau_{op}}{\tau_{op} + \tau}\right)^2 \cdot \left((0.5) \cdot \frac{\tau}{\tau_{op}} - Z_1 - 2 \cdot Z_2\right) \cdot (0.5 - Z_1) \cdot e^{\gamma_1 \frac{\theta - 1}{\theta}} \tag{e}$$

$$r_2 = k_2(T_0) \cdot C_0^2\left(\frac{2 \cdot \tau_{op}}{\tau_{op} + \tau}\right)^2 \cdot \left((0.5) \cdot \frac{\tau}{\tau_{op}} - Z_1 - 2 \cdot Z_2\right)^2 \cdot e^{\gamma_2 \frac{\theta - 1}{\theta}}. \tag{f}$$

We select Reaction 1 as the leading reaction; hence, the characteristic reaction time is

$$t_{cr} = \frac{1}{k_1(T_0) \cdot C_0} = \frac{1}{(0.02) \cdot (3)} = 16.67 \text{ min}. \tag{g}$$

Substituting (e), (f), and (g) into (a) and (b), the design equations reduce to

$$\frac{dZ_1}{d\tau} = \left(\frac{2 \cdot \tau_{op}}{\tau_{op} + \tau}\right) \cdot \left((0.5) \cdot \frac{\tau}{\tau_{op}} - Z_1 - 2 \cdot Z_2\right) \cdot (0.5 - Z_1) \cdot e^{\gamma_1 \frac{\theta - 1}{\theta}} \tag{h}$$

$$\frac{dZ_2}{d\tau} = \left(\frac{k_2(T_0)}{k_1(T_0)}\right)\left(\frac{2 \cdot \tau_{op}}{\tau_{op} + \tau}\right) \cdot \left((0.5) \cdot \frac{\tau}{\tau_{op}} - Z_1 - 2 \cdot Z_2\right)^2 \cdot e^{\gamma_2 \frac{\theta - 1}{\theta}}. \tag{i}$$

a. For isothermal operation, $\theta = 1$, and the design equations are:

$$\frac{dZ_1}{d\tau} = \left(\frac{2 \cdot \tau_{op}}{\tau_{op} + \tau}\right) \cdot \left((0.5) \cdot \frac{\tau}{\tau_{op}} - Z_1 - 2 \cdot Z_2\right) \cdot (0.5 - Z_1) \tag{j}$$

$$\frac{dZ_2}{d\tau} = \left(\frac{k_2(T_0)}{k_1(T_0)}\right) \cdot \left(\frac{2 \cdot \tau_{op}}{\tau_{op} + \tau}\right) \cdot \left((0.5) \cdot \frac{\tau}{\tau_{op}} - Z_1 - 2 \cdot Z_2\right)^2. \tag{k}$$

We solve (j) and (k) subject to the initial conditions that $Z_1(0) = Z_2(0) = 0$ for different values of operating times, τ_{op}, and then determine the final extents for each τ_{op}. The first plot shows the reaction operating curves (Z_m versus τ_{op}) for isothermal semi-batch operations. Using (8.1-20), we determine the dimensionless species operating curves, $N_j(\tau_{op})/(N_{tot})_0$ versus τ_{op}. These curves are shown in figures below. The species operating curves of products V and W are compared to those of isothermal batch operation, derived in (b) below.

b. To compare the operation of semi-batch operation to batch operation, we write the

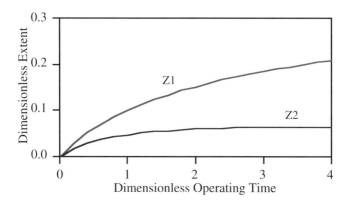

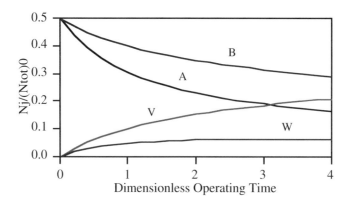

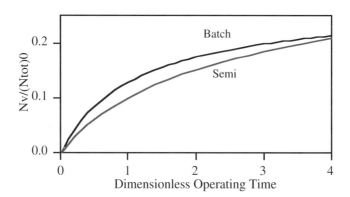

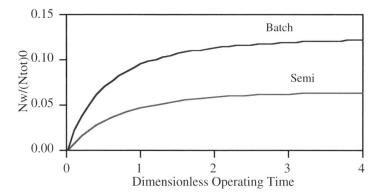

design equations for a constant-volume batch reactor:

$$\frac{dZ_1}{d\tau} = r_1 \cdot \left(\frac{t_{cr}}{C_0}\right) \tag{l}$$

$$\frac{dZ_2}{d\tau} = r_2 \cdot \left(\frac{t_{cr}}{C_0}\right) \cdot \tag{m}$$

When both solutions are charged initially into the reactor, $C_0 = 3.0$ mole/liter and $y_A(0) = y_B(0) = 0.5$. Using (5.1-10) to express the species concentrations, the rate of the reactions are

$$r_1 = k_1(T_0) \cdot C_0^2 \cdot (0.5 - Z_1 - 2 \cdot Z_2) \cdot (0.5 - Z_1) \cdot e^{\gamma_1 \frac{\theta-1}{\theta}} \tag{n}$$

$$r_2 = k_2(T_0) \cdot C_0^2 \cdot (0.5 - Z_1 - 2 \cdot Z_2)^2 \cdot e^{\gamma_2 \frac{\theta-1}{\theta}} \cdot \tag{o}$$

The design equations are

$$\frac{dZ_1}{d\tau} = (0.5 - Z_1 - 2 \cdot Z_2) \cdot (0.5 - Z_1) \cdot e^{\gamma_1 \frac{\theta-1}{\theta}} \tag{p}$$

$$\frac{dZ_2}{d\tau} = \left(\frac{k_2(T_0)}{k_1(T_0)}\right) \cdot (0.5 - Z_1 - 2 \cdot Z_2)^2 \cdot e^{\gamma_2 \frac{\theta-1}{\theta}} \cdot \tag{q}$$

For isothermal operation, $\theta = 1$, and we solve these equations and calculate the dimensionless species operating curves. To compare semi-batch and batch operations, we plot the species operating curves for V and W, shown above.

c. For non-isothermal semi-batch operations, we have to solve design equations (h) and (i) simultaneously with the energy balance equation. Using (8.1-32),

$$\hat{c}_{p0} = \frac{(400 \text{ lit}) \cdot (850 \text{ g/lit}) \cdot (0.9 \text{ cal/g } ^\circ\text{K})}{(1,200 \text{ mole})} = 255 \text{ cal/mole } ^\circ\text{K}.$$

The reference temperature is $T_0 = 333^\circ\text{K}$, and the dimensionless heat of reactions are

$$\text{DHR}_1 = \frac{\Delta H_{R_1}(T_0)}{T_0 \cdot \hat{c}_{p0}} = -0.212 \qquad \text{DHR}_2 = \frac{\Delta H_{R_2}(T_0)}{T_0 \cdot \hat{c}_{p0}} = -0.306.$$

The dimensionless activation energies of the two reactions are

$$\gamma_1 = \frac{E_{a1}}{R \cdot T_0} = 18.14 \qquad\qquad \gamma_2 = \frac{E_{a2}}{R \cdot T_0} = 30.23.$$

For adiabatic operation, $(S/V) = 0$, and, therefore, HTN = 0. Using (8.1-34),

$$\text{CF}(Z_m, \theta) = \frac{\tau_{op} + \tau}{2 \cdot \tau_{op}}. \tag{r}$$

Since $T_{inj} = T_0$, energy balance equation (8.1-33) reduces to

$$\frac{d\theta}{d\tau} = \left(\frac{2 \cdot \tau_{op}}{\tau_{op} + \tau} \right) \cdot \left((0.212) \cdot \frac{dZ_1}{d\tau} + (0.306) \cdot \frac{dZ_2}{d\tau} \right). \tag{s}$$

We solve (s) simultaneously with (h) and (i) subject to the initial conditions that $Z_1(0)$ = $Z_2(0) = 0$, $\theta(0) = 1$, for different values of operating times, τ_{op}, and then determine the extents and θ at each t_{op}. The plot shows the reaction operating curves (Z_m versus τ_{op}) for an adiabatic semi-batch operation. Using stoichiometric relation (8.1-20), we determine the species operating curves, $N_j/(N_{tot})_0$ versus τ_{op}. These curves are shown in figures below. The species operating curves of products V and W are compared to

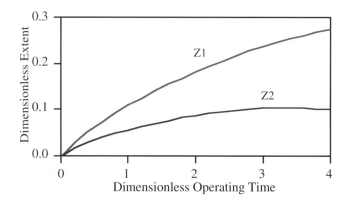

those of isothermal batch operation, derived in (d) below.

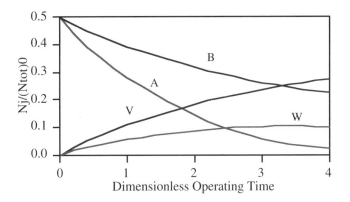

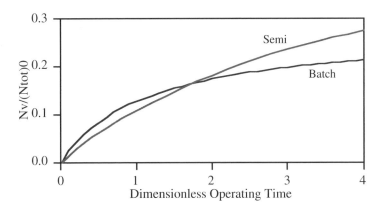

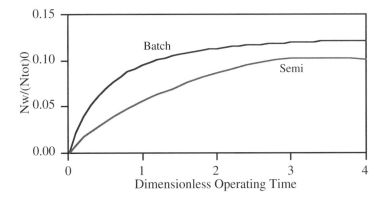

d. For adiabatic batch operation, the energy balance equation is

$$\frac{d\theta}{d\tau} = (0.212) \cdot \frac{dZ_1}{d\tau} + (0.306) \cdot \frac{dZ_2}{d\tau}. \tag{t}$$

We solve (t) simultaneously with (p) and (q) subject to the initial conditions that $Z_1(0)$ = $Z_2(0) = 0$ and $\theta(0) = 1$ and obtain the reaction and species operating curves for batch operation. The two figures below show a comparison of the production of V and W in adiabatic semi-batch and batch operations. An examination of the product operating curves indicates clearly the advantage of selecting the semi-batch mode to reduce the production of undesirable product W.

8.2 PLUG-FLOW REACTOR WITH DISTRIBUTED FEED

In this section, we discuss the operation of a plug-flow reactor with a distributed feed along the reactor length, shown schematically in Figure 8-2. This reactor configuration is used when we want to maintain one reactant at a high concentration while maintaining the second reactant at low concentration to maintain high yield of the desirable product. The operation of a plug-flow reactor with a distributed feed is similar to that of a semi-batch reactor, except that, in the latter, products are generated over time, whereas here, products are generated over space.

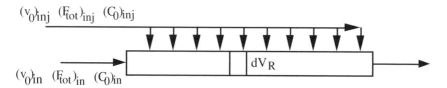

Figure 8-2: Plug-flow Reactor with Distributed Feed

To derive the design equation of a plug-flow reactor with a distributed feed, we write a species balance equation for any species, say species j, that is **not** injected along the reactor. Since the species is not fed or withdrawn, its molar balance equation is

$$dF_j = (r_j) \cdot dV_R. \tag{8.2-1}$$

We follow an identical procedure to the one used in Chapter 3 to derive the reaction-based design equation of a plug-flow reactor and obtain

$$\frac{d\dot{X}_m}{dV_R} = r_m + \sum_{k}^{n_{dep}} \alpha_{km} \cdot r_k, \tag{8.2-2}$$

where $\dot{X}_m$ is the extent per unit time of the m-th independent reaction from the reactor inlet to a given point in the reactor. Eq. (8.2-2) is the differential design equation for a plug-flow reactor with distributed feed, written for the m-th independent chemical reaction. Note that it is identical to the design equation of a plug-flow reactor. The only difference between the two is the way the volumetric flow rate and the concentrations of the species vary along the reactor. Consequently, to solve the design equations, we have to express these quantities in terms of the extents of the independent reactions.

Using stoichiometric relation (1.4-6), the local molar flow rate of species j at a point where the reactor volume is V_R is

$$F_j = (F_j)_{in} + \int_0^{V_R} \left(\frac{d(F_j)_{inj}}{dV_R} \right) \cdot dV_R + \sum_m^{n_{ind}} (s_j)_m \cdot \dot{X}_m, \qquad (8.2\text{-}3)$$

where $(d(F_j)_{inj}/dV_R)$ is the local injection rate of the injected stream, $\dot{X}_m$ is the extent per time of independent reaction m in volume V_R, and $(F_j)_{in}$ is the rate species j is fed at the reactor inlet. The local injection rate is

$$(F_j)_{inj} = (y_j)_{inj} \cdot (C_0)_{inj} \cdot v_{inj} = (y_j)_{inj} \cdot (F_{tot})_{inj}, \qquad (8.2\text{-}4)$$

where v_{inj} is the local volumetric injection rate, and $(F_{tot})_{inj}$ is the total molar injection rate. Differentiating (8.2-4),

$$\frac{d(F_j)_{inj}}{dV_R} = (y_j)_{inj} \cdot (C_0)_{inj} \cdot \left(\frac{d(v_{inj})}{dV_R} \right).$$

In general, the mode of the injection rate along the reactor, $d(v_{inj})/dV_R$, should be specified. For simplicity, we consider here the case where the injection rate is uniform along the reactor. For such operations,

$$\frac{d(v_{inj})}{dV_R} = \frac{(v_0)_{inj}}{V_{R tot}}, \qquad (8.2\text{-}5)$$

where $(v_0)_{inj}$ is the total volumetric flow rate of the injected stream, and V_{Rtot} is the total volume of the reactor. Substituting (8.2-5) into (8.2-3), the local molar flow rate of species j is

$$F_j = (F_j)_{in} + \left(\frac{V_R}{V_{R tot}} \right) \cdot (F_j)_{inj} + \sum_m^{n_{ind}} (s_j)_m \cdot \dot{X}_m. \qquad (8.2\text{-}6)$$

For a species that is not fed in the reactor inlet, $(F_j)_{in} = 0$, and for a species that is not injected along the reactor, $(F_j)_{inj} = 0$. Summing (8.2-6) over all species, the local total molar flow rate is

$$F_{tot} = (F_{tot})_{in} + \left(\frac{V_R}{V_{R tot}} \right) \cdot (F_{tot})_{inj} + \sum_m^{n_{ind}} \Delta_m \cdot \dot{X}_m. \qquad (8.2\text{-}7)$$

To reduce the design equation (8.2-2) to a dimensionless form for all cases, including the case where the entire feed is injected along the reactor, we select a reference stream on the basis of the total molar flow rates of species introduced into the reactor. We define a dimensionless extent of the m-th independent reaction by

$$Z_m = \frac{\dot{X}_m}{(F_{tot})_0} = \frac{\dot{X}_m}{(F_{tot})_{in} + (F_{tot})_{inj}}, \qquad (8.2\text{-}8)$$

where

$$(F_{tot})_{in} = (v_0)_{in} \cdot (C_0)_{in} \qquad (8.2\text{-}9)$$

$$(F_{tot})_{inj} = (v_0)_{inj} \cdot (C_0)_{inj}. \qquad (8.2\text{-}10)$$

We also select a reference volumetric flow rate, v_0, and define the reference concentration by

$$C_0 = \frac{(F_{tot})_{in} + (F_{tot})_{inj}}{v_0}, \qquad (8.2\text{-}11)$$

where

$$v_0 = (v_0)_{in} + (v_0)_{inj}. $$

We also define the dimensionless space time by

$$\tau = \frac{V_R}{v_0 \cdot t_{cr}}. \qquad (8.2\text{-}12)$$

Differentiating (8.2-8) and (8.2-12),

$$d\dot{X}_m = \left[(F_{tot})_{in} + (F_{tot})_{inj}\right] \cdot dZ_m \qquad (8.2\text{-}13)$$

$$dV_R = (v_0 \cdot t_{cr}) \cdot d\tau. \qquad (8.2\text{-}14)$$

Substituting these and (8-2-9) into (8.2-2), the design equation reduces to

$$\frac{dZ_m}{d\tau} = \left(r_m + \sum_{k}^{n_{dep}} \alpha_{km} \cdot r_k\right) \left(\frac{t_{cr}}{C_0}\right). \qquad (8.2\text{-}15)$$

Eq. (8.2-15) is the dimensionless design equation of a plug-flow reactor with a distributed feed, written for the m-th independent reaction. To describe the operation of a plug-flow reactor with a distributed feed with multiple chemical reactions, we have to write (8.2-12) for each of the independent reactions.

To solve the design equations, we should express the rates of all the chemical reactions (r_m's and r_k's) in terms of the extents of the independent reactions. Using stoichiometric relation (8.2-5), the local concentration of species j is

$$C_j = \frac{1}{v}\left((F_j)_{in} + \left(\frac{V_R}{V_{R_{tot}}}\right)\cdot(F_j)_{inj} + \sum_{m}^{n_{ind}}(s_j)_m \cdot \dot{X}_m\right), \qquad (8.2\text{-}16)$$

where v is the local volumetric flow rate. For convenience, we divide the discussion into two cases, liquid-phase reactions and gas-phase reactions.

For **liquid-phase** reactions, we assume that the density of the reacting fluid is constant, and the volumetric flow rate at any point in the reactor is

$$v = (v_0)_{in} + \left(\frac{V_R}{V_{R_{tot}}}\right)\cdot(v_0)_{inj} = (v_0)_{in} + \left(\frac{\tau}{\tau_{tot}}\right)\cdot(v_0)_{inj}, \qquad (8.2\text{-}17)$$

where $\tau_{tot} = V_{Rtot}/(v_0 \cdot t_{cr})$ is the dimensionless total space time. We combine (8.2-16) and (8.2-17) and use (8.2-9) and (8.2-10) to obtain

$$C_j = C_0 \frac{\left(\frac{(F_{tot})_{in}}{(F_{tot})_0}\right)\cdot(y_j)_{in} + \left(\frac{(F_{tot})_{inj}}{(F_{tot})_0}\right)\cdot\left(\frac{\tau}{\tau_{tot}}\right)\cdot(y_j)_{inj} + \sum_{m}^{n_{ind}}(s_j)_m \cdot Z_m}{\left(\frac{(v_0)_{in}}{v_0}\right) + \left(\frac{\tau}{\tau_{tot}}\right)\cdot\left(\frac{(v_0)_{inj}}{v_0}\right)}, \qquad (8.2\text{-}18)$$

where τ is the local dimensionless space time, and τ_{tot} is the total space time of the reactor.

For **gas-phase** reactions, assuming ideal gas behavior, the local volumetric flow rate at any point in the reactor is

$$v = v_0 \cdot \left(\frac{F_{tot}}{(F_{tot})_0}\right)\cdot\left(\frac{T}{T_0}\right)\cdot\left(\frac{P_0}{P}\right). \qquad (8.2\text{-}19)$$

Using the total stream concentrations, for isobaric operation, we obtain

$$v = v_0 \cdot \left[\left(\frac{(F_{tot})_{in}}{(F_{tot})_0}\right) + \left(\frac{(F_{tot})_{inj}}{(F_{tot})_0}\right)\cdot\left(\frac{\tau}{\tau_{tot}}\right) + \sum_{m}^{n_{ind}}\Delta_m \cdot Z_m\right]\cdot\theta. \qquad (8.2\text{-}20)$$

Combining (8.2-20) and (8.2-16), we obtain

$$C_j = C_0 \frac{\left(\frac{(F_{tot})_{in}}{(F_{tot})_0}\right)\cdot(y_j)_{in} + \left(\frac{(F_{tot})_{inj}}{(F_{tot})_0}\right)\cdot\left(\frac{\tau}{\tau_{tot}}\right)\cdot(y_j)_{inj} + \sum_{m}^{n_{ind}}(s_j)_m \cdot Z_m}{\left[\left(\frac{(F_{tot})_{in}}{(F_{tot})_0}\right) + \left(\frac{(F_{tot})_{inj}}{(F_{tot})_0}\right)\cdot\left(\frac{\tau}{\tau_{tot}}\right) + \sum_{m}^{n_{ind}}\Delta_m \cdot Z_m\right]\cdot\theta}. \qquad (8.2\text{-}21)$$

For non-isothermal operations, we have to incorporate the energy balance equation to express variation in the reactor temperature. For a plug-flow reactor with distributed feed, we modify energy balance equation (4.2-29) to account for the side injection,

$$d\dot{Q} = \sum_{m}^{n_{ind}} \Delta H_{R_m}(T_0) \, d\dot{X}_m + (\Sigma F_j \cdot \hat{c}_{p_j}) \, dT - d(\dot{m}_{inj} \cdot h_{inj}), \qquad (8.2\text{-}22)$$

where $d(\dot{m}_{inj} \cdot h_{inj})$ is the rate enthalpy is added to reactor element dV_R by the injection stream. But, for uniform injection rate along the reactor,

$$d(\dot{m}_{inj} \cdot h_{inj}) = (F_{tot})_{inj} \cdot \hat{c}_{p_{inj}} \cdot (T_{inj} - T_0) \cdot \left(\frac{dV_R}{V_{R_{tot}}} \right), \qquad (8.2\text{-}23)$$

where $V_{R_{tot}}$ is the total volume of the reactor. Also,

$$d\dot{Q} = U \cdot (T_F - T_0) \cdot \left(\frac{S}{V} \right) \cdot dV_R, \qquad (8.2\text{-}24)$$

and, using (8.2-6), the local heat capacity of the reacting fluid is

$$(\Sigma F_j \cdot \hat{c}_{p_j}) = \sum_{j}^{all} \left[(F_j)_{in} + \left(\frac{V_R}{V_{R_{tot}}} \right) \cdot (F_j)_{inj} + \sum_{m}^{n_{ind}} (s_j)_m \cdot \dot{X}_m \right] \cdot \hat{c}_{p_j}. \qquad (8.2\text{-}25)$$

Substituting these relations into (8.2-22) and using the reference stream definition (8.2-7), (8.2-8), (8.2-11), and (8.2-12), we obtain

$$\frac{d\theta}{d\tau} =$$

$$= \frac{1}{CF(Z_m, \theta)} \left[HTN \cdot (\theta_F - \theta) + \frac{1}{\tau_{op}} \cdot \left(\frac{(F_{tot})_{inj} \cdot \hat{c}_{p_{inj}}}{(F_{tot})_0 \cdot \hat{c}_{p0}} \right) \cdot (\theta_{inj} - 1) - \sum_{m}^{n_{ind}} DHR_m \cdot \frac{dZ_m}{d\tau} \right]$$

$$(8.2\text{-}26)$$

where HTN is the heat-transfer number, defined by (4.2-23),

$$HTN = \frac{U \cdot t_{cr}}{C_0 \cdot \hat{c}_{p0}} \cdot \left(\frac{S}{V} \right).$$

DHR_m is the dimensionless heat of reaction of the m-th independent reaction, defined by (4.2-24),

$$DHR_m = \frac{\Delta H_{R_m}(T_0)}{T_0 \cdot \hat{c}_{p0}},$$

and $CF(Z_m,\theta)$ is the correction factor of the heat capacity, defined by

$$CF(Z_m,\theta) = \frac{(\Sigma F_j \hat{c}_{pj})}{(F_{tot})_0 \cdot \hat{c}_{p0}}.$$

For **liquid-phase** reactions, the mass-based specific heat capacity, $\bar{c}_p$, is usually specified and assumed constant. For uniform injection rate along the reactor, the specific molar heat capacity of the reference state is

$$\hat{c}_{p0} = \frac{(\dot{m}_{in} + \dot{m}_{inj}) \cdot \bar{c}_p}{(F_{tot})_0} = \frac{(\dot{m}_{in} + \dot{m}_{inj}) \cdot \bar{c}_p}{(F_{tot})_{in} + (F_{tot})_{inj}}, \qquad (8.2\text{-}27)$$

where $(\dot{m}_{in} + \dot{m}_{inj})$ is the total mass injection rate. The correction factor of the heat capacity is

$$CF(Z_m,\theta) = \left(\frac{(F_{tot})_{in}}{(F_{tot})_0}\right) + \left(\frac{(F_{tot})_{inj}}{(F_{tot})_0}\right)\left(\frac{\tau}{\tau_{tot}}\right). \qquad (8.2\text{-}28)$$

Note that although the specific heat capacity is constant, the correction factor varies along the reactor because the mass flow rate changes.

For **gas-phase** reactions, the specific molar heat capacity of the reference stream, $\hat{c}_{p0}$, is

$$\hat{c}_{p0} = \sum_{j}^{all}\left[\left(\frac{(F_0)_{in}}{(F_{tot})_0}\right)\cdot(y_j)_{in} + \left(\frac{(F_0)_{inj}}{(F_{tot})_0}\right)\cdot(y_j)_{inj}\right]\hat{c}_{pj}(1), \qquad (8.2\text{-}29)$$

where $\hat{c}_{pj}(1)$ is the molar heat capacity of species j at the reference temperature. The correction factor of the heat capacity is

$$CF(Z_m,\theta) =$$

$$= \frac{1}{\hat{c}_{p0}}\sum_{j}^{all}\left[\left(\frac{(F_0)_{in}}{(F_{tot})_0}\right)\cdot(y_j)_{in} + \left(\frac{(F_0)_{inj}}{(F_{tot})_0}\right)\cdot\left(\frac{\tau}{\tau_{tot}}\right)\cdot(y_j)_{inj} + \sum_{m}^{n_{ind}}(s_j)_m \cdot Z_m(\tau)\right]\hat{c}_{pj}(\theta). \qquad (8.2\text{-}30)$$

Note that since the reference state is the total feed to the reactor for both gas-phase and liquid-phase reactions, the correction factor is a function of the dimensionless space time.

With these relations at hand, the design formulation of a plug-flow reactor with distributed feed results in a set of first-order, nonlinear differential equations of the form

$$\frac{dZ_{m1}}{d\tau} = G_1(\tau, Z_{m1}, Z_{m2}, \ldots, Z_{nind}, \tau_{tot}, \theta)$$

$$\frac{dZ_{m2}}{d\tau} = G_2(\tau, Z_{m1}, Z_{m2}, \ldots, Z_{nind}, \tau_{tot}, \theta)$$

$$\vdots \qquad (8.2\text{-}31)$$

$$\frac{dZ_{nind}}{d\tau} = G_{nind}(\tau, Z_{m1}, Z_{m2}, \ldots, Z_{nind}, \tau_{tot}, \theta)$$

$$\frac{d\theta}{d\tau} = G_{nind}(\tau, Z_{m1}, Z_{m2}, \ldots, Z_{nind}, \tau_{tot}, \theta).$$

We solve these equations subject to the initial condition that at $\tau = 0$, the dimensionless extents of all independent reactions and the dimensionless temperature are specified. Note that we solve these equations for a given value of τ_{tot}. The solutions describe how the extents and temperature vary along the reactor for the specified reactor. To obtain the operating curves of recycle reactors, we repeat the calculations for different values of τ_{tot} and, for each case, determine the final value of Z_m's and θ. We then construct the reaction operating curves and the dimensionless temperature by plotting $Z_m(\tau_{tot})$'s and $\theta(\tau_{tot})$ versus τ_{tot}.

The example below illustrates the design of a plug-flow reactor with distributed injection.

Example 8-3 A valuable product V is produced in a plug-flow reactor with distributed feed. The following simultaneous, gas-phase chemical reactions take place in the reactor:

$$A + B \rightarrow V \qquad\qquad r_1 = k_1 \cdot C_A \cdot C_B$$

$$2\,A \rightarrow W \qquad\qquad r_2 = k_2 \cdot C_A^{\,2}$$

Two equal streams (50 mole/hr each), one of reactant A and one of reactant B, are to be processed in the reactor. Reactant A is injected uniformly along the reactor, and reactant B is fed into the reactor inlet. The reactor is operated adiabatically, but, to cool the reactor, the temperature of injected stream is 200°C, while the inlet stream is at 300°C. The reactor is operated at 2 atm.

a. Derive the design equations, and plot the reaction and species operating curves.
b. Compare the operation to that of an adiabatic plug-flow reactor.
c. What is the reactor volume, and what are the species production rates at 70% conversion of A?

Data: At 300°C, $k_1 = 0.8$ lit mole^{-1} sec^{-1}; $k_2 = 1.6$ liter mole^{-1} sec^{-1};

$\Delta H_{R1} = -5{,}000$ cal mole^{-1}; $\Delta H_{R2} = -3{,}750$ cal mole^{-1};

$E_{a1} = 12{,}000$ cal mole^{-1}; $E_{a2} = 24{,}000$ cal mole^{-1};

$\hat{c}_{PA} = 30$ cal mole^{-1} °K^{-1}; $\hat{c}_{PB} = 10$ cal mole^{-1} °K^{-1};

$\hat{c}_{PV} = 32$ cal mole^{-1} °K^{-1}; $\hat{c}_{PW} = 40$ cal mole^{-1} °K^{-1}.

Solution The chemical reactions that take place in the reactor are:

Reaction 1: $\qquad A + B \rightarrow V$

Reaction 2: $\qquad 2A \rightarrow W,$

and the stoichiometric coefficients are:

$$s_{A1} = -1; \; s_{B1} = -1; \; s_{V1} = 1; \; s_{W1} = 0; \; \Delta_1 = -1;$$

$$s_{A2} = -2; \; s_{B2} = 0; \; s_{V2} = 0; \; s_{W2} = 1; \; \Delta_2 = -1.$$

Since each reaction has a species that does not appear in the other, the two reactions are independent, and there is no dependent reaction. We select the total feed into the reactor as the reference stream. Hence, $(F_{tot})_0 = (F_A)_0 + (F_B)_0 = 100$ mole/hr. We also select the reference state at $T_0 = 300°C$ and $P_0 = 2$ atm; hence

$$C_0 = \frac{P_0}{R \cdot T_0} = 0.0425 \text{ mole/lit}$$

$$v_0 = \frac{(F_{tot})_0}{C_0} = 2,353 \text{ lit/hr} = 0.653 \text{ lit/sec}.$$

Since only reactant B is fed into the reactor inlet, $(F_{tot})_{in} = 50$ mole/hr, and $(y_A)_{in} = 0$, $(y_B)_{in} = 1$, $(y_V)_{in} = 0$, and $(y_W)_{in} = 0$. Since only reactant A is injected along the reactor, $(F_{tot})_{inj} = 50$ mole/hr, and $(y_A)_{inj} = 1$ and $(y_B)_{in} = (y_V)_{in} = (y_W)_{in} = 0$.

a. We write design equation (8.2-15) for each independent reaction,

$$\frac{dZ_1}{d\tau} = r_1 \cdot \left(\frac{t_{cr}}{C_0} \right) \tag{a}$$

$$\frac{dZ_2}{d\tau} = r_2 \cdot \left(\frac{t_{cr}}{C_0} \right). \tag{b}$$

Using (8.2-18), the concentrations of the reactants at any point in the reactor are

$$C_A = C_0 \frac{(0.5) \cdot \dfrac{\tau}{\tau_{op}} - Z_1 - 2 \cdot Z_2}{\left((0.5) \cdot \dfrac{\tau_{op} + \tau}{\tau_{op}} - Z_1 - Z_2 \right) \cdot \theta} \tag{c}$$

$$C_B = C_0 \frac{0.5 - Z_1}{\left((0.5) \cdot \dfrac{\tau_{op} + \tau}{\tau_{op}} - Z_1 - Z_2 \right) \cdot \theta}. \tag{d}$$

The rates of the two reactions are

$$r_1 = k_1(T_0) \cdot C_0^2 \frac{\left((0.5) \cdot \dfrac{\tau}{\tau_{op}} - Z_1 - 2 \cdot Z_2\right) \cdot (0.5 - Z_1)}{\left((0.5) \cdot \dfrac{\tau_{op} + \tau}{\tau_{op}} - Z_1 - Z_2\right)^2 \cdot \theta^2} \cdot e^{\gamma_1 \frac{\theta-1}{\theta}} \quad \text{(e)}$$

$$r_2 = k_2(T_0) \cdot C_0^2 \frac{\left((0.5) \cdot \dfrac{\tau}{\tau_{op}} - Z_1 - 2 \cdot Z_2\right)^2}{\left((0.5) \cdot \dfrac{\tau_{op} + \tau}{\tau_{op}} - Z_1 - Z_2\right)^2 \cdot \theta^2} \cdot e^{\gamma_2 \frac{\theta-1}{\theta}}. \quad \text{(f)}$$

We select Reaction 1 as the leading reaction; hence, the characteristic reaction time is

$$t_{cr} = \frac{1}{k_1(T_0) \cdot C_0} = 29.41 \text{ sec}. \quad \text{(g)}$$

Substituting (e), (f), and (g) into (a) and (b), the design equations reduce to

$$\frac{dZ_1}{d\tau} = \frac{\left((0.5) \cdot \dfrac{\tau}{\tau_{op}} - Z_1 - 2 \cdot Z_2\right) \cdot (0.5 - Z_1)}{\left((0.5) \cdot \dfrac{\tau_{op} + \tau}{\tau_{op}} - Z_1 - Z_2\right)^2 \cdot \theta^2} \cdot e^{\gamma_1 \frac{\theta-1}{\theta}} \quad \text{(h)}$$

$$\frac{dZ_2}{d\tau} = \left(\frac{k_2(T_0)}{k_1(T_0)}\right) \cdot \frac{\left((0.5) \cdot \dfrac{\tau}{\tau_{op}} - Z_1 - 2 \cdot Z_2\right)^2}{\left((0.5) \cdot \dfrac{\tau_{op} + \tau}{\tau_{op}} - Z_1 - Z_2\right)^2 \cdot \theta^2} \cdot e^{\gamma_2 \frac{\theta-1}{\theta}}. \quad \text{(i)}$$

To set up the energy balance equation, we have to determine several related parameters. Using (8.2-29), the specific molar heat capacity of the reference stream is

$$\hat{c}_{P0} = (0.5) \cdot \hat{c}_{PA}(T_0) + (0.5) \cdot \hat{c}_{PB}(T_0) = 20 \text{ cal/mole}^\circ K.$$

The reference temperature is $T_0 = 573^\circ K$, and the dimensionless heats of reaction are

$$DHR_1 = \frac{\Delta H_{R_1}(T_0)}{T_0 \cdot \hat{c}_{P0}} = -0.4362 \qquad DHR_2 = \frac{\Delta H_{R_2}(T_0)}{T_0 \cdot \hat{c}_{P0}} = -0.3272.$$

The dimensionless activation energies of the two reactions are

$$\gamma_1 = \frac{E_{a1}}{R \cdot T_0} = 10.54 \qquad\qquad \gamma_2 = \frac{E_{a2}}{R \cdot T_0} = 21.08,$$

and $\theta_{inj} = 0.8255$. For adiabatic operation, $(S/V) = 0$, and, therefore, $HTN = 0$. Using (8.2-30), the correction factor of the heat capacity is

$$CF(Z_m, \theta) =$$

$$= \frac{1}{\hat{c}_{P0}} \left[\left((0.5) \cdot \frac{\tau}{\tau_{op}} - Z_1 - 2 \cdot Z_2 \right) \cdot \hat{c}_{PA} + (0.5 - Z_1) \cdot \hat{c}_{PB} + Z_1 \cdot \hat{c}_{PV} + Z_2 \cdot \hat{c}_{PW} \right] =$$

$$= (0.25) + (0.75) \cdot \frac{\tau}{\tau_{tot}} - (0.4) \cdot Z_1 - Z_2. \tag{j}$$

Substituting these parameters and (j) into (8.1-26), it reduces to

$$\frac{d\theta}{d\tau} = \frac{(0.75) \cdot \dfrac{\tau}{\tau_{tot}} \cdot (\theta_{inj} - 1) + (0.43625) \cdot \dfrac{dZ_1}{d\tau} + (0.32725) \cdot \dfrac{dZ_2}{d\tau}}{(0.25) + (0.75) \cdot \dfrac{\tau}{\tau_{tot}} - (0.4) \cdot Z_1 - Z_2}. \tag{k}$$

We solve (k) simultaneously with (h) and (i) subject to the initial conditions $Z_1(0) = Z_2(0) = 0$, $\theta(0) = 1$, for different values of total space time, τ_{tot}, and determine the final extents and θ at each τ_{tot}. The figures below show the reaction operating curves and the temperature profile, respectively. Using stoichiometric relation (8.1-20), we determine the species operating curves, $F_j/(F_{tot})_0$ versus τ_{tot}, shown in the next figure below.

b. For plug-flow reactors, the species concentrations at any point in the reactor are:

$$C_A = C_0 \frac{0.5 - Z_1 - 2 \cdot Z_2}{(1 - Z_1 - Z_2) \cdot \theta} \tag{l}$$

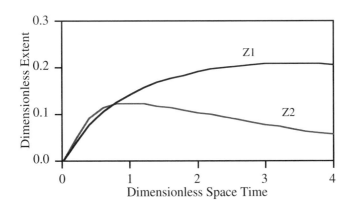

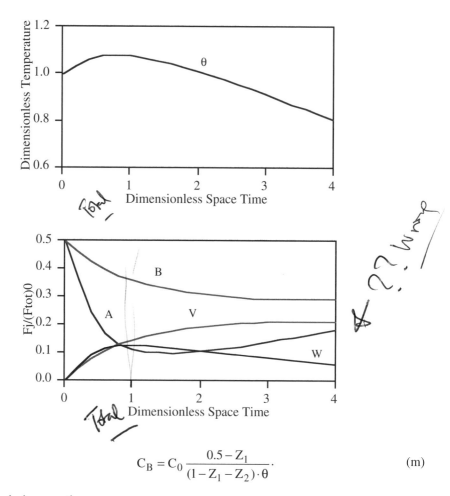

$$C_B = C_0 \frac{0.5 - Z_1}{(1 - Z_1 - Z_2) \cdot \theta} \cdot \tag{m}$$

The design equations are

$$\frac{dZ_1}{d\tau} = \frac{(0.5 - Z_1 - 2 \cdot Z_2) \cdot (0.5 - Z_1)}{(1 - Z_1 - Z_2)^2 \cdot \theta^2} \cdot e^{\gamma_1 \frac{\theta - 1}{\theta}} \tag{n}$$

$$\frac{dZ_2}{d\tau} = \left(\frac{k_2(T_0)}{k_1(T_0)} \right) \cdot \frac{(0.5 - Z_1 - 2 \cdot Z_2)^2}{(1 - Z_1 - Z_2)^2 \cdot \theta^2} \cdot e^{\gamma_2 \frac{\theta - 1}{\theta}} \cdot \tag{o}$$

The correction factor for the heat capacity is

$$CF(Z_m, \theta) = 1 - (0.4) \cdot Z_1 - Z_2. \tag{p}$$

For adiabatic plug-flow operation, HTN = 0, and the energy balance equation is

$$\frac{d\theta}{d\tau} = \frac{(0.4362) \cdot \frac{dZ_1}{d\tau} + (0.3272) \cdot \frac{dZ_2}{d\tau}}{1 - (0.4) \cdot Z_1 - Z_2}. \tag{q}$$

Note that, in the plug-flow operation, the two streams are mixed, and the inlet temperature is 225°C. We solve (n), (p), and (q) simultaneously subject to the initial conditions that $Z_1(0) = Z_2(0) = 0$, $\theta(0) = 0.8691$, and obtain the reaction and species operating curves for a plug-flow reactor. The figures below show comparisons of the reaction and temperature operating curves for the two modes of operation. These figures show the effect of the injection mode and the temperature. The figure above shows the species operating curves for the reactants, A and B. The operating curves for V and W are identical to the operating curves of Z_1 and Z_2, respectively. Note that while the production rate of product V is about the same in both modes, the production rate of W is much lower in the operation with distributed feed.

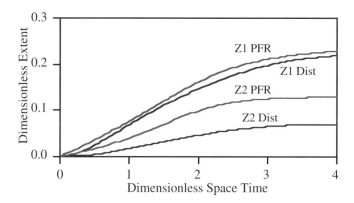

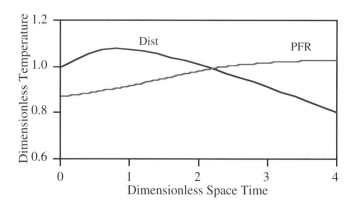

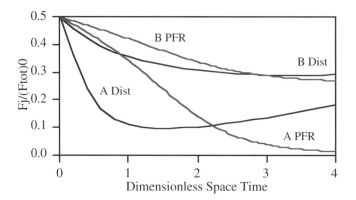

c. For 70% conversion of A, $F_A/(F_A)_0 = 0.30$, and this is obtained at $F_A/(F_{tot})_0 = 0.15$. For the plug-flow reactor with the distributed feed, this is achieved at two values of τ_{tot}: 0.661 and 3.30. At $\tau_{tot} = 0.611$, $F_V/(F_{tot})_0 = 0.114$ and $F_W/(F_{tot})_0 = 0.118$, and at $\tau_{tot} = 3.30$, $F_V/(F_{tot})_0 = 0.21$ and $F_W/(F_{tot})_0 = 0.07$. For the conventional plug-flow reactor, it is achieved at $\tau_{tot} = 1.95$, and, at this point, $F_V/(F_{tot})_0 = 0.158$ and $F_W/(F_{tot})_0 = 0.0960$. Using (g), the corresponding reactor volume in each case is obtained by

$$V_{R_{tot}} = v_0 \cdot t_{cr} \cdot \tau_{tot} = (0.653 \text{ lit/sec}) \cdot (29.41 \text{ sec}) \cdot \tau_{tot} = 19.2 \cdot \tau_{tot} \text{ liter}.$$

For $(F_{tot})_0 = 100$ mole/hr, the species molar flow rates are readily determined using (1.5-5).

8.3 DISTILLATION REACTOR

A distillation reactor is a liquid-phase ideal batch reactor where volatile products are generated and continuously removed from the reactor, as shown schematically in Figure 8-3. Because species are removed, the volume of the reacting fluid reduces during the operation.

To derive a design equation for the reactor, we write a species balance equation for species j that is **not** being removed from the reactor:

$$\frac{dN_j}{dt} = (r_j) \cdot V_R(t). \tag{8.3-1}$$

We use stoichiometric relation (1.4-2) to relate the moles of species j in the reactor at operating time t to the extents of the independent reactions, $X_m(t)$'s, following the procedure described in Chapter 3 for a batch reactor, and obtain

$$\frac{dX_m}{dt} = \left(r_m + \sum_k^{n_{dep}} \alpha_{km} \cdot r_k \right) \cdot V_R(t). \tag{8.3-2}$$

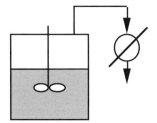

Figure 8-3: Distillation Reactor

This is the differential design equation for a distillation reactor, written for the m-th independent chemical reaction. Note that (8.3-2) is identical to the design equation for an ideal batch reactor. The difference between the two cases is in the variation of the reactor volume and species concentrations during the operation.

To derive a relation for the change in the reactor volume, we write an overall material balance for the reactor. Assuming the mass of the gaseous species inside the reactor is negligible, the reduction in the total mass of the reactor during operating time t is equal to the mass of the volatile species removed,

$$\rho(0) \cdot V_R(0) - \rho(t) \cdot V_R(t) = \sum_{j}^{n_{evap}} MW_j \cdot N_j(t), \qquad (8.3\text{-}3)$$

where MW_j is the molecular weight, and $N_j(t)$ is the mole of gaseous species j formed during the operation. Note that the summation in (8.3-3) is only over all the species that evaporate and removed from the reactor. Assuming the density of the reacting fluid does not vary during the operation, $\rho(t) = \rho(0) = \rho$, and, differentiating (8.3-3), we obtain

$$\frac{dV_R}{dt} = -\frac{1}{\rho} \sum_{j}^{n_{evap}} MW_j \cdot \frac{dN_j}{dt}. \qquad (8.3\text{-}4)$$

To relate the formation rates of these species to the extents of the independent reactions, we differentiate stoichiometric relation (1.4-2),

$$\frac{dN_j}{dt} = \sum_{m}^{n_{ind}} (s_j)_m \cdot \frac{dX_m}{dt}. \qquad (8.3\text{-}5)$$

Substituting this into (8.3-4), we obtain

$$\frac{dV_R}{dt} = -\frac{1}{\rho} \sum_{j}^{n_{evap}} MW_j \cdot \sum_{m}^{n_{ind}} (s_j)_m \cdot \frac{dX_m}{dt}. \qquad (8.3\text{-}6)$$

Multiplying both sides by dt and integrating,

$$V_R(t) = V_R(0) - \frac{1}{\rho} \left(\sum_j^{n_{evap}} MW_j \cdot \sum_m^{n_{ind}} (s_j)_m \cdot X_m(t) \right). \tag{8.3-7}$$

Eq. (8.3-7) provides an expression for the volume of the reactor in terms of the extents of the independent chemical reactions.

To reduce (8.3-2) and (8.3-7) to dimensionless form, we select the initial reactor state as the reference state, $(N_{tot})_0 = N_{tot}(0)$, and define a dimensionless extent of the m-th independent reaction by

$$Z_m = \frac{X_m}{(N_{tot})_0}. \tag{8.3-8}$$

We also define the reference concentration, C_0, by

$$C_0 = \frac{(N_{tot})_0}{V_R(0)}. \tag{8.3-9}$$

To reduce (8.3-7) to a dimensionless form, we divide both sides by $V_R(0)$ and $N_{tot}(0)$, and, using (8.3-8) and (8.3-9),

$$\left(\frac{V_R(\tau)}{V_R(0)} \right) = 1 - \frac{C_0}{\rho} \left(\sum_j^{n_{evap}} MW_j \cdot \sum_m^{n_{ind}} (s_j)_m \cdot Z_m(\tau) \right). \tag{8.3-10}$$

As we did with batch reactors, we define the dimensionless operating time by

$$\tau = \frac{t}{t_{cr}}, \tag{8.3-11}$$

where t_{cr} is the characteristic reaction time, defined by (2.5-1). Differentiating (8.3-8) and (8.3-9),

$$dX_m = (N_{tot})_0 \cdot dZ_m$$

$$dt = t_{cr} \cdot d\tau.$$

Substituting these into (8.3-2) and using (8.3-10), we obtain

$$\frac{dZ_m}{d\tau} = \left(r_m + \sum_k^{n_{dep}} \alpha_{km} \cdot r_k \right) \cdot \left[1 - \frac{C_0}{\rho} \left(\sum_j^{n_{evap}} MW_j \cdot \sum_m^{n_{ind}} (s_j)_m \cdot Z_m(\tau) \right) \right] \left(\frac{t_{cr}}{C_0} \right). \tag{8.3-12}$$

Eq. (8.3-12) is the dimensionless, reaction-based design equation for distillation reactors, written for the m-th independent reaction. To describe the operation, we have to write (8.3-12) for each independent reaction.

To solve the design equations, we should express the species concentrations in terms of the extents of the independent reactions. Using stoichiometric relation (1.5-3) and accounting for changes in the reactor volume, the concentration of reactant j is

$$C_j(t) = C_0 \cdot \left(\frac{V_R(0)}{V_R(t)} \right) \cdot \left(y_j(0) + \sum_m^{n_{ind}} (s_j)_m \cdot Z_m(t) \right). \tag{8.3-13}$$

Substituting (8.3-10) into (8.3-13), the species concentrations are

$$C_j(t) = C_0 \frac{y_j(0) + \sum_m^{n_{ind}} (s_j)_m \cdot Z_m(t)}{1 - \dfrac{C_0}{\rho} \left(\sum_j^{n_{evap}} MW_j \cdot \sum_m^{n_{ind}} (s_j)_m \cdot Z_m(\tau) \right)}. \tag{8.3-14}$$

For non-isothermal operations, we also have to solve the energy balance equation to express variation in the reactor temperature. For a distillation reactor, we modify energy balance equation (4.2-8),

$$\Delta H(t) = Q(t) - \int_0^t \dot{m}_{out} \cdot h_{out} \, dt - W_{sh}(t), \tag{8.3-15}$$

where $\dot{m}_{out} \cdot h_{out}$ is the rate enthalpy is removed from the reactor by the gaseous species. Differentiating (8.3-15),

$$\frac{dH}{dt} = \dot{Q} - h_{out} \cdot \dot{m}_{out} - \dot{W}_{sh}, \tag{8.3-16}$$

where $\dot{W}_{sh}$ is the rate of mechanical (shaft) work done by the reactor on the surrounding, and $\dot{Q}$ is the rate heat is added to the reactor, expressed by (see Chapter 4)

$$\dot{Q} = U \cdot \left(\frac{S}{V} \right) \cdot V_R(t) \cdot (T_F - T). \tag{8.3-17}$$

Using (8.3-5), the term $h_{out} \cdot \dot{m}_{out}$ relates to the extents of the independent reactions by

$$h_{out} \cdot \dot{m}_{out} = \sum_j^{n_{evap}} \hat{H}_j \cdot \sum_m^{n_{ind}} (s_j)_m \cdot \frac{dX_m}{dt}, \tag{8.3-18}$$

where $\hat{H}_j$ is the specific molar-based enthalpy of species j in the vapor phase relative to the reference state. Now, noting that the reacting fluid is liquid and assuming constant density and constant mass-based heat capacity, $\bar{c}_p$, differentiating (4.2-10)

$$\frac{dH}{dt} = \sum_m^{n_{ind}} \Delta H_{R_m}(T_0) \cdot \frac{dX_m}{dt} + M(t) \cdot \bar{c}_p \frac{dT}{dt}. \tag{8.3-19}$$

Combining (8.3-16), (8.3-17), (8.3-17), and (8.3-17) for reactors with negligible shaft work,

$$\frac{dT}{dt} = \frac{1}{M(t) \cdot \bar{c}_p} U \cdot \left(\frac{S}{V}\right) \cdot V_R(t) \cdot (T_F - T) -$$

$$- \frac{1}{M(t) \cdot \bar{c}_p} \left(\sum_j^{n_{evap}} \hat{H}_j \cdot \sum_m^{n_{ind}} (s_j)_m \cdot \frac{dX_m}{dt} + \sum_m^{n_{ind}} \Delta H_{R_m}(T_0) \cdot \frac{dX_m}{dt} \right). \quad (8.3\text{-}20)$$

We define dimensionless temperature by

$$\theta = \frac{T}{T_0}. \quad (8.3\text{-}22)$$

Dividing both sides of (8.3-20) by $(N_{tot})_0\, \hat{c}_{p0}\, T_0$ and using the relations above, we obtain

$$\frac{d\theta}{d\tau} = \frac{1}{CF(Z_m,\theta)} HTN \cdot \left(\frac{V_R(\tau)}{V_R(0)} \right) \cdot (\theta_F - \theta) -$$

$$- \frac{1}{CF(Z_m,\theta)} \left(\sum_j^{n_{evap}} \frac{\hat{H}_j}{\hat{c}_{p0} \cdot T_0} \sum_m^{n_{ind}} (s_j)_m \frac{dZ_m}{dt} + \sum_m^{n_{ind}} DHR_m \cdot \frac{dZ_m}{dt} \right), \quad (8.3\text{-}23)$$

where DHR_m is the dimensionless heat of reaction of the m-th independent reaction defined by (4.2-24), HTN is the heat transfer number, defined by (4.2-23), $\hat{c}_{p0}$ is the specific molar heat capacity of the reference state, and $CF(Z_m,\theta)$ is the correction factor of the heat capacity. For liquid-phase reactions, the specific molar heat capacity of the reference state is defined by

$$\hat{c}_{p0} = \frac{M(0) \cdot \bar{c}_p}{(N_{tot})_0}, \quad (8.3\text{-}24)$$

and, assuming constant specific heat and using (8.3-10), the correction factor is

$$CF(Z_m,\theta) = \left(\frac{V_R(\tau)}{V_R(0)} \right) = 1 - \frac{C_0}{\rho} \left(\sum_j^{n_{evap}} MW_j \cdot \sum_m^{n_{ind}} (s_j)_m \cdot Z_m(\tau) \right). \quad (8.3\text{-}25)$$

We illustrate the design of a distillation reactor in the examples below.

Example 8-4 The dehydration of two heavy organic species, A and B, is carried out in a distillation reactor where the water, product C, is being removed continuously. The reactor is initially charged with a 60 liter organic solution that contains 200 moles of A and 300 moles of B. The reactor is operated isothermally at 200°C, and the following reactions take place in the reactor:

$$A \text{ (liq)} \rightarrow C \text{ (g)} + D \text{ (liq)} \qquad r_1 = k_1 C_A$$

$$B \text{ (liq)} \rightarrow 2 \text{ C (g)} + E \text{ (liq)} \qquad r_2 = k_2 C_B.$$

a. Derive the design equations, and plot the reaction operating curves.
b. Plot the species operating curves.
c. Determine the operating time needed to achieve 90% conversion of B.
d. Determine the heating (or cooling) load during the operation.

Data: at 200°C, $k_1 = 0.05$ min^{-1}, $k_2 = 0.1$ min^{-1},

$\Delta H_{R1} = 30$ kcal/mole min^{-1}; $\Delta H_{R2} = 40$ kcal/mole min^{-1};

$MW_C = 18$ g/mole; $\rho = 800$ g/liter; $\bar{c}_p = 0.9$ cal g^{-1} °K^{-1}

Solution: The chemical reactions taking place in the reactor are:

Reaction 1: $\qquad\qquad A \rightarrow C + D$

Reaction 2: $\qquad\qquad B \rightarrow 2 C + E$,

and the stoichiometric coefficients are:

$$s_{A1} = -1; \; s_{B1} = 0; \; s_{C1} = 1; \; s_{D1} = 1; \; s_{E1} = 0; \; \Delta_1 = 1;$$

$$s_{A2} = 0; \; s_{B2} = -1; \; s_{C2} = 2; \; s_{D2} = 0; \; s_{E2} = 1; \; \Delta_2 = 2.$$

Since each reaction has a species that does not appear in the other, the two reactions are independent, and there is no dependent reaction in this case. We select the initial state as the reference state. Hence, $V_R(0) = 60$ liter,

$$(N_{tot})_0 = N_A(0) + N_B(0) = 500 \text{ mole},$$

$$C_0 = \frac{(N_{tot})_0}{V_R(0)} = \frac{500 \text{ mole}}{60 \text{ liter}} = 8.333 \text{ mole/liter},$$

and $y_A(0) = 0.4$, $y_B(0) = 0.6$. For liquid-phase reactions, using (4.2-31), the specific molar heat capacity of the reference state is

$$\hat{c}_{P0} = \frac{(60 \text{ lit}) \cdot (800 \text{ g/lit}) \cdot (0.9 \text{ cal/g - °K})}{(500 \text{ mole})} = 86.4 \text{ cal/mole°K}.$$

Since only species C is gaseous, (8.3-10) reduces to

$$\left(\frac{V_R(\tau)}{V_R(0)} \right) = 1 - \frac{C_0}{\rho} \cdot MW_C \cdot [Z_1(\tau) + 2 \cdot Z_2(\tau)], \tag{a}$$

and, using (8.3-25), the correction factor of the heat capacity is

$$CF(Z_m, \theta) = \left(\frac{V_R(\tau)}{V_R(0)} \right) = 1 - \frac{C_0}{\rho} \cdot MW_C \cdot (Z_1 + 2 \cdot Z_2) = 1 - 0.1875 \cdot (Z_1 + 2 \cdot Z_2). \tag{b}$$

The reference temperature is $T_0 = 473°K$, and the dimensionless heats of reaction are

$$DHR_1 = \frac{\Delta H_{R_1}(T_0)}{T_0 \cdot \hat{c}_{P0}} = 0.734 \qquad DHR_2 = \frac{\Delta H_{R_2}(T_0)}{T_0 \cdot \hat{c}_{P0}} = 0.979,$$

and the dimensionless specific molar enthalpy of gaseous species C is

$$\frac{\hat{H}_C}{\hat{c}_{P0} \cdot T_0} = \frac{8,300}{(86.4) \cdot (473)} = 0.203. \tag{c}$$

a. We write the design equation for distillation reactors, (8.3-12), for each independent reaction,

$$\frac{dZ_1}{d\tau} = r_1 \cdot \left[1 - \frac{C_0}{\rho} \cdot MW_C \cdot (Z_1 + 2 \cdot Z_2)\right] \cdot \left(\frac{t_{cr}}{C_0}\right) \tag{d}$$

$$\frac{dZ_2}{d\tau} = r_2 \cdot \left[1 - \frac{C_0}{\rho} \cdot MW_C \cdot (Z_1 + 2 \cdot Z_2)\right] \cdot \left(\frac{t_{cr}}{C_0}\right). \tag{e}$$

We use (8.3.14) to express the reactant concentrations in terms of the dimensionless extents,

$$C_A(\tau) = C_0 \frac{y_A(0) - Z_1(\tau)}{1 - \frac{C_0}{\rho} \cdot MW_C \cdot \left[Z_1(\tau) + 2 \cdot Z_2(\tau)\right]} \tag{f}$$

$$C_B(\tau) = C_0 \frac{y_B(0) - Z_1}{1 - \frac{C_0}{\rho} \cdot MW_C \cdot \left[Z_1(\tau) + 2 \cdot Z_2(\tau)\right]}. \tag{g}$$

We select Reaction 1 as the leading reaction and define a characteristic reaction time by

$$t_{cr} = \frac{1}{k_1} = 20 \text{ min}. \tag{h}$$

We substitute (f) and (g) in the rate expressions and the latter together with (h) in the design equations and obtain

$$\frac{dZ_1}{d\tau} = (0.4 - Z_1) \cdot e^{\gamma_1 \frac{\theta - 1}{\theta}} \tag{i}$$

$$\frac{dZ_2}{d\tau} = \left(\frac{k_2(T_0)}{k_1(T_0)}\right) \cdot (0.6 - Z_2) \cdot e^{\gamma_2 \frac{\theta - 1}{\theta}}. \tag{j}$$

a. For isothermal operation, $\theta = 1$, and we solve (i) and (j) subject to the initial conditions that at $\tau = 0$, $Z_1 = Z_2 = 0$. In this case, we obtain analytical solutions:

$$Z_1(\tau) = 0.4 \cdot (1 - e^{-\tau}) \tag{k}$$

$$Z_2(\tau) = 0.6 \cdot \left[1 - e^{-(k_2/k_1)\tau}\right]. \tag{l}$$

The reaction operating curves and the reactor volume curve are shown in the first figure below.

c. Now that the extents of the independent reactions at any operating time are known, we can determine the amount of each species formed or depleted using the stoichiometric relation (1.5-3). The species operating curves are shown in the second figure.

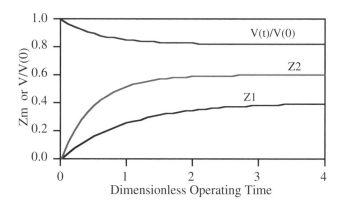

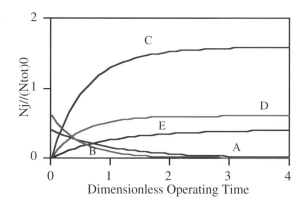

d. For 80% conversion of B, $N_B = 0.2 \cdot N_B(0) = 0.2 \cdot [0.6 \cdot (N_{tot})_0] = 0.12 \cdot (N_{tot})_0$, and, using (1.5-3), $Z_2 = 0.48$. From (l), this value of Z_2 is reached at dimensionless operating time of $\tau = 0.805$, and, at that time, $Z_1 = 0.221$. Using the definition of the dimensionless operating time (8.3-11) and (h), the operating time is

$$t = \tau \cdot t_{cr} = (0.805) \cdot (20 \text{ min}) = 16.09 \text{ min}.$$

From the calculated data, for $\tau = 0.805$, $V_R(t)/V_R(0) = 0.785$. The volume of the reactor when 80% conversion of B is reached is

$$V_R(t) = (0.785) \cdot V_R(0) = 46.71 \text{ liter}.$$

For isothermal operation, $d\theta/d\tau = 0$, and energy balance equation (8.3-21) reduces to

$$\frac{d}{d\tau}\left(\frac{Q}{T_0 \cdot (N_{tot})_0 \cdot \hat{c}_{p0}} \right) =$$

$$= \left[\frac{\hat{H}_C}{\hat{c}_{p0} \cdot T_0}\left(\frac{dZ_1}{d\tau} + 2 \cdot \frac{dZ_m}{d\tau} \right) + DHR_1 \cdot \frac{dZ_1}{d\tau} + DHR_2 \cdot \frac{dZ_2}{d\tau} \right], \tag{m}$$

which, upon integration, becomes

$$Q = (N_{tot})_0 \cdot \left[\hat{H}_C(Z_1 + 2 \cdot Z_2) + \Delta H_{R_1} \cdot Z_1 + \Delta H_{R_2} \cdot Z_2 \right]. \tag{n}$$

For 80% conversion of B, $Z_1 = 0.221$ $Z_2 = 0.48$ and $Q = 17.82 \cdot 10^3$ kcal.

8.4 RECYCLE REACTOR

A recycle reactor is a mathematical model describing a steady plug-flow reactor where a portion of the outlet is recycled to the inlet, as shown schematically in Figure 8-4. Although this reactor configuration is rarely used in practice, the recycle reactor model enables us to examine the effect of mixing on the operations of continuous reactors. In some cases, the recycle reactor is one element of a complex reactor model. Below, we analyze the operation of a recycle reactor with multiple chemical reactions, derive its design equations, and discuss how to solve them.

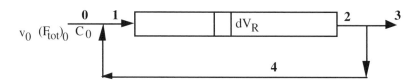

Figure 8-4: Recycle Reactor

To derive the design equation of a recycle reactor, we consider a differential reactor element, dV_R, and write a species balance equation over it for species j,

$$dF_j = (r_j) \cdot dV_R. \tag{8.4-1}$$

We follow the same procedure as the one used to derive the reaction-based design equation of a plug-flow reactor (see Chapter 3) and obtain

$$\frac{d\dot{X}_m}{dV_R} = r_m + \sum_{k}^{n_{dep}} \alpha_{km} \cdot r_k, \tag{8.4-2}$$

where $\dot{X}_m$ is the extent per unit time of the m-th independent reaction from the reactor inlet to a given point in the reactor. This is the differential design equation for a recycle reactor, written for the m-th independent chemical reaction. Note that (8.4-2) is identical to the design equation of a plug-flow reactor. The main difference between this case and a plug-flow reactor is in the way the volumetric flow rate and the concentrations of the species vary along the reactor. Consequently, to solve the design equations, we have to express these quantities in terms of the extents of the independent reactions. Since the reactor inlet is affected by the outlet, let $\dot{X}_{m_{out}}$ denote the extent per unit time of the m-th independent reaction in the entire reactor.

To reduce the design equation (8.4-2) to a dimensionless form, we select the feed stream to the system (Stream 0) as the reference stream and define a dimensionless extent of the m-th independent reaction by

$$Z_m = \frac{\dot{X}_m}{(F_{tot})_0}, \tag{8.4-3}$$

where $(F_{tot})_0 = (v_0 \cdot C_0)$ is the total molar flow rate of the reference system. We define a dimensionless space time by

$$\tau = \frac{V_R}{v_0 \cdot t_{cr}}. \tag{8.4-4}$$

Differentiating (8.4-3) and (8.4-4),

$$d\dot{X}_m = (F_{tot})_0 \cdot dZ_m$$

$$dV_R = (v_0 \cdot t_{cr}) \cdot d\tau,$$

and substituting these in (8.4-2), noting that $(F_{tot})_0 = (v_0 \cdot C_0)$, the design equation reduces to

$$\frac{dZ_m}{d\tau} = \left(r_m + \sum_{k}^{n_{dep}} \alpha_{km} \cdot r_k \right) \cdot \left(\frac{t_{cr}}{C_0} \right). \tag{8.4-5}$$

Eq. (8.4-5) is the dimensionless design equation of a recycle reactor, written for the m-th independent reaction. To describe the operation of a recycle reactor with multiple chemical reactions, we have to write (8.4-5) for each of the independent reactions.

To solve the design equations (8.4-5), we have to express the reaction rates in terms of the dimensionless extents of the independent reactions. We do so by expressing the local volumetric flow rate and the local molar flow rates of all the reactants in terms of Z_m's and calculating the local species concentrations. Using stoichiometric relations (1.4-6) and (1.4-7), the local molar flow rate of species j at any point in the reactor is

$$F_j = F_{j1} + \sum_m^{n_{ind}} (s_j)_m \cdot \dot{X}_m, \tag{8.4-6}$$

and the local total molar flow rate is

$$F_{tot} = (F_{tot})_1 + \sum_m^{n_{ind}} \Delta_m \cdot \dot{X}_m. \tag{8.4-7}$$

The local molar flow rate of species j at the reactor inlet (point 1) is

$$F_{j1} = F_{j0} + F_{j4}. \tag{8.4-8}$$

Using the definition of the recycle ratio, $R = F_{j4}/F_{j3}$, the molar flow rate at point 4 is related to the molar flow rate at point 2 by

$$F_{j4} = \frac{R}{1+R} F_{j2}. \tag{8.4-9}$$

Writing (8.4-6) for the reactor outlet (point 2) and combining it with (8.4-8) and (8.4-9), we obtain

$$F_{j2} = (1+R) \cdot \left(F_{j0} + \sum_m^{n_{ind}} (s_j)_m \cdot \dot{X}_{m_{out}} \right), \tag{8.4-10}$$

where $\dot{X}_{m_{out}}$ is the extent of the m-th reaction in the **entire** reactor per unit time. Substituting (8.4-9) and (8.4-10) into (8.4-8), the molar flow rate of species j at the reactor inlet (point 1) is

$$F_{j1} = (1+R) \cdot F_{j0} + R \sum_m^{n_{ind}} (s_j)_m \cdot \dot{X}_{m_{out}}. \tag{8.4-11}$$

Substituting (8.4-11) into (8.4-6), the local molar flow rate of species j is

$$F_j = (1+R) \cdot F_{j0} + R \sum_m^{n_{ind}} (s_j)_m \cdot \dot{X}_{m_{out}} + \sum_m^{n_{ind}} (s_j)_m \cdot \dot{X}_m. \tag{8.4-12}$$

To obtain the total local molar flow rate, we sum (8.4-12) over all species and obtain

$$F_{tot} = (1+R) \cdot (F_{tot})_0 + R \sum_m^{n_{ind}} \Delta_m \cdot \dot{X}_{m_{out}} + \sum_m^{n_{ind}} \Delta_m \cdot \dot{X}_m. \tag{8.4-13}$$

For the reactor inlet and outlet (points 1 and 2), (8.4-13) reduces, respectively, to

$$(F_{tot})_1 = (1+R) \cdot (F_{tot})_0 + R \sum_m^{n_{ind}} \Delta_m \cdot \dot{X}_{m_{out}} \qquad (8.4\text{-}14)$$

$$(F_{tot})_2 = (1+R) \cdot \left((F_{tot})_0 + \sum_m^{n_{ind}} \Delta_m \cdot \dot{X}_{m_{out}} \right). \qquad (8.4\text{-}15)$$

The concentration of species j at any point in the reactor is

$$C_j = \frac{F_j}{v} = \frac{1}{v} \left((1+R) \cdot (F_{tot})_0 \cdot (y_j)_0 + R \sum_m^{n_{ind}} (s_j)_m \cdot \dot{X}_{m_{out}} + \sum_m^{n_{ind}} (s_j)_m \cdot \dot{X}_m \right). \qquad (8.4\text{-}16)$$

For **liquid-phase** reactions, the density is assumed constant; hence, $v = v_1$, and, using (8.4-8),

$$v_1 = v_0 + v_4 = (1+R) \cdot v_0. \qquad (8.4\text{-}17)$$

Substituting (8.4-17) into (8.4-16) and using dimensionless extent defined by (8.4-3), the local concentration for liquid-phase reactions is

$$C_j = \frac{F_j}{v} = C_0 \left((y_j)_0 + \frac{R}{1+R} \sum_m^{n_{ind}} (s_j)_m \cdot Z_{m_{out}} + \frac{1}{1+R} \sum_m^{n_{ind}} (s_j)_m \cdot Z_m \right). \qquad (8.4\text{-}18)$$

Note that for $R = 0$ (no recycle), (8.4-18) reduces to the local concentration in a plug-flow reactor with liquid-phase reactions, (6.1-9), and, for $R \to \infty$ (very high recycle), (8.4-18) becomes

$$C_j = C_0 \left((y_j)_0 + \sum_m^{n_{ind}} (s_j)_m \cdot Z_{m_{out}} \right),$$

which is the concentration at the reactor outlet, $C_{j_{out}}$. Indeed, at very high recycles, the recycle reactor behaves as a CSTR. For **gas-phase** reactions, assuming ideal gas behavior, the local volumetric flow rate at any point in the reactor is

$$v = v_0 \cdot \left(\frac{F_{tot}}{(F_{tot})_0} \right) \cdot \left(\frac{T}{T_0} \right) \cdot \left(\frac{P_0}{P} \right). \qquad (8.4\text{-}19)$$

Noting that $(F_{tot})_0 = v_0 \cdot C_0$ and substituting (8.4-13),

$$v = v_0 \cdot \left(1 + \frac{R}{1+R} \sum_m^{n_{ind}} \Delta_m \cdot Z_{m_{out}} + \frac{1}{1+R} \sum_m^{n_{ind}} \Delta_m \cdot Z_m \right) \cdot \theta \cdot \left(\frac{P_0}{P} \right). \qquad (8.4\text{-}20)$$

Substituting (8.4-20) into (8.4-18), for isobaric operations, the local concentration of species j is

$$C_j = C_0 \frac{(y_j)_0 + \dfrac{R}{1+R} \sum_m^{n_{ind}} (s_j)_m \cdot Z_{m_{out}} + \dfrac{1}{1+R} \sum_m^{n_{ind}} (s_j)_m \cdot Z_m}{\left(1 + \dfrac{R}{1+R} \sum_m^{n_{ind}} \Delta_m \cdot Z_{m_{out}} + \dfrac{1}{1+R} \sum_m^{n_{ind}} \Delta_m \cdot Z_m\right) \cdot \theta}. \qquad (8.4\text{-}21)$$

Note that, here too, for $R = 0$ (no recycle), (8.4-22) reduces to the local concentration in a plug-flow reactor, and, for $R \to \infty$ (very high recycle), it reduces to the concentration at the reactor outlet.

For non-isothermal recycle reactors, we have to incorporate the energy balance equation to express variation in the reactor temperature. The energy balance equation is the same as that of a plug-flow reactor,

$$\frac{d\theta}{d\tau} = \frac{1}{CF(Z_m, \theta)} \left(HTN \cdot (\theta_F - \theta) - \sum_m^{n_{ind}} DHR_m \cdot \frac{dZ_m}{d\tau} \right). \qquad (8.4\text{-}22)$$

The only difference here is that the temperature at the reactor inlet depends on the outlet temperature and the recycle. Hence, we have to solve (8.4-22) subject to the initial condition that at $\tau = 0$, $\theta(0) = \theta_1$. To determine θ_1, we write an energy balance over the mixing point,

$$(F_{tot})_1 \hat{H}_1 = (F_{tot})_0 \hat{H}_0 + (F_{tot})_4 \hat{H}_4, \qquad (8.4\text{-}23)$$

where $\hat{H}$ is the specific molar enthalpy of the stream. Taking the feed stream to the system as the reference stream, the specific molar enthalpy of the stream is

$$(F_{tot}) \hat{H} = \int_{T_0}^{T} (\Sigma F_j \cdot \hat{c}_{p_j}) \, dT. \qquad (8.4\text{-}24)$$

Using the correction factor of the heat capacity, (8.4-23) reduces to

$$\int_1^{\theta_1} CF(Z, \theta)_1 \, d\theta = \int_1^{\theta_1} CF(Z, \theta)_4 \, d\theta, \qquad (8.4\text{-}25)$$

which relates to the various process variables by

$$\int_1^{\theta_1} \sum_j^{all} \left((1+R) \cdot (y_j)_0 + R \sum_m^{n_{ind}} (s_j)_m \cdot Z_{m_{out}} \right) \hat{c}_{p_j}(\theta_1) \, d\theta =$$

$$= \int_1^{\theta_4} R \sum_j^{all} \left((y_j)_0 + \sum_m^{n_{ind}} (s_j)_m \cdot Z_{m_{out}} \right) \hat{c}_{p_j}(\theta_4) \, d\theta. \qquad (8.4\text{-}26)$$

We solve (8.4-26) to express θ_1 in terms of θ_{4t}. When the heat capacities of the species are independent of the temperature, (8.4-25) reduces to

$$\theta_1 = 1 + \frac{CF(Z_m, \theta)_4}{CF(Z_m, \theta)_1} (\theta_4 - 1). \qquad (8.4\text{-}27)$$

For **liquid-phase** reactions, assuming constant density and heat capacity, $CF(Z_m, \theta) = \dot{m}/\dot{m}_0$, and (8.2-27) becomes, for $\theta_4 = \theta_{out}$,

$$\theta_1 = 1 + \frac{R}{1+R} (\theta_{out} - 1). \qquad (8.4\text{-}28)$$

Note that, when $R = 0$, $\theta_1 = 1$ and when $R \to \infty$, $\theta_1 = \theta_{out}$. For **gas-phase** reactions with constant species heat capacities, (8.4-27) becomes, for $\theta_4 = \theta_{out}$,

$$\theta_1 = 1 + \frac{R \sum\limits_{j}^{all} \left((y_j)_0 + \sum\limits_{m}^{n_{ind}} (s_j)_m \cdot Z_{m_{out}} \right) \hat{c}_{p_j}}{\sum\limits_{j}^{all} \left((1+R) \cdot (y_j)_0 + R \sum\limits_{m}^{n_{ind}} (s_j)_m \cdot Z_{m_{out}} \right) \hat{c}_{p_j}} (\theta_{out} - 1). \qquad (8.4\text{-}29)$$

With the concentration relations and an expression for θ_1 at hand, we can now complete the design formulation of recycle reactors. Substituting the species concentrations and θ in the individual reactions rates, r_m's and r_k's, we obtain a set of first-order, non-linear differential equations of the form

$$\frac{dZ_{m1}}{d\tau} = G_1(\tau, Z_{m_1}, Z_{m_2}, ..., Z_{n_{ind}}, \tau_{tot}, \theta)$$

$$\frac{dZ_{m2}}{d\tau} = G_2(\tau, Z_{m_1}, Z_{m_2}, ..., Z_{n_{ind}}, \tau_{tot}, \theta)$$

$$\vdots \qquad (8.4\text{-}30)$$

$$\frac{dZ_{n_{ind}}}{d\tau} = G_{n_{ind}}(\tau, Z_{m_1}, Z_{m_2}, ..., Z_{n_{ind}}, \tau_{tot}, \theta)$$

$$\frac{d\theta}{d\tau} = G_{n_{ind}}(\tau, Z_{m_1}, Z_{m_2}, ..., Z_{n_{ind}}, \tau_{tot}, \theta).$$

We have to solve these equations simultaneously with the energy balance equation for the initial condition that at $\tau = 0$, Z_m's $= 0$ and $\theta = \theta_1$. Note that we solve these equations for a given value of τ_{tot} corresponding to a given reactor volume The solutions indicate how the extents and temperature vary along the reactor for the specified reactor. To obtain the operating curves of recycle reactors, we repeat the calculations for different values of t_{tot} and for each case determine the final value of Z_m's and θ.

Also note that the species concentrations are expressed in terms of the extents at the reactor outlet, Z_{mout}'s. Therefore, solutions are obtained by an iterative procedure. We first guess the outlet extents, Z_{mout}'s, solve the set of differential equations, and then check if the calculated outlet extents agree with the assumed values.

8.5 SUMMARY

In this chapter, we discussed the design of different reactor configurations whose design formulations are not available in the literature. We showed how the basic approach of reaction-based design formulation is used to describe the operations of these reactors with multiple reactions. We covered in some detail the following reactor models:

a. Semi-batch reactor.
b. Plug-flow reactor with distributed feed.
c. Distillation reactor.
d. Recycle reactor.

The reader is challenged to apply reaction-based design formulation to other reactor configurations and models.

PROBLEMS

8.1₄ Chlorinations of hydrocarbons are notorious for their side, undesirable reactions, where the desired products are further chlorinated to bi- or tri-substitutions. You are on a research team that is assigned to examine the operating mode of a chlorination reactor. The following homogeneous, gaseous reactions take place in the reactor:

$$A + B \rightarrow V \qquad\qquad r_1 = k_1 C_A^{0.5} C_B$$

$$A + V \rightarrow W \qquad\qquad r_2 = k_2 C_A C_V$$

$$A + W \rightarrow P \qquad\qquad r_2 = k_2 C_A C_W,$$

where A is the chlorine, B is the hydrocarbon, and V is the desired product. A constant-volume batch pilot reactor is available for testing. Two members of the team suggest two different approaches. One engineer suggests to operate the reactor as a batch reactor, charging it with an equal amount of A and B. The second engineer suggests to operate the reactor as a semi-batch reactor, charging it initially with B and then feeding an equal amount of A at a constant rate during operating time, t_{op}.

a. Derive the design equations, and plot the operating curves for each mode.
b. Determine the optimal operating time for maximizing the production of V.
c. Suggest the preferred operating mode (check $t_{op} = 10\ t_{cr}$; $t_{op} = 8\ t_{cr}$; $t_{op} = 5\ t_{cr}$).
Data: $k_1 C_0^{0.5} = 0.5$ min⁻¹; $k_2 C_0 = 1.0$ min⁻¹; $k_3 C_0 = 1.2$ min⁻¹;
  $T_0 = 300°K$; $P_0 = 1$ atm.

8.2₄ You are on a research team that is assigned to examine the operating mode of a chemical reactor. The homogeneous, liquid-phase reactions take place in the reactor:

$$A \rightarrow 2\,V \qquad\qquad r_1 = k_1\,C_A$$

$$A + V \rightarrow W \qquad\qquad r_2 = k_2\,C_A\,C_V.$$

One engineer suggests to operate the reactor as a batch reactor, charging it with reactant A. The second engineer suggests to operate the reactor as a semi-batch reactor, feeding it at a constant rate during operating time t_{op}.
a. Derive the design equations, and plot the operating curves for each mode.
b. Determine the optimal operating time for maximizing the production of V.
c. Suggest the preferred operating mode (check $t_{op} = 10\,t_{cr}$; $t_{op} = 8\,t_{cr}$; $t_{op} = 5\,t_{cr}$).
Data: $k_1 = 0.2$ min⁻¹; $k_2\,C_0 = 0.01$ min⁻¹; $C_A(0) = C_0 = 2$ mole/lit.

8.3₄ The elementary gas-phase isomerization reaction $A \rightarrow B$ is carried out in a packed-bed recycle reactor. The recycle ratio is 5 mole recycled per mole taken off in the exit stream. For a volumetric flow rate of 10 dm³/s through the reactor, the corresponding pressure gradient (assumed constant) in the reactor is 0.0025 atm/m. The flow in the reactor is turbulent. What overall conversion can be achieved in a reactor that is 10 m in length and 0.02 m² in cross-sectional area?
Additional information: $C_{A0} = 0.01$ mole/dm³ ($P_0 = 2$ atm). The volumetric flow rate of fresh reactant to the reactor is $v_0 = 10$ dm³/s. The specific reaction rate is $k = 0.25$ sec⁻¹. (Adapted from Fogler's textbook)

8.4₄ In many industrial processes where the conversion per pass through the reactor is low, it may be advantageous to use a recycle reactor . Here, a significant portion of the exit stream is recycled back through the reactor. Calculate the overall conversion $f_A = (F_{A0} - F_{Aout})/F_{A0}$ that can be achieved in a 2-m³ plug-flow reactor when the irreversible, isothermal first-order gas-phase reaction $A \rightarrow B + C$ is carried out at 500°C, and 5 m³ of gas is recycled for every cubic meter of fresh feed.
Additional data: $k = 0.05$ s⁻¹; $P_0 = 1013$ kPa (10 atm).
(Adapted from Fogler's textbook)

8.5₄ Consider the reaction $A + B \rightarrow C + B$ in which $r = k\,C_A C_B$, where B is a catalyst. The reaction is taking place in a semi-batch reactor in which 100 ft³ of a solution containing 2 lbmole/ft³ of A is initially present. No B is present initially. Starting at time $t = 0$, 5 ft³/min. of a solution containing 0.5 lbmole/ft³ of B is fed into the reactor. The reactor is isothermal, and $k = 0.2$ ft³/lbmole min.
a. How many moles of C are present in the reactor after half an hour?
b. Plot the conversion as a function of time.
c. Repeat parts (a) and (b) for the case when the reactor initially contains 100 ft³ of B, and A is fed to the reactor at a concentration of 0.02 lbmole/ft³ and a rate of 5 ft³/min. The feed concentration is 0.5 lbmole/ft³. (Adapted from Fogler's textbook)

8.6₄ The liquid-phase reaction $2A + B \rightarrow C + D$ is carried out in a semi-batch reactor.

The reactor volume is 1.2 m³. The reactor initially contains 20 mole of B at a concentration of 0.03 kmole/m³. A at an aqueous concentration of 0.015 kmole/m³ is fed to the reactor at rate of 4 dm³/min. The reaction is first-order in A and half-order in B with a specific reaction rate of k = 60 (m³/kmole)$^{0.5}$/min. The activation energy is 35 kJ/mole. The feed rate to the reactor is discontinued when the reactor contains 0.53 m³ fluid. Plot the conversion, volume, and concentration as a function of time. Calculate the time necessary to achieve

a. 97% conversion of B and

b. 59% conversion of B.

c. The reaction temperature is to be increased from 25°C to 70°C, and the reaction is carried out isothermally. At this temperature, the reaction is reversible with an equilibrium constant of 10 (m³/kmole)$^{0.5}$. Plot the conversion of A and B and the equilibrium conversion of A as functions of time.

(Adapted from Fogler's textbook)

8.7$_4$ Aqueous feed containing reactant A (C_{A0} = 2 mole/lit) enters a plug-flow reactor (10 lit) which has a provision for recycling a portion of the flowing stream. The reaction kinetics and stoichiometry are:

$$A \rightarrow R \qquad (-r_A) = k\, C_A\, C_R \text{ mole/lit min.,}$$

and we wish to get 96% conversion. Should we use the recycle stream? If so, at what value should we set the recycle flow rate so as to obtain the highest production rate, and what volumetric feed rate can we process to this conversion in the reactor?

(Adapted from Levenspiel's Omnibook)

CHAPTER NINE

ECONOMIC-BASED OPTIMIZATION

In the preceding chapters, we discussed how to design chemical reactors – how to determine the reactor size (or operating time) to obtain a specified production rate and how to determine the production rate attainable on an existing reactor. However, we have not addressed the question: is it desirable to achieve high reaction extents (and use large reactors) or to utilize smaller reactors and recycle unconverted reactants? Further, when more than one reactant is used, we have to address the question: what is the most economical proportion of reactants fed to the reactor? Also, we saw that by selecting different schemes to heat (or cool) chemical reactors, we affect their performance. The answer to all these questions is not straightforward; it depends on the value of the products, the cost of the reactants, the cost of operating the reactor, as well as the cost of recovering unconverted reactants. The basis to the design and operation of chemical reactors lies in their economic performance. The dimensionless reaction operating curves tie (by stoichiometric relations) the species production rates to the reactor size (through dimensionless operating or space time, τ). The dimensionless energy balance equation ties the utilities (heating/cooling) needed for the operation to the extents and τ. Together, they provide the means to conduct an economic-based optimization of reactor operations. In this chapter, we discuss briefly how to apply these relations to optimize operations with chemical reactors.

9.1 ECONOMIC-BASED PERFORMANCE OBJECTIVE FUNCTIONS

Optimization of chemical processes is based on the objective of maximizing the profit of the entire process. For processes involving chemical reactions, we can write the objective function as

$$\begin{Bmatrix} \text{profit} \\ \text{rate} \\ (\$/\text{time}) \end{Bmatrix} = \begin{Bmatrix} \text{value of} \\ \text{products} \\ \text{generated} \\ (\$/\text{time}) \end{Bmatrix} - \begin{Bmatrix} \text{cost of} \\ \text{species} \\ \text{fed} \\ (\$/\text{time}) \end{Bmatrix} - \begin{Bmatrix} \text{rate of} \\ \text{operating} \\ \text{expenses} \\ (\$/\text{time}) \end{Bmatrix}. \tag{9.1-1}$$

For convenience, the operating expenses are divided into different categories,

$$\begin{Bmatrix} \text{rate of} \\ \text{operating} \\ \text{expenses} \\ (\$/\text{time}) \end{Bmatrix} = \begin{Bmatrix} \text{cost of} \\ \text{utilities} \\ \text{used} \\ (\$/\text{time}) \end{Bmatrix} + \begin{Bmatrix} \text{cost of} \\ \text{operating} \\ \text{equipment} \\ (\$/\text{time}) \end{Bmatrix} + \begin{Bmatrix} \text{other costs} \\ \text{-- overhead,} \\ \text{labor, etc.} \\ (\$/\text{time}) \end{Bmatrix} + \begin{Bmatrix} \text{amortized} \\ \text{capital} \\ \text{investment} \\ (\$/\text{time}) \end{Bmatrix}. \tag{9.1-2}$$

The revenue portion of (9.1-1) depends on the composition of the reactor outlet stream. A higher purity product is more valuable, and the economics of the process depends on whether unconverted reactants and undesirable by-products are separated. When species are separated, the separation cost should be incorporated into the analysis of the reactor operation. To maximize the profit of the process, an engineer can adjust, both in design and operation, several reactor operating parameters:

- the feed rate per reactor volume (expressed in terms of dimensionless operating or space time),
- proportion of reactants (expressed in terms of y_{j0}'s), and
- the heating (or cooling) rate (by adjusting the temperature of the heating fluid, θ_F).

The profit objective function (9.1-1) is then expressed in terms of these (and other) operating parameters, and we determine the optimal values of the parameters by solving the following equations,

$$\frac{\partial\{\text{profit}\}}{\partial\tau} = 0, \quad \frac{\partial\{\text{profit}\}}{\partial(y_j)_0} = 0, \quad ..., \quad \frac{\partial\{\text{profit}\}}{\partial\theta_F} = 0. \tag{9.1-3}$$

For most operations with single chemical reactions, the profit function is expressed in terms of relatively simple functions of these parameters, and the equations in (9.1-3) can be solved analytically. For operations with multiple chemical reactions, the optimal parameters are usually determined numerically.

Next, we derive the gross revenue function of the process, expressed in terms of the extents of the independent reactions. Let V_j denote the value of species j (expressed in \$/mole). When all the species in the reactor outlet are separated, using stoichiometric relation (1.5-5), the value of the product stream is

$$\sum_{j=1}^{\text{all}} F_{j_{\text{out}}} \cdot V_j = (F_{\text{tot}})_0 \sum_{j-1}^{\text{all}} V_j \cdot \left((y_j)_0 + \sum_m^{n_{\text{ind}}} (s_j)_m \cdot Z_{m_{\text{out}}} \right), \tag{9.1-4}$$

and the value of the feed stream is

$$\sum_{j=1}^{all} F_{j_{in}} \cdot V_j = (F_{tot})_0 \sum_{j=1}^{all} V_j \cdot y_{j0}. \tag{9.1-5}$$

Hence, when the inlet stream is selected as the reference stream, $Z_{min} = 0$, and the gross revenue of the process is

$$\begin{Bmatrix} \text{gross} \\ \text{revenue} \\ \text{rate} \\ \text{($/time)} \end{Bmatrix} = (F_{tot})_0 \sum_{j=1}^{all} V_j \cdot \sum_{m}^{n_{ind}} (s_j)_m \cdot Z_{m_{out}}. \tag{9.1-6}$$

In many instances, the separation expense is expressed in terms of cost per mole of product recovered, SC_j (expressed in $/mole of j); hence, the separation expense rate is

$$\begin{Bmatrix} \text{rate of} \\ \text{separation} \\ \text{expense} \\ \text{($/time)} \end{Bmatrix} = (F_{tot})_0 \sum_{j=1}^{sep} SC_j \cdot \left(y_{j0} + \sum_{m}^{n_{ind}} (s_j)_m \cdot Z_{m_{out}} \right). \tag{9.1-7}$$

Combining (9.1-6) and (9.1-7), the gross income rate of the reactor operation (without accounting for reactor operation expense) is

$$\begin{Bmatrix} \text{gross} \\ \text{income} \\ \text{rate} \\ \text{($/time)} \end{Bmatrix} = (F_{tot})_0 \left(\sum_{j=1}^{all} (V_j - SC_j) \cdot \sum_{m}^{n_{ind}} (s_j)_m \cdot Z_{m_{out}} - \sum_{j=1}^{all} SC_j \cdot y_{j0} \right). \tag{9.1-8}$$

When the species in the reactor effluent stream are **not** separated (unconverted reactants are discarded with the product), the gross revenue rate is

$$\begin{Bmatrix} \text{gross} \\ \text{revenue} \\ \text{rate} \\ \text{($/time)} \end{Bmatrix} = (F_{tot})_0 \sum_{j=1}^{sold} V_j \cdot \left(y_{j0} + \sum_{m}^{n_{ind}} (s_j)_m \cdot Z_{m_{out}} \right) - (F_{tot})_0 \sum_{j=1}^{all} V_j \cdot y_{j0}. \tag{9.1-9}$$

Note that the summation in the first term on the right of (9.1-9) is over the valuable products that are being sold. For example, unconverted reactants that are discarded with the product have no value. Also, note that polluting species in the product stream have a negative value (the cost of removing them to meet environmental specifications).

To obtain the net profit of the operation, we have to substitute into (9.1-1) the operating expenses of the reactor. These are divided into two categories: utility costs that are directly related to reaction extents and the total flow rate, and other expenses such as labor, maintenance, capital expenditure, etc. The latter is usually expressed on the basis of expenses per unit reactor volume.

The example below illustrates an economic-based optimization procedure.

Example 9-1 A valuable product V is produced in a plug-flow reactor by the liquid-phase, elementary chemical reaction

$$A + B \rightarrow V.$$

Under the operating conditions, V reacts with reactant B according to the elementary reaction

$$V + B \rightarrow W$$

to form toxic, undesirable product W. A stream of reactant A (C_{A0} = 16 mole/lit, r = 800 g/lit) and a stream of reactant B (C_{B0} = 20 mole/lit, r = 800 g/lit) are available in the plant, and we want to utilize an available 200 liter tubular reactor (plug-flow). The reactor is operated isothermally at 190°C. Based on the data below, determine the optimal total feed rates of the two feed stream (operation at maximum yield of product V). There is no separation of reactor outlet stream.
b. Operation to maximize profit (no separation of reactor outlet).
c. Operation to maximize profit (with separation of reactor outlet stream).
Data: At the reactor temperature, k_1 = 9.20·10⁻⁴ lit mole⁻¹ min⁻¹,
  k_2 = 1.84·10⁻³ lit mole⁻¹ min⁻¹.
  Values of the species: A = 1, B = 2, V(raw) = 30, W(raw) = -3 $/mole,
  Species separation costs: A = 0.2, B = 0.3, V = 0.3, W = 0.3 $/mole.
Solution This is an example of a series (consecutive) chemical reaction. The chemical reactions taking place in the reactor are:

Reaction 1: $A + B \rightarrow V$

Reaction 2: $V + B \rightarrow W,$

and the stoichiometric coefficients are:

$$s_{A1} = -1; \ s_{B1} = -1; \ s_{V1} = 1; \ s_{W1} = 0; \ \Delta_1 = -1;$$

$$s_{A2} = 0; \ s_{B2} = -1; \ s_{V2} = -1; \ s_{W2} = 1; \ \Delta_2 = -1.$$

Since each reaction has a species that does not participate in the other, the two reactions are independent as a reference stream and denote its flow rate by v_0, where, $v_0 = v_1 + v_2$. The concentration of the reference stream is

$$C_0 = \frac{(F_{tot})_0}{v_0} = \frac{v_1 \cdot C_{A0} + v_2 \cdot C_{B0}}{v_1 + v_2} = w_1 \cdot C_{A0} + (1 - w_1) \cdot C_{B0}, \qquad (a)$$

where $w_1 = v_1/(v_1 + v_2)$ is the volume fraction of Stream 1. The molar fractions of the species in the reference stream are

$$y_{A0} = w_1 \frac{C_{A0}}{C_0}, \qquad y_{B0} = (1 - w_1) \frac{C_{B0}}{C_0}, \tag{b}$$

$y_{V0} = y_{W0} = 0$, and $Z_{1in} = Z_{2in} = 0$. We write the dimensionless design equation of plug-flow reactors, (6.1-1), for each of the independent reactions,

$$\frac{dZ_1}{d\tau} = r_1 \cdot \left(\frac{t_{cr}}{C_0} \right) \tag{c}$$

$$\frac{dZ_2}{d\tau} = r_2 \cdot \left(\frac{t_{cr}}{C_0} \right). \tag{d}$$

Using stoichiometric relation (1.5-5), the local species molar flow rates are

$$F_A = (F_{tot})_0 (y_{A0} - Z_1) \tag{e}$$

$$F_B = (F_{tot})_0 (y_{B0} - Z_1 - Z_2) \tag{f}$$

$$F_V = (F_{tot})_0 (Z_1 - Z_2) \tag{g}$$

$$F_W = (F_{tot})_0 Z_2. \tag{h}$$

For liquid-phase reactions, we use (6.1-11) to express the concentrations, and the rates are

$$r_1 = k_1 \cdot C_0^2 \cdot (y_{A0} - Z_1) \cdot (y_{B0} - Z_1 - Z_2) \tag{i}$$
$$r_2 = k_2 \cdot C_0^2 \cdot (Z_1 - Z_2) \cdot (y_{B0} - Z_1 - Z_2). \tag{j}$$

We define the characteristic reaction time on the basis of Reaction 1; hence,

$$t_{cr} = \frac{1}{k_1 \cdot C_0}. \tag{k}$$

Substituting (i), (j), and (k) into (c) and (d), the design equations reduce to

$$\frac{dZ_1}{d\tau} = (y_{A0} - Z_1) \cdot (y_{B0} - Z_1 - Z_2) \tag{l}$$

$$\frac{dZ_2}{d\tau} = \left(\frac{k_2}{k_1} \right) \cdot (Z_1 - Z_2) \cdot (y_{B0} - Z_1 - Z_2). \tag{m}$$

a. We solve (l) and (m) numerically subject to the initial condition that at $\tau = 0$, $Z_1 = Z_2$

= 0, for different values of y_{B0} (note that $y_{A0} = 1 - y_{B0}$). Once we obtain Z_1 and Z_2 as functions of τ, we use (g) to calculate the dimensionless operating curve of the desirable product V as a function of the dimensionless space time. The figure below shows the curve for different values of y_{B0}. An examination of the curves indicates that

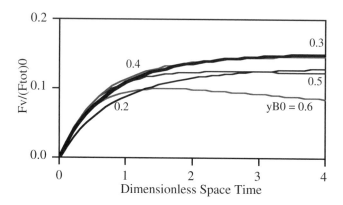

maximum value of $F_V/(F_{tot})_0$ is achieved for $y_{B0} = 0.3$ and is achieved at about $\tau = 4$. Since the unconverted reactants are not recovered, the profit of the operation is calculated by (9.1-9),

$$\{\text{profit rate}\} = (F_{tot})_0 \left[V_V \cdot (Z_1 - Z_2) + V_W \cdot Z_2 - (y_{A_0} \cdot V_A + y_{B_0} \cdot V_B) \right]. \quad (n)$$

The curves of the profit rates for different values of y_{B0} are shown in the figure below.

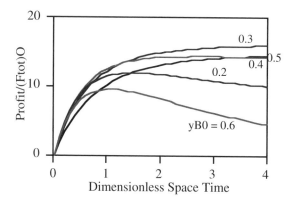

From the figure, the highest profit rate is achieved for $y_{B0} = 0.3$. From the solutions of the design equations, for $y_{B0} = 0.3$ and $\tau = 4$, $Z_1 = 0.218$ and $Z_2 = 0.068$. Using (n), the operation profit rate is

$$\{\text{profit rate}\} = (F_{tot})_0\left[30\cdot(Z_1 - Z_2) + (-3)\cdot Z_2 - (0.70)\cdot 1 - (0.30)\cdot 2\right]. \quad (o)$$

To determine the optimal feed rate, $(F_{tot})_0$, we use the definition of the dimensionless space time,

$$v_0 = \frac{V_R}{\tau\cdot t_{cr}} = \frac{200}{4}k_1\cdot C_0. \quad (p)$$

Hence, to determine the volumetric flow rate, we have to determine the value of C_0. Dividing the first equation in (b) by the second, for $y_{B0} = 0.3$, $w_1 = 0.744$. Using (a), the concentration of the reference stream is $C_0 = 17.02$ mole/lit, and from (p), the volumetric flow rate of the reference stream is $v_0 = 0.783$ lit/min. The volumetric flow rate of stream 1 is 0.583 lit/min, and the rate of stream 2 is 0.200 lit/min. Using (o), the optimal profit rate of the operation is

$$\{\text{profit rate}\} = (0.783 \text{ lit/min})\cdot(17.40 \text{ mole/lit})\cdot(3.00) = 40.87 \text{ \$/min}.$$

b. When the reactor is connected to a separation unit, the gross income rate of the operation is calculated by (9.1-8),

$$\{\text{profit rate}\} =$$

$$= (F_{tot})_0\left[\begin{matrix}(1-0.2)\cdot(-Z_1) + (2-0.3)\cdot(Z_1 - Z_2) + (30-0.3)\cdot(Z_1 - Z_2) + \\ (-3-0.3)\cdot(Z_2) - (0.2)\cdot y_{A0} - (0.3)\cdot y_{B0}\end{matrix}\right] =$$

$$= (F_{tot})_0\left[(27.2)\cdot Z_1 - (34.7)\cdot Z_2 - (0.2)\cdot y_{A0} - (0.3)\cdot y_{B0}\right]. \quad (q)$$

The figure below shows the profit curve for different values of y_{B0} for this case. Maximum profit is obtained with $y_{B0} = 0.3$ at $\tau = 4$, and solutions of the design equations

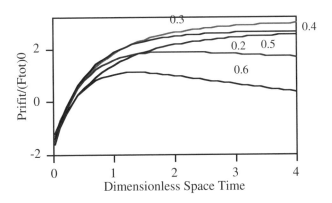

are: $Z_1 = 0.218$ and $Z_2 = 0.068$. Using (q), the profit rate of the operation is $3.34 \cdot (F_{tot})_0$, or 46.5 \$/min. Note that the profit of operation with separation is higher than that when unconverted reactants are discarded.

9.2 BATCH AND SEMI-BATCH REACTORS

The designs of batch and semi-batch reactors covered in Chapter 5 and 8 addressed only the operating time needed to obtain a given conversion; the size of the batch was not considered. The latter is determined by the required production rate of the unit. The operation of batch reactors usually consists of four steps: (i) preparation and filling, in duration time, t_f; (ii) the reaction time (or the operating time), t_r; (iii) discharging time, t_d; and (iv) idle time, t_i. The size of the batch and the duration of the reaction time are determined on the basis of economic-based optimization. Usually, the production of a species is specified, and the size of the batch, expressed in terms of $(N_{tot})_0$, is

$$F_j = \frac{(N_{tot})_0}{t_f + t_r + t_d + t_i} \left(y_{j0} + \sum_m^{n_{ind}} (s_j)_m \cdot Z_m \right). \tag{9.2-1}$$

The extents of the independent reactions are determined by the design equations for a given operating time. Below, we discuss the determination of the optimal operating time.

Let's denote the respective operating costs per unit time of these steps by W_f, W_r, W_d, and W_i. The total cost of the operation is therefore

$$\text{Cost}_{tot} = t_f \cdot W_f + t_r \cdot W_r + t_d \cdot W_d + t_i \cdot W_i. \tag{9.2-2}$$

When the species in reactor discharge are **not** separated, the gross revenue of a batch is

$$\left\{ \begin{array}{c} \text{gross} \\ \text{revenue} \\ \text{per batch} \\ (\$) \end{array} \right\} = (N_{tot})_0 \sum_{j=1}^{sold} V_j \cdot \left(y_{j0} + \sum_m^{n_{ind}} (s_j)_m \cdot Z_m \right) - (N_{tot})_0 \sum_{j=1}^{all} V_j \cdot y_{j0}. \tag{9.2-3}$$

Note that, as in (9.1-9), the summation in the first term on the right is over the species that have any value. Combining (9.1-1) and (9.1-2), the net profit rate is

$$\left\{ \begin{array}{c} \text{net} \\ \text{profit} \\ \text{rate} \\ (\$/\text{time}) \end{array} \right\} = \frac{(N_{tot})_0}{t_f + t_r + t_d + t_i} \left[\sum_{j=1}^{sold} V_j \cdot \left(y_{j0} + \sum_m^{n_{ind}} (s_j)_m \cdot Z_m \right) - \sum_{j=1}^{all} V_j \cdot y_{j0} \right] -$$

$$-\frac{t_f \cdot W_f + t_r \cdot W_r + t_d \cdot W_d + t_i \cdot W_i}{t_f + t_r + t_d + t_i} . \tag{9.2-4}$$

Note that the reaction time is expressed by the dimensionless operating time, $t_r = t_{cr} \cdot \tau$, and relates to the extents by the design equations and to $(N_{tot})_0$ by (9.2-1). To optimize the operation, we determine the values of the parameters by solving the following equations,

$$\frac{\partial\{profit\}}{\partial \tau} = 0, \quad \frac{\partial\{profit\}}{\partial (y_j)_0} = 0, \, ... \tag{9.2-5}$$

For operations with multiple chemical reactions, the optimal parameters are usually determined numerically. For most operations with single chemical reactions, the profit function is expressed in terms of relatively simple functions of these parameters, and the optimal operating time can be determined analytically. This is illustrated in the example below.

Example 9-2 The liquid-phase, first-order reaction $A \rightarrow B + C$ is carried out in a batch reactor. The value of reactant A is 0.5 \$/mole, the value of product B is 1 \$/mole, and the value of product C is 1.5 \$/mole. Determine:
a. the optimal operating (reaction) time,
b. the optimal extent, and
c. the reactor volume if the desired production rate of B is 5 mole/min.
Data: $C_A(0) = 2$ mole/liter; $k = 0.1$ min^{-1}.
Solution The reaction is $A \rightarrow B + C$, and its stoichiometric coefficients are: $s_A = -1$, $s_B = 1$, and $s_C = 1$. We select the initial state as the reference state; hence, $C_0 = C_A(0) = 2$ mole/liter, and $y_{A0} = 1$ and $y_{B0} = y_{C0} = 0$. For a first-order reaction, the characteristic reaction time is

$$t_{cr} = \frac{1}{k} = 10 \text{ min}. \tag{a}$$

Using stoichiometric relation (1.5-3),

$$N_A = (N_{tot})_0 (1 - Z) \tag{b}$$
$$N_B = N_C = (N_{tot})_0 \cdot Z. \tag{c}$$

The dimensionless design equation, in this case, is

$$\frac{dZ}{d\tau} = 1 - Z. \tag{d}$$

Solving (d) subject to the initial condition that $Z(0) = 0$, we obtain

$$Z(\tau) = 1 - e^{-\tau}. \tag{e}$$

a. Using (9.2-4), the net profit rate of the operation is

$$
\left\{
\begin{array}{c}
\text{net} \\
\text{profit} \\
\text{rate} \\
(\$/\text{time})
\end{array}
\right\}
= (N_{tot})_0 \frac{V_B \cdot Z + V_C \cdot Z - V_A \cdot y_{A0} - t_r \cdot W_r}{t_f + t_r + t_d + t_i} =
$$

$$
= (N_{tot})_0 \frac{(2.5) \cdot Z - 0.5 - t_r \cdot 0.1}{30 + t_r}, \tag{f}
$$

where $t_r = t_{cr} \cdot \tau$ is the operating time to be determined. Substituting (e) into (f) and taking the derivative, we find that the optimal dimensionless time is $\tau = 0.895$. Hence, the optimal operating time is $t_r = t_{cr} \cdot \tau = 8.95$ min, and the duration of a cycle is 38.95 min.

b. Using (e), the optimal extent is $Z = 0.591$.

c. Using (9.2-1), the specified production rate of B is

$$
5 \text{ mole/min} = \frac{(N_{tot})_0}{t_f + t_r + t_d + t_i} \cdot Z = \frac{(N_{tot})_0}{38.95} \cdot (0.591). \tag{g}
$$

Solving (g), $(N_{tot})_0 = 329.53$ mole, and the required reactor volume is

$$
V_R(0) = \frac{(N_{tot})_0}{C_0} = \frac{329.53 \text{ mole}}{2 \text{ mole/lit}} = 164.76 \text{ liters}.
$$

9.3 FLOW REACTORS

The economic-based optimization may be based on other criteria than those discussed in Section 9.1. In some cases, the production rate is specified, and the value of the product is fixed. In such cases, we would like to minimize the total operating cost. This is illustrated in the example below.

Example 9-3 We want to produce 1,000 mole of product R per hour from an aqueous solution of reactant A ($C_{A0} = 2$ mole/liter) in a CSTR. Product R is formed by the first-order chemical reaction $A \rightarrow 2\,R$. The cost of the feed stream is 0.4 $/mole A, and the cost of operating the reactor (labor, recovery of capital cost, overhead, etc.) is 0.1 $/liter·hr. The unconverted reactant A is being discarded. Based on the data below,
a. determine the optimal operating conditions (V_R, Z_{out}, $(F_{tot})_0$, and
b. calculate the cost of producing R under these conditions.
c. Plot the production cost as a function of the extent.
Data: At the operating conditions, $k = 0.125$ hr^{-1}.

Solution The chemical reaction is $A \rightarrow 2\,R$, and the stoichiometric coefficients are: $s_A = -1$, $s_R = 1$, and $\Delta = 1$, and the rate expression is $r = k \cdot C_A$. We select the inlet stream as the reference stream; hence, $C_0 = C_{A0}$, $y_{A0} = 1$, and $Z_{in} = 0$. The characteristic reaction time is

$$t_{cr} = \frac{1}{k} = 8 \text{ hr}. \tag{a}$$

Using stoichiometric relation (1.5-3),

$$F_A = (F_{tot})_0(1 - Z) \tag{b}$$

$$F_R = (F_{tot})_0 \cdot 2 \cdot Z. \tag{c}$$

The design equation of a CSTR, for this case, is

$$\tau = \frac{Z_{out}}{1 - Z_{out}}, \tag{d}$$

where the dimensionless space time is

$$\tau = \frac{V_R}{v_0 \cdot t_{cr}}. \tag{e}$$

a. The cost of the operation is

$$\text{Cost (\$/hr)} = (0.4) \cdot (F_{tot})_0 \cdot y_{A0} + (OC) \cdot V_R, \tag{f}$$

where OC is the reactor operating cost, $OC = 0.1$ \$/liter·hr. Based on the specified production rate of R and using (c),

$$(F_{tot})_0 = \frac{1,000}{2 \cdot Z} = \frac{500}{Z}. \tag{g}$$

From (e) and (d),

$$V_R = v_0 \cdot t_{cr} \frac{Z_{out}}{1 - Z_{out}}. \tag{h}$$

Substituting (g) and (h) into (f) and noting that $(F_{tot})_0 = v_0 \cdot C_0$, the cost of the operation is

$$\text{Cost (\$/hr)} = (0.4)\frac{500}{Z_{out}} + (0.1) \cdot \left(\frac{500}{C_0} \cdot t_{cr} \cdot \frac{1}{1 - Z_{out}} \right). \tag{i}$$

The figure below shows the cost of the operation as a function of the extent. To determine the minimum cost for the given production rate of R, we take the derivative of the cost with respect to Z and equate it to zero,

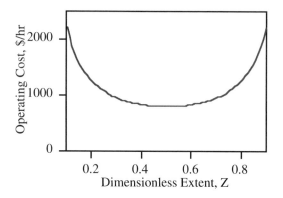

$$\frac{d(\text{Cost})}{dZ_{out}} = -\frac{200}{Z_{out}^{2}} + \left(\frac{50 \cdot t_{cr}}{C_0}\right)\frac{1}{(1-Z_{out})^{2}} = 0. \tag{j}$$

Substituting numerical values, we obtain $Z_{out} = 0.50$, and, from (g), $(F_{tot})_0 = 1,000$ mole/hr, and

$$v_0 = \frac{(F_{tot})_0}{C_0} = \frac{1,000 \text{ mole/hr}}{2 \text{ mole/liter}} = 500 \text{ liter/hr}.$$

From (h), the optimal volume of the reactor is

$$V_R = (500 \text{ liter/hr}) \cdot (8 \text{ hr})\frac{0.5}{1.0.5} = 4,000 \text{ liter}.$$

b. Using (b), the cost of producing R under optimal conditions is
$$\text{Cost ($/hr)} = 0.4 (1,000) + 0.1 (4,000) = 800 \text{ $/hr}.$$

9.4 SUMMARY

In this chapter, we discussed the optimization of chemical reactor operations on the basis of economic criteria. We showed how stoichiometric relations are combined with economic data and the design equations to describe the profitability of the opera-tions. We covered in some detail the following topics:
a. Economic objective function of reactor operations.
b. Economic objective function of reactor-separation operations .
c. Economic-based optimization .
d. Sizing and optimizing batch reactor operations.

The reader is challenged to apply the methods described in this chapter to other applications.

BIBLIOGRAPHY

More detailed treatment of optimizing the operation of batch reactors can be found in

Aris, R. B., *An Introduction to Chemical Reactor Analysis*. Englewood Cliffs, New Jersey: Prentice-Hall, 1960.
   A concise review of reactor optimization is provided by
Levenspiel, O., *Chemical Reactor Omnibook*, Second Edition. Corvallis, Oregon: Oregon State University, 1989.
   A general treatment of optimizing reactor operations and chemical processes can be found in
Biegler, L. T., I. E. Grossman, and A. W. Weterberg, *Systematic Methods of Chemical Process Design*. Englewood Cliffs, New Jersey: Prentice-Hall, 1997.

PROBLEMS

9-1$_2$ The plug-flow reactor is to produce 1,000 mole R/hr from an aqueous feed of A (C_{A0} = 1 mole/lit). The reaction is $2\,A \rightarrow R$, and its rate expression is $(-r_A) = 2 \cdot k \cdot C_A^2$. The cost of reactant stream is 0.50 \$/mole A, and the cost of operating the reactor comes to 0.20 \$/lit hr. Find V_R, f_A, F_{A0} for optimum operations under the following conditions:

a. The unconverted A is discarded.
b. What is the cost of producing R in (a)?
c. The unconverted A is recovered and recycled at a loss of 0.10 \$/mole A.
d. What is the cost of producing R in (c)?
Data: k = 1 liter/mole hr. (Adapted from Levenspiel's Omnibook)

9-2$_2$ Aqueous feed (C_{A0} = 1 mole/lit, v_0 = 1,000 lit/hr) is available to us at 1.00 \$/mole A from our friendly neighbor. The second-order chemical reaction is $2\,A \rightarrow R$, and the rate constant is 2 lit/mole hr. We can sell the product at 3.70 \$/mole R. A CSTR and a product purification unit cost 0.20 \$/lit hr of reactor (labor, utilities, value of money). Unconverted A is destroyed when the product is purified. Find the best way to operate such a system (V_R, f_A, F_R, hourly profit \$) and then recommend a course of action. (Adapted from Levenspiel's Omnibook)

9-3$_2$ Species A, present in a waste stream (v_0 = 1,000 lit/hr, C_{A0} = 1 mole/lit), can be converted into product R by the chemical reaction $2\,A \rightarrow R$. This product is useful and can be sold at 1.00 \$/mole R. We have to pay 0.10 \$/lit to get rid of the waste stream. The cost of operating a CSTR, including all charges, comes to 0.20 \$/lit hr. Find the best way of running this recovery operation (V_R, f_A, F_R, hourly profit \$) and then decide whether to proceed with this venture. The rate of reaction of A to R is given by

C_A, mole/lit	0.1	0.2	0.3	0.4	0.6	0.8	1.0
$(-r_A)$, mole/lit hr	0.15	0.5	1.0	1.7	3.6	6.0	9.1

(Adapted from Levenspiel's Omnibook)

9-4$_2$ We want to produce 500 mole/hr of R (value if 3 $/mole) by the chemical reaction $A + B \rightarrow R$. The rate expression is $r = k\,C_A\,C_B$. The cost of reactant A is 0.8 $/mole, and that of reactant B is 0.2 $/mole. The unconverted A is recovered and recycled to the reactor at a cost of 0.1 $/mole, and unconverted B is discarded. The cost of operating the reactor is 0.4 $/lit hr. Find:

a. The optimal conversion and feed ratio F_{B0}/F_{A0}.
b. The size of the reactor (CSTR).
c. The profit ($/hr) of the operation.

Data: $k = 2$ lit/mole hr; $C_{A0} = 1$ mole/lit; $v_0 = 500$ lit/hr; $F_{B0} = ?$
(Adapted from Levenspiel's Omnibook)

APPENDIX

NUMERICAL METHODS

A.1 NUMERICAL DIFFERENTIATION

For equally-spaced points, the first derivative of function $f(x)$ is approximated (to error order of Δx^2) as follows:

Forward differentiation

$$\left(\frac{df}{dx}\right)_i = \frac{-3 \cdot f(x_i) + 4 \cdot f(x_{i+1}) - f(x_{i+2})}{2 \cdot \Delta x}$$

Central differentiation

$$\left(\frac{df}{dx}\right)_i = \frac{f(x_{i+1}) - f(x_{i-1})}{2 \cdot \Delta x}$$

Backward differentiation

$$\left(\frac{df}{dx}\right)_i = \frac{3 \cdot f(x_i) - 4 \cdot f(x_{i-1}) + f(x_{i-2})}{2 \cdot \Delta x}.$$

A.2 NUMERICAL INTEGRATION

Trapezoidal rule. The trapezoidal rule provides a first-order approximation of the area of a function between two points:

$$I = \int_{x_1}^{x_2} f(x)dx = \frac{x_2 - x_1}{2}\left[f(x_1) + f(x_2)\right]$$

Simpson's rule. This method is based on a second-order polynomial approximation of the function. For equally-spaced points, the integral of the function between x_0 and x_2 is

$$I = \int_{x_0}^{x_2} f(x)dx = \frac{\Delta x}{3}\left[f(x_0) + 4 \cdot f(x_1) + f(x_2)\right]$$

 NOTATION

All quantities are defined in their generic dimensions (length, time, mass or mole, energy, etc.). Symbols that appear in only one section are not listed. Numbers in parentheses indicate the equations where the symbol is defined or appears for the first time.

A	Cross-section area, area.
a	Species activity coefficient
C	Molar concentration, mole/volume.
CF	Correction factor of heat capacity, dimensionless, (4.2-19).
c_p	Mass-based heat capacity at constant pressure, energy/mass °K.
$\hat{c}_p$	Molar-based heat capacity at constant pressure, energy/mole °K.
D	Reactor (tube) diameter, length.
DHR	Dimensionless heat of reaction, dimensionless, (4.2-24).
d_p	Particle diameter, length.
E	Total energy, energy.
E_a	Activation energy, energy/mole extent.
e	Specific energy, energy/mass.
F	Molar flow rate, mole/time.
f	Conversion of a reactant, dimensionless, (1.5-7a) and (1.5-7b).
f	Friction factor, dimensionless.
G	Mass velocity, mass/time area.
G_j	Generation rate of species j in a flow reactor, moles j/time.
g	Gravitational acceleration, length/time^2.
H	Enthalpy, energy.
$\hat{H}$	Molar-based specific enthalpy, energy/mole.
ΔH_R	Heat of reaction, energy/mole extent.
h	Mass-based specific enthalpy, energy/mass.
HTN	Dimensionless heat-transfer number, dimensionless, (4.2-23).
J_j	Molar flux of species j, mole j/(time area).
K	Equilibrium constant.
KE	Kinetic energy, energy.
k, k(T)	Reaction rate constant.

L	Length, length.
M	Mass, mass.
MW	Molecular weight, mass/mole.
$\dot{m}$	Mass flow rate, mass/time.
N, N(t)	Molar content in a reactor, moles.
n	Number of chemical reactions.
n	Unit outward vector.
$(n_j - n_{j0})_i$	Moles of species j formed by the i-th reaction, moles of j.
OC	Operating cost,
P	Total pressure, force/area.
PE	Potential energy, energy.
Q(t)	Heat added to the reactor in time t, energy.
$\dot{Q}$	Rate heat added to the reactor, energy/time.
R	Gas constant, energy/temperature mole.
r	Volume-based rate of a chemical reaction, mole extent/time volume.
(r_j)	Volume-based rate of formation of species j, mole j/time volume.
$(r_j)_s$	Surface-based rate of formation of species j, mole j/time surface area of catalyst.
$(r_j)_w$	Mass-based rate of formation of species j, mole j/time catalyst mass.
S	Surface area, area.
s_j	Stoichiometric coefficient of species j, mole j/mole extent.
SC	Separation cost
T	Temperature, °K or °R.
t	Time, time.
t_{cr}	Characteristic reaction time, time, (2.5-1).
U	Internal energy, energy.
U	Heat-transfer coefficient, energy/time area °K.
u	Mass-based specific internal energy, energy/mass.
u	Velocity, length/time.
V	Volume, volume.
V_R	Reactor volume, volume.
V_j	Value of species j, $/mole
v	Volumetric flow rate, volume/time.
W	Mass, mass.
W	Work, energy.
X, X(t)	Extent of a chemical reaction, mole extent, (1.2-1).
$\dot{X}$	Reaction extent per unit time, mole extent/time, (1.8-7).
y	Molar fraction, dimensionless.
Z	Dimensionless extent, $X/(N_{tot})_0$ or $\dot{X}/(F_{tot})_0$, dimensionless, (1.5-1) and (1.5-2).
z	Vertical location, length.

Greek Symbols

α	Order of the reaction with respect to species A, dimensionless.
α_{km}	Multiplier factor of m-th independent reaction for k-th dependent reaction, (1.6-9).
β	Order of the reaction with respect to component B, dimensionless.
γ	Dimensionless activation energy, $E_a/R \cdot T_0$, dimensionless, (2.3-5).
Δ	Change in the number of moles per unit extent, mole, (1.1-6).
ε	Void of packed-bed, dimensionless.
η	Yield, dimensionless, (1.8-3) and (1.8-4).
θ	Dimensionless temperature, T/T_0, dimensionless.
μ	Viscosity, mass/length time.
ρ	Density, mass/volume.
σ	Selectivity, dimensionless, (1.8-9).
τ	Dimensionless operating time, t/t_{cr}, or space time, $V_R/v_0 \cdot t_{cr}$, dimensionless, (3.5-3) and (3.5-8).
Φ	Particle sphericity, dimensionless

Subscripts

0	Reference state or stream
A	The limiting reactant
all	All
cr	Characteristic reaction
dep	Dependent
eq	Equilibrium
F	Heating (or cooling) fluid
gas	Gas-phase
I	Inert
i	The i-th reaction
in	Inlet, inlet stream
ind	Independent
inj	Injected stream
j	The j-th species
k	Index number for dependent reactions
liq	Liquid-phase
m	Index number for independent reactions
op	Operation
out	Outlet
R	Reactor
S	Surface
sh	Shaft work (mechanical work)
sp	Space
sys	System

tot	Total
V	Volume basis
vis	Viscous
W	Mass basis

INDEX